ROUTLEDGE HANDBOOK OF SOCIAL PSYCHOLOGY OF TOURISM

The impacts of tourism, an increasingly crucial area of study amongst researchers, are primarily investigated through economic, socio-cultural or environmental perspectives. The social psychological effects of tourism have not been adequately researched despite often being much more important for many destinations, especially where conflicts among different stakeholders exist. This book investigates the social psychological effects of tourism within the scope of social psychology theory.

This book introduces the concept of social psychology, as distinct from psychology and sociology, and its relationship to tourism, examines tourism within various theoretical frameworks, e.g. career ladder theory and Maslow's 7 hierarchy, explores the ways in which tourism changes attitudes and finally investigates social psychological issues in tourism business.

It is an important resource for advanced undergraduates, graduate students and relevant practitioners in the field of tourism, and in some cases for a broader public in the field of social psychology.

Dogan Gursoy, PhD is the Taco Bell distinguished professor in Hospitality Business Management at Washington State University in the School of Hospitality Business Management and the editor of the *Journal of Hospitality Marketing & Management*. He is also the recipient of the 2021 ICHRIE Lifetime Research Achievement Award and the 2019 University of Delaware's Michael D. Olsen Research Achievement Award.

Sedat Çelik, PhD is Associate Professor in the Department of Tourism and Hotel Management at Şırnak University, Turkey. He completed his doctorate at Akdeniz University. His research interests are destination, tourism management, organizational behavior and the social psychology of tourism. He is referee for many important journals. He is also Editor-in-Chief of the *Journal of Hospitality and Tourism Issues (JOHTI)*.

ROUTLEDGE HANDBOOK OF SOCIAL PSYCHOLOGY OF TOURISM

Edited by
Dogan Gursoy and Sedat Çelik

Routledge
Taylor & Francis Group
LONDON AND NEW YORK

Cover image: Getty Images

First published 2022
by Routledge
4 Park Square, Milton Park, Abingdon, Oxon OX14 4RN

and by Routledge
605 Third Avenue, New York, NY 10158

Routledge is an imprint of the Taylor & Francis Group, an informa business

British Library Cataloguing-in-Publication Data
A catalogue record for this book is available from the British Library

Library of Congress Cataloging-in-Publication Data
A catalog record has been requested for this book

ISBN: 978-0-367-75287-3 (hbk)
ISBN: 978-0-367-75289-7 (pbk)
ISBN: 978-1-003-16186-8 (ebk)

DOI: 10.4324/9781003161868

Typeset in Bembo
by codeMantra

CONTENTS

FIGURES

TABLES

ABOUT THE CONTRIBUTORS

Fulya Acikgoz is a PhD candidate at University of Bristol, UK. Her research focuses on technology marketing, information management, technology adoption, social media, and tourism. She published in journals including *Journal of Business & Industrial Marketing, International Journal of Contemporary Hospitality Management, Behaviour & Information Technology, Journal of Marketing for Higher Education, International Journal of Human-Computer Interaction.*

Manisha Agarwal is a full-time staff at James Cook University Singapore as the associate lecturer and senior manager under Deputy Vice Chancellor office. She is currently enrolled at a PhD program in tourism at College of Business, Law and Governance at the James Cook University, Townsville, Australia. Her research interests include tourist behaviour and understanding of underlying psychological processes in tourists, in particular motivation, experience, satisfaction, and loyalty. She has published several works in tourism journals.

Sakib Bin Amin, PhD, is an associate professor in the School of Business and Economics (SBE) and the director of the Accreditation Project Team (APT) at North South University (NSU), Bangladesh. He holds a PhD in economics from Durham University (UK). He was the Commonwealth Rutherford Fellowship (2017–2018) receiver for conducting his postdoctoral research at Durham University (UK). He also holds a master's degree in international economics from the University of Essex (UK). He is a fellow of the Higher Education Academy (HEA). His research focuses on energy and tourism policy in developing countries. He has published a book, *The Economy of Tourism in Bangladesh: Prospects, Constraints, and Policies* (Palgrave Macmillan).

Sebastian Amrhein is a tourism professional with several years of practical experience in tourism planning, management and working with indigenous communities in Asia, Europe and Latin America. Currently, he is employed as a research associate and lecturer for the BA Sustainable Tourism course at the Rhine-Waal University of Applied Sciences in Kleve, Germany. He is a PhD candidate at the Radboud University in Nijmegen, Netherlands. His research interests include the interrelations and relationships of tourism and neoliberalism and the resulting consequences on societies and the environment.

Nagihan Cakmakoglu Arici holds a PhD in tourism management. She is currently working for the Turkish Consulate General, Stuttgart, Germany. Her research interests include consumer behaviors, tourist experiences, e-marketing and business administration. She carried out a number of international projects funded by EU. Her work has been published in international journals such as *German Journal of Human Resource Management* and *Leadership & Organization Development Journal.*

Serkan Aylan is the associate professor within the tourism guidance department of tourism faculty at Selçuk University in Turkey. He received his PhD in tourism management, master of science degree in tourism management education and a bachelor of science degree in travel management and tourism guidance teaching from Gazi University. He worked in tourism sector in various functional areas and in roles as waiter, travel consultant and licensed tourist guide. Dr. Aylan's teaching areas include tourism management, tourism guidance, tourism geography, alternative tourism, management of travel agencies and tour operators and recreation management. His research areas include overtourism, tourism management, qualitative research methods and sport tourism. Dr. Aylan has academic publications (refereed journal articles, one book, several book chapters and conference presentations) related to tourism.

Galina Berjozkina is a senior lecturer at City Unity College Nicosia. She is a PhD student in the University of Strathclyde, Department of Work, Employment and Organisation, engaged in research on seasonal employees' work performance in the tourism industry. She lectures on courses like tourism planning and development, hospitality animation and introduction to hospitality, among others. Her academic interests include tourism, hospitality and management. She has attended several academic conferences, and has published a book on destination marketing.

Erhan Boğan, PhD, received his PhD from Sakarya University, Turkey. He is an associate professor at Adıyaman University, Turkey. His research interests include corporate social responsibility, halal tourism and behavioral integrity. He has published articles in top-tier tourism and hospitality journals such as *International Journal of Hospitality Management*, *Journal of Hospitality Marketing & Management* and *Tourism Management Perspectives*. He has also been serving as a reviewer for international tourism and hospitality journals.

P. Monica Chien, PhD, is a senior lecturer at the University of Queensland Business School. Her research focuses on consumer behavior in marketing, tourism and sport, specializing in examining information processing and behavioral change using experimental studies.

Erhan Coşkun, PhD, assistant professor, completed his undergraduate education in Department of Tourism and Hotel Management, School of Tourism and Hotel Management, Düzce University Akçakoca in 2009; his master's degree in Department of Tourism and Hotel Management, Institute of Social Sciences, Düzce University; and his doctorate in the Department of Tourism Management at Institute of Social Sciences, Aydın Adnan Menderes University. He is currently a lecturer at Department of Hotel, Restaurant and Catering Services, Davutlar Vocational School, Aydın Adnan Menderes University. His research interests include financial strategy, strategic management, innovation, alternative tourism types and tourism investments.

Ali Doğantekin, PhD, is currently working as a lecturer at the Department of Hotel, Restaurant and Catering Services at Yozgat Bozokz University. He earned his bachelor's degree from Akdeniz University and master's degree and a doctoral degree from Eskişehir Osmangazi University. He has served as a reviewer for *Tourism Academic Journal*. His research interests include organizational behavior, sustainable tourism and halal tourism.

Filiz Gümüş Dönmez is working as assistant professor in the Department of Tourism Management, Tourism Faculty, Muğla Sıtkı Koçman University, Turkey. She holds a PhD and MS in tourism management and a bachelor's degree in tourist guiding. She worked for 10 years as a tourist guide, restaurant manager and saleswoman in the tourism industry before venturing into academia. Her research areas include tourism management, workplace loneliness, organizational behavior, sustainable tourism, local residents and diversification in tourism. She teaches social psychology, behavioral science, organizational behavior and human resources management. She has several papers in academic journals and national and international conferences.

Erol Duran is a professor of the Faculty of Tourism at Çanakkale Onsekiz Mart University. He holds a PhD in tourism management from the Graduate School of Social Sciences at Dokuz Eylül University, İzmir. His expertise includes sociology of tourism, consumer complaining behavior, tourist behavior, sustainable tourism and culture. He has published around 70 journal articles, book chapters and conference papers. His papers have appeared in respected refereed journals such as *International Journal Culture, Tourism and Hospitality Management*; *Career Development International*; *International Journal of Event and Festival Management*; *Journal of Human Sciences*; and *Journal of Awareness*.

Aysen Ercan Iştin, PhD, completed her undergraduate education in 2010 at Department of Tourism and Hotel Management, Mustafa Kemal University. She completed her master's degree in 2014 at Department of Tourism and Hotel Management, Akdeniz University, and her PhD degree in 2018 at Department of Tourism Management, Mersin University. She started to work as a research assistant in 2017 and was eventually appointed as an assistant professor in 2018 at Department of Tourism and Hotel Management, Şırnak University, where she is working currently. Her main fields of study are organizational behavior in tourism enterprises, social media, alternative tourism types, ecotourism/sustainable tourism and other areas related to tourism.

Çağrı Erdoğan graduated from Akçakoca School of Tourism and Hotel Management, Düzce University (BA in 2011). In the same year, he was appointed as a research assistant at Department of Tourism Management, Faculty of Management, Sakarya University, within the scope of the Academic Member Training Program known as ÖYP. He completed his master's degree (MSc) in 2013 and his PhD with the dissertation titled "The Relationship between Touristic Experiences and Self-Concept" in 2019 at Department of Tourism Management, Institute of Social Sciences, Sakarya University. In 2021, he earned a BA from Sociology Program, Faculty of Open and Distance Education, İstanbul University. Erdoğan, whose academic interests are tourist psychology, tourism sociology, tourism history and tourism theory, has been continuing his academic activities at Department of Tourism Management, Faculty of Tourism, Sakarya University of Applied Sciences since May 2018.

Bekir Eşitti is a researcher at the Department of Travel Management and Tourism Guidance, Tourism Faculty, Çanakkale Onsekiz Mart University, where he earned a PhD on tourism management. Earlier he also earned an MBA degree from the University of Wales Institute Cardiff (2008). He started to work at Department of Hotel, Restaurant and Catering Services, Aksaray University (2013). He is an associate professor in the field of social and administrative sciences, and in "tourism" science. His research focuses on tourism management, behavioral sciences and gastronomy. Particularly, Bekir studies the phenomenon of behavioral tourism by providing links between hospitality businesses, employee values, culture and tourist experiences.

Buket Buluk Eşitti is currently working as a research assistant at Department of Management, Faculty of Tourism, Tourism Çanakkale Onsekiz Mart University. She graduated from Department of Tourism Management, Faculty of Business, Akdeniz University, Alanya. She received her master's degree from Department of Tourism Management, Graduate School of Social Sciences, Çanakkale Onsekiz Mart University. She received her PhD degree from Department of Tourism Management, Institute of Graduate Studies, Çanakkale Onsekiz Mart University. Her main areas of study are behavioral sciences, tourism marketing, sustainable tourism, food and beverage management, media and tourism, travel agency and e-tourism.

Zanete Garanti is a PhD holder and associate professor in City Unity College Nicosia, Cyprus. She lectures on marketing and management courses and is actively involved in researching branding, social media marketing and influencer marketing. Her recent studies have been on travel and tourism influencers, brand personality, loyalty and equity on social media networks, e-referral, brand image and personality of Iran as a destination, among others. Her work has been published in various books and internationally recognized journals.

Volkan Genç, PhD, is an assistant professor at the School of Tourism and Hotel Management at Batman University. He graduated with a PhD in Tourism Management from Eskisehir Anadolu University in Turkey. He has published some articles in prestigious journals such as *International Journal of Contemporary Hospitality Management*. He continues to work on emotional and social competence, emotional labor, aesthetic labor, and the role of employee's emotions in resistance to change. He has also served as a reviewer in international tourism and hospitality journals.

Seray Gülertekin Genç, PhD, is an assistant professor at the School of Tourism and Hotel Management at Batman University. She graduated with a PhD in Tourism Management from Eskisehir Eskisehir Osmangazi University in Turkey. Her academic background is reflected in her main research interests, which include the aesthetics of destinations, tourism behavior, and tourism marketing. She has also served as a reviewer in international tourism and hospitality journals.

Selami Gültekin is currently working as instructor at Department of Recreation Management, School of Tourism Management and Hospitality, Siirt University. He earned his bachelor's degree from Department of Tourist Guiding, School of Tourism Management and Hospitality, Balıkesir University. He holds a master's degree from Department of Tourism Management, Institute of Social Sciences, at Akdeniz University. He's pursuing his PhD at Department of Tourism Management Institute of Social Sciences, Akdeniz University. His research interests include social psychology, sociology of tourism, emotional intelligence and tour guidance.

Hacer Harlak, PhD, is a professor of social psychology and earned her bachelor's degree in psychology from Department of Psychology, Aegean University in İzmir. She continued her graduate training in the same university and received her MSc and PhD in social psychology. She worked in Dokuz Eylül University and then in Aydın Adnan Menderes University. She had taught as an assistant and associate professor in the departments of sociology and psychology for many years. Currently she is a full professor at the Department of Psychology in Aydın Adnan Menderes University. Her research interests include health behavior, intergroup prejudice, interpersonal communication skills, clinical social psychology and social psychological aspects of tourism.

Gert-Jan Hospers is professor of urban and regional transition at Radboud University, Nijmegen, the Netherlands. Besides, he is director of Stichting Stad en Regio, a foundation promoting urban and regional development at a human scale. In the slipstream of his PhD thesis *Regional Economic Change in Europe: a Neo-Schumpeterian Vision* (2004), he has published extensively on transformation processes in territories across Western Europe such as the Ruhr Area, Greater Copenhagen and several Dutch regions. He is Fellow of the Regional Studies Association and member of the scientific committee of Cittaslow International.

Irene Huertas-Valdivia, PhD, is a lecturer in business administration at the University Rey Juan Carlos, Madrid. From 2012 to 2014, she served as lecturer in tourism and hospitality management at the University of Guadalajara (Mexico). For over nine years, she worked as a middle manager in leading hospitality corporations. Her main research focuses on issues related to human resource management, engagement, empowerment and leadership in the hospitality industry. The results of her work have been published in *Tourism Management, International Journal of Hospitality Management* and *International Journal of Contemporary Hospitality Management*.

Üzeyir Kement graduated from Ordu Persembe Anatolian Hotel Management and Tourism Vocational High School in 2005. He graduated from Gazi University Tourism Management Education Department in 2009. He completed his master's degree in Hacettepe University, Institute of Social Sciences, Tourism Management program. He completed his doctorate in Gazi University, Institute of Social Sciences, Department of Recreation Management in 2015. In 2020, he was entitled to receive an associate professorship in the field of Tourism. From 2014 to 2021, he worked in the Tourism and Hotel Management program and Recreation Management departments at Bingöl University. In 2021, he started to work in the Department of Gastronomy and Culinary Arts, Tourism Faculty, Ordu University. He has written papers, books, and articles in the fields of recreation management, sustainable tourism and gastronomy and culinary arts. He also carried out scientific research projects on eco-recreation and gastronomic culture. Üzeyir Kement has worked in the tourism sector in positions such as waiter, assistant cook and food and beverage teacher.

Farhan Khan is currently working as a graduate assistant at North South University (NSU). He completed his bachelor of science (BS) in economics from NSU with magna cum laude honors. Currently he is enrolled in master of science (MS) in economics program at NSU. He is an advocate of ecological economics. Farhan believes that economic efficiency and good economic decisions are not possible if all of the costs as well as benefits are not included in prices. He has published several scientific papers in international and national reputed journals.

Burhan Kılıç, PhD, is an associate professor with Faculty of Tourism, Mugla Sıtkı Kocman University, Turkey. He completed his PhD in Gazi University, Ankara. His area of study includes tourism marketing and management. He studies tourist behavior mostly in relation to the field of gastronomy.

Erdogan Koc is professor of marketing Faculty of Economics and Administrative, and Social Sciences, Bahcesehir University, Turkey. He has extensively published in top-tier tourism, hospitality and services marketing and management journals. He is the author/editor of several books (by publishers such as Routledge and CABI) and book chapters. He serves as a member on the editorial boards of a number of journals (such as *Journal of Hospitality Marketing and Management, International Journal of Intercultural Relations, International Journal of Hospitality and Tourism Administration, Journal of Promotion Management*) and has acted as an ad hoc referee for over 15 respectable journals. His research primarily focuses on the human element (both as consumers and employees) in tourism, hospitality, services marketing and management. He provides consultancy and training services for a wide variety of service sector businesses.

Maximiliano E. Korstanje is editor in chief of *International Journal of Safety and Security in Tourism* (UP Argentina) and editor in chief emeritus of *International Journal of Cyber Warfare and Terrorism* (IGI-Global US). Besides being a senior researcher in the Department of Economics at University of Palermo, Argentina, he is a global affiliate of Tourism Crisis Management Institute (University of Florida), Centre for Ethnicity and Racism Studies (University of Leeds), The Forge (University of Lancaster and University of Leeds UK) and The International Society for Philosophers, hosted in Sheffield UK. With more than 1,200 published papers and 35 books, in 2015, he was awarded as Visiting Research Fellow at School of Sociology and Social Policy, University of Leeds UK and the University of La Habana Cuba. In 2017 he was elected as foreign faculty member of AMIT, Mexican Academy in the study of Tourism, which is the most prominent institutions dedicated to tourism research in Mexico. He had a vast experience in editorial projects working as advisory member of Elsevier, Routledge, Springer, IGI Global and Cambridge Scholar Publishing. Korstanje had visited and given seminars in many important universities worldwide, a great distinction given by Marquis Who's Who in the world.

Elif Yolbulan Okan, after graduating from Middle East Technical University with a BSc degree in business administration (1995), earned her MSc in marketing from University of Salford, Manchester (1997) and her PhD in marketing from Yeditepe University (2007). She decided to pursue her academic studies after working in the banking sector for a few years. She was a faculty member of Department of Business Administration, Yeditepe University between 2002 and 2016. Since February 2017, she has been working as an associate professor of marketing and vice dean at Faculty of Economics, Administrative and Social Sciences, Bahçeşehir University. Her research interests include marketing, brand management, marketing theory and consumer behavior. Her research has appeared as book chapters, case studies in books and articles published by reputable national and international publishers. Besides her academic studies, she also lectures on brand management and integrated marketing communication courses at TURQUALITY program, whose goal is to facilitate and support the success of Turkish brands in the international arena.

Hera Oktadiana earned her PhD from the School of Hotel and Tourism Management the Hong Kong Polytechnic University and CHE (Certified Hospitality Educator) from the American Hotel & Lodging Educational Institute. She has been the Head of Hotel Management/Hospitality and Tourism Departments at several Indonesia tourism educational institutions. She is an adjunct senior lecturer at James Cook University Australia and assistant professor at Trisakti School of Tourism Indonesia. Her research interests include tourists' behaviour (particularly the Muslim and emerging markets) and tourism and hospitality education. She serves as an editorial board of *Asia Pacific Journal of Tourism Research, International Journal of Tourism Cities*, and a number of Indonesian tourism journals. She is also a guest editor for a special issue of *International Journal of Tourism Cities*. Currently, she is the Regional Vice President of the International Tourism Studies Association (ITSA).

Yakup Kemal Özekici, PhD, earned his PhD degree from Department of Tourism Management, Gazi University, and now works in the Adıyaman University (Turkey). He has worked at Gaziantep University, Gazi University and Ankara Hacı Bayram Veli University. His research interests and published articles focus on the concepts of acculturation, cultural change, global culture and restaurant management.

Emrah Özkul is a professor in Faculty of Tourism, Kocaeli University, Turkey. He completed his undergraduate education at Akçakoca School of Tourism and Hotel Management, Bolu Abant İzzet Baysal University. He completed his master's degree in tourism management at Institute of Social Sciences, Balıkesir University, and his PhD in tourism management at Dokuz Eylül University. He serves on the editorial boards of several national/international journals. His field of study for both his doctorate and associate professorship is tourism. His research primarily focuses on the tourism sociology, tourism marketing and CRM in tourism.

Emrah Öztürk is an assistant professor at Akçakoca School of Tourism and Hotel Management, Düzce University, Turkey. He earned his master's degree in Department of Tourism and

Hotel Management, Institute of Social Sciences, Düzce University in 2012 and his doctorate from Department of Tourism Management, Sakarya University in 2017. His main research interests include alternative tourism types, strategic management, competitive strategies and innovation in tourism enterprises.

Jessica Mei Pung holds a PhD in business and economics from the University of Cagliari (Italy). During her PhD program, she also conducted her research as a visiting scholar at the Department of Tourism at University of Otago (New Zealand) and the Department of Tourism, Sport and Hotel Management at Griffith University (Australia). Her research interests lie in tourist experiences and behavior, peer-to-peer accommodation and innovation. She has published in leading academic journals such as *Annals of Tourism Research*, *International Journal of Hospitality Management* and *Current Issues in Tourism*. She is currently a tutor, research assistant and project coordinator at the University of Otago.

Shah Zahidur Rahman is a recent graduate from North South University (NSU). He completed his bachelor of science (BS) in economics in 2018 with magna cum laude honors. He has worked as a teaching assistant in the Department of English and Modern Languages of North South University for a semester. He is currently working as a customer relations executive at Getco Business Solutions. His areas of interest are international trade and macroeconomics. When he isn't busy with work and study, he spends his time writing fictions.

Dirk Reiser is professor for sustainable tourism management at the Rhine-Waal University of Applied Sciences in Kleve, Germany. His research interests are sustainable tourism, in particular wildlife tourism, CSR, marketing and environmental management. He is a member of the International Association of Scientific Experts in Tourism (AIEST).

Gaunette Sinclair-Maragh is an associate professor at the University of Technology, Jamaica. She holds a doctorate of philosophy in business administration with specialization in hospitality and tourism management from the Washington State University in the United States. Her research interests span across tourism planning and development, sustainable tourism development, hospitality management and marketing, event planning and management and service management. She has a number of journal and book chapter publications and has presented at many local, regional and international conferences. She is also a member of the editorial team for several hospitality- and tourism-related journals and also serves as a reviewer for a number of academic journals.

Seda Sökmen is a research assistant in Faculty of Tourism at Anadolu University, Turkey. She is pursuing her PhD. Her research interests include museology, social psychology and event experiences.

Birsen Bulut Solak graduated from Ankara University with a BSc in dairy technology. She obtained her MSc and PhD in food engineering from Selcuk University (TR). She was awarded an international postdoctoral research fellowship by the Scientific and Technical Research Council of Turkey (TUBITAK). She fulfilled her postdoctoral research in food and nutrition sciences at University College Cork (IRL) under the supervision of Dr. Seamus Anthony O'Mahony between 2013 and 2014. She was a university lecturer in the Programme of Dairy Technology at Selcuk University between 2002 and 2013. Dr. Bulut-Solak is currently an assistant professor in the Department of Gastronomy and Culinary Arts at Selcuk University, Konya (TR). She has been an ambassador in Global Harmonization Initiative for Turkey since 2016.

Wanting Sun is a PhD candidate at the University of Queensland Business School. Her research interests encompass the areas of tourist behavior, destination marketing and sport fandom.

Gozde Turktarhan is a visiting research scholar at University of South Florida Muma College of Business M3 Center Technology and Innovation for the Hospitality. Her master thesis has been awarded The Best Master Thesis in 2014. She completed her doctorate degree in the area of destination marketing. After completing her doctorate degree in 2019, she went to University of South Florida for her post doctorate researches by the support of Scientific and Technological Research Council of Turkey. Her research interests are in the areas of destination marketing, smart destinations and tourism technologies. She serves as assistant editor of the *Journal of Global Business Insights*. She is an active member of International Federation for IT and Travel & Tourism (IFITT), Hospitality Financial and Technology Professionals (HFTP), American Hotel & Lodging Educational Institute (AHLEI) and Association of North America Higher Education International (ANAHEI).

Medet Yolal, PhD, is professor of marketing in Faculty of Tourism at Anadolu University, Turkey, where he mainly teaches issues related to destination management and marketing, tourism marketing and consumer behavior. He has authored or coauthored several articles, book chapters and conference papers on hospitality marketing, consumer behavior, management of small- and medium-sized enterprises in tourism and event management. His research interests mainly focus on tourism marketing, consumer behavior, tourist experience, event management, tourism development and quality of life research in tourism.

Nisan Yozukmaz, PhD, is a researcher from Faculty of Tourism, Pamukkale University, Turkey. She earned her bachelor's degree from Department of American Culture and Literature, Ege University, and her master's degree in the field of tourism management and studied tourism marketing. Her doctoral thesis was on social psychology of tourism, more specifically existential authenticity and heterotopic destinations. She is currently working as a research assistant in the Department of Tourism Management.

1

PSYCHOLOGY, SOCIOLOGY, AND SOCIAL PSYCHOLOGY

Hacer Harlak

Introduction

Three friends went to Rome to spend their holiday. They had booked an economic hotel room for three persons for five nights by using one of the online internet booking sites. The hotel was not far from the city center by bus. It was an old apartment transformed into tourist accommodation. There was no reception, room service, restaurant, etc. They found the keys as written in the directions in the message they had received from the booking site. They went into the room. It was a large room with only one small window. There were three single beds and simple furniture. However, as soon as they entered the room they felt a bad smell and when they opened the bathroom door the smell became unbearable. One of them called the contact number and asked for a solution for the bad smell to the woman on the phone. The woman said that a plumber would repair the washbasin the next morning. However, the next day nobody arrived to repair it, and they had to sleep with that disgusting smell. The next morning, they called the contact person again. In the daytime, they were out to visit tourist attractions in Rome but when they arrived at the room in the night the same bad smell continued. One of them proposed to change the hotel, while the other two rejected the idea because it was high season and said that it would be very difficult to find an available room for a low price. They managed to spend five nights despite the horrible smell in the room.

Many travelers with a low budget could have experienced a scenario similar to the one above. Incidents like this and somewhat different others could be analyzed from various aspects. An economist, a psychologist, a sociologist, a social psychologist, and other social scientists can have diverse perspectives for various situations in which such social interactions occur. Focusing on psychology, sociology, and social psychology, this chapter deals with the theoretical and methodological approaches of the three disciplines to understand and explain human behavior and relationships. Each area has its collection of topics, perspectives, history, and research methodologies. In this chapter, first, the disciplines of psychology, sociology, and then social psychology will be introduced. In the final section, they will be compared and contrasted with one another in relation to some aspects.

Definition, Main Concerns, and Subject Areas of Psychology

Many people ask such questions as what kind of relations are there between the human brain and behavior? Why do we dream? Do animals have intelligence? Why do many people become nervous before an important exam? Why do some people hurt themselves? How does a newborn human baby who has a 3.5 kilogram weight and 48–50 centimeters height and is not able to speak

DOI: 10.4324/9781003161868-1

and walk become a capable person who can do almost whatever she/he wants? How can we learn anything? How do we remember many things but sometimes forget some important ones? How do we make decisions? etc. You can add your questions to the list. These questions are not new. Namely, many thinkers asked them and more others similar to them before. However, searching for scientifically valid answers to these questions is relatively new.

The questions above are examples of which psychology researchers deal with for a long time. Psychology as a scientific area is defined as "the scientific investigation of mental processes (thinking, remembering, feeling, etc.) and behavior" (Griggs, 2012; Kowalski & Westen, 2010). Behaviors are generally defined as observable responses or actions to the environment. They vary from smiling, speaking loudly, running, reading a poem, etc. to hitting someone, talking in sleep, complaining about something, crying, etc. Mental processes include cognitive processes such as thinking, remembering, information processing, and affective processes such as feeling sad, joy, angry, or anxious.

Psychologists are interested in both the nature of specific behaviors, emotions, and mental processes and their underlying causes. Many psychological theories were developed to explain behavior and related processes by utilizing psychological research methods ranging from experiments carried out with limited participants in laboratory settings to surveys with many participants from various social backgrounds.

Psychology is a social science as well as natural science. Many psychology studies focus on human beings; however, there are many investigations on animal behavior too. Psychologists do research on the nature of humans and animals which questions the effects of biological factors such as the roles of evolutionary processes, physiological processes (inner states, brain activation, nervous system, etc.) on behavior, sensation, affection, and mental health. Also, they study the environmental or sociocultural effects (child-rearing practices, social norms, and beliefs) on behavior.

Let us consider psychological research on stress. Stress is generally defined as "the response to the events that threaten or challenge a person" (Feldman, 1997, p. 307). The reasons for stressful experiences of a person might be various such as unemployment, serious health problems, important life events (divorce, car accident, losing a beloved one, moving to another country, etc.), or other real-life situations such as going for a holiday and asking for a room change at a hotel, like in the example at the beginning of this chapter which seems not as important as or as negative as "big" life events. When a person is under stress, several physiological reactions and changes are experienced in the body. A psychology researcher (e.g., a neuropsychologist) is concerned with exploring these physiological processes or mechanisms to explain the behavior or the relations between the mechanisms and behavior. However, another psychologist (e.g., personality psychologist) concentrates on the effects of personality characteristics on behavior to explain why there are individual differences in reactions to stressful situations. Their theories, research methods, and levels of explanations would differ from one another to some degree.

To gain an understanding of the main focus of psychology we can look closely into the history of psychology.

Major Developments in Psychology from a Historical Perspective

In psychology textbooks, the following phrase written by German psychologist Hermann Ebbinghaus (Ebbinghaus, 1908) is frequently quoted: "psychology has a long past but a short history." Namely, the questions and answers which exist even in today's psychology were argued by ancient philosophers (e.g., Plato, Aristotle, Hippocrates) and later by others (e.g., R. Descartes, T. Hobbes, I. Kant, J. Locke) long before psychology has been founded as an institutionalized scientific discipline. However, the beginning of psychology that many textbooks seem to agree on is 1879, when Wilhelm Wundt founded a laboratory in Leipzig, although recent studies showed that there were early developments in Germany before that (see Greenwood, 2015).

Table 1.1 Milestones in the History of Psychology

F.J. Gall – 1st volume of *Anatomie et Physiologie du Système Nerveux* (Anatomy and Physiology of the Nervous System), 1810
J. Müller – *Handbüch des Physiologie des Menschen* (Handbook of Human Physiology), 1834
C. Darwin – *On the Origin of Species*, 1859 and *The Descent of Man*, 1871
G. Fechner – *The Elements of Psychophysics*, 1860
Wundt – *Principles of Physiological Psychology*, 1873
E. Kraepelin – *Compendium der Psychiatrie*, 1883
H. Spencer – *Principles of Psychology*, 1855
F. Brentano – *Psychology from an Empirical Standpoint*, 1874
W. James – *What is an Emotion?* 1884 and *The Principles of Psychology*, 1889
H. Ebbinghaus – *On Memory*, 1885
O. Külpe – *Outlines of Psychology*, 1893
W. Wundt – *Outlines of Psychology*, 1897
E.B. Titchener – *The Postulates of a Structural Psychology*, 1898
E.L. Thorndike – *Animal Intelligence*, 1898

The historical roots of psychology are in philosophy and physiology. For a long time philosophers searched for the answers to various questions about the nature of human thought, feeling, and behavior by using philosophical methods like logic and argumentation. On the other hand, throughout the nineteenth century studies in physiology constituted a basis of psychology. Some major developments in the foundation of modern psychology are presented in Table 1.1.

In the last quarter of the nineteenth century, the first journals (*Mind* in 1876 and *American Journal of Psychology* in 1887) in the field of psychology were founded and the American Psychological Association (APA) was founded with 42 members in the USA in 1892.

The long past of psychology above presents the picture of the background of this discipline. At the beginning of the twentieth century, there were two schools of thought in psychology: structuralism and functionalism. Structuralism was developed by Edward Titchener who had been Wilhelm Wundt's student in his laboratory. Wundt believed that consciousness could be investigated in the laboratory using introspection, that is, asking people about their subjective experience during a perceptual process; however, he never accepted that experimentation was the only route to psychological knowledge (Kowalski & Westen, 2010). On the contrary, Titchener, who was the founder of structuralism in psychology, accepted that laboratory experimentation was the only way to study consciousness. Functionalism, on the other hand, was interested in how our minds adapt to a changing environment rather than the structure of consciousness (Plotnik & Kouyoumdjian, 2011).

Throughout the twentieth century, there were two main distinct approaches to behavior. One emphasized the effects of the environment on behavior and the other concentrated on the effects of internal processes on behavior. The first one was behaviorism which was founded by E. Thorndike, J.B. Watson, and then B.F. Skinner. The other approaches were psychoanalysis and cognitive perspective which concentrated on internal processes. Freud and followers studied the unconscious and suggested several concepts to investigate the unconscious processes. However, cognitive perspective concentrated on the person's cognitive processes between given stimuli and making a response to these stimuli. These approaches exist as distinct perspectives in contemporary psychology.

Main Perspectives in Psychology

Today's psychologists approach main questions of psychology from various perspectives. These perspectives can be classified into two as internal and external in terms of the type of causes that they emphasize (Griggs, 2012). The approaches focusing on internal factors are biological and cognitive, and the others explaining psychological phenomena as caused by external factors are behavioral and sociocultural.

Psychodynamic Perspective

The founder of psychodynamic perspective, S. Freud (1856–1939), was a neurologist who lived mostly in Vienna, Austria. He treated many patients with hysteria by using first hypnosis, then psychoanalysis, which he developed as a new way of treatment for psychological disorders. He described the unconscious as a repository of aggression and sexual motivations. Human actions are primarily directed by these motives unconsciously. In his theory, early childhood experiences are very important in the formation of personality. The phases what he called psychosexual developmental stages (oral, anal, phallic, latent, and genital stages) shape one's personality. The unconscious expresses itself in indirect ways, for instance through dreams. A. Adler (1870–1937) and C.G. Jung (1875–1961), who were followers of Freud, contributed to psychodynamic theory, emphasizing factors other than unconscious sexual motives. Freud's theories echoed in the USA in the 1920s. Karen Horney (1885–1952) and Erik Erikson (1902–1994) are well-known neo-Freudians who developed their own theories although kept the basic premises of psychodynamic perspective. Today there are diverse psychoanalytic schools of thought (J. R. Greenberg & Mitchell, 1983).

Behaviorist Perspective

The behaviorist perspective focuses on how the external stimuli create behavior. I.P. Pavlov (1849–1936) was a Russian researcher who introduced the concept of a conditional reflex while studying the same on the digestion of dogs and extended the domain of neurophysiology to cover non-natural responses (Harré, 2006). In the USA, E.L. Thorndike (1874–1949) suggested "law of effect" which meant that the connection between a stimulus and response is strengthened when the response is followed by a "satisfying outcome" (e.g., reward) and weakened when followed by a "non-satisfying result" (Sheehy, 2004). J.B. Watson (1878–1958) emphasized, as many later behaviorists, the great effects of environmental stimuli on behavior. Behaviorists explained behavior as responses to environmental stimuli and thus gave great importance to the process of learning. B.F. Skinner (1904–1990) thought that solely observable behavior must be studied for scientific psychology (Harré, 2006). From Skinner's viewpoint, a person was a learning machine controlled by the environmental stimuli through operant conditioning, that is, the schedules of reinforcement. Behaviorists used laboratory experiments as research methodology and accepted that generalizations can be made on human behavior from experiments on animals. Today, historically behavioristic thought is named radical behaviorism, and the way of thinking that non-observable processes also have effects on behavior is named neo-behaviorism.

Cognitive Perspective

In contrast to behaviorism, the cognitive perspective concentrated on mental processes in between exposure to stimulus and action. These processes are thinking, reasoning, deciding, planning, calculating, and remembering. The cognitive perspective is based on F.C. Bartlett's (1886–1969) book published in 1932 titled *Remembering* and J.S. Bruner's (1915–2016) studies on how cognitive processes shape perception. The effects of cognitive perspective are so wide and strong in psychology that the subject issue of psychology is defined as cognitive processes. Today this perspective is not only an approach but also a way of thought for a variety of fields ranging from cognitive neuroscience which includes the brain connections and the neural correlates of cognitive processes and artificial intelligence to cognitive behavioral therapy.

Biological Perspective

The idea that biological factors such as genes, brain, nervous system, and endocrine system have effects on mental processes, affection, and behavior existed for a long time. It was suggested that there are two debates on the effects of biology on psychology (Harré, 2006). One of them is related to the question of how much neurophysiological concepts can describe and explain human social and cognitive activity, namely, whether or not psychology can be reduced to neurophysiology. The other is concerned with the effects of heredity and culture on human behavior, that is, to what degree behavior can be determined by genetics. Today behavioral neuroscience, cognitive neuroscience, and evolutionary psychology are the fields in which psychological processes are studied from a biological perspective.

There are also other perspectives in psychology. From the sociocultural perspective, the behavior and mental processes have been discussed, concentrating on the similarities and diversities between societies and cultures. The humanistic perspective suggests that every person is unique, can make choices freely, and self-actualization needs direct him or her. Some humanistic theorists focus on the people's way of finding meaning in life. Particular concepts and theories are developed from these diverse perspectives for various topics of psychology.

Topics and Subfields of Psychology

Psychology has a wide array of topics. Some of these are sensation, perception, learning, thinking, motivation, memory, language, intelligence, emotions, developmental processes, personality, distress and coping, mental health, psychological disorders, psychotherapy, and applying psychological knowledge to daily life. APA has 56 subdivisions for separate areas of psychological studies such as environmental psychology, psychology of women, psychology of religion, health psychology, and industrial and organizational psychology (for the full list of these subdivisions see *https://www.apa.org/about/division*).

There are a variety of subfields in psychology. Each of these has its main areas of study, theories, and even research methodologies. Some examples of these areas are as follows: Neuropsychology is an area in which researchers are interested in physiological processes of the nervous system and its normal functions as well as dysfunctional processes for behavior and cognition. Cognitive psychology is associated with the way human thinks, learns, remembers, and makes decisions. Developmental psychology is an area related to how people grow and adapt to their environment in the course of their lives. Clinical psychology is an area of specialization in the research, assessment, diagnosis, evaluation, prevention, and treatment of emotional and behavioral disorders. Social psychology is related to the social influences on thoughts, emotions, and behavior. Environmental psychology centers on human-environment interactions. Health psychology is an area using the scientific knowledge of psychology to promote health, prevent illness, and improve health care. Industrial and organizational psychology is an applied area in which researchers study human behavior and relations in organizations and the workplace.

Research Methods in Psychology

The objectives of psychology are to describe, explain, predict, and change behavior (Shaughnessy et al., 2012) and mental processes. For instance, a psychology researcher studying adolescents' behavior deals with finding answers to such questions as to what kind of behavior adolescents typically do? Why do they behave in certain ways? How would they respond to diverse environmental conditions (e.g., school settings)? How do they change their behavior? The goals underlying these questions are describing, explaining, predicting, and changing, respectively.

Psychologists research various subjects using scientific research methods ranging from surveys to experiments. The scientific method is a somewhat different way of thinking, explaining, and problem-solving than the ones which a layperson uses in daily life. It is based on an empirical approach, that is, it includes *systematic and controlled observation, unbiased and objective reporting, clear operational definitions of constructs, accurate and precise instruments, valid and reliable measures, and testable hypotheses* (Shaughnessy et al., 2012, p. 40). Research methods used in psychology might be classified into two groups as experimental and non-experimental. There is also another method called quasi-experimental.

Researchers follow certain steps in the process of making scientific study whatever their research method is. First, they formulate research questions derived from the theory. Second, they decide the method they will use based on the research question they have formulated. Third, they collect data and then analyze it by utilizing the various statistical and qualitative analysis techniques according to the type of data (quantitative and/or qualitative) they have collected. Finally, they report their findings and discuss them based on the theory which they have chosen by making comparisons to previous research results.

Experimental Method

Knowledge of psychology overwhelmingly is based on the results revealed in the experiments. The experimental method is usually considered the best way to explore the cause and effect relations between variables. Researchers choose a variable to test the effect of it (*independent variable*) on another variable (*dependent variable*) in a controlled setting (usually laboratory). There are two types of experimental settings: laboratory and field (real-life situations). In a *laboratory experiment*, the independent variable (cause) is manipulated by the experimenter. *Experimental manipulation* means creating more than one condition of an independent variable. For example, consider the following research question: What is the effect of being sleepless on attention? In an experiment being sleepless (independent variable) is manipulated, that is, at least two levels of it are generated. One is being sleepless (e.g., participants are not permitted to sleep for 24 hours) and the other not being sleepless (eight hours sleep in last 24 hours). These are called *conditions*. At least two conditions should be created in an experiment: One is the *experimental condition* in which the independent variable is manipulated and the other is the *control condition* in which there is no manipulation for the independent variable. The participants are randomly assigned in one of the two conditions. Then the degrees of attention are measured and compared between conditions. The researcher should certainly have *experimental control* over the factors that are likely to have effects on attention other than sleeplessness. If the participants in the experimental condition have lower attention scores than those in the control condition, then the researcher concludes that sleeplessness causes lower attention.

Psychology experiments can also be conducted in real-life settings (in a school, a workplace, a restaurant, etc.) as well as in a laboratory. This kind of method is called the *field experiment*. In the laboratory, possible *confounding variables* which are likely to have effects on the relationship between independent and dependent variable can be controlled at some degree by the experimenter; however, in field experiments, experimental control is a problematic issue.

Non-experimental methods

Many psychology researchers use non-experimental methods to seek an answer to research questions. One of the non-experimental methods is *case study*. This method is based on making observations and obtaining detailed information from a person or a group of people who has/have certain characteristics related to the research question. Case studies are more frequently used in clinical psychology than in other areas of psychology.

Another non-experimental method is *surveying*. Researchers seek answers to both descriptive and correlational research questions by using this method. In surveys, questionnaires are used to collect data from a large sample of people. Questionnaires usually consist of standard *self-report scales* and closed- as well as open-ended questions related to the research subject. The most preferred self-report scales are Likert-type with a five-point response set (from "completely agree" to "completely disagree") that is generally developed to measure the attitudes. Closed-ended questions are presented with answering options to the respondents. The options can be yes/no type as well as a checklist.

Sampling is an important issue in surveying methodology. To provide the generalizability of results the representativeness of respondents in the research sample should be sufficiently well planned. Nevertheless, the respondents of many psychology studies are university students, mostly introductory psychology students. There are many criticisms associated with this point (e.g., Hanel & Vione, 2016).

Whatever the research methodology, developing a psychological measurement tool is a very important issue in psychology. Measures vary from very simple ones such as some personal characteristics (gender, likes/dislikes, etc.) to very complex phenomena such as intelligence, personality, and motivational orientations. There are several tests (*standardized measurements*), inventories, and scales developed by psychometrists and other psychology researchers. Two features are particularly important, which all kinds of standard measures are required to have. One of them is *validity*, that refers to how well a measure actually measures the construct it is intended to measure (Netemeyer et al., 2003). The other is *reliability*, that refers to the consistency between test items and between measures taken using the instrument from time to time.

In sum, psychology is a field of study concentrated on understanding, explaining, and sometimes changing behavior and mental processes by using the scientific method. In lots of subfields, psychology researchers seek answers to the questions related to humans (sometimes animals) in an enormous variety. However, psychology is not the only field concerned with human behavior and relations. Sociology also asks questions on the same area of study by concentrating on different aspects.

Definition, Main Concerns, and Subject Areas of Sociology

Sociologists seek answers to the questions concerning human societies and collectivities. However, the interaction between individuals as actors or agents and society, that is, how society is formed collectively, is the main focus in sociology. Human life continues in societies. In many societies, people socialize in a family and groups, live in a neighborhood, attend schools, work in diverse workplaces, get married and experience intimate relationships, watch media, and are convinced that there are sacred things. Sociologists carry out studies on many aspects of human life. Hence, sociology is the scientific study of human life, social groups, whole societies, and the human world as such (Giddens, 2009).

In everyday life, people often talk about topics that might be considered in the field of sociology. For example, think about the situation at the beginning of this chapter. You can ask why people do (or do not) complain about any problem when they are on a vacation and make some explanations for the situation. So, one can ask what are the differences between a layperson and a sociologist in their thoughts and explanations about sociological issues. The answer is in sociology's broader view of people's lives to explain why people act as they do (Giddens, 2009). Sociologists think sociologically (Bauman, 1990). Namely, they try to understand the social world and make explanations (cause and effect relationships) based on sociological knowledge. However, the ultimate goal of sociology is not to find an absolute truth, but to improve the adequacy of sociological knowledge (Bauman & May, 2001, p. 167) and to play a role in the process of changing society.

In a society, people are members of several social groups which are formed in terms of gender, race, ethnicity, class, age, etc. There are status differences or social inequalities between them. That is, they are not equal in the allocation of rewards, opportunities, and other social resources (Renzetti & Curran, 1998). Sociologists seek to understand the relations between these groups and to explore the cause and effect relationships with a point of view far from ethnocentrism that leads people to judge a situation or a relation in a way that creates a good or bad judgment of value. Also, the process of social (economical, political, and cultural) change is one of the core contents of sociology. The history of sociology reflects the main sociological thoughts.

Major Developments in Sociology from a Historical Perspective

Sociology emerged in the nineteenth century in Europe though there had been some thoughts and explanations on society by thinkers or philosophers long before that. However, the industrial revolution changed people's way of life deeply and led some thinkers to think and to ask questions about society such as how and why do societies change (Giddens, 2009, p. 12). Hence, sociological knowledge gradually increased from the mid-nineteenth century on.

Some illustrious pioneers founded sociology as a science. Auguste Comte (1798–1857) gave sociology its name. Comte was strongly influenced by positivism that was dominant in his age. The positivistic understanding is based on the idea that there is an order in the world and one can find answers related to it through direct observation of the entities, and can infer the laws underlying the relations between these entities. Comte believed that sociology would become a positive science. Another founder is Émile Durkheim (1858–1917) who established sociology as an independent scientific discipline by distinguishing it from the other fields of study. Durkheim focused on social order. For him, a social order could be possible through collective conscience which he defined as a set of rules that give the members of society the shared feeling that they belong to something larger than themselves as individuals (Renzetti & Curran, 1998). He emphasized that sociology is the study of social facts, that is, aspects of social life that shape our actions as individuals such as the state of the economy or the influence of religion (Giddens, 2009).

On the other hand, Karl Marx (1818–1883), as the founder of conflict theory in sociology, believed that societies changed through the struggles between two classes in society: the ruler and the ruled or the exploiters and the exploited. In contemporary capitalist societies, these two antagonistic classes are the bourgeoisie and the proletariat. He predicted that society will be transformed from capitalism which is based on inequalities to communism where all inequalities are abandoned. Another founder Max Weber (1864–1920) worked on the Western type of capitalism, protestant ethics, and democracy. He suggested an ideal type of bureaucracy for modern societies, which he defined as a hierarchically organized structure designed for rational coordination of working people who strive for achieving administrative duties and organizational goals (Slattery, 2003).

Besides the European founders, some early sociologists in the USA made significant contributions to sociology in the beginning. After the first sociology department established at the University of Chicago, a group of sociologists such as Robert Park (1864–1944), Ernest Burgess (1886–1966), W.I. Thomas (1863–1947), and Florian Znaniecki (1882–1958) which was named as Chicago School studied important topics. They were interested in social problems such as social deviance, urbanization, and cultural change. In the 1940s and 1950s, Durkheim's ideas were transported to the USA and thus had effects on Talcott Parsons (1902–1979) and Robert Merton (1910–2003) who founded the functionalist perspective in sociology. C. Wrights Mills (1916–1962) suggested the term "sociological imagination," a wider perspective that linked personal situations to the circumstances of society.

All the ideas and perspectives of these pioneering sociologists gave direction to recent sociological theory and research. The questions and answers related to the social causes and consequences are considered from these diverse perspectives.

Main Perspectives in Sociology

There are several theories in sociology. Perspectives are approaches underlying theories. One of the perspectives is structural functionalism that was influenced by Durkheim's ideas. Functionalism assumes that society is a stable and ordered system made up of structures. These structures or institutions, that is, economy, government, education, family, religion, and the health care system, have a function or role to play in this system. Parsons, developing Durkheim's approach, suggested a theory of social systems for the relations between the whole system and its structures. For him, every society is a system in which various subsystems are interrelatedly and interdependently connected (Slattery, 2003). The subsystems contribute to the fundamental needs that have to be satisfied for society's survival. These needs are adaptation, goal attainment, integration, and latency. Another functionalist sociologist, Merton, distinguished two types of functions as manifest (observable consequences or outcomes that are intended) and latent (observable consequences or outcomes that are intended and not expected).

Another perspective in sociology is conflict theory derived from Marx's ideas. The main emphasis of this perspective is on social inequality. As each individual or group struggles to attain the maximum benefit there is always social conflict and consequently the need for change in society. The processes of dominance, competition, upheaval, and social change are some of the topics that conflict theory concentrates on.

Some sociologists accept that people interact with one another collectively within a particular structure and give collective meanings to their interactions, and that sociology should study these interactions (Renzetti & Curran, 1998). Symbolic interactionism is a somewhat different perspective in sociology. Contrary to functionalism and conflict theory, symbolic interactionism focuses on interpersonal interactions in everyday life rather than on society as a whole. The basics of this perspective were founded by G.H. Mead (1863–1931). For Mead, human beings have a specific mental mechanism. Owing to this mechanism they can plan and adjust their behavior to the goals they adopt; to communicate through symbols, that is, language; and to be aware of themselves as well as others' feelings, thoughts, motivations, roles, etc., and therefore have a self-image. So, a person has two selves at the same time, one is based on her/his ideas about self (I), and the other consists of the significant others' reflections about one's self (me). Social life is solely possible through interpersonally communicating the meanings.

Sociologists study various subjects and generate theories adopting one of these perspectives.

Topics and Subfields of Sociology

Contemporary sociology is a wide field of study. The topics studied in sociology spread out over a wide range of areas. Socialization, family, social interaction, urban life, social class, poverty, inequality, social movements, health, illness, disability, sexuality and gender, ethnicity, migration, education, religion, media, work and labor, deviance, and environment are some of the topics of sociology. *American Sociological Association* listed 52 sections that focus on separate specialty areas varied from evolution, biology and society, sociology of emotions, sociology of human rights to global and transnational sociology.

Sociological knowledge depends on the findings and explanations revealed in sociological studies in which scientific research methodology is used as in psychology.

Research Methods in Sociology

Research methodology in sociology is the most important subject to generate sociological knowledge. Sociologists use either *qualitative* or *quantitative* methods or both in accordance with the research question that they choose to answer. The theoretical perspectives that they adopt determine their approach in doing research. Qualitative methods are appropriate for a researcher who wants to gather detailed data about an individual action in the context of social life, whereas quantitative methods comprise the measurements of related social phenomena and statistical analyses (Giddens, 2009).

Research methods in sociology are surveys, fieldwork, secondary data analysis, and experiments. Surveying can be carried out using in-person interviews as well as questionnaires. Interviewing can be in structured or semi-structured form. It provides the researcher to produce new and in-depth questions to the respondents besides giving the possibility of clarifying both questions and answers. However, it requires much time and labor. Using questionnaires the researcher can gather both quantitative and qualitative data according to the way the questions have been formed (open- and closed-ended questions). Questionnaires shorten the time for data collection and thus economical.

Another qualitative method is fieldwork. Fieldworks can be conducted by making observations, ethnography, and case studies. In fieldwork, researchers only observe and take notes without participation in a situation or personally participate in the situation on which they collect data. The observation method has the advantage of obtaining in-depth information about the subject that researchers study. On the other hand, ethnography means collecting data about a group or a community in which the researcher lives for some time. In addition, the case study is used, like in psychology, as an in-depth analysis of a single individual, event, or situation.

In some studies, secondary analysis or documentary research is also used. This qualitative method allows the researcher to collect or analyze historical and longitudinal data in relation to the research question. Various records (written, visual, or tape recording) could be data for this type of study. Usually, content analysis can be applied to data to interpret it.

Besides the non-experimental methods, experimentation is used by sociologists particularly to explore the cause and effect relations between social phenomena. Experiments can be in the laboratory as well as in the field, that is, real-life social settings like in psychology. For experimentation, a researcher chooses one or more factors (*independent variable*) to explore its effect on an outcome (*dependent variable*), mostly aimed at testing a hypothesis derived from a theory. Researchers first create *experimental* (in which the manipulation of independent variable occurs) and *control conditions* (no manipulation), then *randomly* assign the participants to the conditions and subsequently apply treatment and compare results for the conditions. In laboratory experiments, the researcher can have control over the research conditions and can afford the possibility of carrying out a precise study. Nevertheless, the generalizability of findings to a real-life context (*external validity*) is limited because of the artificial conditions in the laboratory.

Field experiments are recommended to overcome the generalizability problem in laboratory experiments. In this case, both experimenter and participant bias may occur because double-blind and randomization procedures cannot be applied in real-life situations. However, randomization could be possible in studies that are called "social experiments" (D. Greenberg et al., 1999). Furthermore, there are quasi-experimental designs in which experimental manipulation is applied in survey questionnaires (vignettes) to participate in larger sample sizes, though they are very rare.

The use of the experimental method in sociology is quite restricted (Giddens, 2009), even though there was an invitation to experimental sociology as early as the 1930s (Brearley, 1931). However, a review revealed that experimental designs in sociology increased from the 1990s to 2010s (Baldassarri & Abascal, 2017; Jackson & Cox, 2013).

In sum, sociology and psychology are complementary fields of social sciences. To gain a comprehensive understanding about persons and their social organizations, that is, the social life of humans, we need both disciplines. At this point, one might ask that why another field of study is needed when there already exist psychology and sociology which are distinct yet complementary fields. Answering this question requires understanding the subject matter of social psychology. Many people think about several questions at least once in their lifetime such as why do people behave similarly to one another in social environments? Why do people treat each other in a friendly or hostile manner? Why do some people give help to another person while others do not? How does a person attract one another? How does a person become a leader while some others prefer to be followers? How does being a member of a group affect perceiving the other groups? How do the other's behavior and thoughts influence a person? How does a single person influence many people? These queries bring us to social psychology.

Definition, Main Concerns, and Subject Areas of Social Psychology

Social psychology is a field of social science that is closely related to both psychology and sociology. To clarify the status of social psychology, it would be useful to refer to Allport's definition of the field. Social psychology is a scientific investigation of how the thoughts, feelings, and behaviors of individuals are influenced by the actual, imagined, or implied presence of others (Allport, 1954). That is to say, social psychology's main concern is on the interactions between persons in various social situations. These situations consisted of person-to-person relations, relationships between members in social groups, and social processes in intergroup contexts.

Social psychologists develop concepts and theoretical explanations and carry out research to test their hypotheses derived from various social psychological theories on the wide range of topics that vary from interpersonal to intergroup interactions. Investigating the social and cognitive processes in human interactions can serve to comprehend why people act the way they do, and also to guide social psychologists to make suggestions to solve some important social problems.

Most social psychologists are interested in human social interactions rather than animals in contrary to some psychologists because making generalizations from animal to human is quite difficult (Hogg & Vaughan, 2011). Also, social psychologists focus on cognitive and social processes rather than observed behavior. Let us consider the situation at the beginning of this chapter as an example of those kinds of social situations. A social psychologist might ask the following questions: How do people influence one another in small groups (at least three persons) especially when there is a problem to be solved? Is it possible for a single person to persuade a majority to change their mind? How do people negotiate and make decisions in groups? How do people resolve the cognitive dissonance that they experience after they make a decision? How are social interactions in groups affected by the social and cultural backgrounds of the people involved in the situation? Also, what are the effects of the features of certain situations (being on vacation in this case) on group problem-solving?

Social psychologists focus on social interactions, that is, the immediate processes where the behavior and mental processes are experienced, and the effects of social group membership and wider sociocultural contexts on all social interactions. Social psychology is concerned with how social interactions occur and how it does affect people to make changes in their behavior, thoughts, or emotions. In other words, the social psychological concepts and theories explain the effects of *the other* on the individual. The other varies depending on the context. It means another person, a group of people, a group that the person is being membered, or other groups that are not being membered (sometimes hostile group) in different social situations. Individuals are influenced by the other continuously, even if it does not exist physically in the situation. This emphasis on social influence distinguishes social psychology from both psychology and sociology.

Examining the historical developments in social psychology could be useful to comprehend the main focus of social psychology.

History of Social Psychology

Ebbinghaus's phrase "a long past and short history" is true also for social psychology (Farr, 1996). Questions related to the social psychological phenomena have been inquired for a long time. However, social psychology as a scientific area emerged in Europe toward the end of the nineteenth century. The name of social psychology was first used in an article published in Italy in 1864 (Jahoda, 2007). In France, Gabriel Tarde (1843–1904) published a book titled *Etudes de Psychologie Sociale* (1898) in that he discussed the need for social psychology. Gustave Le Bon (1841–1931) wrote on a person's unexpected behaviors in crowds. In Germany, M. Lazarus (1824–1903) and H. Steinthal studied on folk psychology (*Völkerpsychologie*) like W. Wundt. In the USA, W. James (1842–1910) wrote on the self and the distinction between *I* (personal self) and *me* (social self) which has been still frequently referred to today (e.g., Leary, 1999; Leary & Tangney, 2012).

In the last decade of the nineteenth century, some studies which emphasized the social influences on human performance were published. Binet and Henri studied children and found some evidence for social influence as early as 1894 (Stroebe, 2012). An agricultural engineer Max Ringelmann carried out some studies on maximum performance in group and individual performance conditions in France between 1882 and 1887. However, his studies were published as late as 1913. Triplett (1898) found that motor performance enhanced by the effects of others' existence (social facilitation, presence of others increases one's performance) in his study with cyclists which was often noted as the first social psychological experiment.

On the other side, some sociologists such as G.H. Mead (1863–1931) and W. Thomas (1863–1947) thought on the connection of society with the individual. Their ideas created symbolic interactionist social psychology. C.H. Cooley (1864–1929) and G. Simmel (1858–1918) were major contributors to sociological social psychology. Cooley theoretically formulated the distinction of primary and secondary groups. In primary groups, people have face-to-face contact, whereas secondary groups consist of more people and less intimate relations. Simmel considered society as social interaction networks between persons.

At the beginning of the twentieth century, two textbooks were published in the USA in the same year as 1908. One of them was written by a psychologist, W. McDougall (1871–1938), and the other was published by a sociologist, E.A. Ross (1866–1951). These books reflect the status of social psychology as a scientific discipline in which contains approaches from both psychological and sociological perspectives.

During the twentieth century both World War I and II, and the rise of Nazism in Europe affected the history of social psychology. Many social psychologists moved to the USA and suggested social psychological theories and developed research programs on various social psychological topics. One of the most eminent researchers was Kurt Lewin (1890–1947). Lewin and his research group studied group dynamics, leadership, and attitude change in the 1930s and 1940s. A group of social psychologists, T.W. Adorno (1903–1969), E. Frenkel-Brunswik (1908–1958), D. Levinson, and R. Nevitt Sanford (1905–1995), studied the roots of prejudice and published the theory of authoritarian personality in 1950. M. Sherif (1906–1988) who emigrated from Turkey to the USA by the indirect effects of World War II investigated the effects of social norms on human judgment in 1936.

In the 1940s and 1950s, some studies influenced the development of social psychology. Some of them were the following: S.A. Stouffer (1949) laid the foundations of relative deprivation theory by his study on adjustment in army life. C.I. Hovland, I.L. Janis, and H.H. Kelley (1953) studied the effects of propaganda in Yale University and developed the Yale model of persuasive communication. S. Asch (1951) conducted his well-known study on conformity in groups. L. Festinger (1957) published his book on cognitive dissonance. F. Heider (1958) founded the fundamentals of attribution theory.

In the 1970s, the studies on social influence particularly minority influences were conducted in Europe. Moscovici (1972) suggested the concept of social representations adapted from Durkheim's collective representations. However, in the USA cognitive perspective had become dominant in mainstream social psychology by the 1970s. Social psychology's focus had shifted from real-life social problems to individual processes, and laboratory experimentation had become the dominant research method. After all, criticisms were pronounced toward both theory and method of social psychology (e.g., Moscovici, 1972; Tajfel, 1972). One of the main points in criticisms was that social psychology researchers mostly preferred the experimental method, and the participants of laboratory experiments were college students attending introductory psychology courses. Another point was related to the environmental conditions in which the laboratory experiments were conducted because they were far from being similar to real-life situations. Also, the research focus was the individual not the social. Some social psychologists (Greenwood, 2004; Moscovici, 1972; Stam, 2006; Tajfel, 1972), inclined to a *more social* psychology, discussed the theorization of social psychology (Abrams & Hogg, 2004), and sought new methodologies (Wetherell & Potter, 1988).

From the 1980s social psychology in Europe differed from the USA concerning the emphasis on the social. Social identity perspective became pervasive in Europe while approaching social psychology from the cognitive perspective and considering it with personality was dominant in the USA. The prominent journal of the field was the *Journal of Personality and Social Psychology* founded in 1965. Nevertheless, graduate programs in the USA gradually accepted more students from countries in Asia and other parts of the world. Consequently, cross-cultural studies of which the findings based on data obtained from various cultural groups (ethnic and linguistically different) increased, and hence the importance of cultural context for social psychological phenomena was increasingly recognized. From the 1990s, social psychological studies continued to expand into diverse topics such as self-esteem, causal attributions, social identity, group processes, intergroup contact, values, morality, communication, emotions, romantic love, ideology, terrorism, stress, religiosity, altruism, migration, minority issues, and social change.

Main Perspectives in Social Psychology

There are various theories of social psychology. However, approaches underlying these theories can be grouped into social cognitive, evolutionary, social neuroscience, and sociocultural perspectives.

Social Cognitive Perspective

Individuals are continuously exposed to social information concerned with themselves, other persons, and social groups or events. They try to make sense of these stimuli in an enormous variety. They process social information in a way that it is simplified and adjusted to it. Hence, people are information processors who think, remember, and reason about social stimuli or social world, that is, other persons or groups and also themselves. Impression formation, social categorization, social schemas, causal attributions, stereotyping, biases in information processing, and heuristics are some of the social cognitive concepts. Social psychologists who adopt the social-cognitive perspective utilize the methods of cognitive psychology. Namely, they use laboratory experimentation as a research method.

Evolutionary Perspective

Evolutionary perspective emerged in social psychology in the 1990s. It is an approach in that evolutionary theory is applied to social psychology. From this perspective, many social processes can be explained based on the basic assumptions of evolutionary theory. If a given behavior exists in people of today's world, it is determined genetically either for survival or for the continuation of generations. Evolutionary explanations are often used to understand several questions related to the topics such as interpersonal attraction, close relations, group membership, social behavior in groups, aggression, helping behavior, and intergroup hostility.

Social Neuroscience

Social neuroscience is a relatively new perspective in social psychology. Social neuroscientists study the neural bases of social psychological phenomena. That is, neurophysiological mechanisms in the brain and nervous system are used to explain the underlying processes of social behavior. Therefore, social neuroscience is an integrative field that examines how nervous (central and peripheral), endocrine, and immune systems are involved in sociocultural processes (Harmon-Jones & Winkielman, 2007). This perspective is neither reductionist (reducing something sociocultural to biological) nor dualist (making difference between these two), but rather emphasizes the mutual dependence between the sociocultural and the biological. Social neuroscience researchers make use of neuroimaging techniques such as fMRI (functional magnetic resonance imaging). Several issues in social psychology, for example, prejudice, intergroup emotions, moral beliefs, and social perception, have been studied from the social neuroscience perspective.

Sociocultural Perspective

From the sociocultural perspective, the social psychological phenomena are considered in their sociocultural environment, that is, in the context of social norms, beliefs, and values. The social relations in different (person-to-person, social group, and intergroup) contexts are mostly formed by the effects of cultural norms or values that are commonly held in society. So, many social interactions or social behavior can be understood and explained in reference to social norms, beliefs, and roles. Social psychologists who adopt the sociocultural perspective study family relations, socialization, and the impact of cultural beliefs and norms socially in various social situations including prosocial behavior, aggression, and intergroup encountering.

Topics and Applied Fields of Social Psychology

Topics of social psychology are spread over a wide area. Some of them are social influence, conformity, obedience, attitudes, impression formation, persuasion, impression management, self, self-presentation, social identity, collective behavior, crowd behavior, emotions, prejudice, discrimination, stress, friendship, social dilemmas, family, group decision making, love, stereotyping, collective violence, bargaining, sexuality, sexism, communication, aggression, small groups, power, leadership, language, violence, social categories, speech, altruism, intergroup relations, attraction, prosocial behavior, and social change. Also, social psychological knowledge is used in many applied areas such as health, clinical and counseling psychology, media, sports, education, environment, political science, consumer behavior, intervention to social problems (e.g., alcohol and substance use), and criminal and justice system (e.g., Schneider et al., 2012).

Research Methods in Social Psychology

Social psychologists use both experimental and non-experimental research methods to test their theories related to diverse topics. Experimental methods include laboratory and field experiments. Non-experimental methods are mostly surveys and qualitative techniques.

Experimental Methods

Social psychological knowledge depends on the results of experimental studies to a large extent. Social psychological experimentation is similar to that in psychology. The experimenter manipulates the *independent variable(s)*, that is, creates one or more conditions of it, allocates participants randomly to these conditions, and exposes them to one or two independent variables. Then he/she observes (measures) the changes in *the dependent variable(s)* between the conditions, consequently compares the measures of dependent variables in all conditions, and makes conclusions in a way that consists of the cause and effect relations. The experimentation process is similar to that in psychology; however, there is an important difference between a psychology experiment and a social psychology experiment. In social psychology experiments independent variable(s) are mostly the elements in social situations, in contrast to those in psychology. These aspects require some arrangements before the experiment. For instance, imagine an experimenter studying social influence in a group and the research question is what does a person do in a group if there is an obvious reality but other people do not tell the truth?

S. Asch (1951) sought an answer to this question in his well-known experiment on conformity. He created one experimental condition in which one participant found himself contradicted by a group of seven. Several cards that consisted of one line (x) with three others (a, b, c) were presented to the participants and asked to judge which lines of three were the same with line x. There were 18 trials similar to this. 50 men in the experimental and 37 men in the control group participated in the experiment. The experimenter noted participants' answers in each trial. In fact, the people except for one participant in each experimental group condition were *confederates*, that is, they were not actual participants, but Asch's assistants. Apparently, the experimenter has to deceive participants to create a majority consensus in the group. Also, the truth about the aim of the study was not known by the participants. However, ethical rules were developed concerning the issue. Researchers must display powerful reasons to ethical research boards in universities for using deception if it is necessary. On the other hand, social psychologists do not always use deception in experiments.

Most experiments in social psychology are conducted in laboratories, that is, artificially created environments in university departments. Therefore, their *ecological (external) validity* is low. Researchers use field experiments to overcome artificiality. Field experiments are conducted in real-life settings. They are rare in social psychology because of the difficulties concerning experimental control. However, Sherif's (Sherif et al., 1988) summer camp studies on intergroup prejudice and conflict with boys 13–14 years of age are good examples of field experiments.

In laboratory experiments measuring dependent variable(s) is an important issue. Researchers generally use *self-report* scales or questionnaires. However, self-reports are criticized due to the biases related to social desirability. *Implicit techniques* were developed to overcome these biases. These measurements depend on reaction time as an indicator of the power of association between a stimulus and the activation of the information in memory.

Non-experimental Methods

Surveys are very frequently used in social psychology for descriptive, correlational, and cause and effect studies. According to research aims, researchers obtain data through questionnaires including open- and closed-ended questions and scales developed in terms of psychometric assumptions, as in psychology. Data were analyzed by statistical techniques and the results of the statistical analyses reveal the situation, and correlations or cause and effect relations between the variables. Also, conducting quasi-experimental research is possible by using questionnaires.

Qualitative data analyses include content analysis, conversation analysis, and discourse analysis. These methods are very rare and they are usually utilized for descriptive purposes.

Comparing Social Psychology to Psychology and Sociology

As you can see psychology, sociology, and social psychology are very close disciplines and share some subjects, concepts, and methods. Yet, each discipline has its own focus which distinguishes it from the other one. We can have a full understanding if we compare them by focuses, levels of analysis, and methodological specifications (see Table 1.2).

From the very beginning psychology concentrated on the individual and intraindividual processes. Social psychology's emphasis is on the, connection between the individual and the other, although it shares many concepts with both psychology and sociology. Social psychology comprehends individuals in a social context. Individuals influence others and are influenced by others. They construct a cognitive representation of the social world that they are also a part of it, interact with others, and develop social relations. On the other hand, sociology's focus is the society and the larger collectivities and social processes. The sociological perspective is a consideration of society to understand, explore, and explain the structure and functions of institutions and the process of change. The status of social psychology has been considered frequently in between psychology and sociology. However, social psychology is neither sociology nor psychology, as pointed out by major contributors of the field (M. Sherif, S. Moscovici, and H. Tajfel).

A close look into social psychology reveals the distinctions between the three disciplines, though it seems that it is difficult to differentiate social psychology from the other two. We can direct our attention to their subjects. Remember that psychology studies behavior and mental processes. The questions of social psychology are related to both behavior and mental processes too; however, they are also somewhat different from general psychology's questions like how do individuals have different personalities? Or how does a person learn dancing? Why do some people have extraordinary memories? A psychologist does not need to focus on social interactions but rather on personal characteristics that make a person different from one another in trying to find answers to such questions.

On the other hand, sociology is related to the structure and functioning of society. The main concern of sociology is how the institutions (religion, education, health systems, politics, economy, etc.) in a society function and change. Its focus is on the larger groups and institutions rather than the individual. Sociologists attempt to understand the forces that operate throughout society—forces that mold individuals, shape their behavior, and thus determine social events (Tischler, 2011). However, sociologists differ in their approach to society. Some of them tend to consider society as a total of individuals whereas some others think that society has its transcendent existence beyond individuals.

Social psychology has generally presented a subdiscipline of both psychology and sociology. However, social psychologists from these two scientific backgrounds adopt so diverse approaches and develop their social-psychological knowledge in so diverse environments that it is mentioned that there are two social psychologies, one is psychological social psychology (PSP) and the other is sociological social psychology (SSP) (e.g., House, 1977; McMahon, 1984; Stryker, 1977). Psychological social psychology's main concern is the effects of social and cognitive processes on the way that individuals perceive, influence, and relate to others (Smith & Mackie, 2007). In other words, the focus is often on how aspects of self, attitudes, and interpersonal perception influence behavior (Delamater, 2006). On the other hand, sociological social psychology is rather concerned with social groups or collectivities and families, organizations, communities, and social institutions.

Table 1.2 Comparisons of Psychology, Sociology, and Social Psychology

Fields	*Common Definition*	*Main Perspectives*	*Levels of Analysis*	*Main Questions*	*Research Methods*
Psychology	A scientific investigation of mental processes (thinking, remembering, feeling, etc.) and behavior	Psychodynamic Behaviorist Cognitive Biological	Intraindividual	What is the main cause of behavior, nature or nurture? Do people have free will? How accurate the human information processing? Which factors determine behavior, conscious or unconscious? How do people differ from each other?	Experimentation Case study Surveying
Sociology	A scientific study of human society and social interactions (social groups)	Structural functionalism Conflict theory Symbolic interactionism	Society (Macro level) Interpersonal interactions (Micro level)	What are the functions and dysfunctions of institutions and society? How do social inequalities produce conflict and who benefits from particular social arrangements? How do people give meaning to the world in which they participate and reproduce them through interactions?	Surveys Fieldwork Secondary data analysis Qualitative research Experimentation
Social psychology	A scientific investigation of how the thoughts, feelings, and behaviors of individuals are influenced by the actual, imagined, or implied presence of others	PSP Behaviorist Cognitive Social neuroscience Sociocultural SSP Symbolic interaction Social structure and personality Group processes	Intraindividual Interindividual Positional Ideological Intersocietal Neurological	How do people influence and influenced by other? How does an individual construct a representation of the social world? How do social and neural processes interact? How social and cognitive processes affect the way that individuals perceive, influence, and relate to others? How do people make and share meanings as members of a social group? How do social structures influence individuals? What are the processes that occur in group contexts?	Experimentation Surveying Qualitative research

PSP: Psychological Social Psychology, SSP: Sociological Social Psychology.

Levels of Analysis

It would be useful to examine the levels of analyses in the three fields to differentiate social psychology from psychology and sociology. In psychology generally, the level of explanations is intra-individual. The concepts and theories of psychology direct us to the inner world of the individual. Learning, memory, motivation, hormones, mood, etc. are investigated as individual processes, and explanations and theories concerning these processes focus on the individual, though they often emphasize environmental effects. In sociology, three levels of analyses are distinguished as micro, meso, and macro units (e.g., Turner, 2006). The focuses are face-to-face interactions, group dynamics, and societal indicators, respectively.

In social psychology, four levels of explanations were pointed out as intraindividual, interindividual/situational, positional, and ideological (Doise, 1980, 1986). Intrapersonal level refers to individuals' experience related to perceptions of social environment and the way they organize their experience and behavior (e.g., research on cognitive dissonance, cognitive balance). In the interindividual/situational level of analysis, the focus is interpersonal relations in a given situation. Research on communication networks in groups, interpersonal attributions are at this level of analysis. When interpersonal relations in specific situations are considered by taking into account the relative status and power of two parts the level of explanations is positional (e.g., research on social identity and power). The level of analysis becomes ideological when researchers make explanations on social interactions by considering the role of social beliefs, sociocultural norms, and values. Furthermore, an intersocietal level and a neurological level were added (Doise & Valentim, 2015). The intersocietal level of analysis refers to considerations on some social variables across cultures or societies. Finally, when some researchers investigate the social or societal variables (social beliefs, norms, moralities, etc.) by reference to evolutionary or neurological processes the level of analysis is neurological. Social psychologists both in Europe and in the USA use these levels of analysis; however, the individual and interindividual levels may have been more favored by leading American social psychologists than they were by their European colleagues (Doise & Valentim, 2015).

Methodological Issues

The methods used in psychology, sociology, and social psychology are quite similar. Nevertheless, the experimental method is used often in most psychology and social psychology research whereas in sociology studies this method is used very rarely. On the other hand, surveying is common in psychology and social psychology as well as sociology though there are some technical differences. In the questionnaires used in both psychology and social psychology, some standardized measurements (scales, e.g., five-point Likert type, from completely agree to completely disagree) and implicit techniques (both in the laboratory and in the self-report questionnaires) were very often used. Therefore, quantitative analyses or statistical analyses that differ in complexity (e.g., Student's t-test and multivariate statistical tests such as regression analysis, path analysis, and structural equation modeling) are used. Besides, qualitative analyses are more common in sociology than in psychology and social psychology.

Why Do We Need Social Psychology besides Psychology and Sociology to Understand Tourism?

Psychology, sociology, and social psychology are concerned with the same field of study in a large sense, human behavior and relations. The questions that they seek answers are quite similar to a great extent. They can be thought of as complementary areas. They contribute to the scientific knowledge of people, human societies, and the relations between them. However, the levels of analyses and explanations are significantly different (see Table 1.2). Psychology explains behavior rather by reference to inner processes, whereas sociology tends to search society and its institutions. Social psychology has its unique perspective distinct from psychology and sociology. Society cannot be reduced into individual and individual cannot be reduced into inner processes. To gain a comprehensive understanding of both individual and society, and the interactions between the individual and the social, social psychology is needed.

Understanding and explaining tourist behavior and the social processes during touristic travels requires psychology, sociology, and social psychology. Let us think about the example given at the beginning of this chapter. Concerning these kinds of situations, one can ask the same question, "why do people avoid making complaints about some problems during their vacation?" The answers could be formulated from psychological, sociological, and social psychological viewpoints. Namely, if you direct your attention to personality characteristics or effects of past experiences, motivational orientations, or emotional states that a person experienced in this kind of situation your viewpoint is psychological. If you concentrate on group dynamics, conformity, majority influence, persuasion, group decision-making processes, negotiations in small groups yours is a social psychological perspective. Also, you can have a sociological approach. You can inquire about the differences between tourists from high class and low class concerning approaches to complaining about something while staying in five-star hotels and bed and breakfast accommodations. Or you would like to analyze the effects of neoliberal economy politics on the systems of organizations (e.g., tourist accommodations) served to travelers to explain the patterns of tourist complaints in hotels and reveal how the social system continuously reproduces itself by making customer complaints difficult.

Acknowledgement

I want to thank Prof. Dr. Şerife Geniş for her helpful suggestions especially in organizing the part of the chapter related to sociology.

References

Abrams, D., & Hogg, M. (2004). Metatheory: Lessons from social identity research. *Personality and Social Psychology Review, 8*(2), 98–106.
Allport, G. W. (1954). The historical background of modern social psychology. In G. Lindzey (Ed.), *Handbook of social psychology* (pp. 3–56). Addison-Wesley.
Asch, S. E. (1951). Effects of group pressure upon the modification and distortion of judgments. In H. Guetzkow (Ed.), *Groups, leadership and men; Research in human relations* (pp. 177–190). Carnegie Press.
Baldassarri, D., & Abascal, M. (2017). Field experiments across the social sciences. *Annual Review of Sociology, 43*, 41–73. https://doi.org/10.1146/annurev-soc-073014-112445
Bauman, Z. (1990). *Thinking sociologically.* Basil Blackwell.
Bauman, Z., & May, T. (2001). *Thinking sociaologicallay.* Blackwell Publishing Ltd.
Brearley, H. C. (1931). Experimental sociology in the United States. *Social Forces, 10*(2), 196–199. https://www.jstor.org/stable/2570247
Delamater, J. (2006). Preface. In J. Delamater (Ed.), *Handbook of social psychology.* Springer.
Doise, W. (1980). Levels of explanation in the European Journal of Social Psychology. *European Journal of Social Psychology, 10*(3), 213–231. https://doi.org/10.1002/ejsp.2420100302
Doise, W. (1986). *Levels of explanation in social psychology.* Cambridge University Press.
Doise, W., & Valentim, J. P. (2015). Levels of analysis in social psychology. In J. D. Wright (Ed.), *International encyclopedia of the social & behavioral sciences* (2nd ed., pp. 899–903). Elsevier.
Ebbinghaus, H. (1908). *Psychology: An elementary text-book.* D C Heath & Co Publishers.
Farr, R. M. (1996). *The roots of modern social psychology: 1872–1954.* Wiley-Blackwell.
Feldman, R. S. (1997). *Essentials of understanding psychology.* McGraw Hill.
Festinger, L. (1957). *A theory of cognitive dissonance.* Stanford University Press.
Giddens, A. (2009). *Sociology.* Polity Press.

Greenberg, D., Shroder, M., & Matthew, O. (1999). The social experiment market. *Journal of Economic Perspectives, 13*(3), 157–172.
Greenberg, J. R., & Mitchell, S. A. (1983). *Object relations in psychoanalytic theory*. Harvard University Press.
Greenwood, J. D. (2004). *The disappearance of the social in American social psychology*. Cambridge University Press.
Greenwood, J. D. (2015). *A conceptual history of psychology: Exploring the tangled web* (2nd ed.). Cambridge University Press.
Griggs, R. A. (2012). *Psychology: A concise introduction*. Worth Publishers.
Hanel, P. H. P., & Vione, K. (2016). Do student samples provide an accurate estimate of general public? *PLoS ONE, 11*(12). https://doi.org/10.1371/journal.pone.0168354
Harmon-Jones, E., & Winkielman, P. (Eds.). (2007). *Social neuroscience: Integrating biological and psychological explanations of social behavior*. The Guilford Press.
Harré, R. (2006). *Key thinkers in psychology*. Sage.
Heider, F. (1958). *The psychology of interpersonal relations*. John Wiley & Sons Inc.
Hogg, M. A., & Vaughan, G. M. (2011). *Social psychology* (5th ed.). Pearson.
House, J. S. (1977). The three faces of social psychology. *Sociometry, 40*(2), 161–177.
Hovland, C. I., Anis, I. L., & Kelley, H. H. (1953). *Communication and persuasion*. Yale University Press.
Jackson, M., & Cox, D. R. (2013). The principles of experimental design and their application in sociology. *Annual Review of Sociology, 39*, 27–49. https://doi.org/10.1146/annurev-soc-071811-145443
Jahoda, G. (2007). *A history of social psychology: From the eighteenth-century enlightenment to the second world war*. Cambridge University Press.
Kowalski, R. M., & Westen, D. (2010). *Psychology*. John Wiley & Sons,.
Leary, M. R. (1999). Making sense of self-esteem. *Current Directions in Psychological Science, 8*(1), 32–35.
Leary, M. R., & Tangney, J. P. (2012). The self as an organizing construct in the behavioral and social sciences. In M. R. Leary & J. P. Tangney (Eds.), *Handbook of self and identity* (2nd ed., pp. 1–18). Guilford Press.
McMahon, A. M. (1984). The two social psychologies: Postcrises directions. *Annual Review of Sociology, 10*, 121–140.
Moscovici, S. (1972). Society and theory in social psychology. In J. Israel & H. Tajfel (Ed.), *The context of social psychology: A critical assessment*. Academic Press.
Netemeyer, R. G., Bearden, W. O., & Sharma, S. (2003). *Scaling procedures: Issues and applications*. Sage Publications.
Plotnik, R., & Kouyoumdjian, H. (2011). *Introduction to psychology* (9th ed.). Wadsworth.
Renzetti, C. M., & Curran, D. J. (1998). *Living sociology*. Allyn & Bacon.
Schneider, F. W., Gruman, J. A., & Coutts, L. M. (Eds.). (2012). *Applied social psychology: Understanding and addressing social and practical problems*. Sage Publications.
Shaughnessy, J. J., Zechmeister, E. B., & Zechmeister, J. S. (2012). *Research methods in psychology*. McGraw-Hill Companies, Inc.
Sheehy, N. (2004). *Fifty key thinkers in psychology*. Routledge.
Sherif, M., Harvey, O. J., White, B. J., Hood, W. R., & Sherif, C. W. (1988). *The Robbers Cave experiment: Intergroup conflict and cooperation*. Wesleyan University Press.
Slattery, M. (2003). *Key ideas in sociology*. Nelson Thornes Ltd.
Smith, E. R., & Mackie, D. M. (2007). *Social psychology* (3rd ed.). Psychology Press.
Stam, H. J. (2006). Introduction: Reclaiming the social in social psychology. *Theory & Psychology, 16*(5), 587–595.
Stouffer, S. A. (1949). *The American soldier*. Princeton University Press.
Stroebe, W. (2012). The truth about triplett (1898), but nobody seems to care. *Perspectives on Psychological Science*, 7(1), 54–57. https://doi.org/10.1177/1745691611427306
Stryker, S. (1977). Developments in "Two Social Psychologies": Toward an appreciation of mutual relevance. *Sociometry, 40*(2), 145. https://doi.org/10.2307/3033518
Tajfel, H. (1972). Experiments in a vacuum. In J. Israel & H. Tajfel (Eds.), *The context of social psychology: A critical assessment* (pp. 69–119). Academic Press.
Tischler, H. L. (2011). Introduction to sociology. In *Introduction to sociology* (10th ed.). Wadsworth Cengage Learning.
Triplett, N. (1898). The dynomogenic factors in peacemaking and competition. *American Journal of Psychology, 9*, 507–533.
Turner, J. H. (2006). The state of theorizing in sociological social psychology: A grand theorists view. In P. J. Burke (Ed.), *Contemporary social psychological theories* (pp. 353–374). Stanford Social Sciences.
Wetherell, M., & Potter, J. (1988). Discourse analysis and the identification of interpretative repertoires. In C. Antaki (Ed.), *Analysing everyday explanation: A casebook of methods* (pp. 168–183). Sage Publications, Inc.

2

THE RELATIONSHIP BETWEEN TOURISM AND SOCIOLOGY, PSYCHOLOGY, AND SOCIAL PSYCHOLOGY

Çağrı Erdoğan

Introduction: The Emanation of Travel and Tourism Need's Innate Nature from a Historical Perspective

In the same way as any other species, humans seek to satisfy their needs naturally to ensure their survival. Unlike plants with fixed roots, a person's basic needs cannot be met from a fixed position. Humans are mobile from the moment they *start walking*, and in that sense, travel[1] is an essential part of life. As Gosch and Stearns (2007, pp. 2–3) point out, travel has always played a predominant role in sustaining the necessities of everyday life in pre- and post-agricultural times. In accordance with it, Zuelow (2016, p. 3) states that after villages and cities were established, the walls surrounding them did not stop the flow of humans and commodities. Instead, they helped promote a sense of integrity, keep strange(r) out, and promote safety. The world beyond the walls was not beyond the reach, trade has existed for centuries and made many parts of the world reachable through the travel required to trade. Engaging with others beyond the city walls has connected different cultures and ideas found a ground to meet. From travel required to survive to touristic purposes, the forms of travel widely differ in time. Regardless of form, travel – like an innate feature – can be seen as a human condition (Brodsky-Porges, 1981, p. 176).

There is little evidence of tourism activity until more modern times, and most early travels were not undertaken for that purpose. Still, tourism activities can be traced back for almost two millennia to the times of ancient Rome (Butler, 2015, pp. 16–17; Towner & Wall, 1991, p. 73). No matter how limited in scale tourism activities were until the age of modern tourism, most of its motivational and behavioral basics remained somewhat similar. Therefore, it could be said that this similarity is not only an essential link to understand modern tourism which is a significant social, economic, environmental, and cultural force (Butler, 2015, pp. 16–18) but also a flare pointing to its relation with the human condition.

Adopting a historical perspective sheds light on the transformation occurring on the surface of the tourism practices along with their partially stable core, and how they adapted over time. Also helps to reveal its interwoven and inseparable structure with psychology and with social psychology since a tourist as an individual is a social being by nature at the first place (Murphy, 1939; Rogoff, 2003). The sociological aspect of tourism also becomes more visible in time. While the scope of "aristocratic" tourism activities of the past was mostly limited to individuals, especially after the eighteenth century, the impacts of travel and tourism became an unignorable force that demands the attention of society. In this regard, it is understood that examining the tourism

DOI: 10.4324/9781003161868-2

phenomenon without psychological, sociological, and social-psychological lenses can only offer a partial understanding.

Relationship between Tourism and Sociology: The Transforming Step, from the Individual to Social Structure

Human mobility has existed throughout history, so has travel. As a structured way of travel, the origins of tourism can be traced historically too, which has been addressed above. All the touristic activities done so far, the increasing multilayered impacts caused by the intensifying tourist movement (E. Cohen, 1978; Doğan, 1989; Shepherd, 2002; Zhou et al., 1997), and from individual to destination, the transformative effects taking place (Baniya et al., 2018; Higgins-Desbiolles, 2006; Saarinen, 2004; Soulard et al., 2021), all started with embracing *touristhood* (Jafari, 1987), thus being the focal point of tourism (Leiper, 1979, p. 396) – a step taken towards outside the usual environment[2] turning the individual into a tourist. Before taking the step, even when considering to take a step, there are some internal and external factors – such as demographics, psychological elements, personal values, previous related purchases, influences from family, friends, and reference groups, marketing influences, and cognitive processes that process the mentioned information – affecting the consideration as it can be followed from Woodside and King's (2001) purchase-consumption system applied to leisure and travel behavior. While deciding from the possible choices and before taking the first step, tourism product purchasing starts. However, it is neither easy for a business to produce a highly experience-based and intangible-driven tourism products nor possible to establish a total control over the weather, landscape, cultural and natural values and attractions in a destination that are part of a touristic product, so that the production process demands businesses to sustain good relations not only with tourists but also with other businesses (Aarstad et al., 2015; Sfandla & Björk, 2013). While these circumstances necessitate companies forming supply chains (Slusarczyk et al., 2016), from a wider industrial look, they also necessitate being a part of a functional integration focused on a product-based value creation – global value chains (Humphrey & Schmitz, 2001; Romero & Tejada, 2011).

In both the origin and destination countries, the tourism global value chain contains a multitude of parties. They include tour operators, travel agents, global distribution systems, airlines, cruise companies, travel agents, and destination management organizations. Also, transportation services, accommodation, local guides, national parks, retailers, restaurants, natural assets, flora and fauna, and historical sites all form parts of the value chain (Christian, 2013, p. 46). Considering the comprehensive coverage of the tourism global value chain and the numerous operations and transactions it puts into practice along with the multiplier effect (Var & Quayson, 1985), pointing out the indirect and induced impacts generating economic benefits originated from the visitors' direct spending, the major wheel systems energized via the steps visitors take, the mobility made under the tourism roof can be understood better.

It has long been fixed on that inbound visitors are seen as foreign exchange generators. However, besides the direct revenue the industry provides, its labor-intensive structure and openness to low-skilled (Ladkin, 2011), young (Robinson et al., 2019) and female employees (Cave & Kilic, 2010) not only transforms the economy but also how the urban environment is developed with tourism growth (Gospodini, 2001), and the general image of the urban environment (Ashworth & Page, 2011). In addition to this, making natural and cultural values visible as touristic products brings some positive outcomes such as cultural rejuvenation (Čaušević & Tomljenović, 2003) as well as negative ones like overcrowding (Namberger et al., 2019) and commodification of culture (Shepherd, 2002), not to mention things arise from the issues discussed as *pseudo-events* (Boorstin, 1992), *modern cannibalism* (MacCannell, 1990), *McDisneyization* (Ritzer & Liska, 2004), and such.

Studies from different perspectives shed light on the relationship between the tourism activity – which has an unignorable impact on the entire social structure – and the sociocultural space (E. Cohen, 1972, 1979b, 1984, 1988; Dann & Cohen, 1991). Still, Forster (1964, p. 227) reveals no consistent theoretical view focusing on tourism studies. Also, he points to the rapid economic development, changes in labor force distribution, work habits, conflict and tension, social change in general, and social-psychological issues as something to focus on. Although related research has increased in time and E. Cohen and Cohen (2012, p. 2195) state that topics like urban, space, dark tourism and tourism's relationship to global financial crises, social media, crime, and prostitution are under-explored in contemporary sociological studies of tourism. In addition to this, they underline the importance of non-Western research to acquire a complete comprehension. S. A. Cohen and Cohen (2019, p. 167) warn about excessively focusing on issues in a narrow scope and abandoning sociology and broader emergent issues that affect tourism, such as the aging of the global population, the effects of automatization, and the rapid global expansion of urban living.

Overall, the complex process brought by tourism has a significant part in shaping the sociocultural space with its positive and negative aspects, especially when tourism has a vital position for the countries and administrative zones. Furthermore, in line with the increasing global interdependence and connectivity, tourism considerably influences the social structure and, therefore, social change,[3] a reminder that the relationship between tourism and sociology should never be overlooked.

Relationship between Tourism and Psychology: Geographic Mobility Reflected on the Mental Space

It is commonly said that "we are living in strange times". It is "strange" not just because it is different to 10,000 years ago when the Neolithic or agricultural revolution took place or different from the mid-eighteenth century when the Industrial Revolution was encountered. It can be easily observed the apparent difference spans just 50 years back. With each passing decade becoming more "distinctive" because the homogeneous and static structure of the pre-modern period falls behind, and the void left is filled with heterogeneity and dynamism instead.

The mobility of both people and goods has reached a previously untouched level. Tourism plays a predominant role in this human traffic, which forms a basis for the constant interaction between strangers. Recalling what Bauman (2014) highlights as beneficial, even if one never travels far nor at all, it would not be possible to protect sociocultural fixity and integrity since it would be eroded anyway by the stranger influence (pp. 20–26). We are in the middle of a constant flow of the unfamiliar in terms of people and commodities. Considering the formal social structure and the dynamic manufactured/cultural environment we live in, it is not easy to take root or remain stable. Furthermore, people are less in touch with the natural environment than ever before. Since these inorganic changes happen so fast, and we cannot keep up with the rapid change as organically slow adapters, conditions that escalate alienation become widespread.

Coping with the restricting and dedifferentiating effect brought about by the traditional and professional norms of everyday life (Schouten, 2005; Stronza, 2001, p. 266 as cited in Aktaş Polat, 2016, p. 107), perceived mundane existence (S. A. Cohen, 2010, p. 30), and the aforementioned alienating effect of both fluctuant and inorganically mechanized world, as Aktaş Polat and Polat (2016) emphasized, tourism experiences could be a valuable asset. At the center of alienation, the authors separate the daily life and touristic life in a specific way (p. 243); while people lose their value against objects in everyday life; objects are used for people's happiness in touristic life. While people try to make the world happy in daily life, people try to (or tried to) be happy in touristic life. While spiritual deprivation takes place in everyday life, returning spirituality with the fulfillment of the need for authenticity becomes reality in touristic life; while people become the instrument of economic ends in daily life, economic instruments become the instrument of human reality and authenticity in touristic life. With these and similar propositions the dichotomy they form suggests how the codes of alienation could be reversed through tourism.

After examining the relationship between tourism and sociology, the link between tourism and psychology is now discussed, specifically concerning the role played by tourism in overcoming the alienating effect of the existing social structure on the individual. Before moving to tourist role and experience in examining this relationship, the issue is touched upon in a business context. Pine and Gilmore's (1998) experience economy is worth remembering; clients turn into guests, and intangible customized services that provide benefits turn into memorable personal experiences. Within this economy, issues related to psychology such as symbolic consumption, the self, and experience become a priority for marketing research and product design. When it comes to the tourism industry, it is known that consumption or experiences are spread across the destination and exceed the boundaries of businesses. Therefore, the core of the tourist experience and the psychological outcome of tourism activity is not limited to what companies provide. As the co-producer of the product/experience, the individual/tourist plays an active role since the destination-wide experience neither can be controlled externally from one point nor the psychological foundations specific to the individual that are effective at every stage of the interaction process with the elements such as environment, employees, and locals (Guthrie & Anderson, 2010; Lugosi & Walls, 2013). Beyond any doubt, within a *psychologized economy*, the psychological part of the *experience product* that is turned into *experiencescape* by the tourism industry, individuals'/consumers'/tourists' role on the psychological ground reaches its peak (Gelter, 2010; O'Dell & Billing, 2005; Toffler, 1971). However, experience-related and psychology-focused research depth is not yet at the desired level, despite its high importance level (Otto & Ritchie, 1996; Ritchie et al., 2011).

Coinciding with the fact that travel provides an ample opportunity to see beyond the defining sociocultural matrix (Leed, 1991), as mentioned before, tourism has a share in shaping the material world and the perceptual one (Chambers, 2014). In this regard, by emphasizing the psychological processes, Volo (2009) highlights the reflections of the tourism mobility carried out in the geographical space on the mental space. On the other hand, although tourism supports cross-cultural understanding and raises awareness of global issues like poverty, migration, and power imbalances, such outcomes are also criticized for supporting prejudices and because the effects are temporary (Lean, 2012, p. 152). Nevertheless, since touristic experiences that are temporary in nature could eventuate in permanent acquisitions, especially when they are considered in cumulative terms (Erdoğan & Kıngır, 2021), and not all tourists are mass tourists (Wickens, 2002) nor lifestyle travelers (S. A. Cohen, 2011), they contain various qualities which should be evaluated from different perspectives (Graburn & Barthel-Bouchier, 2001).

Some perspectives shed light on excessive drinking, unrestrained gambling, visiting prostitutes, and other activities performed in tourist-related settings and categorized under deviant tourist behavior. So as not to ignore the internal and unconscious psychological processes in the context of their relationship with social order, Uriely, Ram, and Malach-Pines (2011) discussed the subject within psychoanalytic sociology. "The analysis suggests that various unconscious drives can either be gratified by normative tourist activities that involve adaptive defense mechanisms or lead to deviant tourist behaviors that entail distorting defense mechanisms". Although the experience lived at the destination is important, its psychological processes exceed the destination's limits. Therefore, psychological parts of the experience that occur before and after the trip are within its scope, for example, cognitive, affective, psychomotor, and personal development factors (Cutler & Carmichael, 2010). Eventually, as is evident from what has been considered, comprehension of the tourism phenomenon without considering psychology and psychological processes points to an effort that cannot yield beneficial results.

Relationship between Tourism and Social Psychology: Conceptual and Practical Unity

Sociologists like George Herbert Mead's (1972) and Charles Horton Cooley's (1902, 1926) thoughts and works on the social self together with psychologists like Floyd Henry Allport's (1919) and Gordon Willard Allport's (1943) studies on the social part of personality and psychology relating to social influence (F. H. Allport & Allport, 1921) shaped the cornerstones of social psychology.[4] It is possible to state that this goes parallel with the interrelatedness highlighted throughout the chapter and reminds the importance of scientific ground they provide for tourism studies especially when considering tourism beyond the solely industrial scope. The symbiotic relationship between tourism and social psychology was demonstrated by Stringer and Pearce (1984). As the authors underline, topics related to group dynamics, attitude formation, and change are part of tourism activities and happen within temporal processes, different environmental conditions, and sociocultural contexts. This demonstrates the need to study tourism from a social psychology perspective to clarify the issues like tourist behavior and experience that are interactive and contextual, along with the tourism phenomenon as a whole.

Besides the tourism activities' "call for social-psychological magnifier" and pull the perspective on itself, social psychology also relates notable push factors leading one to join touristic activities from individual to sociocultural levels. As it can be recalled, the innate nature of the need for travel and tourism and the ambiguous and fluctuant structure of the sociocultural space was mentioned. All kinds of self-related information become diffused within this sociocultural space, unlike its given form in traditional settings. Therefore, where the individual can see and test him/herself on the different natural and cultural backgrounds, interact with the strange and authentic via touristic activities, it increases their objective self-awareness and self-perception under Bem's theory (1972). Also, it promotes the gaining of self-confidence and self-respect. It allows people to explore self-identification sources, collect self-related information, and finally form a stable and reliable self-concept. Tourism experiences are also beneficial in terms of personal and social identity. Considering Festinger's (1954) social comparison theory, they offer ample opportunities for the individual to compare his/her personal and social identity elements with various related entities. At the same time, interacting with others via touristic experiences could help obtain evidence-based knowledge instead of the previous groundless opinions. This learning can play a role in overcoming prejudices, stereotypes, and dehumanization factors (Erdoğan, 2019).

Numerous social psychology and tourism subjects are associated with conceptual and practical terms that Stringer and Pearce (1984, p. 13) discuss. Although there are some dominant themes as attitudinal outcomes of international travel and contacts, tourist-guide relations, and the effects of culture shock, the researches need to be deepened to overcome its fragmentary existence (do consider the cited study's date). Focusing specifically on tourist motivation, Harrill and Potts (2002) state that there are some social psychology-focused tourism studies, yet sociological perspectives were more common during the 1970s and 1980s. Additionally, the authors note that continuity and consensus should be secured if the related social-psychological tourism studies end the current superficiality. To clarify significant social psychology-based tourism trends and encourage additional related scientific research to cover a broader range and reach deeper areas, Tang (2014) reviewed 12 selected hospitality and tourism journals from 1999 to 2012. It is determined that social exchange theory, equity theory, and cognitive dissonance are some of the theories and concepts concentrated on. Since this concentration constituted some well-developed areas with low marginal utility in terms of scientific knowledge, it is advised that generating cumulative knowledge is vital. Even though it is not easy and beneficial in the short term – step by step and in patience, clarification on the subjects should be sought.

In conclusion, the relationship between tourism and social psychology can be seen explicitly. Whenever a tourism-related topic is considered, the set of intersections it shares with social psychology summarily becomes apparent. Concordantly, this study puts a sincere effort forward to highlight the scientific bridge between the two research areas and the much effort required to develop a more thorough understanding of tourism as a whole by investigating factors that have not been heavily touched upon in previous research.

Conclusion

Travel is an inseparable part of the human condition, and tourism can be seen as a form of structured travel. Although tourism activities can be traced with ease thousands of years in the past, especially after the eighteenth century, they became widespread in ordinary life and every corner of the world. This ubiquitousness and the mass that tourism activities involve naturally strengthen the shaping force of tourism in every aspect. To prevent tourism from destroying areas and develop an understanding that maximizes its benefits, the factors that exceed the naturally visible impacts of mass tourism need to be demonstrated (E. Cohen, 1973). Also, the perception towards tourists must not be confined to a superficial and straightforward state by associating the tourists with pure pleasure and entertainment-driven travels (E. Cohen, 1979a; Graburn & Barthel-Bouchier, 2001). The entire tourism and touristic experience-related outcomes should not be reduced to a suntan and souvenirs (Pearce, 2010, p. 257). The experience comprises the individual traveling and interacting with different natural and sociocultural settings. Tourism is naturally indissociable with social psychology which aims to study the individual without detaching it from its sociocultural context while unifying the mind and social world by integrating "psychological social psychology and sociological social psychology". Within this context, it can be confidently stated that the general comprehension of tourism and tourism-related phenomena based on research shaped by ignoring the psychological and sociological foundations, and also without considering the relationship with social psychology, cannot be described as comprehensive.

In tourism research conducted on the basis of psychology, Pearce and Packer (2013) stated that a notable level of focus is placed on the themes of decision making, consumer behavior, motivation, satisfaction and attitudes, and social interaction. Nevertheless, studies shed light on the "human aspect" of tourism in the social-psychological context can still be described as far from saturation. In this respect, rather than highlighting certain topics one by one for researchers who are likely to contribute to this field, in line with the relatively implicit nature of the related subject and its high level of relationality, the need to create a consensus by expanding, deepening and sustaining the research together with overcoming the fragmented existence of social-psychological tourism studies are emphasized. Despite the fact that touristic experiences have aspects that differ from everyday life with their episodic structure and less familiar content, it is apparent that a way to clarify both the effects of touristic experiences on daily life and its place in the integrity of individual life, to make a common definition and categorization of tourism, and to make a significant contribution to a clearer form of tourism paradigm cannot be thought of independently from the social psychology perspective (Iso-Ahola, 1983; Pearce, 1982, 2011).

Notes

1 "Travel refers to the activity of travellers. A traveller is someone who moves between different geographic locations for any purpose and any duration" (UNDESA/*United Nations Department of Economic and Social Affairs* Statistics Division, 2010, p. 9).

2 Remembering the tourist definition of UNWTO/*United Nations World Tourism Organization* (2021) at this point could be beneficial:

A traveler taking a trip that includes an overnight stay to a main destination outside his/her usual environment, for less than a year, for any main purpose other than to be employed by a resident entity in the country or place visited.

3 "Social structure referring to basic characteristics of social life, those demonstrating a lasting and permanent quality" (Form & Wilterdink, 2020) and "social change, in sociology, the alteration of mechanisms within the social structure, characterized by changes in cultural symbols, rules of behaviour, social organizations, or value systems" (Wilterdink & Form, 2020).

4 "Social psychology has been defined as a social science that aims to examine and comprehend the influence of the actual, imagined, or implied presence of others on an individual' thoughts, experience, and behaviour" (Allport, 1984, as cited in Tang, 2014, p. 188).

References

Aarstad, J., Ness, H., & Haugland, S. A. (2015). Innovation, uncertainty, and inter-firm shortcut ties in a tourism destination context. *Tourism Management, 48*(1), 354–361. https://doi.org/10.1016/j.tourman.2014.12.005

Aktaş Polat, S. (2016). *Turizm Antropolojisi [Tourism Anthropology]* (1st ed.). Gazi Kitabevi.

Aktaş Polat, S., & Polat, S. (2016). Turizm perspektifinden yabancılaşmanın sosyo-psikolojik analizi: Günlük yaşamdan turistik yaşama yabancılaşma döngüsü [Socio-psychological analysis of alienation in tourism perspective: The alienation cycle from daily life to touristic life]. *Sosyoekonomi, 24*(28), 235–253. https://doi.org/10.17233/se.29769

Allport, F. H. (1919). Behavior and experiment in social psychology. *The Journal of Abnormal Psychology, 14*(5), 297–306. https://doi.org/10.1037/h0073020

Allport, F. H., & Allport, G. W. (1921). Personality traits: Their classification and measurement. *The Journal of Abnormal Psychology and Social Psychology, 16*(1), 6–40. https://doi.org/10.1037/h0069790

Allport, G. W. (1943). The ego in contemporary psychology. *Psychological Review, 50*(5), 451–478. https://doi.org/10.1037/h0055375

Ashworth, G., & Page, S. J. (2011). Urban tourism research: Recent progress and current paradoxes. *Tourism Management, 32*(1), 1–15. https://doi.org/10.1016/j.tourman.2010.02.002

Baniya, R., Shrestha, U., & Karn, M. (2018). Local and community well-being through community based tourism – A study of transformative effect. *Journal of Tourism and Hospitality Education, 8*, 77–96. https://doi.org/10.3126/jthe.v8i0.20012

Bauman, Z. (2014). *Küreselleşme: Toplumsal sonuçları [Globalization: The human consequences] (A. Yılmaz, Trans.; 5th ed.).* Ayrıntı.

Bem, D. J. (1972). Self-perception theory. In L. Berkowitz (Ed.), *Advances in experimental social psychology* (Vol. 6, Issue C, pp. 1–62). Academic Press. https://doi.org/10.1016/S0065-2601(08)60024-6

Boorstin, D. J. (1992). *The image: A guide to pseudo-events in America* (25th Anniv). Vintage Books.

Brodsky-Porges, E. (1981). The grand tour travel as an educational device 1600–1800. *Annals of Tourism Research, 8*(2), 171–186. https://doi.org/10.1016/0160-7383(81)90081-5

Butler, R. (2015). The evolution of tourism and tourism research. *Tourism Recreation Research, 40*(1), 16–27. https://doi.org/10.1080/02508281.2015.1007632

Čaušević, S., & Tomljenović, R. (2003). World Heritage site, tourism and city's rejuvenation: The case of Poreč, Croatia. *Tourism (Zagreb), 51*(4), 417–426.

Cave, P., & Kilic, S. (2010). The role of women in tourism employment with special reference to Antalya, Turkey. *Journal of Hospitality Marketing and Management, 19*(3), 280–292. https://doi.org/10.1080/19368621003591400

Chambers, I. (2014). *Göç, kültür, kimlik [Migrancy, culture, identity -1994]. Türkmen, İ. & Beşikçi, M. (trans.).* (2nd ed.). Ayrıntı Yayınları.

Christian, M. (2013). Global value chains, economic upgrading, and gender in the tourism industry. In C. Staritz & J. G. Reis (Eds.), *Global value chains, economic upgrading, and gender case studies of the horticulture, tourism, and call center industries* (Issue January). The World Bank: International Trade Department Gender Development Unit.

Cohen, E. (1972). Toward a sociology of international tourism. *Social Research, 39*(1), 164–182.

Cohen, E. (1973). Nomads from affluence: Notes on the phenomenon of drifter-tourism. *International Journal of Comparative Sociology, 14*(1–2), 89–103. https://doi.org/10.1163/156854273X00153

Cohen, E. (1978). The impact of tourism on the physical environment. *Annals of Tourism Research, 5*(2), 215–237.

Cohen, E. (1979a). A phenomenology of tourist experiences. *Sociology, 13*(2), 179–201. https://doi.org/10.1177/003803857901300203

Cohen, E. (1979b). Rethinking the sociology of tourism. *Annals of Tourism Research, 6*(1), 18–35. https://doi.org/10.1016/0160-7383(79)90092-6

Cohen, E. (1984). The sociology of tourism: Approaches, issues, and findings. *Annual Review of Sociology, 10*(1), 373–392.

Cohen, E. (1988). Traditions in the qualitative sociology of tourism. *Annals of Tourism Research, 15*(1), 29–46. https://doi.org/10.1016/0160-7383(88)90069-2

Cohen, E., & Cohen, S. A. (2012). Current sociological theories and issues in tourism. *Annals of Tourism Research, 39*(4), 2177–2202. https://doi.org/10.1016/j.annals.2012.07.009

Cohen, S. A. (2010). Searching for escape, authenticity and identity: Experiences of "Lifestyle travellers." In M. Morgan, P. Lugosi, & J. R. B. Ritchie (Eds.), *The tourism and leisure experience: Consumer and managerial perspectives* (pp. 27–42). Channel View Publication.

Cohen, S. A. (2011). Lifestyle travellers: Backpacking as a way of life. *Annals of Tourism Research, 38*(4), 1535–1555. https://doi.org/10.1016/j.annals.2011.02.002

Cohen, S. A., & Cohen, E. (2019). New directions in the sociology of tourism. *Current Issues in Tourism, 22*(2), 153–172. https://doi.org/10.1080/13683500.2017.1347151

Cooley, C. H. (1902). *Human nature and the social order* (1st ed.). Charles Scribner's Sons.

Cooley, C. H. (1926). The roots of social knowledge. *American Journal of Sociology, 32*(1), 59–79.

Cutler, S. Q., & Carmichael, B. A. (2010). The dimensions of the tourist experience. In Michael Morgan, P. Lugosi, & J. R. B. Ritchie (Eds.), *The tourism and leisure experience: Consumer and managerial perspectives* (pp. 3–26). Channel View Publication. https://doi.org/10.21832/9781845411503-004

Dann, G., & Cohen, E. (1991). Sociology and tourism. *Annals of Tourism Research, 18*(1), 155–169. https://doi.org/10.1016/0160-7383(91)90045-D

Doğan, H. Z. (1989). Forms of adjustment: Sociocultural impacts of tourism. *Annals of Tourism Research, 16* (2), 216–236. https://doi.org/10.1016/0160-7383(89)90069-8

Erdoğan, Ç. (2019). *Turistik deneyimlerin benlik kavramı ile ilişkisi [The relationship between touristic experiences and self-concept]* [Unpublished doctoral dissertation]. Sakarya University of Applied Sciences.

Erdoğan, Ç., & Kıngır, S. (2021). Liminal evrenin ardından: kümülatif turistik deneyimler temelinde kalıcı davranışsal edinimler [After the liminal stage: The permanent behavioral acquisitions based on cumulative touristic experiences]. *MANAS Sosyal Araştırmalar Dergisi [MANAS Journal of Social Studies], 10*(1), 591–607. https://doi.org/10.33206/mjss.781401

Festinger, L. (1954). A theory of social comparison processes. *Human Relations*, 7(2), 117–140. https://doi.org/10.1177/001872675400700202

Form, W., & Wilterdink, N. (2020). Social structure. In *Encyclopedia Britannica*. https://www.britannica.com/topic/social-structure

Forster, J. (1964). The sociological consequences of tourism. *International Journal of Comparative Sociology, 5*(-2), 217–227.

Gelter, H. (2010). Total experience management – a conceptual model for transformational experiences within tourism. *Conference Proceedings The Nordic Conference on Experience 2008. Research, Education and Practice in Media*, 46–66.

Gosch, S., & Stearns, P. (2007). *Premodern travel in world history* (e-book). Routledge.

Gospodini, A. (2001). Urban design, urban space morphology, urban tourism: An emerging new paradigm concerning their relationship. *European Planning Studies, 9*(7), 925–934. https://doi.org/10.1080/09654310120079841

Graburn, N. H. H., & Barthel-Bouchier, D. (2001). Relocating the tourist. *International Sociology, 16*(2), 147–158. https://doi.org/10.1177/0268580901016002001

Guthrie, C., & Anderson, A. (2010). Visitor narratives: Researching and illuminating actual destination experience. *Qualitative Market Research: An International Journal, 13*(2), 110–129. https://doi.org/10.1108/13522751011032575

Harrill, R., & Potts, T. D. (2002). Social psychological theories of tourist motivation: Exploration, debate, and transition. *Tourism Analysis*, 7(2), 105–114. https://doi.org/10.3727/108354202108749989

Higgins-Desbiolles, F. (2006). More than an "industry": The forgotten power of tourism as a social force. *Tourism Management, 27*(6), 1192–1208. https://doi.org/10.1016/j.tourman.2005.05.020

Humphrey, J., & Schmitz, H. (2001). Governance in global value chains. *IDS Bulletin, 32*(3), 19–29. https://doi.org/10.1111/j.1759-5436.2001.mp32003003.x

Iso-Ahola, S. E. (1983). Towards a social psychology of recreational travel. *Leisure Studies, 2*(1), 45–56. https://doi.org/10.1080/02614368300390041

Jafari, J. (1987). Tourism models: The sociocultural aspects. *Tourism Management, 8*(2), 151–159. https://doi.org/10.1016/0261-5177(87)90023-9
Ladkin, A. (2011). Exploring tourism labor. *Annals of Tourism Research, 38*(3), 1135–1155. https://doi.org/10.1016/j.annals.2011.03.010
Lean, G. L. (2012). Transformative travel: A mobilities perspective. *Tourist Studies, 12*(2), 151–172. https://doi.org/10.1177/1468797612454624
Leed, E. J. (1991). *The mind of the traveller: From Gilgamesh to global tourism*. Basic Books.
Leiper, N. (1979). The framework of tourism: Towards a definition of tourism, tourist, and the tourist industry. *Annals of Tourism Research, 6*(4), 390–407. https://doi.org/10.1016/0160-7383(79)90003-3
Lugosi, P., & Walls, A. R. (2013). Researching destination experiences: Themes, perspectives and challenges. *Journal of Destination Marketing and Management, 2*(2), 51–58. https://doi.org/10.1016/j.jdmm.2013.07.001
MacCannell, D. (1990). Cannibal tours. *SVA Review, 6*(2), 14–24.
Mead, G. H. (1972). *Mind, self, and society: From the standpoint of a social behaviorist* (18th ed.). The University of Chicago Press.
Murphy, A. E. (1939). Concerning Mead's the philosophy of the act. *The Journal of Philosophy, 36*(4), 85–103.
Namberger, P., Jackisch, S., Schmude, J., & Karl, M. (2019). Overcrowding, overtourism and local level disturbance: How much can munich handle? *Tourism Planning and Development, 16*(4), 452–472. https://doi.org/10.1080/21568316.2019.1595706
O'Dell, T., & Billing, P. (Eds.). (2005). *Experiencecapes: Tourism, culture and economy*. Copenhangen Business School Press.
Otto, J. E., & Ritchie, J. R. B. (1996). The service experience in tourism. *Tourism Management, 17*(3), 165–174. https://doi.org/10.4324/9780080519449-38
Pearce, P. L. (1982). Tourists and their hosts: Some social and psychological effects of inter-cultural contact. In S. Bochner (Ed.), *Cultures in contact: Studies in cross-cultural interaction* (pp. 199–221). Pergamon Press.
Pearce, P. L. (2010). New directions for considering tourists' attitudes towards others. *Tourism Recreation Research, 35*(3), 251–258. https://doi.org/10.1080/02508281.2010.11081641
Pearce, P. L. (2011). Preparation and prospects. In P. L. Pearce (Ed.), *Study of tourism: Foundations from psychology* (pp. 3–22). Emerald Group Publishing Limited.
Pearce, P. L., & Packer, J. (2013). Minds on the move: New links from psychology to tourism. *Annals of Tourism Reseacrh, 40*, 386–411. https://doi.org/10.1016/j.annals.2012.10.002
Pine, J. B., & Gilmore, J. H. (1998). Welcome to the experience economy. *Harvard Business Review, 76*(July–August), 97–105.
Ritchie, J. R. B., Tung, V. W. S., & Ritchie, R. J. B. (2011). Tourism experience management research: Emergence, evolution and future directions. *International Journal of Contemporary Hospitality Management, 23*(4), 419–438. https://doi.org/10.1108/09596111111129968
Ritzer, G., & Liska, A. (2004). "McDisneyization" and "post-tourism": Complementary perspectives on contemporary tourism. In S. Williams (Ed.), *Tourism: Critical concepts in the social sciences – Volume IV: New directions and alternative tourism* (pp. 65–82). Routledge – Taylor and Francis Group.
Robinson, R. N. S., Baum, T., Golubovskaya, M., Solnet, D. J., & Callan, V. (2019). Applying endosymbiosis theory: Tourism and its young workers. *Annals of Tourism Research, 78*(June), 1–12. https://doi.org/10.1016/j.annals.2019.102751
Rogoff, B. (2003). *The cultural nature of human development*. Oxford University Press.
Romero, I., & Tejada, P. (2011). A multi-level approach to the study of production chains in the tourism sector. *Tourism Management, 32*(2), 297–306. https://doi.org/10.1016/j.tourman.2010.02.006
Saarinen, J. (2004). 'Destinations in change': The transformation process of tourist destinations. *Tourist Studies, 4*(2), 161–179. https://doi.org/10.1177/1468797604054381
Schouten, F. (2005). The process of authenticating souvenirs. In S. Melanie and R. Mike (Eds.), *Cultural Tourism in a Changing World Politics, Participation and (Re)presentation* (pp. 191–202). Channel View Publications.
Sfandla, C., & Björk, P. (2013). Tourism Experience Network: Co-creation of experiences in interactive processes. *International Journal of Tourism Research, 15*, 495–506. https://doi.org/10.1002/jtr.1892
Shepherd, R. (2002). Commodification, culture and tourism. *Tourist Studies, 2*(2), 183–201.
Slusarczyk, B., Smolag, K., & Kot, S. (2016). The supply chain of a tourism product. *Actual Problems of Economics, 179*(5), 197–207.
Soulard, J., McGehee, N. G., Stern, M. J., & Lamoureux, K. M. (2021). Transformative tourism: Tourists' drawings, symbols, and narratives of change. *Annals of Tourism Research, 87*, 103–141. https://doi.org/10.1016/j.annals.2021.103141

Stringer, P. F., & Pearce, P. L. (1984). Toward a symbiosis of social psychology and tourism studies. *Annals of Tourism Research*, *11*(1), 5–17. https://doi.org/10.1016/0160-7383(84)90093-8

Stronza, A. (2001). Anthropology of tourism: Forging new ground for ecotourism and other alternatives. *Annual Review Anthropology, 30*, 261–283.

Tang, L. R. (2014). The application of social psychology theories and concepts in hospitality and tourism studies: A review and research agenda. *International Journal of Hospitality Management*, *36*, 188–196. https://doi.org/10.1016/j.ijhm.2013.09.003

Toffler, A. (1971). *Future shock*. Bantam Books.

Towner, J., & Wall, G. (1991). History and tourism. *Annals of Tourism Research*, *18*(1), 71–84. https://doi.org/10.1016/0160-7383(91)90040-I

UNDESA Statistics Division. (2010). *Role of the international recommendations for tourism statistics 2008*. United Nations Publication.

UNWTO. (2021). *Glossary of tourism terms | UNWTO*. https://www.unwto.org/glossary-tourism-terms

Uriely, N., Ram, Y., & Malach-Pines, A. (2011). Psychoanalytic sociology of deviant tourist behavior. *Annals of Tourism Research*, *38*(3), 1051–1069. https://doi.org/10.1016/j.annals.2011.01.014

Var, T., & Quayson, J. (1985). The multiplier impact of tourism in the okanagan. *Annals of Tourism Research*, *12*(4), 497–514. https://doi.org/10.1016/0160-7383(85)90074-X

Volo, S. (2009). Conceptualizing experience: A tourist based approach. *Journal of Hospitality Marketing and Management*, *18*(2–3), 111–126. https://doi.org/10.1080/19368620802590134

Wickens, E. (2002). The sacred and the profane: A tourist typology. *Annals of Tourism Research*, *29*(3), 834–851. https://doi.org/10.4324/9780203711941-12

Wilterdink, N., & Form, W. (2020). Social change. In *Encyclopedia Britannica*. https://www.britannica.com/topic/social-change

Woodside, A. G., & King, R. I. (2001). An updated model of travel and tourism purchase-consumption systems. *Journal of Travel and Tourism Marketing*, *10*(1), 3–27. https://doi.org/10.1300/J073v10n01_02

Zhou, D., Yanagida, J. F., Chakravorty, U., & Leung, P. S. (1997). Estimating economic impacts from tourism. *Annals of Tourism Research*, *24*(1), 76–89. https://doi.org/10.1016/s0160-7383(96)00035-7

Zuelow, E. (2016). *A history of modern tourism* (1st ed.). Palgrave, Macmillan Publishers.

3

THE EFFECTS OF TOURISM

Economic, Environmental, Social, Cultural, Social Psychological

Emrah Öztürk

Introduction

The tourism movement, which started in 1841 for the first time in a modern sense, with a group of 570 people from Leicester to Loughborough, was organized by Thomas Cook in England (Lickorish & Jenkins, 1997; Williams, 1998; Mason, 2003; Goeldner & Ritchie, 2009; Page, 2014; Özdemir & Büyükkuru, 2019), and since the start of civil aviation service after World War II and due to technological innovations, it has developed rapidly.

In tourism, which is one of the fastest and steadily growing sectors in the world (Duran, 2011, p. 302), the number of tourists traveling for tourism purposes in the world was 25 million in 1950 (UNWTO, 2008), while the number of tourists traveling for tourism in the world reached 1.5 billion in 2019 (UNWTO, 2020).

One of the fastest-growing industries worldwide (Steyn et al., 2004; Golzardi et al., 2012; Mbagwu Felicia et al., 2016), tourism covers almost all aspects of society as it relates to other academic subjects such as language, psychology, marketing, business and law as well as economic changes, sociocultural activities and environmental development, apart from geography, economy, and history (Education Bureau, 2013, p. 11). In this context, as a result of the continuous increase in the number of participants in tourism (Mason, 2003; Cunha, 2012), it is an essential human activity and creates economic, social, cultural, environmental (Duran & Özkul, 2012; Hammad et al., 2017; Türker, 2020), psychological (Pizam, 1978; Rhama, 2019; Salazar & Cardoso, 2019), and social-psychological (Jurowski et al., 1997; Frent, 2016; Cianga & Sorocovschi, 2017; Hammad et al., 2017; Castela, 2018; Alipour et al., 2019; Çelik, 2019a) effects in every destination.

In the literature, economic, social, cultural, environmental, and social-psychological effects of tourism (Kim, 2002; Çelik, 2019a) are discussed with both positive and negative aspects (Jackson, 2008; Cianga & Sorocovschi, 2017; Hammad et al., 2017; Castela, 2018; Alipour et al., 2019). In this context, the first studies on the effects of tourism focused on the economic effects of tourism in the 1960s (Pizam, 1978; Kim, 2002; Türker, 2020). Studies on the social-cultural effects of tourism started in the 1970s (Kim, 2002; Castela, 2018; Türker, 2020). After the 1980s, studies were conducted on the environmental effects of tourism (Pender & Sharpley, 2005; Paul, 2012; Türker, 2020) and its social-psychological effects (Iso-Ahola, 1980; Iso-Ahola, 1982; Pearce, 1982; Stringer & Pearce, 1984; Harrill & Potts, 2002).

Tourism has such positive effects as employment, foreign exchange income, contribution to government revenues, regional development (Kim, 2002; Mason, 2003), effect on other sectors (Pender & Sharpley, 2005; Bahar & Kozak, 2015), effect on infrastructure and superstructure (Ince

DOI: 10.4324/9781003161868-3

et al., 2020), and the balance of payments (Ardahaey, 2011; Zaei & Zaei, 2013; Bahar & Kozak, 2015) but it also has negative economic effects such as inflation, opportunity cost, excessive dependence on tourism (Mason, 2003), seasonality effect (Kim, 2002; Salazar & Cardoso, 2019), and import effect (Bahar & Kozak, 2015). Moreover, tourism has certain positive social effects such as improving lifestyle, reducing negative stereotypes between the host community and tourists, enabling different communities to develop positive attitudes towards each other (host community and tourists), increasing the reputation and visibility of the host community against foreigners (tourists) and social relations (Education Bureau, 2013). However, there are also adverse social effects of tourism such as hostility and conflict between local people and tourists (Chang et al., 2018); smuggling, alcohol and drug use (Ap, 1990; Castela, 2018); prostitution, gambling, bullying, congestion, and other excesses (Goeldner & Ritchie, 2009). When it comes to the cultural effects of tourism, the positive ones might be listed as preserving cultural heritage (Loureiro et al., 2019); cultural interaction; preserving the cultural identity of the host population (Ap, 1990); reviving customs and traditions; strengthening cultural values; and understanding, maintaining, and preserving local culture, arts, crafts, and traditions (Ap, 1990; Shahzalal, 2016), while the adverse cultural effects seem to be the breakdown of the family (Kreag, 2001); changes in traditions, religion, and language; loss of cultural identity (Pizam, 1978); and commercialization of culture (Lickorish & Jenkins, 1997; Shahzalal, 2016). Besides, tourism has positive environmental effects including environmental protection and development (Almeida-García et al., 2016; Segota et al., 2017); protecting wildlife, encouraging natural tourism education (Kim, 2002); increasing environmental awareness, and improving environmental quality (Nematpour & Faraji, 2019), but it has also certain negative ones including pollution (air, water, noise, solid waste, and visual) (Pizam & Milman, 1986; Kreag, 2001; Jackson, 2008); erosion and physical damage (Williams, 1998); crowds, garbage, destruction of heritages, unappealing appearance of temporary and permanent holiday homes (Jackson, 2008); and depletion of natural resources (Paul, 2012).

Finally, in the related literature, tourism is thought to have social-psychological effects as well as economic, social, cultural, and environmental effects (Pizam et al., 1991; Pizam et al., 2002; Çelik, 2019a). Social psychology focuses on leadership, motivation, conflict, compliance, obedience, persuasion, control, surveillance, prejudice, and so on, which have effects on human behavior as a result of positive and negative interactions of the individual with others, the individual with the society and the society with other societies (Göksu, 2007) focuses on issues (Zencirkıran, 2017, p. 3). In this context, it can be said that tourism has positive social psychological effects, including reducing hostility, social distance, discrimination (Çelik, 2019a), prejudice, conflict, and tensions (Çelik, 2019b) in addition to increasing benevolence and cooperation. However, it also might lead to negative behaviors such as aggression (Göksu, 2007), hostility, prejudice, violence, and gender roles (Stringer & Pearce, 1984).

In this section, both positive and negative aspects of tourism's economic, social, cultural, environmental, and social-psychological effects have been mentioned.

Economic Effects of Tourism

From a global perspective, tourism can positively and negatively affect regions and destinations in economic, social, cultural, and environmental dimensions (Vanhove, 2011, p. 223). The effects of tourism have historically been the most studied tourism area, and the economic effects have been studied more than any other aspect. (Mason, 2003). This is because the most significant benefit of tourism for a region or country is economic. It provides job opportunities and income opportunities at international, national, regional, and local levels. Therefore, tourism also benefits economies at the regional and local levels, as money flows into urban and rural areas. This creates a more positive image in the region, thereby encouraging new commercial ventures (Zaei & Zaei,

2013, p. 16). In this sense, tourism can be considered an economic activity that produces trade opportunities and jobs and is one of the global economies (Castela, 2018).

Tourism has various economic effects (Ardahaey, 2011) because tourism is a mighty economic power that provides employment, foreign exchange, income, and tax revenue. The factors that create an economic impact for a city, country, or destination are visitors, visitor spending, and multiplier effect. Therefore, the economic impact of tourism expenditures is a function of numbers and expenditures resulting from domestic and foreign visitors (Goeldner & Ritchie, 2009, p. 380).

In general, the economic dimension of tourism is the most crucial reason for the positive attitude of the local people towards tourism. However, this economic dimension needs consideration as to the evaluation of positive and negative aspects depending on the level of development of the countries (Nematpour & Faraji, 2019, p. 262). Thus, the economic effects of tourism are categorized under positive and negative headings in the literature (Jurowski et al., 1997; Kreag, 2001; Mason, 2003).

Various studies on the economic effects of tourism have shown that tourism, which is a growing industry, has positive effects on economic conditions (Ince et al., 2020, p. 176). The positive economic effects of tourism on the economy of a country or region can be listed as follows (Lickorish & Jenkins, 1997; Vanhove, 2011; Zaei & Zaei, 2013; Bahar & Kozak, 2015):

- Tourism contributes to the balance of payments.
- Tourism has an income-generating effect.
- Tourism provides employment opportunities
- Tourism has an impact on interregional development.
- Tourism has an impact on other sectors.

In addition to these positive economic effects, there are also the following positive effects (Ap, 1990; Kreag, 2001; Kim, 2002; Marzuki, 2011; Vanhove, 2011; Nematpour & Faraji, 2019; Ince et al., 2020):

- Tourism contributes to an increase in income and living standards
- Tourism improves the local economy
- Tourism improves investment, development, and infrastructure spending in the economy
- Tourism increases tax revenues
- Tourism improves the infrastructure of public services
- Tourism improves transport infrastructure
- Tourism improves shopping opportunities
- Tourism encourages widespread direct and indirect expenditures
- Tourism provides job diversity and new job opportunities.

In addition to the positive economic effects of tourism mentioned above, tourism helps the host community and country earn foreign exchange. It increases the gross national product (Kim, 2002; Marzuki, 2011; Vanhove, 2011; Nematpour & Faraji, 2019). It increases government revenue (Lickorish & Jenkins, 1997; Kim, 2002; Mason, 2003; Vanhove, 2011). It enables the creation and growth of local markets (Nematpour & Faraji, 2019; Salazar & Cardoso, 2019). It helps to reduce poverty and distribute wealth in better ways (Zaei & Zaei, 2013; Nematpour & Faraji, 2019). It increases investments and investment incentives (Goeldner & Ritchie, 2009; Marzuki, 2011; Zaei & Zaei, 2013; Salazar & Cardoso, 2019). It diversifies the economy (Kim, 2002). In addition, thanks to tourism, real estate, and other asset values increase (Vanhove, 2011; Salazar & Cardoso, 2019).

The economic effects of tourism are generally discussed with its positive aspects. However, there are also adverse economic effects caused by tourism activities in a region or country in which they are carried out (Frent, 2016). These adverse economic effects are generally listed as follows (Kreag, 2001; Kim, 2002; Bahar & Kozak, 2015; Frent, 2016):

- Tourism has an import effect.
- Tourism has an effect on inflation.
- Tourism has an opportunity cost effect.
- Tourism has a seasonal effect.
- Tourism has a foreign capital effect.

In addition to the negative economic effects of tourism mentioned above, increases in prices in a tourism region increase the cost of living there (Ap, 1990; Kreag, 2001; Nematpour & Faraji, 2019). Tourism causes acquired revenues to go out of the region or country economy, that is, leakage (Goeldner & Ritchie, 2009; Ardahaey, 2011; Vanhove, 2011; Education Bureau, 2013; Frent, 2016; Nematpour & Faraji, 2019). Due to overdependence, national economies might become dependent on tourism revenues (Mason, 2003; Goeldner & Ritchie, 2009; Ardahaey, 2011; Frent, 2016; Nematpour & Faraji, 2019). Tourism attracts foreign labor due to a lack of knowledge and expertise (Nematpour & Faraji, 2019; Salazar & Cardoso, 2019). Tourism causes government spending (Vanhove, 2011) and infrastructure costs to increase (Kreag, 2001; Frent, 2016). As tourism increases the demand for local products, it raises the price of products (Kim, 2002). In addition, shortages in some primary products during the tourism season are among the negative economic effects of tourism (Nematpour & Faraji, 2019). As a result, it can be concluded that tourism shows economic effects on both regional and countrywide levels positively and negatively.

Social Effects of Tourism

Tourism is a social process that provides social interaction by bringing tourists and host people together (Brunt & Courtney, 1999). In this process, tourists, directly and indirectly, affect the culture and social life of destinations. These effects can bring changes to traditional lifestyles, families, and relationships (Alipour et al., 2019, p. 282). Therefore, tourism has great importance for society as a whole and its importance in human life because tourism is a social event that affects a society's worldview, understanding, and thoughts about people of other countries (Civelek, 2010, p. 342). Thanks to tourism, there is a social interaction between the host people and tourists from different countries and regions. Mutual interaction can be between individuals as well as between an individual and social groups. This shows that tourism has social effects on social groups such as individuals, families, and society (Avcıkurt, 2009).

Social effects of tourism are described as the ways tourism contributes to changes in value systems, individual behavior, family relationships, collective lifestyles, levels of safety, moral behavior, creative expression, traditional ceremonies, and community organizations" (Mathieson & Wall, 1982; Ap, 1990). In other words, the social effects of tourism are defined as the factors that affect the traditions, habits, social lives, beliefs, and values of the local people in tourism destinations (Nematpour & Faraji, 2019). So, the social effects of tourism have a faster impact on both tourists and the host public in terms of quality of life (Brunt & Courtney, 1999) since the social effects of tourism include more immediate changes in the social structure of the society, regulations for the economy and industry of the destination (Haralambopoulos & Pizam, 1996).

Since tourism mostly occurs from developed countries to developing countries, it affects developing countries by spreading values, behavioral patterns, and organizational structures of developed societies (Doğan & Üngören, 2010, p. 397). Therefore, social interactions between local people and tourists can lead to new social and cultural opportunities for both parties. However, these interactions can also create emotions of distress, pressure, and congestion (Nematpour &

Faraji, 2019). As a result, benefiting from a tourism region's natural and cultural heritage creates both positive and negative social effects on that region (Cianga & Sorocovschi, 2017). In general, the positive social effects of tourism are as follows (Ap, 1990; Kreag, 2001; Gürbüz, 2002; Goeldner & Ritchie, 2009; Kozak et al., 2014):

- It fosters positive relations and an atmosphere of tolerance between local people and guests.
- It accelerates the urbanization of rural areas.
- It gives women more rights and opportunities.
- It develops the habits of spending leisure time.
- It helps to develop the awareness of cleanliness.
- It leads to the emergence of new social institutions.
- It creates new professions.
- It increases the desire of local people and guests to learn foreign languages.
- It strengthens family ties by allowing families to have a holiday together.
- It affects the world view of local people and guests (dressing, consumption habits, etc.).

In addition to the positive social effects of tourism mentioned above, tourism provides an incentive for local people to stay in the area by reducing migration pressures (Nematpour & Faraji, 2019). It enables local recreational activities to expand due to tourist visits (Golzardi et al., 2012). Tourism improves the image of the country and the region positively (Kim, 2002; Hammad et al., 2017). It provides peace, trust, and harmony between different groups of people in society (Ince et al., 2020). It provides an opportunity for education (Kim, 2002). It provides security level improvement (Nematpour & Faraji, 2019).

The negative social effects of tourism, in general, are as follows (Ap, 1990; Kreag, 2001; Kim, 2002; Avcıkurt, 2009; Goeldner & Ritchie, 2009; Marzuki 2011; Education Bureau, 2013; Frent, 2016; Nematpour & Faraji, 2019; Salazar & Cardoso, 2019):

- Bad habits such as gambling, prostitution, smuggling, alcohol and drug use, and increased crime rates.
- Too many people in public places cause overcrowding, congestion, traffic jams, and noise.
- The difference between the social and moral values of tourists and local people causes some conflicts.
- Differences in the wealth level of tourists and local people cause resentment and hostility.
- It causes displacement of indigenous people for the development of tourism.
- It causes undesirable changes in lifestyle.
- It causes negative changes and conflicts in values and traditions.
- It causes families to break up.
- It causes the spread of some infectious diseases, especially AIDS, and an increase in health problems.

In addition to the negative social effects of tourism mentioned above, tourism increases xenophobia (Kozak et al., 2014). It causes the social carrying capacity to be exceeded (Salazar & Cardoso, 2019). It causes the exploitation of local people (Ap, 1990). It also causes law and social order changes and commercialization in local norms (Gu & Wong, 2006).

Cultural Effects of Tourism

Culture consists of behavioral patterns, knowledge, and values that are acquired and transmitted over generations. In other words, culture is a complex whole that includes knowledge, belief, art, moral law, tradition, and all other abilities and habits acquired by a person as a member of society.

Therefore, culture is about interacting through social interaction, social relations, and material artifacts (Mason, 2003, pp. 40–41).

Tourism is the most crucial sector that mediates cultural exchange and cultural dissemination (Avcıkurt, 2009, p. 66) because tourism is an activity that involves the movement of people from one place to another. In this process, tourists, on the one hand, bring the traditions and cultures of their own countries to the host countries. On the other hand, they learn the customs and traditions of the host countries and bring them back to their countries (Education Bureau, 2013, p. 174).

Among the cultural attractions related to tourism are handicrafts, language, religion, tradition, architecture, gastronomy, art and music, historical/visual reminders of the region, livelihoods of local people, education systems, clothing, and leisure activities (Mason, 2003). With these cultural values, local people have high wealth, and preserving this unique culture is critical to gain more benefits from tourism activities (Ince et al., 2020). In this context, the cultural effects of tourism focus on the long-term changes that will gradually emerge in a society's norms and standards, social relations, and works (Haralambopoulos & Pizam, 1996, p. 503). However, the cultural effects of tourism include the relationship between local people and tourists, which is not always beneficial for the local people (Cianga & Sorocovschi, 2017). Therefore, when it comes to the cultural effects of tourism, the results point to two situations. While the first states that tourism can preserve local culture, the second argues the opposite (Pizam & Milman, 1986). Many studies in the literature support the idea that tourism has both positive and negative cultural effects (Lickorish & Jenkins, 1997; Kim, 2002; Mason, 2003; Doğan & Üngören, 2010; Education Bureau, 2013; Shahzalal, 2016; Cianga & Sorocovschi, 2017; Hammad et al., 2017; Türker, 2020). The positive effects of tourism on culture are listed as follows (Ap; 1990; Lickorish & Jenkins, 1997; Kreag, 2001; Long & Kayat, 2011; Nematpour & Faraji, 2019):

- Tourism enhances the understanding and image of different cultures.
- Tourism increases the demand for historical and cultural exhibitions.
- Tourism causes cultural exchange between tourists and local people.
- Tourism encourages local people to engage in various cultural activities.
- Tourism keeps the culture of the host people alive and preserves their cultural identity.
- Tourism causes positive values and traditions (honesty, courtesy, etiquette, mutual trust).

In addition to the positive cultural effects of tourism mentioned above, tourism increases local people's pride in their culture. It preserves cultural heritage (Kim, 2002; Long & Kayat, 2011; Education Bureau, 2013). It drives local culture to gain importance and local people to learn more about their own culture. Local people learn about the culture of foreigners and increase their knowledge (Avcıkurt, 2009).

The negative cultural effects of tourism are as follows (Kreag, 2001; Avcıkurt, 2009; Doğan & Üngören, 2010; Duran, 2011; Education Bureau, 2013; Hammad et al., 2017):

- It causes changes in religious beliefs and behaviors.
- It causes changes in moral values and behaviors.
- The lifestyle of foreigners is adopted.
- It causes an increase in the rate of foreign words in the language of the local people.
- It causes the function and meaning of native art to change.
- It causes tourism to become a part of the local people's own culture.

In addition to the negative cultural effects of tourism mentioned above, tourism might harm local culture. It encourages local people to imitate the behavior of tourists and abandon cultural traditions (Kim, 2002; Long & Kayat, 2011). Because of tourism, culture is seen by local people as a commercial resource (Brunt & Courtney, 1999; Shahzalal, 2016; Salazar & Cardoso, 2019). Existing tourism activities offered to tourists do not reflect the region's authentic culture (Marzuki, 2011; Hammad et al., 2017). In addition, tourism has negative cultural effects such as cultural diffusion, cultural degeneration, cultural change (acculturation), hybrid culture, and cultural delay (Avcıkurt, 2009; Shahzalal, 2016).

Environmental Effects of Tourism

When it comes to the effects of tourism, one of the most emphasized factors is the effects of tourism on the environment (Türker, 2020, p. 1482) because the tourism industry is one of the most significant components of the service industry and has an important ability to affect the environmental quality (Shakouria et al., 2017). In addition, the environment is a fundamental element of the tourism product, and tourism is an environment-dependent activity (Pender & Sharpley, 2005).

The concept of environment refers to the physical environment in which tourism takes place and provides incentives for travel (Zaei & Zaei, 2013). The tourism environment is "the physical and human environment consisting of all-natural and human factors, including air, water, soil, flora and fauna, built facilities, landscape, color, sound, and other environmental factors" (Zhong et al., 2011). So this environment can be said to consist of both natural and human characteristics. The natural environment is often referred to as the physical environment. The natural or physical environment includes certain land features such as rivers, rock outcrops, beaches, plants, and animals (flora and fauna) (Mason, 2003). Therefore, the environmental dimension covers the physical environment of a destination. Destinations with invaluable natural resources (e.g., beaches, lakes, mountains), museums, historical buildings, or heritage sites are attractive to tourists. Tourism can generate income to preserve such natural places, restore monuments or historical sites, or even improve the visual and aesthetic appearance (Loureiro et al., 2019).

Tourism is more dependent on the environment than any other field of activity since the primary source on which the development, quality, and components of various tourism activities depend is the tourist attraction factor. Thus, tourism development is facilitated or hindered by the environment (Cianga & Sorocovschi, 2017). Therefore, tourism development in a region with an attractive natural or artificial environment attracts tourists (Zaei & Zaei, 2013). Accordingly, tourists seek attractive, different, or distinctive environments that can support certain tourist activities. Therefore, for the long-term success of tourism (Pender & Sharpley, 2005), maintaining and improving the quality of a destination's tourism environment is essential (Zhong et al., 2011).

Most of the tourism activities occur in nature, and consequently, the tourism activities that take place affect the physical environment in various ways (Kozak et al., 2014). The environmental impacts of tourism occurring in the physical environment mainly depend on local conditions and planning practices such as locality, type of activity, and tourist infrastructure (Chang et al., 2018). Thus, when tourism develops in a region, environmental protection and improvement also develop positively. In addition, an unplanned and unhealthy touristic development can lead to deterioration of the environment and environmental quality, which can negatively affect the environment (Avcıkurt, 2009, p. 43). In this context, it can be said that tourism generally creates both negative and positive effects on the environment in which it develops (Pizam & Milman, 1986; Lickorish & Jenkins, 1997; Zaei & Zaei, 2013). The positive environmental effects of tourism are generally as follows (Lickorish & Jenkins, 1997; Kreag, 2001; Mason, 2003; Long & Kayat, 2011; Education Bureau, 2013; Hammad et al., 2017; Nematpour & Faraji, 2019; Salazar & Cardoso, 2019; Nematpour & Faraji, 2019):

- Tourism ensures the protection of the environment, wildlife, and natural areas.
- Tourism protects archaeological and historic sites, historic buildings, and monuments.

- Tourism improves environmental quality.
- Tourism contributes to the wellness of the environment.
- Tourism provides improvement of infrastructure.
- Tourism increases environmental awareness.

In addition to the positive environmental effects of tourism mentioned above, tourism ensures that water conservation measures are taken (Salazar & Cardoso, 2019). It provides management of waste products (Nematpour & Faraji, 2019). It ensures the survival of places with scarce natural resources. It enables the use and exploitation of low-yielding agricultural lands through the implementation of appropriate tourist facilities (Cianga & Sorocovschi, 2017).

The increasing number of tourists resulting from tourism development can cause negative effects by increasing the pressure on the environment over time (Baysan, 2001). In this context, the most typical negative effects of tourism on the environment are (Pizam & Milman, 1986; Lickorish & Jenkins, 1997; Kreag, 2001; Mason, 2003; Long & Kayat, 2011; Golzardi et al., 2012; Paul, 2012; Frent, 2016; Nematpour & Faraji, 2019; Salazar & Cardoso, 2019):

- Tourism causes visual, noise, soil, air, and water pollution.
- Tourism causes traffic congestions and overcrowding.
- Tourism causes the destruction of settlements, agricultural and forest lands, and their use in tourism.
- Tourism causes ecosystem degradation and the destruction of vegetation.
- Tourism causes environmental hazards such as erosion, landslides, flooding, and the destruction of beaches.
- Tourism causes damage and destruction of historical and archaeological sites and monuments.
- Tourism results in increased waste and improper disposal.
- Tourism causes excessive urbanization and concretization.
- Tourism causes excessive consumption of natural resources.

In addition to the negative environmental effects of tourism mentioned above, tourism also causes the reduction of open spaces due to uncontrolled construction (Kreag, 2001; Nematpour & Faraji, 2019; Salazar & Cardoso, 2019). In addition, tourism causes buildings that do not fit the local architecture and causes architectural pollution (Mason, 2003; Marzuki, 2011; Nematpour & Faraji, 2019).

Social Psychological Effects of Tourism

The beginning of the development process of social psychology goes back to Plato. However, it was not until the 1900s that social psychology began to develop as a branch of science (Güney, 2009, p. 1). In this process, social psychology, a newly developing branch of science, has been examining people's social behavior by using the concepts and approaches of both sociology and psychology (Tutar, 2013, p. 43). Sociology focuses on social factors, and psychology focuses on individual factors; social psychology, on the other hand, examines the individual in the group, the effects of the group on the individual, and the group-individual interaction (Zencirkıran, 2017, p. 3). In other words, social psychology deals with the behavior of people who are influenced by their environment and other people, rather than just individual (field of psychology) or social (field of sociology) behaviors. Thus, social psychology was born out of the need to explain interactions, relationships, situations, and phenomena that sociology and psychology fail to explain (Güney, 2009, p. 3).

Social psychology is the science of examining the behavior of an individual or individuals in a society (Kağıtçıbaşı, 2014, p. 22). In other words, social psychology is a discipline that scientifically examines people's opinions about other people and the way they interact with them (Tutar, 2013, p. 45). In addition, social psychology is also expressed as a field of science that tries to understand the nature and causes of individual behavior in social environments (Gnoth, 2014). In this sense, social psychology examines the individual and his/her environment and the social groups consisting of the group and its environment (Küçün, 2019, p. 43).

Tourism is a mind-opening experience that teaches people that the world does not consist of a single life model. There are other life models (Wintersteiner & Wohlmuther, 2014) because it is believed that tourism positively affects world peace. When people travel from place to place with a sincere desire to learn more about their global neighbors, their knowledge and understanding increases. Thus, a start is made to develop world communication, which is very important in building bridges of mutual appreciation, respect, and friendship through tourism (Goeldner & Ritchie, 2009).

Among the areas of interest of social psychology, which is about understanding individual behavior in society (Gnoth, 2014), we can include topics including group conformity behavior, persuasion, power, social influence, obedience, prejudice, reduction of prejudice, discrimination, stereotypes, social cognition, social perception, social categories, aggression, altruistic behaviors, interpersonal attraction, attitudes and attitude change, communication, impression making, small groups, leadership, mass behavior, and intergroup relations (Tutar, 2013, p. 45), and learning (Güney, 2009). It is possible to see both positive and negative aspects of these issues in the behaviors that occur due to the mutual interaction of an individual with another individual or society and society with another society. The social-psychological effects of tourism, which is a social event that occurs as a result of the mutual relations of tourists with tourists, local people or employees, and other individuals/groups, can emerge with the positive and negative aspects of these issues. Therefore, tourism may cause the emergence of the above-mentioned social psychological issues among individuals and groups in the region where it takes place. For example, thanks to tourism, different societies that are prejudiced against each other can get to know each other closely and break their prejudices. Çelik (2019b) also stated in his study that social psychology tries to explain the relationships between society, groups, and individuals. The interaction arising from the relationship between tourists and local people can also be approached from a social psychological perspective. On the other hand, the concept of learning, which is included in the subjects of social psychology (Güney, 2009); It is the long-term permanent changes that occur in the behavior of the individual as a result of experiences (experience, education-training, observation) (Zencirkıran, 2017, p. 87). Transformational learning, one of the learning theories, was first put forward by Jack Mezirow in 1978 (Akpınar, 2010, p. 186). Transformational learning is learning to purposefully question one's own assumptions, beliefs, feelings and perspectives for personal development and maturation (Akçay, 2012, p. 7). Therefore, at the center of the transformative learning theory, there are processes of people to critically research themselves, evaluate their experiences, and interpret and rename these experiences (Çimen & Yılmaz, 2014, p. 341). In this context, the research of transformational learning in tourism recently supports the existence of social psychological effects of tourism. For example, Pitman et al. (2010) in the research on transformational learning in educational tourism; educational tourism has been found to be characterized by intentional and structured learning experiences that provide opportunities for the teacher to immerse himself in experiences that have the potential to challenge previously held beliefs and prejudices. Coghlan & Gooch (2011) reconceptualized volunteer tourism as a form of transformative learning in which all participants (including members of the host community) learn and change as a result of their experience. More specifically, aspects of voluntary tourism that are compatible with the theory and practice of transformative learning have been identified. Stone & Duffy (2015), in their research to systematically review travel and tourism research and identify strategies using transformational learning; in addition to a better understanding of the theoretical basis of transformational learning, it is emphasized that scientific research that emphasizes the deliberate, creative and effective uses of transformational learning in tourism is necessary. Müller et al. (2020), in their research on the effects of volunteer tourism on transformational learning and career, argue that volunteer tourism contributes to the practitioner's career sustainability and employability, as well as career reassessments of new behaviors and worldviews developed during their experience. Cavender et al. (2020), in their research on transformational learning through study abroad in a niche tourism destination within the scope of transformational travel, explored how travel can be transformative for the traveler within the framework of Mezirow's transformational learning theory.

In the literature, it is seen that there is a limited number of studies on the social-psychological effects of tourism. Some of these are listed below.

Iso-Ahola (1982) carried out a study titled "Toward a Social Psychological Theory of Tourism Motivation." Pearce's (1982) *The Social Psychology of Tourist Behavior* is a seven-part book describing tourists, tourism, and tourist psychology. The book includes explicitly economic, geographical, anthropological, and sociological studies of tourism. Fedler (1986) used various studies to document the diversity of tourist roles, the diversity of tourist motivations at tourist-host social contact levels, and the uniqueness of tourist environments. However, the development and analysis of these concepts and their incorporation into a new or existing social psychology theory or conceptual framework are often insufficient. Many social, psychological and social psychological theories are mentioned throughout the book to fill this gap. For example, Freud's psychoanalytic theory, Hull's drive reduction theory, Maslow's hierarchy of needs, and achievement motivation were compiled from attribution theories for social contact and environment learning theory. The article by Harrill & Potts (2002) aims to review and evaluate the dominant social psychological, and related tourist motivation models by reviewing the conceptual development of tourist motivation from the early 1970s to the present. This study argues that an integrated social-psychological approach to tourist motivation has not yet been achieved. The study by Wolfe and Hsu (2008) aimed to empirically test Iso-Ahola's (1982) Social Psychological Model of Tourism Motivation. Tomljenovic (2010), to identify examples of the positive relationship between tourism and peace and to learn from them; to provide and promote more academic and scientific research output focused on tourism and peace proposition; starting from the original question of whether tourism contributes to peace, it has sought to find ways in which tourism can be managed and conducted to meet the goal of peace. The tourism experience model (TEM), developed by Gnoth (2014), is a meta-analytic, phenomenological model that shows how tourists experience destinations. This article argues that social and cultural psychology is only part of analyzing how the tourist filters consciousness interactions. TEM adds to the scope of social psychology, describing how and why tourists may experience social interactions, considering the existential self instead of the role-authentic self of social psychology. The model thus formed goes beyond both the exclusionary social focus and the self-centered concept, which is the dichotomy of individualism-collectivism. Tang (2014) examined articles on applying social psychology theories and concepts presented in 12 leading academic journals published between 1999 and 2012. A study was conducted by Çelik (2019b) to evaluate the tourist-local people interaction in Allport's theory of intergroup contact. This study examined the relationship between tourism and prejudice in the context of intergroup contact theory. As a result, the positive and negative effects of tourism could not be revealed in the studies on social psychology in tourism. In addition, since the number of studies on social psychology in tourism is low, its conceptual and theoretical framework has not been fully developed.

Conclusion and Suggestions

In this study, the positive and negative aspects of tourism's effects are discussed from general, economic, social, cultural, environmental, and social psychological perspectives. Therefore, it is possible to see these effects at both macro and micro levels in the regions, destinations, and countries where tourism activities occur.

When tourism first started to develop, the importance of tourism began to be perceived in terms of the economy in general because tourists going to the destinations created an income effect. Later, these positive economic effects were followed by some other economic effects such as employment, the effect on the balance of payments, interregional development, and other sectors. However, tourism has negative economic effects such as inflationary effect, import effect, opportunity cost effect, seasonality effect and foreign capital effect. On the other hand, one of the first studies to realize that the economic effects of tourism alone do not give a comprehensive perspective to the phenomenon of tourism was made by Pizam (1978). The researcher examined the negative effects of tourism on the social structure in this study (Brida et al., 2011, p. 360). Therefore, the effect of the social environment is very important in terms of the tourism sector. Situations that the social structure cannot accept and tolerate may end a business or local tourism. Therefore, it is possible to examine the subjects such as human groups, relationships, demographic and cultural characteristics, customs and traditions, lifestyles, folklore, and value judgments that are directly or indirectly affected or affecting, within the concept of social structure (Civelek, 2010, p. 333). Depending on this situation, tourism affects the cultural structure both positively and negatively. It causes the destruction of culture and tangible and intangible values of culture, deterioration of the moral structure, and positive effects such as cultural interaction and recognition of different cultures, protection of cultural heritage.

Tourism has both positive and negative effects on the environment. The environment is where we see the worst effects of tourism because tourism activities usually occur in physical environments with natural or cultural resources. Due to the increasing number of tourists, tourism creates negative effects that endanger the future of regions and destinations, such as pollution, destruction of natural and cultural resources. However, tourism also has positive environmental effects in protecting the environment protecting and improving natural and cultural resources.

In addition to the effects of tourism mentioned above, the social-psychological effects of tourism have also been mentioned recently. However, the lack of sufficient studies on the positive and negative effects of tourism on this subject has caused the conceptual and theoretical framework of social psychological effects not to be fully formed. In addition, the confusion of social psychological effects with the social or cultural effects of tourism makes it difficult to reveal what the social-psychological effects of tourism are.

When the studies on the effects of tourism in the literature are examined, it is seen that, apart from the economic effects of tourism, studies on the environmental, social, cultural, and social-psychological effects of tourism have been around since 1980. The increase in the number of studies on the effects of tourism is due to the increase in the number of tourists participating in tourism activities and the number of businesses opened in the sector. In future studies on the effects of tourism, tourism's sociological effects and psychological effects can be investigated separately. Therefore, it can be suggested that tourism researchers contribute to this field of tourism by conducting studies on sociology and psychology.

References

Akçay, C. (2012). Dönüşümsel öğrenme kurami ve yetişkin eğitiminde dönüşüm. *Milli Eğitim Dergisi,* 42(196), 5–19.

Akpınar, B.(2010). Transformatif öğrenme kurami: dönüşerek ve değişerek öğrenme. *Anadolu Üniversitesi Sosyal Bilimler Dergisi,* 10(2), 185–198.

Alipour, H., Rezapouraghdam, H. & Hasanzade, B. (2019). A holistic analysis of the second-home tourism impacts in the Caspian Sea region of Iran. *Experiencing Persian Heritage (Bridging Tourism Theory and Practice),* 10, 275–293.

Almeida-García, F., Peláez-Fernández, M. Á., Balbuena-Vázquez, A. & Cortés-Macias, R. (2016). Residents' perceptions of tourism development in Benalmádena (Spain). *Tourism Management,* 54, 259–274.

Ap, J. (1990). Residents' perceptions research on the social impacts of tourism. *Annals of Tourism Research,* 17(4), 610–616.

Ardahaey, F.T. (2011). Economic Impacts of tourism industry. *International Journal of Business and Management,* 6(8), 206–215.

Avcıkurt, C. (2009). *Turizm sosyolojisi, genel ve yapısal yaklaşim.* Ankara: Detay Yayıncılık.

Bahar, O. & Kozak, M. (2015). *Turizm ekonomisi* (7th edn). Ankara: Detay Yayıncılık.

Baysan, S. (2001). Perceptions of the environmental impacts of tourism: A comparative study of the attitudes of German, Russian and Turkish tourists in Kemer, Antalya. *Tourism Geographies,* 3(2), 218–235.

Brida, J.G., Osti, L. & Faccioli, M. (2011). Residents' perception and attitudes towards tourism impacts, A case study of the small rural community of Folgaria (Trentino – Italy). *Benchmarking: An International Journal,* 18(3), 359–385.

Brunt, P. & Courtney, P. (1999). Host perceptions of sociocultural impacts. *Annals of Tourism Research,* 26(3), 493–515.

Castela, A. (2018). Impacts of tourism in an urban community: The case of Alfama, Athens. *Journal of Tourism,* 5(2), 133–148.

Cavender, R., Swanson, J. R. & Wright, K. (2020). Transformative travel: Transformative learning through education abroad in a niche tourism destination. *Journal of Hospitality, Leisure, Sport & Tourism Education,* 27, 1–14.

Çelik, S. (2019a). Does tourism reduce social distance? A study on domestic tourists in Turkey. *Anatolia,* Routledge, 30(1), 115–126.

Çelik, S. (2019b). Social psychological effects of tourism Evaluation of the tourist–local people interaction within the context of Allport's intergroup contact theory, In D. Gürsoy & R. Nunkoo (Eds.), *Handbook of tourism impacts theoretical and applied perspectives* (pp. 242–251). London and New York: Routledge.

Chang, K.G., Chien, H., Cheng, H. & Chen, H. (2018). The impacts of tourism development in rural indigenous destinations: An investigation of the local residents' perception using choice modeling. *Sustainability,* 10(4766), 2–15.

Cianga, N. & Sorocovschi, V. (2017). The impact of tourism activities. A point of view. *risks and catastrophes Journal,* 20(1), 25–40.

Çimen, O. & Yılmaz, M. (2014). Dönüşümsel öğrenme kuramina dayali çevre eğitiminin biyoloji öğretmen adaylarinin çevre sorunlarina yönelik algilarina etkisi. *Bartın Üniversitesi, Eğitim Fakültesi Dergisi,* 3(1), 339–359.

Civelek, A. (2010). Turizmin sosyal yapiya ve sosyal değişmeye etkileri. *Selçuk Üniversitesi Sosyal Bilimler Meslek Yüksekokulu Dergisi,* 13(1–2), 331–350.

Coghlan, A. & Gooch, M. (2011). Applying a transformative learning framework to volunteer tourism. *Journal of Sustainable Tourism,* 19(6), 713–728.

Cunha, L. (2012). The definition and scope of tourism: A necessary inquiry. *COGITO Journal of Tourism Studies,* 5, 91–114.

Doğan, H. & Üngüren, E. (2010). Alanya halkinin turizme sosyo-kültürel açidan bakişi. *e-Journal of New World Sciences Academy,* 5(4), 396–415.

Duran, E. (2011). Turizm, kültür ve kimlik ilişkisi; turizmde toplumsal ve kültürel kimliğin sürdürülebilirliği. *İstanbul Ticaret Üniversitesi, Sosyal Bilimler Dergisi,* 10(19), 291–313.

Duran, E. & Özkul, E. (2012). Yerel halkin turizm gelişimine yönelik tutumlari: Akçakoca örneği üzerinden bir yapisal model. *International Journal of Human Sciences (Online),* (9)2, 500–520.

Education Bureau, (2013). Manual on Module I Introduction to Tourism (Fine-tuned version), Personal, Social and Humanities Education Section, The Government of the Hong Kong Special Administrative Region, Tourism and Hospitality Studies, https://www.edb.gov.hk/attachment/en/curriculum-development/kla/pshe/references-and-resources/tourism/Tourism_English_19_June.pdf.

Fedler J. A. (1986). The social psychology of tourist behaviour. Philip L.Pearce. *Journal of Leisure Research,* 18(3), 213–214.

Frent, C. (2016). An overview on the negative impacts of tourism. *Journal of Tourism – Studies & Research in Tourism,* 22, 32–37.

Gnoth, J. (2014). *The role of social psychology in the tourism experience model (tem), tourists' perceptions and assessments* (Advances in Culture, Tourism and Hospitality Research, Vol. 8) (pp. 61–69). Bingley: Emerald Group Publishing Limited.

Goeldner, R.C. & Ritchie, J.R.B. (2009). *Tourism: Principles, practices, philosophies.* Hoboken: John Wiley and Sons.

Göksu, T. (2007). *Sosyal psikoloji* (1st ed.). Ankara: Seçkin Yayınları.

Golzardi, F., Sarvaramini, S., Sadatasilan, K. & Sarvaramini, M. (2012). Residents attitudes towards tourism development: A case study of Niasar, Iran. *Research Journal of Applied Sciences, Engineering and Technology,* 4(8), 863–868.

Gu, M. & Wong, P.P. (2006). Residents' perception of tourism impacts: A case study of homestay operators in Dachangshan Dao, North-East China. *Tourism Geographies,* 8(3), 253–273.

Güney, S. (2009). *Sosyal psikoloji* (1st ed.). Ankara: Nobel Yayıncılık.

Gürbüz, A. (2002). Turizmin sosyal çevreye etkisi üzerine bir araştirma. *Teknoloji,* 5(1–2), 49–59.

Hammad, N.M., Ahmad, S.Z. & Papastathopoulos, A. (2017). Evaluating perceptions of residents' towards impacts of tourism development in Emirates of Abu Dhabi, United Arab Emirates. *Tourism Review,* 72(4), 448–461.

Haralambopoulos, N. & Pizam, A. (1996). Perceived impacts of tourism, the case of Samos. *Annals of Tourism Research,* 23(3), 503–526.

Harrill, R. & Potts, T. D. (2002). Social psychological theories of tourist motivation: Exploration, debate, and transition. *Tourism Analysis,* 7(2), 105–114.

Ince, E., Iscioglu, D. & Ozturen, A. (2020). Impacts of cittaslow philosophy on sustainable tourism development. *Open House International,* 45(1/2), 173–193.

Iso-Ahola, S. E. (1980). *The social psychology of leisure and recreation.* Dubuque, IA: William C. Brown Company Publishers.

Iso-Ahola, S. E. (1982). Toward a social psychological theory of tourism motivation: A rejoinder. *Annals of Tourism Research,* 9(2), 256–262.

Jackson, L. A. (2008). Residents' perceptions of the impacts of special event tourism. *Journal of Place Management and Development,* 1(3), 240–255.

Jurowski, C., Uysal, M. & Williams, D. R. (1997). A theoretical analysis of host community resident reactions to tourism. *Journal of Travel Research,* 36(3), 3–11.

Kağıtçıbaşı, Ç. (2014). *Yeni insan ve insanlar: Sosyal psikolojiye giriş.* İstanbul: Evrim Yayınları.

Kim, K. (2002). The effects of tourism impacts upon quality of life of residents in the community. (Unpublished PhD. dissertation). Virginia Polytechnic Institute and State University, Blacksburg, Virginia.

Kozak, N., Kozak, M.A. & Kozak, M. (2014). *Genel turizm ilkeler ve kavramlar.* Ankara: Detay Yayıncılık.

Kreag, G. (2001). *The impacts of tourism.* New York: Minnesota Sea Grant.

Küçün, N. T. (2019). Sosyal Psikoloji Çerçevesinden Satın Alma Sürecinin Nöropazarlama Yöntemleri İle İncelenmesi, (*Yayınlanmamış Doktora Tezi*). Trakya Üniversitesi Sosyal Bilimler Enstitüsü, Edirne.

Lickorish, L.J. & Jenkins, C.L. (1997). *An introduction to tourism.* Boston: Butterworth-Heinemann.

Long, P. H. & Kayat, K. (2011). Residents' perceptions of tourism impact and their support for tourism development: The case study of Cuc Phuong National Park, Ninh Binh Province, Vietnam. *European Journal of Tourism Research,* 4(2), 123–146.

Loureiro, S.M.C., Sarmento, E.M. & Rosário, J.F. (2019). *Overview of underpinnings of tourism impacts from.* Ed. Dogan Gursoy & Robin Nunkoo. The Routledge Handbook of Tourism Impacts, Theoretical and Applied Perspectives Routledge.

Marzuki, A. (2011). Resident attitudes towards impacts from tourism development in Langkawi Islands, Malaysia. *World Applied Sciences Journal,* 12 (Special Issue of Tourism & Hospitality), 25–34.

Mason, P. (2003). *Tourism impacts, planning and management.* Boston: Butterworth-Heinemann.

Mathieson, A. & Wall, G. (1982). *Tourism: Economic, physical and social impacts.* Harlow: Longman.

Mbagwu Felicia, O., Bessong Columbus, D. & Anozie Okechukwu, O. (2016). Contributions of tourism to community development. *Review of European Studies,* 8(4), 121–130.

Müller, C. V., Scheffer, A.B.B. & Closs, L.Q. (2020). Volunteer tourism, transformative learning and its impacts on careers: The case of Brazilian volunteers. *International Journal of Tourism Research,* 22(5), 726–738.

Nematpour, M. & Faraji, A. (2019). Structural analysis of the tourism impacts in the form of future study in developing countries (case study: Iran). *Journal of Tourism Futures,* 5(3), 259–282.

Özdemir, E.G. & Büyükkuru, M. (2019). *Turizmin tarihsel gelişimi.* Ed. Şule Aydın & Duygu Eren. Alternatif Turizm, Ankara: Detay Yayıncılık.

Page, S.J. (2014). *Tourism management.* London: Routledge.

Paul, B.D. (2012). The impacts of tourism on society, annals of faculty of economics, University of Oradea. *Faculty of Economics,* 1(1), 500–506.

Pearce, P. L. (1982). Tourists, tourism and tourist psychology. In *The social psychology of tourist behavior* (pp. 1–25). New York: Pergamon Press.

Pender, L. & Sharpley, R. (2005). *The management of tourism.* SAGE Publications.

Pitman, T., Broomhall, S., Majocha, E. & McEwan, J. (2010). Transformative learning ineducational tourism. In Educating for sustainability. *Proceedings of the 19th AnnualTeaching Learning Forum, 28–29 January 2010.* Perth: Edith Cowan University.
Pizam, A. (1978). Tourism's impacts: The social costs to the destination community as perceived by its residents. *Journal of Travel Research,* 16(8), 8–12.
Pizam, A., Fleischer, A. & Mansfeld, Y. (2002). Tourism and social change: The case of Israeli ecotourists visiting Jordan. *Journal of Travel Research*, November, 177–184.
Pizam, A., Jafari, J. & Milman, A. (1991). Influence of Tourism on attitudes: US students visiting USSR. *Tourism Management*, 12(1), 47–54.
Pizam, A. & Milman, A. (1986). The social impacts of tourism. *Tourism Recreation Research,* 11(1), 29–33.
Rhama, B. (2019). Psychological costs on tourism destination. *Journal of Advanced Management Science,* 7(3), 100–106.
Salazar, A. & Cardoso, C. (2019). Tourism planning: Impacts as benchmarks for sustainable development plans. *Worldwide Hospitality and Tourism Themes*, 11(6), 652–659.
Segota, T., Mihalic, T. & Kuscer, K. (2017). The impact of residents' informedness and involvement on their perceptions of tourism impacts: The case of Bled. *Journal of Destination Marketing & Management*, 6, 196–206
Shahzalal, M. (2016). Positive and negative impacts of tourism on culture: A critical review of examples from the contemporary literature. *Journal of Tourism, Hospitality and Sports*, 20, 30–34.
Shakouria, B., Yazdia, S. K. & Ghorchebig, E. (2017). Does tourism development promote CO2 emissions? *Anatolia*, 28(3), 444–452.
Steyn, S., Saayman, M. & Nienaber, A. (2004). The impact of tourist and travel activities on facets of psychological wellbeing: Research article. *South African Journal for Research in Sport, Physical Education and Recreation*, 26(1), 97–106.
Stone, G.A. & Duffy. L.N. (2015). Transformative learning theory: A systematic review of travel and tourism scholarship. *Journal of Teaching in Travel & Tourism*, 15(3), 204–224, DOI: 10.1080/15313220.2015.1059305.
Stringer, P. F. & Pearce, P. L. (1984). Toward a symbiosis of social psychology and tourism studies. *Annals of Tourism Research*, 11(1), 5–17.
Tang, L.R. (2014). The application of social psychology theories and concepts in hospitality and tourism studies: A review and research agenda. *International Journal of Hospitality Management,* 36,188– 196.
Tomljenovic, R. (2010). Tourism and intercultural understanding or contact hypothesis revisited. In O. Moufakkir & I. Kelly (Eds.), *Tourism, progress and peace* (pp. 17–34). Wallingford, UK: CABI.
Türker, G.Ö. (2020). Are the effects of tourism positive or negative? A review of the studies in Turkey. *Journal of Turkish Tourism Research*, 4(2), 1477–1492.
Tutar, H. (2013). *Davranış bilimleri, kavramlar ve kuramlar* (1. Baskı). Seçkin Yayınları.
UNWTO (2008). *Tourism highlights, 2008 Edition, 1–12.*
UNWTO (2020). *World tourism barometer* January 2020 EXCERPT, 18(1), 1–6.
Vanhove, N. (2011). *The economics of tourism destinations* (2nd ed.). Routledge.
Williams, S.(1998).*Tourism geography.* Routledge.
Wintersteiner, W. & Wohlmuther, C. (2014). Peace sensitive tourism: How tourism can contribute to peace. In C. Wohlmuther & W. Wintersteiner (Eds.), *International handbook on tourism and peace* (p. 31). Avusturya: Drava.
Wolfe, K. & Hsu, H.C.C. (2004). An Application of the social psychological model of tourism motivation. *International Journal of Hospitality & Tourism Administration*, 5(1), 29–47.
Zaei, M. E. & Zaei, M. E. (2013). The impacts of tourism industry on host community. *European Journal of Tourism Hospitality and Research,* 1(2), 12–21.
Zencirkıran, M. (2017). Davranış bilimleri üzerine. In M. Zencirkıran (Ed.), *Davranış bilimleri* (ss.1–10). Dora Publish.
Zhong, L., Deng, J., Song, Z. & Ding, P. (2011). Research on environmental impacts of tourism in China: Progress and prospect. *Journal of Environmental Management*, 92, 2972–2983.

4

TOURISM, PREJUDICE, STEREOTYPES, AND PERSONAL CONTACT

Gordon Allport's Contributions to Tourism Research

Maximiliano E Korstanje

Introduction

One of the aspects that define tourism seems to be associated not only to the right to travel everywhere to enjoy other cultures and landscapes, that is, the curiosity for something new, but also to the acculturation process behind the host-guest meeting (Graburn, 2012; Smith, 2012; Van der Duim, Peters & Wearing, 2005). The industry of leisure and business travels has created a cosmopolitan spirit straddling cultures, economies, and nations. Democratic institutions, as well as political stability, are of vital importance for the prosperity and growth of the tourism industry (Sandell, 2005). What is equally important for the popular parlance tourism is being considered an activity of peace and prosperity which helps to placate inter-ethnic conflicts (Farmaki, 2017). Over recent years, some voices have questioned the assumption that tourism promotes peace in the hosting communities (Comaroff & Comaroff, 2009). The literature does not explain with clarity if tourism contributes or not to political stabilities (Litvin, 1998). Having said this, some studies show amply how a covert sentiment of racism, xenophobia, or prejudice still remains stable in host-guest relations (Çelik, 2019; Korstanje, 2011). As John Dovidio and Samuel Gaertner (1986) brilliantly observed, prejudice never disappears but mutates towards new (cultural) versions. The classic (ethnic) racism which characterized the American life in the 1950s or 1960s sets the pace to a new type of discrimination where the cultural superiority of Western rationality occupies a central position. In a seminal book titled*Prejudice, Discrimination, and Racism*, Dovidio and Gaertner are motivated to expand the current understanding how white Americans repress internally their prejudices when they are before a stranger and liberate them when they feel safe within the rules of the in-group.

The present book chapter intends to discuss critically to what extent prejudice and stereotypes are present in the host-guest relation, as well as what are the preconditions for prejudice to be eradicated in the service sectors. In so doing, we review critically the contributions of one of the modern fathers of prejudice research, Gordon Willow Allport. *The Nature of Prejudice*, one of his bestsellers, marked the pace of social psychology to study the correlation of social contact and discrimination. For some reason very hard to be precise here and now, neither prejudice nor Allport was widely worked out in the constellations of tourism research. As discussed, scholars embraced enthusiastically the belief that tourism encourages peace and mutual understanding, disarticulating radical sentiments and xenophobic expressions. One might speculate that laypeople accustomed to deal with foreign tourists are more open to the cultural difference than other professions, but to what extent is this case is the main goal the current conceptual essay-review will answer.

DOI: 10.4324/9781003161868-4

What Are the Prejudice, Discrimination and Racism?

The studies on prejudice and discrimination started from the urgency to understand the nefarious effects of Nazi Germany in the Western imaginary. What were the real factors that ushered the most civilized nation of Europe in the nightmare of Nazism?

Many of the theorists who were originally motivated by prejudice studies forcedly migrated from Europe during World War II (Noelle-Neumann, 1974). Although prejudice has been widely studied by social sciences for more than five decades, experts failed to reach consensus about the specific causes that leads a layperson to develop a radical sentiment against others. As Derrida and Dufourmantelle (2000) explains, the stranger who interrogates furtherly the laws of hosts (reign of dogmatism) should respond about its intentions to the hosting state. In this vein, Zygmunt Bauman (2017) called the attention to the difficult relationship between vagabonds and rich tourists. While the former are immobilized, scrutinized, and even imprisoned as *an undesired guests,* the latter are encouraged to travel, enjoy and discover the world. The armchair anthropologist Lord George Frazer (1951) argued that the outlanders are always valid reasons for fear and anxiety in the hosting communities. The fear of strangers motivates the tribal sorcerers to organize religious rituals to exonerate them of bad spirits. This happens because the stranger seems to be an object of mistrust. Some studies have reflected that the regulation of self-esteem in the in-group needs from an alter-id to regulate the internal social cohesion (Hewstone, 2015; Jones, Dovidio & Vietze, 2013; Markus, 2016). In this respect, those theories discussing prejudice and discrimination failed to form an all-encompassing model that describes how the phenomenon evolves. Far from this, prejudice-related theories speak us of a multidimensional issue very hard to grasp. The produced knowledge may be very well divided into four academic theories: *social identity theory, authoritarian personality theory, social structuralism theory and social contact theory.*

In perspective, social identity theory resulted from the advances in economics in the fields of the deprivation model in the 1930s. These works marked a direct correlation between the economic downturn and the racial riots in the US. Authors of the caliber of Dollard J et al. (1939) or Hovland and Sears (1940) validated a firm correlation between lynching and long-lasting economic deprivations. Per their viewpoints, psychological frustration is commonly rechanneled against the stranger confirming an uncanny interplay between social upward and racism. In consonance with this obtained outcome, Bruno Bettelheim and Morris Janowitz (1950) review conceptually different works which prove the social upwards is directly proportional to the reduction of inter-class and ethnic conflicts. However, they caution that in some cases where ethnic minorities gain further benefits and privileges of racism in the dominant group becomes particularly strong. In 1986, Tajfel and Turner (1986) evince how the economic factor, as well as the economic crises, does not suffice to explain discrimination unless through the lens of positive esteem of the group. Normally, there are additional but not for this less powerful factors as the success or failure which potentiate or deteriorate the social identity of the group. When members feel in danger or a stranger interpellates us, hostility re-draws the natural borders of the in-group. Social theory outcomes, anyway, are contradictory or rest in shaky foundation simply because sometimes the psychological frustration – far from being directed against a stranger – is expressed against the in-group. Philomena Essed (1991), in her book *Understanding Everyday Racism,* contradicts the thesis that frustrated people are prone to express racism against ethnic minorities. Under some conditions, racism goes through the veins of successful professionals who take advantages of their privilege positions to ridicule their subordinates. This happens because of the denial of racism, which operates in subtle *ethnicism* denoting a multicultural pluralism, where all groups are placed in egalitarian conditions but dominated under the control of the dominant cultural group.

The second theory discusses the effects of authoritarian regimes in the formation of psychological characters. Authoritarian personality theory was originally a discussion ignited by

Frankfurt School's member Theodor Adorno, who jointly Else Frenkel-Brunswick, Daniel Levinson, and Nevitt Stanford, develop an instrumental scale (F-Scale) to measure how prone to an authoritarian submission a person is. They start from the premise that fascism resulted from the fascination of a collective character dubbed as "authoritarian personality". Based on Freudian and Frommian theories, they hold the thesis that the authoritarian personality keeps a strict superego that had problems to deal with the intensity of the Id. The intrapsychic conflicts generate internal tensions and insecurities which lead the subject to venerate conventional – and externally designed – norms. Given the things in these terms, the subject falls on an authoritarian submission while a leader administers unilaterally the rules. An emerging ego-defense mechanism orients the anxiety-producing drive of the Id towards ethnic minorities of society. As the authors agree, authoritarian personality undermines the value of mankind and the "Other", an emotion associated with the needs of wielding power while concentrating authority (Adorno et al., 2019). Although illustrative – if not eloquent – the theory, as well as the F scale was widely criticized by colleagues because the theory overemphasizes of ultra-right movements overlooking the left-wind ones. For Adorno and colleagues, the authoritarian spirit was linked to Fascism and anti-Semitism.

In third, social structuralism focuses on prejudice and discrimination as an evolving mechanism proper of capitalism to maximize profits minimizing costs. As stable in social reproduction, racism facilitates the social reproduction of those rules and norms which would forge the economic means of production. The inter-ethnic conflict is feasible only when the conditions of production are changing. In this way, as Balibar and Wallerstein (1991) remark, "the figure of the Non-Western Other" legitimated the expansion of European powers paving the ways to the rise of global capitalism to a politics of world economy. Some scholars have exerted a caustic critique of this position for some theoretical simplifications. In their landmark book, *Racism,* Robert Miles and Malcolm Brown (2003) remind that discrimination as a liminoid process should be rethought as something more complex than an act of violence which is finely enrooted in the capitalist production. Discrimination would be fruitful for the dominant but dangerous for the dominated groups. As radical ideologies, dissociated from capitalism, chauvinism and racism mine negatively the social trust which is the core of capitalist production. Any form of discrimination, as authors add, very well ignites a complex sentiment of discontent – above all in discriminated classes or groups – where the dominant position of status quo is being gradually placed in jeopardy.

Last but not least, the social contact theory has its origin in the ink of Gordon W. Allport through the 1950s. Although Allport was a prominent American psychologist, he was reluctant to accept passively the ideas of psychoanalysis and the behavioral sciences. For him, the history and the biography historically constituted were two key factors that explain racism and prejudice. Each person reacts differently to a stranger, sometimes with empathy but sometimes aggressively. Prejudices derive from our needs of framing the external world through the use of stereotypes, which serve as simplifications of the environment. These stereotypes are normally neutral but they can evolve to negative or positive dynamics when emotional mechanisms arouse. Allport knows that social contact – in equal conditions – reduces and controls racism but far from disappearing, prejudice is always present in human relations. In Allport's works, prejudice should be defined as general feelings favorable or unfavorable towards a person or a group, which are not based on rational validation. Much the contradictory evidence is presented to change the mind stronger the prejudice is. Rather, discrimination seems to be the verbal or physical manifestation of prejudice. In the same token, racism is a subtype of discrimination. Allport, in *The Nature of Prejudice,* proffers an erudite work that aims at explaining the formation and evolution of a simple biased idea (prejudice) to a violent manifestation (racism). This moot point will be continued in the next sections.

The Tourism Industry and Prejudice

It is unfortunate that the specialized literature, which certainly examines empirical study cases of racism in tourism and hospitality, does not abound. Scholars have not given a prominent place to racism and prejudice in their respective applied research. As stated, they strongly believed in the peaceful nature of tourism and hospitality. The fact is that racism is esteemed as a barrier to the sustainable development of some destinations (Ruhanen & Whitford, 2018). Some recent studies have widely showed not only how nationality molds inter-group conflicts but also social distance as well as racism (Bai & Chang, 2021). Far from being placated, racism remains hidden or neglected in the modernity. As Gustav Jahoda (2001) highlighted, there is no reason to suppose a black tourist would not be accepted at a hotel, of course, if he pays for the services, but what is more important, he may find all rooms reserved at the time the booking is requested. Racism – above all in the service sector – goes on subtle forms, which far from disappearing, has durable effects in daily life. Richard La Pierre (1934) conducted more than interesting research. He questions the dissociation between ideas and behavior. La Pierre plans a leisure journey jointly a Chinese couple across the US. They found no problems to be lodge at hotels in the touring, while in one case the Chinese tourists were overtly rejected. Six months later, he delivered a questionnaire asking owners if they would welcome Chinese guests. Almost 92% of participants said they would not accept Chinese guests at their establishments. The evidence suggests that prejudices – which are ideal construes – do not determine behavior when economic interests are at stake. In an innovative method in 1948 confirmed a similar hypothesis. Sydney Lawrence Wax (1948) wrote and posted two letters to more than 50 hotels for the same dates. One letter was signed on behalf of Mr. Greenberg and another in the name of Mr. Lockwood. The former obtained only 52% of acceptance in the reply while the latter more than 90%. La Pierre shows widely that racism not only exists but keeps a firm engaged in the host-guest relations. In 1952, Kutner, Wilkins, and Yarrow (1952) replicated similar outcomes in their studies. They hold the thesis prejudice and racism coexist in an earlier stage but both dissociate in contexts of status or economic subordination. Per their earlier interviews, restaurant owners manifested their aversion to serving black tourists while in the real life they did it. All these studies were conducted during the 1950s when racism was stronger, in which case we have no evidence in the modern tourism industry the same behavior remains active. What is equally important, such hostility is still silenced – if not denied – given the current legal jurisprudence. In his classic book *Sociology*, Anthony Giddens (1979) cites the case of American citizen Katherine Dunham who is discriminated in a tourist establishment in Brazil. The Senate promptly passed a law sanctioning those who express racist manifestation because of religion, ethnicity, or genre. On 26 May of 1995, *The New York Times* printed a newspaper article entitled “German Accuses Tourist Office of prejudice”. The event takes a room at the tourist office of the German embassy in New York when a staff member told the reporter he was fired after being public an order to reject the solicitude of visas of Jews, blacks, and Latin American tourists who opt to travel to Germany. The event escalated to Ulrich Geisendorf, a senior official of economic ministry in Berlin, who expressed his concerns and worries about these discriminatory practices (Cowell, 1995; Milton, 1995). In this vein, research with a focus on some ethnic minorities living in the UK shows that Asians lack the necessary skills and training to get a job in tourist operators. With a focus on Yorkshire, Northern England, Kelmm (2002) conclude that a whole portion of ethnic minorities lacks opportunities to be recruited in the tourism industry. Although they are educated at prestigious universities earning their graduate tourism bachelor, British Asians are not being recruited as permanent staff in tourist establishments. This moot point contradicts other national realities where the staff is mainly formed by ethnic minorities which are daily subject to adverse working conditions such as excessive working hours, undeclared work or low-paid salaries (Lazaridis & Wickens, 1999; Mbaiwa, 2005; Yang, 2011). In doctoral research,

Evelyn Newman Phillips (1994) finds robust evidence that describes the political structure of racism which impedes foreigner tourists to have direct contact with some Afro-Americans in St. Petersburg, Florida. As she stresses, the roots of discrimination are not only limited to undemocratic nations but also finely ingrained in established democracies. She centers her outcomes on an ethnography conducted over the Afro-American community which evinces some signs of social maladies resulting from years of segregation from the economic prosperity produced by tourism. In this context, not only did tourism fell short to improve the black and white relations in the city, but it also reinforced the submission of one group to the other. Under some conditions, as Newman Phillips puts it, tourism acts as an ideological mechanism of social control that accelerates the previous material asymmetries. Tourism and racism have been persistent factors in St. Petersburg but as soon as social conditions are altered the authorities adapt their strategies so that Afro-Americans keep away from the wealth tourism amasses. Whatever the case may be, the denial of racism seems to be a condition stable in the host-guest relations, probably covered in the dynamics of economic production.

Over recent years, digital technology offered a fertile ground to racist expressions and xenophobic discourses in the website. Some researchers try to evaluate tourists' sentiments in sites such as Tripadvisor, Yelp, Flickr, or even Twitter. Dataset often obtained from these digital platforms gives fresh alternative insight to issues which are widely denied (for example racism). Li et al. (2020) inspect various racism-related reviews of tourists to revisit the impact of racism in a post-tourist experience. Per their findings, tourism websites reflect some racism-associated reviews which were certainly aggravated after a bad tourist-experience. For reasons of skin color these tourists say they were impeded to access some places or even asked for paying more to visit the site. In other cases, they claimed of pour services by the white staffs. The gathered reviews punctuate that racism is still active in the tourism industry though differently distributed in the geography of the planet. While Europe concentrates a more large percentage of racism-related reviews (40%) followed by the US (30%), Asia seems to be in a third position with 10% (less than the rest of the world comprising Oceania, Latin America and Africa, 20%). In consonance with this, Markus Stephenson (2004) argues convincingly that the UK Afro-Caribbean diaspora is shaped by migrants descended from various islands of the former British colonies as Jamaicans, Barbadians and Vincentians. This migration represents the second largest of all groups of non-European ethnicities living in the UK. The different subgroups were geographically allocated in different cities and neighborhoods. The nostalgia for returning to their ancestral homes occupies an important place in the social imaginary of many "Black British citizens". This Afro-Caribbean Diaspora rests on mobile racialized landscapes – impregnated of racial alienation – where travelers internalize their own race as the only precondition to dialogue with others. To put the same bluntly, racialized realities at Caribbean destinations are designed and packaged by whites, but also continues the old stereotypes of the colonial period. To wit, Elizabeth Hoppe, goes beyond and unearths Frantz Fannon's discontent about the future of tourism as an ideological instrument of domination to affirm the hegemony of a global north. She discusses critically how tourism cements a previously forged dependency between the colonial center and its periphery but behind its economic growth, racism re-organizes structurally new social relations to relegate some local communities from the prosperity of tourism. As Fannon recognizes, Hoppe clarifies, tourism does not need of firm strains to prosper whereas the economic future former colonies are dominated by the economic interests of a global (European) marketplace. Tourism today, like colonialism yesterday, stimulates consumption through *exoticism*. It is important not to lose the sight of the fact that exoticism is one aspect of cultural ethnocentrism, preventing any genuine dialogue with the alterity, in the exoticism there is no confrontation with the "Other", but a monologue characterized by the prone not to take the Other's culture seriously. For Fannon, it is naïve to research racism in tourism simply because tourism is a nuanced form of racism, a new opportunity to dominate the world again through the cultivation of travels and cultural consumption. Tourism provides tourists and locals with a romanticized story of the past perpetuating the dependency. In this vein, M. Korstanje (2012) brings some reflection on the fact that there are ideological labels, stereotypes and concepts invented by the ruling elite which needs to dominate other ethnicities. The label is imposed on the "Other" at the same time the elite avoids to be circumstantially marked. This exhibits the dark side of racism. Beyond the Western curiosity of cultural tourism lies a long-dormant discourse forged during the colonial process. The Western civilization has the divine mandate to educate the "Non-Western Other" through the stimulation of trade, literature and travels. The "Other" should be considered as a child-like, primitive or superstitious people who, if dully educated, may very well embrace the mainstream values of European culture. At the same time, when we go to an aboriginal reservation located in Arizona, we are doing "cultural tourism" but if a Navajo who lives in a reservation want to travel Chicago he is making tourism. This suggests two important assumptions in this hot-debate. On one hand, ethnic segregation starts with the language – citing Derrida. On the other hand, the imposed label demonstrates the ethnic inter-group tension which otherwise remains in secrecy. Another point of entry in this discussion appears to be the difficult tension between hosts and guests after the warfare, an ethnic conflict or a geopolitical dispute. The nation-state creates and circulates powerful narratives oriented to cause a sentiment of empathy with the "Other", or simply a common feeling of rejection. Since the borders mark the difference between us and them, or here-safer and there-unsafe, no less true seems to be that nationality plays a leading role in the configuration of racism. As Maximiliano Korstanje and colleagues acknowledge, the history of discrepancies and disputes between Argentina and Chile for the large borders both share have led both parties on the brink of the warfare, however, both avoided a real war-making. Paradoxically, Argentineans and Chileans have developed negative stereotypes with each other in the threshold of time. After the crisis of 2001 and in the auspices of the end of convertibility, thousands of Chilean tourists arrived in Argentina looking for enjoying gastronomy and tourist attractions. An experimental survey administered on a sample of 150 tour operators, hotels and rent-a-car companies confirmed the hypothesis that Argentineans tourism professionals had negative stereotypes and racist attitudes against Chilean tourists. Although covered by the subordination of professionals to their clients, interviewees showed serious prejudices – even racism – against Chilean tourists. This probes not only how the prejudice follows a much deeper emotional dynamic but also how it keeps covered over many years (Korstanje, 2011). Additional research gives some hint that the prejudice is overtly manifested in conditions of rivalry and competition such as sport events or labor competition (Korstanje, George & Amorin, 2016). This begs a more interesting question, To what extent it is feasible to say social contact – as Allport emphasizes – undermines prejudice and racism?

Gordon Allport's Model: Contributions and Limitations

Without any doubt, racism and prejudice were the tugs of the war of psychology since the end of World War II. Other disciplines such as sociology and anthropology have paid attention to these topics but marginally. While anthropology focused on the construction of the "Non-Western Other", sociology devotes time and efforts in deciphering the structural conditions for the rise of social maladies as Anomie, drug addiction, and criminalities in the largest cities. Gordon Willow Allport was a pioneering voice in analyzing the individual and sociological background that predisposes a person towards racism. In his trailblazing work, *The Nature of Prejudice,* he toys with the belief people are not born with prejudice, but rather it is culturally learned. Allport defines prejudices "as any hostile attitude or idea against a person solely by the fact that he or she belongs to a group which has been marked with questionable aspects or qualities. Unlike other simplifications as stereotypes, prejudice can be equated to a poison which blinds the subject from the principle of reality. Laypeople needs stereotypes to understand the complexity of the world, and reportedly they exist in daily life, but once these ideas become impermeable to empathy the prejudice makes stronger. Having said this, discrimination is a manifestation that appeared when the negative prejudice invites the person to act. Negative prejudice and discrimination escalate to more sophisticated forms of violence against ethnic minorities. Hence, social scientists should give with accurate diagnosis and instrument to control discrimination. Otherwise, the story we lived in World War II will be surely repeated (Allport, 1979).

As the previous argument is given, the escalating levels – Allport mentions in his book – include *spoken abuse, avoidance, discrimination, violence against people*, and *extermination*. When the stage of spoken abuse takes place, the person shows their antagonism with stranger freely. Most likely, a whole portion of people will stay at this stage over years and never pass to the second facet. The avoidance associates to the reluctance of a person to have closer contact with the unwanted group. The third stage (discrimination) speaks us of active steps to exclude the "undesired Other" from residential housing, employment or even public life. The physical attack is part of the fourth stage. Under some circumstances, this stage can be substantially radicalized towards a ghettoization process. When this happens we are a few steps to extermination. Nazi Germany seems to be a clear example of the evolution of daily political turmoil to the radicalization of this facet, but it is not limited to. Lynching, massacres and incursions to flagellate Afro-Americans revived the curiosity and disgust of historians. As Allport discusses, the stranger, who does not look like, us emulates the archetype of a scapegoat, a ritual which condenses all the violence for the group to gain the lost political stability. In this respect, social contact plays a crucial role in impeding the proliferation of racism and the gradual evolution of the five above-commented stages (Allport, 1979).

To the philosophical question posited in the earlier sections, Allport goes on respond that social interaction – when it progresses in uneven and asymmetrical relations based on power, wealth or status – does not reduce the prejudice but aggravates it or mutates in ultra-violent contexts. Since the hostility is centered in self-love, Allport accepts that negative stereotypes operate in the same levels of positive stereotypes. When two comrades or soldiers fight hand to hand at the same trench they generate a sentiment of reciprocity which transcends any ethnicity or religious affiliation. Social interaction should be done in egalitarian conditions or symmetric relations of status and power. In a nutshell, Allport's legacy helps to understand prejudice, as well as negative stereotypes, are part of our human nature. They vary on culture and time but transcend all human organizations. Social contact enhances inter-personal reciprocity when the spirit of interaction takes a cooperative – not competitive logic. What Allport adheres, the personality of persons as well as their psychological character is significant to explain why some classes are less open than others to the cultural difference. The genesis of prejudice can be – at least – explained by the rise of ignorance and the lack of introspection, which means in a manifest impossibility to place oneself in the "Other's place". Two contrasting groups can work together – with opportunities to diminish their hostilities – if both share the same goals. In consequence, the ends and rules of in-groups should be legalized into a coherent framework. To validate this thesis, he collects a series of experiments and research that point out those classes or groups with lower inter-group interaction are more prone to develop racist attitudes than those who have higher levels of interaction. At a closer look, the inter-group incompatibilities when are legally regulated have little possibilities of inter-ethnic conflict. In the same way, the geographical proximities between two clans, groups or tribes dispose people to hostilities when they have a low familiarity with the "Other". Rather, when neighbors meet in a frank dialogue the hostility lessens. Although Allport shed light on the studies of prejudice and racism to the next generations, proffering one of the most erudite and robust editorial projects of the history of psychology, some of the limitations in his theory should be carefully examined.

At a first glimpse, Lewis Coser (1998) has clearly explained that when people are close-up interacting regularly, negative events engender more mutual hostility, simply because the institutions fail to regulate the conflict. Indifference, in some conditions, seems to be of paramount importance to undermine social conflict. Secondly, some historical events, which often pit a nation against its neighbor, awaken long-dormant hostilities. For Coser, prejudice and the resulting violence is a natural instrument to keep in control the psychological borders of the group when trespassed or violated. This suggests that prejudice is a counter-balance reaction when the self feels threatened. In a nutshell, detractors of Allport's model start from the premise that social contact is not enough to deter prejudice and xenophobia, in fact under some circumstances, contact aggravates the inter-group hostilities. These studies delve into the lack of interest and familiarity of Allport for history, as well as the structural means of production which organize social relations (Gaines & Reeds, 1995). As children of their epoch, humans are unable to break the rules of their societies, the prejudices and expectances. Sometimes, racism and prejudice are encrypted in the discourse which is culturally determined (Billig, 1988; Riggins, 1997; Van Dijk, 1984).

What are the challenges for tourism research in a post COVID world where the Alterity is neglected, a new normality where the "Other", the foreign tourist situates as an "undesired guest"?

Conclusion

Unfortunately, tourism-related scholars have not paid attention to the problem of racism in tourism and hospitality. They strongly believe tourism promotes peace and political stability reducing the possibilities of inter-ethnic or inter-class conflicts. In this chapter, we have shown precisely the opposite. Prejudice and racism take a hidden form in service sectors where the hosts are subordinated to guest's desires. Methodologically speaking, the application of intruding methods as interviews or administered scales to measure racism does not describe the phenomenon with accuracy. Not only people hide their real emotions, but racism is also legally punished. This leads to what Essed dubbed as "the denial of prejudice". Nowadays, the classic racism, which was based on so-called biological differences, set the pace to subtle cultural racism camouflaged of cultural diversity. In perspective, citing Essed, while globalization and tourism evolve in a cosmopolitan climate of diversity and toleration, racism remains in the core of industrialized societies. In this context, Allport's contributions to tourism research are manifold. He dissects masterfully the effects of racism and discrimination on daily life. The end of World War II and the beginning of the Cold War led Allport and many other social scientists to assess the negative effects of racism in society. In the present chapter, we have reviewed the main contributions and limitations of the main theories that take prejudice as their main object of study. These theories alert on the economic downturns, the political instability and the authoritarian governments as fertile ground to the rise of racism. With a focus on the social contact theory generally and Gordon Allport's model, in particular, we assess the emergence of a new cultural prejudice orchestrated to mark host guests relations. Far from disappearing, prejudice and racism persist in the tourism and hospitality industries. In a post-COVID world plagued of uncertainty and anxieties, Allport's book *The Nature of Prejudice* seems to be more relevant today than ever before.

References

Adorno, T., Frenkel-Brenswik, E., Levinson, D. J., & Sanford, R. N. (2019). *The authoritarian personality.* London, Verso Books.

Allport, G. W. (1979) *The nature of prejudice.* Reading, Addison Wesley.

Bai, S & Chang, H. (2021). Effects of tourist-to-tourist encounter: Increased conflict or reduced social distance?. *Journal of Hospitality & Tourism Research,* ahead of print. DOI:10.1177/10963480211014938

Balibar, E.& Wallerstein, I. M. (1991). *Race, nation, class: Ambiguous identities.* London, Verso.

Bauman, Z. (2017). Tourists and vagabonds: Or, living in postmodern times. In J. David (Ed.), *Identity and social change* (pp. 19–32). Abingdon, Routledge.

Bettelheim, B., & Janowitz, M. (1950). *Dynamics of prejudice.* New York, Harper.
Billig, M. (1988). The notion of "prejudice": Some rhetorical and ideological aspects. *Text-Interdisciplinary Journal for the Study of Discourse, 8*(1–2), 91–110.
Çelik, S. (2019). Does tourism change tourist attitudes (prejudice and stereotype) towards local people?. *Journal of Tourism and Services, 10*(18), 35–46.
Comaroff, J. L., & Comaroff, J. (2009). *Ethnicity, Inc.* Chicago, University of Chicago Press.
Coser, L. A. (1998). *The functions of social conflict* (Vol. 9). Abingdon, Routledge.
Cowell, A. (1995). German accuses tourist office of prejudice. *The New York Times, 26*(05).
Derrida, J., & Dufourmantelle, A. (2000). *Of hospitality.* Stanford, Stanford University Press.
Dollard, J., Miller, N. E., Doob, L. W., Mowrer, O. H., & Sears, R. R. (1939). Frustration and aggression. New Haven, Yale University Press.
Dovidio, J. F., & Gaertner, S. L. (1986). *Prejudice, discrimination, and racism.* New York, Academic Press.
Essed, P. (1991). *Understanding everyday racism: An interdisciplinary theory* (Vol. 2). London, Sage.
Farmaki, A. (2017). The tourism and peace nexus. *Tourism Management, 59,* 528–540.
Frazer, J. G., (1951). *The golden bough: A study in magic and religion. Spirits of the Corn and of the Wild, 1.* London, Macmillan.
Gaines Jr, S. O., & Reed, E. S. (1995). Prejudice: From Allport to DuBois. *American Psychologist,* 50(2), 96–105.
Giddens, A. (1979). *Central problems in social theory.* London, Macmillan.
Graburn, N. (2012). Tourism: the sacred journey. In V. Smith (Eds.), *Hosts and guests: The anthropology of tourism.* Philadelphia, University of Pennsylvania Press, pp. 21–36.
Hewstone, M. (2015). Consequences of diversity for social cohesion and prejudice: The missing dimension of intergroup contact. *Journal of Social Issues, 71*(2), 417–438.
Hovland, C & Sears, R. (1940) Minor studies in aggression: Correlation of lynching with economic indices. *Journal of Psychology,* 9(1), 301–310
Jahoda, G. (2001). Beyond stereotypes. *Culture & Psychology,* 7(2), 181–197.
Jones, J. M., Dovidio, J. F., & Vietze, D. L. (2013). *The psychology of diversity: Beyond prejudice and racism.* New York, John Wiley & Sons.
Klemm, M. S. (2002). Tourism and ethnic minorities in Bradford: The invisible segment. *Journal of Travel Research,* 41(1), 85–91.
Korstanje, M. E. (2011). Influence of history in the encounter of guests and hosts. *Anatolia, 22*(2), 282–285.
Korstanje, M. (2012). Reconsidering cultural tourism: An anthropologist"s perspective. *Journal of Heritage Tourism,* 7(2), 179–184.
Korstanje, M., George, B. P., & Amorin, E. (2016). Chile decime que se siente: Sports, conflicts, and chronicles of miscarried hospitality. *Event Management, 20*(3), 457–462.
Kutner, B., Wilkins, C., & Yarrow, P. R. (1952). Verbal attitudes and overt behavior involving racial prejudice. *The Journal of Abnormal and Social Psychology,* 47(3), 649–659.
La Pierre, R. T. (1934). Actitudes versus actions. *Social Forces,* 1(13), 230–237.
Lazaridis, G., & Wickens, E. (1999). "Us" and the "Others": ethnic minorities in Greece. *Annals of Tourism Research, 26*(3), 632–655.
Li, S., Li, G., Law, R., & Paradies, Y. (2020). Racism in tourism reviews. *Tourism Management, 80,* 104100.
Litvin, S. W. (1998). Tourism: The world"s peace industry? *Journal of Travel Research, 37*(1), 63–66.
Markus, A. B. (2016). *Mapping social cohesion: The scanlon foundation surveys 2016.* Monash, Monash University.
Mbaiwa, J. E. (2005). Enclave tourism and its socio-economic impacts in the Okavango Delta, Botswana. *Tourism Management, 26*(2), 157–172.
Miles, R. & Brown, M. (2003). *Racism.* London, Routledge.
Milton, S. (1995). An ambivalent memory: The American perception of Nazis and Germans. *German Politics & Society, 13*(3 (36), 89–94.
Newman Phillips, E. (1994), An ethno-historical analysis of the political economy of ethnicity among African American in St. Petersburg, Florida, May, Doctoral dissertation directed by Susan Greenbaum, Ph. D., Florida, University of St. Petersburg.
Noelle-Neumann, E. (1974). The spiral of silence a theory of public opinion. *Journal of Communication, 24*(2), 43–51.
Riggins, S. H. E. (1997). *The language and politics of exclusion: Others in discourse.* London, Sage Publications, Inc.
Ruhanen, L., & Whitford, M. (2018). Racism as an inhibitor to the organisational legitimacy of Indigenous tourism businesses in Australia. *Current Issues in Tourism, 21*(15), 1728–1742.
Sandell, K. (2005). Access, tourism and democracy: A conceptual framework and the non-establishment of a proposed national park in Sweden. *Scandinavian Journal of Hospitality and Tourism, 5*(1), 63–75.

Smith, V. L. (Ed.). (2012). *Hosts and guests: The anthropology of tourism*. Philadelphia, University of Pennsylvania Press.

Stephenson, M. L. (2004). Tourism, racism and the UK Afro-Caribbean diaspora. In T. Coles & D. Timothy (Eds.), *Tourism, diasporas and space* (pp. 62–77). Abingdon, Routledge.

Tajfel, H. & Turner, J. C. (1986). The social identity theory ofintergroup behavior. In S. Worchel & W. G. Austin (Eds.), *The Psychology ofIntergroup Relations*. Chicago: Nelson-Hall.

Van Dijk, T. A. (1984). *Prejudice in discourse: An analysis of ethnic prejudice in cognition and conversation*. London, John Benjamins Publishing.

Van der Duim, R., Peters, K., & Wearing, S. (2005). Planning host and guest interactions: Moving beyond the empty meeting ground in African encounters. *Current Issues in Tourism, 8*(4), 286–305.

Wax, S. L. (1948). A Survey of restrictive advertising and discrimination by summer resorts in the province of Ontario. *Canadian Jewish Congress: Informations and Comments*, 2(7), 10–15.

Yang, L. (2011). Minorities, tourism and ethnic theme parks: employees'' perspectives from Yunnan, China. *Journal of Cultural Geography, 28*(2), 311–338.

5

MERE EXPOSURE EFFECT AND TOURISM RELATIONSHIP

Erhan Coşkun

Introduction

People are faced with a series of stimulant that constantly stimulates them in their lives. It has great importance that only the most important stimulants for human prosperity and health be put into process in human consciousness. Emotional reactions are one of the most critical factors that human attention focuses on the environment's most relevant objects. These reactions differ according to the nature of the stimulant that occurs. For example, stimulants that are considered positive and stimulants containing threats are processed rapidly, automatically, and primarily in human consciousness compared to neutral stimulants. In addition, it is difficult to leave these stimulants (Young & Claypool, 2010: 424). If people constantly encounter these stimulants in time, their liking for a stimulant that is encountered again and again increases compared to the stimulants that have never been encountered (Becker & Rinck, 2016: 153). The development of an individual's attitude towards that object resulting from repeated exposure to a stimulating object is explained by the effect of mere exposure (Zajonc, 1968: 23). The mere exposure effect expressing a stimulus preference increase with exposure to repeated stimuli is a robust and vital social psychology phenomenon (Monahan et al., 2000: 462).

The effects of TV commercials on customers' purchase intention and purchasing decisions are associated with the mere exposure effect (Carreon et al., 2019: 1339). For example, seeing the photograph of an advertising product repeatedly can increase the preference for that product and positive attitudes towards that product (Köroğlu & Cebeci, 2019: 119). The mere exposure effect is associated not only with the advertising industry but also with the tourism industry. Lin and Kuo (2016) associate the mere exposure effect with the formation of the tourist experience in their studies. Accordingly, when tourists go to a new destination, the culture, history, religion, nature, activities, shopping, architecture, hospitality, accommodation, and transportation of that destination trigger tourist experiences. These features of the destination are a stimulant of constant exposure for tourists. As a result of the constant exposure of tourists to these stimulants, a tourist experience begins to form, and a search for a memorable experience begins.

This chapter defines the concept of mere exposure effect, and explains its emergence and development in the historical process. Afterward, it aims to associate the concept with the tourism sector and reveal in which areas it is used in the tourism sector. Finally, the chapter deals with planning in relation to transforming the concept of mere exposure effect into a more understandable form by reflecting on application examples in the tourism literature.

DOI: 10.4324/9781003161868-5

The Meaning of Mere Exposure Effect

The mere exposure effect can be explained by two fundamental theories such as cognitive and affective models. Cognitive models suppose that repeated presentation of stimuli first leads to a change in cognitive experience, such as increased familiarity with the stimulant. Affective models are concerned with the assessment of cognitive change, such as an increased liking for the stimulant (Becker & Rinck, 2016: 153). The mere exposure effect is also known as the familiarity effect, as repeated stimuli increase familiarity. People exposed to repeated stimuli are more familiar with repeated stimuli than other stimuli. Familiarity also affects preference tendency. The mere exposure effect is a psychological phenomenon in which people tend to prefer stimuli they are familiar with (Falkenbach et al., 2013: 1255).

The mere exposure effect is expressed as an increase in a person's emotional preference for a stimulant to which he or she is repeatedly exposed (Pugnaghi et al., 2019: 1). Even if there is no awareness, a positive emotional reaction occurs when people are exposed to a stimulant again. The mere exposure effect beyond explicit memory, measured by traditional measures of recall and recognition of communication effects, reaches the implicit memory, which includes one's abilities. Implicit memory refers to the effects of latent communication for people who are exposed to certain stimulants such as love. The mere exposure effect represents implicit memory when the absence of awareness (Ye & Raaij, 1997: 629). In the mere exposure effect, unusual stimulants are presented to the participants repeatedly. The degree of liking for that stimulant increases with the repeated presentation of previously unfamiliar stimulants (Dechene et al., 2009: 1117).

The finding that exposure to a stimulant repeatedly improves attitudes towards it is a rooted phenomenon. However, it is difficult to determine the effects of exposure to products as people may have been exposed to stimulants before (Hekkert et al., 2013: 411). According to this phenomenon, being exposed to a new stimulus one or more times causes stimulants to be liked and preferred more than other stimulants that one is not exposed to (Köroğlu & Cebeci, 2019: 119).

Emergence and Historical Development of Mere Exposure Effect

When the emergence and development of the exposure effect is examined, it is seen that it goes back to Gustav Fechner, who made studies on the fundamental problems of psychology with empirical research methods and experiments. In 1876, he researched the effect and wrote the book *Preschool of Aesthetics*. He determined the basic principles of likes and dislikes and developed these principles as the fundamental laws of aesthetics. These consist of six laws such as primary, secondary, quantitative, qualitative, formal, and content. According to the qualitative law, due to the inability to resist two or more visualizations of the same object, a feeling of liking, and in the case of resistance and opposition, a feeling of dislike occurrences (Neumann & Selim, 2010: 97).

Edward Titchener has also researched the effect. In 1910, Titchener documented the effect and defined it as the glow of warmth felt from the presence of something familiar (Falkenbach et al., 2013: 1255). For example, listening to a well-known song from childhood days on the radio often creates a warm and positive feeling (Jakesch & Carbon, 2012: 1). When we look at the classical studies, it has been concluded that as a result of exposure to a stimulus, personal preference can also lead to increased familiarity with the stimulus (Kwan et al., 2015: 49).

In 1885, a study was conducted by Ebinghaus in which he was the only subject. This research in the field of psychology explores the relationship between repetition and learning. Ebinghaus, who exposed himself to previously unfamiliar words, revealed the relationship between repetition and learning. Accordingly, he found that as a result of the words he repeatedly read, remembering the words he repeated increased, and the level of forgetting decreased (Şahin, 2012: 99).

Some other scientists, such as Robert Zajonc, went on to explore this effect. In 1968, Zajonc revealed that more exposure to a specific stimulus increases the rate of recognition of the stimulus. In addition, he determined that the frequency of exposure to the stimulus also caused a change in attitude towards the stimulus. Zajonc produced a broader theory stating that more exposure to the stimulus will increase the likelihood of stimulus recognition and that as a result of repeated exposure, the attitude change is positive (Falkenbach et al., 2013: 1255).

Areas of Use of Mere Exposure Effect

The mere exposure effect is used in many fields. It can be said that it is used significantly in the advertising sector and it contributes to the achievement of this sector's goals. In addition, it is seen that it is used in fields such as games, health, politics, and tourism.

The main reason why the same products are constantly advertised in TV commercials is considered to be the mere exposure effect. The exposure effect is so strong that the subconscious presentation of the product before choosing from among the options is sufficient for that product to be preferred. For example, the more you are exposed to Coca-Cola's advertisements, the more it is considered the preferred option (Kauffman & Radin, 2021: 12).

Gledhill, Smith, and Medway's (2020) study state that the products embedded in video games affect the implicit memories of the people who play the game, and they find evidence that this may affect their purchasing decisions. In the study, in which the brands of tires, soft drinks, airlines, banks, and game developers are used, it is stated that the mere exposure effect is used as a communication strategy and this is a part of the brand strategy.

In a study conducted on the effects of news and discussion programs about Covid-19 on people's cognitive, emotional and behavioral responses, the focus was on understanding the psychological impact mechanisms that encourage the application of human health-protective rules such as wearing masks and physical distance in the Covid-19 epidemic. Accordingly, positive behaviors that prevent contamination in public spaces appear to emerge as a result of exposure to TV content related to positive attitudes towards maintaining physical distance and the fear of contamination (Scopelliti et al., 2021: 11).

Kim (2021) conducted a research on the effects of the public, who were exposed to the social media posts of political candidates in the 2018 local elections in South Korea, on the election decision. Accordingly, he stated that the people exposed to the posts of the candidate, who will interact with the local people and share their personal feelings and thoughts about various social and political issues and events, evaluate that candidate positively, and this may lead to votes in favor of the candidate.

Mere Exposure Effect in Tourism Industry

The mere exposure effect is associated with destination selection, food and beverage, travel and transportation, as well as marketing and promotional activities for the tourism sector. Visual components such as pictures and videos play an important role in giving a positive direction to tourists' perceptions of the destination. A positive destination image can be created through visuals and can affect destination selection. Tourists are constantly exposed to information, television and radio advertisements, websites and other marketing stimuli. Apart from that, they have to constantly interact with people and the world around them. Their brains cannot process all stimuli and they focus on their basic priorities and include these priorities in their mental capacity. In this process, tourists have to be directed information, advertisements, websites, brochures, etc. stimulants that only about tourism and perception should be affected (Scott et al., 2019: 1244).

Destination marketers carry out effective advertising activities for the destination's promotion and marketing to get a share from the market in an increasingly competitive environment (Jiang et al., 2020: 1). Positive evaluations for that destination can occur by repeatedly watching the advertisements about the destination by potential visitors (Wang & Yao, 2020: 268). When potential tourists choose a destination, they collect information such as accommodation, transportation, restaurants, and travel activities. They can access this information from information sources such as brochures, the internet, and travel agencies. They have a positive evaluation of the destination in the destination choosing due to continuous exposure to these sources of information (Jeong & Holland, 2012: 503). On the other hand, tourists are exposed to secondary sources of information about destinations they have not visited yet. These resources, prepared by others, depend on many factors such as the type of resource, the level of attention paid to the destination, and subjective judgments. Tourists shape the image of the destination in their own minds by being exposed to different sources. Destination image plays an important role in the success of the destination. Because a positive image attracts potential visitors who will visit the destination for the first time, providing a positive competitive advantage. In addition, it also increases the likelihood of the visitor to revisit and recommend the destination (Palacio & Santana, 2020: 2572).

In the study of Björk and Raisanen (2017) on food and travel satisfaction, it is stated that the intensity and extent of the meals that travelers are exposed to affect the travel experience. Accordingly, intensity refers to being exposed to similar experiences during travel. Travelers who choose the all-inclusive system are exposed to the same meals at the same time of the day and at the same place in every hotel they visit during the whole holiday. An example would be breakfast between 8.30 and 10.00 in the morning and dinner between 17.00 and 21.30 in the evening. Extend refers to the range of experiences at the facility. The scope of the dining experience ranges from casual hotel breakfasts, snacks offered by locals at the local food market, to food served by beachfront restaurants and fine-dining restaurants. The mere exposure effect claims that repeated exposure causes intensity and extensity. The exposure effect, which states that liking occurs with intense exposure, emphasizes that extensive exposure affects memory.

Since people generally prefer familiar foods, the mere exposure effect is seen as an essential factor influencing tourist food consumption. Tourists are getting familiar with the local cuisine of the destination they visit day by day. Thus, the preferences for the local cuisine of the destination may increase. In other words, with the constant exposure of tourists to certain foods, their preferences for these foods increase. Tourists' past experiences of exposure can also increase familiarity and guide their potential choices. For example, when tourists are exposed to the local cuisine in the destination they have visited before, familiarity increases, and preferences increase too when they are exposed to the local cuisine in their next visit. (Mak et al., 2012: 933).

According to Kim et al. (2019), repeated exposure to agricultural products during an agrotourism experience can evoke positive feelings towards products among agro-tourists. Accordingly, participation in agro-tourism will affect the familiarity and attractiveness of agricultural products for agro-tourists. In contrast, from the point of view of agro-tourism service providers, the aim of agrotourism can be an effective promotion strategy for agricultural products.

Conclusion

In this section, where the mere exposure effect is explained, the subject of the mere exposure effect is introduced, the concept's meaning is discussed, information on its emergence and development is given, and finally, it is associated with the tourism sector.

One of the critical issues in the mere exposure effect is seen as stimulants. Young and Claypool (2010) state in their study that these stimuli are processed in human consciousness and reveal emotional reactions. Becker and Rinck (2016) emphasize that due to the repetition of these stimuli, a situation of appreciation for these stimuli occurs in human consciousness.

Another important issue in the mere exposure effect is the studies of researchers on the explanation of the concept. Trying to explain the mere exposure effect, Becker and Rinck (2016) refer to two basic models, cognitive and affective. Accordingly, there is an increase in familiarity with the repeated stimulus in the cognitive model, while there is an increase in the emotional model. The mere exposure effect is expressed as an increase in emotional preference for repeated stimuli, according to Pugnaghi et al. (2019). According to Zajonc (1968), who theorized the pure exposure effect, mere exposure effect is the situation in which an individual's attitude towards the object develops as a result of repeated exposure to a stimulating object.

When Neumann and Selim's (2010) study is examined, it is seen that the foundations of the mere exposure effect go back to Gustav Fechner with his book *Preschool of Aesthetic* in 1876. Afterward, it is known that Edward Titchener's studies on the effect (Falkenbach et al., 2013) contributed to the emergence of the mere exposure effect. The relationship between repetition and learning was investigated with the experiments that Ebinghaus conducted on himself in 1885 (Şahin, 2012). Finally, Zajonc developed the theory that the number of stimulus exposures increases recognition, and the frequency of exposure causes a change in attitude towards the stimulus.

Although the mere exposure effect is used extensively in the advertising sector, it is possible to associate this effect with the tourism sector. According to Wang and Yao (2020), potential tourists have a positive evaluation of the destination due to constantly watching the advertisements for the destination. Mak et al. (2012) state that tourists' preferences for those flavors increase due to exposure to certain local flavors.

When the literature about the mere exposure effect is examined, it is seen to be emphasized that the effect of recognition and liking that the exposed stimulants reveal in humans. When this stimulant is accepted as a good or service to be purchased, it is recommended to researchers who are interested in this subject to investigate how the positive or negative evaluation of the consumer after experiencing that product has an effect when he is exposed to that product again.

It can be accepted as a suggestion for the tourism sector to determine to what extent the promotional tools such as TV advertisements and brochures offered to tourists by tourism enterprises affect tourists and what are the errors or deficiencies of these tools, if there are, by receiving feedback from tourists.

References

Becker, E. S. & Rinck, M. (2016). Reversing the mere exposure effect in spider fearfuls: Preliminary evidence of sensitization. *Biological Psychology*, 121(2016), 153–159.

Björk, P. & Räisänen, H. K. (2017) Interested in eating and drinking? How food affects travel satisfaction and the overall holiday experience. *Scandinavian Journal of Hospitality and Tourism*, 17(1), 9–26.

Carreon, E. C. A., Nonaka, H., Hentona, A. & Yamashiro, H. (2019). Measuring the influence of mere exposure effect of TV commercial adverts on purchase behavior based on machine learning prediction models. *Information Processing and Management*, 56 (2019), 1339–1355.

Dechene, A., Stahl, C., Hansen, J. & Wanke, M. (2009). Mix me a list: Context moderates the truth effect and the mere-exposure effect. *Journal of Experimental Social Psychology*, 45(2009), 1117–1122.

Falkenbach, K., Schaab, G., Pfau, O., Ryfa, M. & Birkan, B. (2013). Mere exposure effect. European University Viadrina Frankfurt. https://www.wiwi.europa-uni.de/de/lehrstuhl/fine/mikro/bilder_und_pdfdateien/WS0910/VLBehEconomics/Ausarbeitungen/MereExposure.pdf, (13.3.2018).

Gledhill, M., Smith, P. & Medway, D. (2020). I like you, but don't remember you-Mere exposure effects in videogames and e-Sports. *Proceedings of the European Marketing Academy*, 49, 1–11.

Hekkert, P., Thurgood, C. & Whitfield, T. W. A. (2013). The mere exposure effect for consumer products as a consequence of existing familiarity and controlled exposure. *Acta Psychologica*, 144(2013), 411–417.

Jakesch M. & Carbon C. C. (2012). The mere exposure effect in the domain of haptics. *Plos One*, 7(2), 1–8.

Jeong, C. & Holland, S. (2012). Destination image saturation. *Journal of Travel & Tourism Marketing*, 29(6), 501–519.

Jiang, H., Tan, H., Liu, Y., Wan, F. & Gursoy, D. (2020). The impact of power on destination advertising effectiveness: The moderating role of arousal in advertising. *Annals of Tourism Research*, 83(2020), 1–13.

Kauffman, S. & Radin, D. (2021). Is Brain-Mind Quantum? A theory and supportive evidence. https://arxiv.org/abs/2101.01538.

Kim, H. (2021). The mere exposure effect of tweets on vote choice. *Journal of Information Technology & Politics*, 18(4), 1–11.

Kim, S., Lee, S. K., Lee, D., Jeong, J. & Moon, J. (2019). The effect of agritourism experience on consumers' future food purchase patterns. *Tourism Management*, 70(2019), 144–152.

Köroğlu, A. E. ve Cebeci, U. (2019). Arka plan değişikliklerinin salt maruz bırakma etkisi üzerindeki etkisi. *Niğde Ömer Halisdemir Üniversitesi Sosyal Bilimler Enstitüsü Dergisi*, 1(2), 118–124.

Kwan, L. Y. Y., Yap, S., & Chiu, C. Y. (2015). Mere exposure affects perceived descriptive norms: Implications for personal preferences and trust. *Organizational Behavior and Human Decision Processes*, 127 (2015), 48–58.

Lin, C. H. & Kuo, B. Z. L. (2016). The behavioral consequences of tourist experience. *Tourism Management Perspectives*, 18(2016), 84–91.

Mak, A. H. N., Lumbers, M., Eves, A. & Chang, R. C. Y. (2012). Factors influencing tourist food consumption. *International Journal of Hospitality Management*, 31(2012), 928–936.

Monahan, J. L., Murphy, S. T. & Zajonc, R. B. (2000). Subliminal mere exposure: Specific, general, and diffuse effects. *Psychological Science*, 11(6), 462–466.

Neumann, E. & Selim, Ö (2010). Gustav Theodor Fechner'de ampirik estetiğin temellendirilmesi. *Atatürk Üniversitesi Türkiyat Araştırmaları Enstitüsü Dergisi*, 0 (12), 97–101.

Palacio, A. B. & Santana, J. D. M. (2020). Explaining the gap in the image of tourist destinations through the content of and exposure to secondary sources of information. *Current Issues in Tourism*, 23(20), 2572–2584.

Pugnaghi, G., Memert, D. & Kreitz, C. (2019). Examining effects of preconscious mere exposure: An inattentional blindness approach. *Consciousness and Cognition*, 75(2019), 1–10.

Scopelliti, M., Pacilli, M. G. & Aquino, A. (2021). TV news and COVID-19: Media influence on healthy behavior in public spaces. *International Journal of Environmental Research and Public Health*, 18 (1879), 1–15.

Scott, N., Zhang, R., Le, D. & Moyle, B. (2019). A review of eye-tracking research in tourism. *Current Issues in Tourism*, 22(10), 1244–1261.

Şahin, A. (2012). Televizyon reklamlarının tekrarlanma etkisi. *Istanbul University Faculty of Communication Journal*, 0(4), 97–111.

Wang, Y. M. & Yao, M. Z. (2020). Did you notice the ads? Examining the influence of telepresence and user control on the effectiveness of embedded billboard ads in a VR racing game. *Journal of Interactive Advertising*, 20(3), 258–272.

Ye, G. & Raaij, W. F. (1997). What inhibits the mere-exposure effect: Recollection or familiarity? *Journal of Economic Psychology*, 18(1997), 629–648.

Young, S. G. & Claypool, H. M. (2010). Mere exposure has differential effects on attention allocation to threatening and neutral stimuli. *Journal of Experimental Social Psychology*, 46(2010), 424–427.

Zajonc, R. B. (1968). Attitudinal effects of mere exposure. *Journal of Personality and Social Psychology*, 9(2), 1–27.

6

SOCIAL EXCHANGE THEORY AND TOURISM

Ali Doğantekin

Introduction

Social exchange theory (SET) is a dominant theory in explaining attitudes of local people (Gursoy et al., 2019; Hadinejad et al., 2019) as well as being a common theory used in explaining organisational behaviour (Cropanzano & Mitchell, 2005). SET is more than the other theories as it can explain positive or negative perceptions as well as offering the ability to study relationships or interactions at both an individual and a collective level (Ap, 1992). Therefore, tourism researchers often use SET because it can clearly express perceived individual perceptions of benefit and cost, that it is strong in explaining the relationship between local people and tourists, due to the structure and consequences of the exchange (Hadinejad et al., 2019).

Previous research (Nunkoo et al., 2013) reveals that most research in the field of tourism has no theoretical basis. Previous research examining local people's reactions to the impacts of tourism has been criticised for lacking a theoretical basis. For this reason, researchers have begun to use many theoretical frameworks, including SET, to eliminate uncertainty about how, why and under what conditions local people will react to the impacts of tourism (Nunkoo, 2016). In filling this gap in tourism research, researchers have adopted the theories used in other social science disciplines, such as sociology and psychology (Gursoy et al., 2019).

The widespread use of SET in tourism research has led to an increase in criticism of SET over time (Gursoy et al., 2019). Therefore, the purpose of this section is to identify the major role of SET in the relationship between local people and tourists, which is often used by tourism researchers, and to present evaluations regarding SET. To achieve this goal, this section presents details under the following titles; introduction, the definition of SET, the main categories of social exchange in tourism, tourism studies based on the social exchange theory, and conclusion and directions for future tourism studies, respectively.

The Definition of Social Exchange Theory

SET is *"a general sociological theory concerned with understanding the exchange of resources between individuals and groups in an interaction situation"* (Ap, 1992, p. 668). In another study, social exchange is defined as *"voluntary actions of individuals that are motivated by the returns are expected to bring and typically do bring from others"* (Blau, 1964, p. 91). In interpersonal exchange, the person who rewards another person by helping him/her puts that person under obligation. However, this obligation, unlike economic obligations, is specifically unspecified obligations. In this case, the individual who feels

DOI: 10.4324/9781003161868-6

an obligation engages in rewarding behaviour towards the individual who helps him or her. In interpersonal exchange, the person who rewards another person by helping him puts that person under obligation. If both sides are happy with their gains, this interpersonal exchange continues. What is important is that this interpersonal change begins with the trust in the other party as the return of the favour done is not guaranteed. Therefore, social exchange occurs as a result of a sense of personal obligation, gratitude and trust (Blau, 1964). In addition, according to SET, individuals' perceptions are affected by the exchange between individuals. In other words; the perception of the person who considers this situation beneficial will not be the same as that of the person who considers it harmful (Gursoy et al., 2002).

Social behaviour refers to the exchange of material or nonmaterial goods between individuals. The value and cost of what an individual gives and receives vary based on the number of gains obtained and the cost borne. At this point, individuals aim to obtain maximum gain at minimum cost (Homans, 1958). A resource is defined as "any item, concrete or symbolic, which can become the object of exchange among people" (Foa & Foa, 1980, p. 78). In other words, the source refers to the possessions and abilities owned by an individual and considered valuable by other actor/s (Molm, 2003). Therefore, the source/s may have some material, social or psychological characteristics (Ap, 1992). In this exchange, resources can be material goods or services, or some abstract values such as approval or status (Molm, 2003). Another study claim that the sources could be "intrinsic" (love, affection, respect, etc.) and "extrinsic" (money, labour) (Ritzer, 2010).

In exchange, actors refer to exchange parties, and these actors could be individuals or corporate groups (Molm, 2003). Gergen (1980) highlights the basic logic of exchange as follows: "To say that people behave in such a way as to achieve maximum rewards at a minimum cost indeed has the ring of universal truth about it… people are bent on achieving what is valuable and desirable to them" (Gergen, 1980, p. 266). Thus, it is seen that in the traditional logic of social exchange, the exchange is based on the idea of an economic swap, and the traded resources are considered "objects". In other words, according to traditional exchange theorists, logical and economic principles constitute the basis of exchange (Mitchell et al., 2012). Adams (1965) who focussed on equality in social exchange highlights that individuals tend to calculate the resources they acquire and give in the process of social exchange. In addition, individuals try to perceive whether the occurring social exchange is equal when compared to the behaviour between the parties (Adams, 1965). There are two forms of interpersonal exchange as negotiated and reciprocal relationships. In the negotiated exchange, the exchange terms that the parties agree on apply. Most economic exchanges could be given as examples of this type of exchange. In reciprocal exchange, the parties do not agree on any object subject to exchange. In this process, individuals do a favour without knowing whether or when the other party will do a favour to them (Molm, 2003; Molm et al., 1999).

The Main Categories of Social Exchange Process in Tourism

Individuals' perception regarding the impact of tourism depends on the impact of tourism on their resources (Jurowski et al., 1997). In other words, local people support the development of tourism as long as the benefit that they receive from tourism is more than the cost that they bear (Yoon et al., 2001). According to SET, local people support the development of tourism if they think that the economic, environmental and socio-cultural impacts of tourism are positive (Paraskevaidis & Andriotis, 2017). In previous research (Andereck et al., 2005; Jurowski et al., 1997; Kuvan & Akan, 2005), resources that are subject to exchange among local people and tourists are divided into three main categories as economic, social and environmental, this research has considered these categories.

Local people, who believe that they will get good results from tourism and that the benefits they will gain will be higher than the cost borne, tend to support tourism and its development (Jurowski et al., 1997). From an economic perspective, tourism creates jobs, generates tax revenue and personal income (Abdollahzadeh & Sharifzadeh, 2014; Andereck et al., 2005; Haralambopoulos & Pizam, 1996; Yoon et al., 2001), and increases production in other sectors such as agriculture and animal husbandry (Özel & Kozak, 2017) causes local people to perceive the impacts of tourism positively. In addition, those who are economically dependent on income from tourism (Chuang, 2010; Deccio & Baloglu, 2002; Ko & Stewart, 2002) also have positive perceptions of tourism. In this context, the economic benefit is the most important factor affecting local people's perceptions and support for tourism (Abdollahzadeh & Sharifzadeh, 2014). However, the fact that tourism has increased the cost of living (Perdue et al., 1990) also increases the prices of goods and services (Haralambopoulos & Pizam, 1996; Johnson et al., 1994), which negatively affects local people's perception of tourism.

Local people consider the economic impact of tourism as well as its environmental impact. Because of this, local people adopt a positive or negative perception of tourism. On the one hand, while improvements in infrastructure along with tourism (Abdollahzadeh & Sharifzadeh, 2014; Özel & Kozak, 2017), local recreational opportunities (Perdue et al., 1990) increase, tourism also plays a role in reducing demand for resource use and environmental pollution (Dwyer et al., 2009). On the other hand, although tourism is defined as "a clean industry", it is clear that it causes significant environmental damage (Andereck et al., 2005). Research results supporting this also reveal that the development of tourism causes crowds, noise, pollution, destruction of natural life (Johnson et al., 1994; Yoon et al., 2001) as well as environmental degradation (Choi & Sirakaya, 2005). These negative environmental impacts reduce local people's support for tourism (Yoon et al., 2001). Another element that affects local people's perception of tourism is the positive or negative social impacts of tourism. Tourism contributes to the development of communities as well as improving the quality of life in the community (Andereck & Vogt, 2000). In addition, tourism provides cultural exchange that brings local people and tourists together, and thus it provides a valuable experience and cultural activities (Yoon et al., 2001). However, tourism has some negative social effects, such as overcrowding, overuse of recreational resources, distracting local people, declining quality of life (Choi & Sirakaya, 2005), changing local culture (Johnson et al., 1994) and sexual permissiveness (Teye et al., 2002),.

In short, local people's perceptions of the economic, environmental and socio-cultural elements that make up the exchange elements between the parties are a determinant of local people's reactions to tourism and tourism development (Abdollahzadeh & Sharifzadeh, 2014). In this context, the tourism perception of the people who interact directly with the tourist and are involved in tourism planning (Jani, 2018; Rasoolimanesh et al., 2015) is more positive than other individuals. Therefore, individuals who benefit from tourism positively perceive the impacts of Tourism (Andereck et al., 2005). In other words, the more people benefit from tourism, the greater their support for the development of tourism (McGehee & Andereck, 2004). However, individuals who consider the exchange between local people and tourists a problem are opposed to the development of tourism (Abdollahzadeh & Sharifzadeh, 2014).

Tourism Studies Based on Social Exchange Theory

Ap explains the reason for the adoption of SET in tourism as follows:

> There is limited understanding of why residents respond to the impacts of tourism as they do, and under what conditions residents respond to these impacts. The lack of explanatory research limits the current literature on understanding residents' behaviour toward the impacts.
>
> *(Ap, 1992, p. 666)*

Currently, SET is widely used by tourism researchers (Abdollahzadeh & Sharifzadeh, 2014; Andereck et al., 2005; Jurowski & Gursoy, 2004; Rasoolimanesh et al., 2017). While some of the studies that use SET as a theoretical framework offer results that support this theory (Andereck & Vogt, 2000; Ap, 1992), others, unlike these results, offer unclear or controversial findings (Andereck et al., 2005; Cropanzano et al., 2017; Paraskevaidis & Andriotis, 2017). For example, economic gains obtained from tourism in line with SET are considered as an important factor in supporting tourism development; the fact that individual benefits from tourism are not a premise of tourism planning does not correspond to the principles of SET (McGehee & Andereck, 2004). Therefore, it is seen that tourism studies have revealed contradictory results about SET.

The main reason why SET is used by tourism researchers is the belief that local people will support the development of tourism to gain economic interests (Burns & Fridman, 2011). Interestingly, the principle of rationality, which is an important element of the economic point of view, comes to the fore in explaining this relationship between local people and tourism (Abdollahzadeh & Sharifzadeh, 2014). Similarly, Chang (2018) and McCool and Martin (1994) found that the economic impact of tourism is the most important factor affecting the attitude of local people towards tourism. The main reason for this is that those who economically benefit from tourism have low perceptions of the social and environmental impacts of tourism (Getz, 1994). Besculides et al. (2002) found that the cultural impact of tourism is the most important element that shapes the attitude of local people towards tourism. Individuals' perceptions of the economic, environmental and socio-cultural effects of tourism vary based on the people's socio-demographic characteristics. In particular, individuals who cannot economically benefit from tourism are more critical about the impacts of tourism (Kuvan & Akan, 2005). At this point, the perception of local people towards tourism depends on the economic, environmental and social impacts of tourism. The dimensions that local people care most about are, in turn, economic gains and socio-cultural benefits (Gursoy et al., 2019).

Previous studies claim that SET provides an appropriate theoretical basis for explaining local people's attitudes towards tourism (Andereck & Vogt, 2000; Ap, 1992; Gursoy et al., 2002). However, new research reveals that, unlike SET, the "Altruistic Surplus Phenomenon"(ASP) provides a more appropriate theoretical basis for explaining local people's perceptions of tourism (Paraskevaidis & Andriotis, 2017). The main reason for this is that SET focusses on individual interests, but ASP considers collective interests superior over individual interests (Paraskevaidis & Andriotis, 2017). In line with these findings, a recent study claims that tourism researchers use some new theories, such as "institutional theory and bottom-up spillover theory" although SET is the most widely used theory (Hadinejad et al., 2019).

According to SET, individuals who make up the local community should positively perceive tourism as their gains from tourism increase (Andereck et al., 2005; Kang & Lee, 2018). However, individuals who experience negative interaction with tourists have a higher perception of the cost caused by tourism than those without such a negative experience. At this point, the nature of the interaction between the individual and the tourist comes to the fore. Therefore, SET may be insufficient in explaining the special relationship between such an individual and a tourist (Andereck et al., 2005). In another study, SET was used to identify the elements that influence the Airbnb hosts and guest relationship. Since the formation of social exchange is based on the exchange of resources between individuals, SET is considered an appropriate theoretical basis for determining the elements affecting customers' quality perception (Priporas et al., 2017).

Conclusion and Directions for Future Tourism Studies

This research aims to identify the major role of SET in the relationship between local people and tourists and to reveal the evaluations regarding SET. For this purpose, tourism research using SET as a theoretical framework has been examined. In the light of the studies reviewed, evaluations and recommendations regarding the use of SET in tourism research have been presented. It seems that the interaction between local people and tourism has been studied by tourism researchers in much research. However, it is noteworthy that previous research on this subject lacks a theoretical basis. This has led tourism researchers to adapt theories used in different fields such as sociology and psychology. In this process, many theories, including SET, have been used by tourism researchers to explain the attitudes of local people towards tourism. This suggests that tourism researchers conduct theory-based research to explain the attitudes of local people, taking into account the results of "systematic reviews" research. It seems that this trend will continue among researchers in the future (Hadinejad et al., 2019). In addition, the widespread use of SET as a theoretical framework in the research conducted by tourism researchers is noteworthy. However, although SET is the most commonly used theoretical framework in this research, this study reveals results compatible with SET as well as being non-compatible.

In the research, which examined the perception of local people about tourism and its effect, the main reason for choosing SET as a theoretical framework is based on economic gains because rationality plays a big role in the relationship between tourism and local people. Therefore, individuals who economically benefit from tourism tend to support tourism development in their region. In this context, the general view is that as individuals' benefits obtained from tourism increase, their support for the development of tourism also increases. However, local people's perceptions regarding the impacts of tourism are not limited to its economic dimension. In addition to this economic dimension, environmental and socio-cultural dimensions also positively or negatively affect local people's support for tourism.

When the research conducted is considered, it is seen that the elements affecting the perception of local people about tourism and its impacts are examined in basically three dimensions: economic, environmental and socio-cultural and that local people value economic gains the most and then socio-cultural benefits come out of these dimensions (Gursoy et al., 2019). In addition, local people's perceptions of the economic, environmental and socio-cultural impacts of tourism differ based on their socio-demographic characteristics. For this reason, the differentiation of socio-demographic characteristics of individuals and societies in different destinations where tourism research is carried out also differentiates the degree of importance given to such dimensions.

With the employment of SET by tourism researchers, previous research reveals that SET provides an adequate theoretical basis for explaining the relationship between tourism and local people. However, recent tourism research reveals that SET is insufficient in explaining the relationship between these structures. Therefore, most studies using SET seem to make behaviour estimates without theoretically providing an adequate infrastructure (Cropanzano et al., 2017). Especially in recent tourism research, it is noteworthy that theories such as "altruistic surplus phenomenon", "institutional theory" and "bottom-up spillover theory" are preferred by researchers. At this point, SET is the most commonly used theory in the research examining local people's perceptions of tourism, but its failure to explain the special relationship between individuals and tourists has led to the use of already existing or new theories. In addition, the fact that studies based on SET present contradictory results pushes tourism researchers to other theories.

Comparing existing or new theories together with SET in future research to explain local people's attitudes towards tourism and its impacts will be more useful in determining the strengths and weaknesses of the theories (e.g. Paraskevaidis & Andriotis, 2017). In this way, tourism researchers could choose the theoretical framework most capable of explaining relationships or structures, taking into account the focus of their research. In addition, when the studies explaining the relationship between tourism and local people using SET are considered, they generally seem to have given a wide range of results compatible with SET. However, most studies that produce results incompatible with SET do not adequately explain the reason and reasons for SET's inability to explain relationships.

References

Abdollahzadeh, G., & Sharifzadeh, A. (2014). Rural residents' perceptions toward tourism development: A study from Iran. *International Journal of Tourism Research, 16*(2), 126–136. https://doi.org/10.1002/jtr.1906

Adams, J. S. (1965). Inequity in social exchange. In L. Berkowitz (Ed.), *Advances in experimental social psychology* (Vol. 2, pp. 267–299). Academic. https://doi.org/10.1016/S0065-2601(08)60108-2

Andereck, K. L., Valentine, K. M., Knopf, R. C., & Vogt, C. A. (2005). Residents' perceptions of community tourism impacts. *Annals of Tourism Research, 32*(4), 1056–1076. https://doi.org/10.1016/j.annals.2005.03.001

Andereck, K. L., & Vogt, C. A. (2000). The relationship between residents' attitudes toward tourism and tourism development options. *Journal of Travel Research, 39*(1), 27–36. https://doi.org/10.1177/004728750003900104

Ap, J. (1992). Residents' perceptions on tourism impacts. *Annals of Tourism Research, 19*(4), 665–690. https://doi.org/10.1016/0160-7383(92)90060-3

Besculides, A., Lee, M. E., & McCormick, P. J. (2002). Resident's perceptions of the cultural benefits of tourism. *Annals of Tourism Research, 29*(2), 303–319. https://doi.org/10.1016/S0160-7383(01)00066-4

Blau, P. M. (1964). *Exchange and power in social life*. New York: John Wiley and Sons. https://doi.org/10.4324/9780203792643

Burns, P. M., & Fridman, D. (2011). Actors' perceptions of the newly designated South Downs National Park: Social exchange theory and framework analysis approach. *Tourism Planning and Development, 8*(4), 447–465. https://doi.org/10.1080/21568316.2011.628808

Chang, K. C. (2018). The affecting tourism development attitudes based on the social exchange theory and the social network theory. *Asia Pacific Journal of Tourism Research*, 1–16. https://doi.org/10.1080/10941665.2018.1540438

Choi, H. S. C., & Sirakaya, E. (2005). Measuring residents' attitude toward sustainable tourism: Development of sustainable tourism attitude scale. *Journal of Travel Research, 43*(4), 380–394. https://doi.org/10.1177/0047287505274651

Chuang, S. T. (2010). Rural tourism: Perspectives from social exchange theory. *Social Behavior and Personality, 38*(10), 1313–1322. https://doi.org/10.2224/sbp.2010.38.10.1313

Cropanzano, R., Anthony, E. L., Daniels, S. R., & Hall, A. V. (2017). Social exchange theory: A critical review with theoretical remedies. *Academy of Management Annals, 11*(1), 1–38. https://doi.org/10.5465/annals.2015.0099

Cropanzano, R., & Mitchell, M. S. (2005). Social exchange theory: An interdisciplinary review. *Journal of Management, 31*(6), 874–900. https://doi.org/10.1177/0149206305279602

Deccio, C., & Baloglu, S. (2002). Nonhost community resident reactions to the 2002 winter Olympics: The spillover impacts. *Journal of Travel Research, 41*(1), 46–56. https://doi.org/10.1177/0047287502041001006

Dwyer, L., Edwards, D., Mistilis, N., Roman, C., & Scott, N. (2009). Destination and enterprise management for a tourism future. *Tourism Management, 30*(1), 63–74. https://doi.org/10.1016/j.tourman.2008.04.002

Foa, E. B., & Foa, U. G. (1980). Resource theory. *Social Exchange*, 77–94. https://doi.org/10.1007/978-1-4613-3087-5_4

Gergen, K. J. (1980). Social exchange. In K. J. Gergen, M. S. Greenberg, & R. H. Willis (Eds.), *Social exchange: Advances in theory and research* (pp. 261–280). Plenum. https://doi.org/10.1007/978-1-4613-3087-5

Getz, D. (1994). Residents' attitudes towards tourism: A longitudinal study in Spey Valley, Scotland. *Tourism Management, 15*(4), 247–258. https://doi.org/10.1016/0261-5177(94)90041-8

Gursoy, D., Jurowski, C., & Uysal, M. (2002). Resident attitudes: A structural modelling approach. *Annals of Tourism Research, 29*(1), 79–105. https://doi.org/10.1016/S0160-7383(01)00028-7

Gursoy, D., Ouyang, Z., Nunkoo, R., & Wei, W. (2019). Residents' impact perceptions of and attitudes towards tourism development: A meta-analysis. *Journal of Hospitality Marketing and Management, 28*(3), 306–333. https://doi.org/10.1080/19368623.2018.1516589

Hadinejad, A., D. Moyle, B., Scott, N., Kralj, A., & Nunkoo, R. (2019). Residents' attitudes to tourism: A review. *Tourism Review, 74*(2), 157–172. https://doi.org/10.1108/TR-01-2018-0003

Haralambopoulos, N., & Pizam, A. (1996). Perceived impacts of tourism: The case of samos. *Annals of Tourism Research, 23*(3), 503–526. https://doi.org/10.1016/0160-7383(95)00075-5

Homans, G. C. (1958). Social behavior as exchange. *American Journal of Sociology, 63*(6), 597–606.

Jani, D. (2018). Residents' perception of tourism impacts in Kilimanjaro: An integration of the social exchange theory. *Tourism, 66*(2), 148–160.

Johnson, J. D., Snepenger, D. J., & Akis, S. (1994). Residents' perceptions of tourism development. *Annals of Tourism Research, 21*(3), 629–642. https://doi.org/10.1016/0160-7383(94)90124-4

Jurowski, C., & Gursoy, D. (2004). Distance effects on residents' attitudes toward tourism. *Annals of Tourism Research, 31*(2), 296–312. https://doi.org/10.1016/j.annals.2003.12.005

Jurowski, C., Uysal, M., & Williams, D. R. (1997). A theoretical analysis of host community resident reactions to tourism. *Journal of Travel Research, 36*(2), 3–11. https://doi.org/10.1177/004728759703600202

Kang, S. K., & Lee, J. (2018). Support of marijuana tourism in Colorado: A residents' perspective using social exchange theory. *Journal of Destination Marketing and Management, 9*(February), 310–319. https://doi.org/10.1016/j.jdmm.2018.03.003

Ko, D. W., & Stewart, W. P. (2002). A structural equation model of residents' attitudes for tourism development. *Tourism Management, 23*(5), 521–530. https://doi.org/10.1016/S0261-5177(02)00006-7

Kuvan, Y., & Akan, P. (2005). Residents' attitudes toward general and forest-related impacts of tourism: The case of Belek, Antalya. *Tourism Management, 26*(5), 691–706. https://doi.org/10.1016/j.tourman.2004.02.019

McCool, S. F., & Martin, S. R. (1994). Community attachment and attitudes toward tourism development. *Journal of Travel Research, 32*(3), 29–34. https://doi.org/10.1177/004728759403200305

McGehee, N. G., & Andereck, K. L. (2004). Factors predicting rural residents' support of tourism. *Journal of Travel Research, 43*(2), 131–140. https://doi.org/10.1177/0047287504268234

Mitchell, M. S., Cropanzano, R. S., & Quisenberry, D. M. (2012). Social exchange theory, exchange resources, and interpersonal relationships: A modest resolution of theoretical difficulties. In *Handbook of social resource theory* (Issue September, pp. 99–118). Springer. https://doi.org/10.1007/978-1-4614-4175-5_6

Molm, L. D. (2003). Theoretical comparisons of forms of exchange. *Sociological Theory, 21*(1), 1–17.

Molm, L. D., Peterson, G., & Takahashi, N. (1999). Power in negotiated and reciprocal exchange Author (s): Linda D. Molm, Gretchen Peterson and Nobuyuki Takahashi Source : American Sociological Review, Vol. 64, No. 6 (Dec., 1999), pp. 876–890 Published by : American Sociological Association St. *American Sociological Association, 64*(6), 876–890.

Nunkoo, R. (2016). Toward a more comprehensive use of social exchange theory to study residents' attitudes to tourism. *Procedia Economics and Finance, 39*, 588–596. https://doi.org/10.1016/S2212-5671(16)30303-3

Nunkoo, R., Smith, S. L. J., & Ramkissoon, H. (2013). Residents' attitudes to tourism: A longitudinal study of 140 articles from 1984 to 2010. *Journal of Sustainable Tourism, 21*(1), 5–25. https://doi.org/10.1080/09669582.2012.673621

Özel, Ç. H., & Kozak, N. (2017). An exploratory study of resident perceptions toward the tourism industry in Cappadocia: A social exchange theory approach. *Asia Pacific Journal of Tourism Research, 22*(3), 284–300. https://doi.org/10.1080/10941665.2016.1236826

Paraskevaidis, P., & Andriotis, K. (2017). Altruism in tourism: Social exchange theory vs altruistic surplus phenomenon in host volunteering. *Annals of Tourism Research, 62*, 26–37. https://doi.org/10.1016/j.annals.2016.11.002

Perdue, R. R., Long, P. T., & Allen, L. (1990). Resident support for tourism development. *Annals of Tourism Research, 17*(4), 586–599. https://doi.org/10.1016/0160-7383(90)90029-Q

Priporas, C. V., Stylos, N., Rahimi, R., & Vedanthachari, L. N. (2017). Unravelling the diverse nature of service quality in a sharing economy: A social exchange theory perspective of Airbnb accommodation. *International Journal of Contemporary Hospitality Management, 29*(9), 2279–2301. https://doi.org/10.1108/IJCHM-08-2016-0420

Rasoolimanesh, S. M., Jaafar, M., Kock, N., & Ramayah, T. (2015). A revised framework of social exchange theory to investigate the factors influencing residents' perceptions. *Tourism Management Perspectives, 16*, 335–345. https://doi.org/10.1016/j.tmp.2015.10.001

Rasoolimanesh, S. M., Roldán, J. L., Jaafar, M., & Ramayah, T. (2017). Factors influencing residents' perceptions toward tourism development: Differences across rural and urban world heritage sites. *Journal of Travel Research, 56*(6), 760–775. https://doi.org/10.1177/0047287516662354

Ritzer, G. (2010). *Sociological theory* (8th Ed.). McGraw-Hill.

Teye, V., Sönmez, S. F., & Sirakaya, E. (2002). Residents' attitudes toward tourism development. *Annals of Tourism Research, 29*(3), 668–688. https://doi.org/10.1016/S0160-7383(01)00074-3

Yoon, Y., Gursoy, D., & Chen, J. S. (2001). Validating a tourism development theory with structural equation modelling. *Tourism Management, 22*(4), 363–372. https://doi.org/10.1016/S0261-5177(00)00062-5

7

SOCIAL REPRESENTATION THEORY AND TOURISM

Selami Gültekin

Defining Social Representations

Social representation (the abbreviation "SR" will be used hereinafter) theory is concerned with making sense of and explaining what and how people think in their daily lives, and how social reality influences these thoughts. Representations broadly refer to metasystems that include values, benefits, and common understandings about how the world operates, besides embodying shared knowledge (Andriotis & Vaughan, 2003: 173). In this approach, the common sense of "individual" and "social" cannot be regarded as separate entities. This concept is used to explain the genesis of representations in different societies in terms of values, ideas, and practices (Duveen & Lloyd, 2013b: 157).

SR has become yet a more important issue as the unifying elements such as science, religion, ideology, and the state became mutually incompatible. The means of mass communication accelerated this trend, and the facilitation of communication increased both the need to combine the theoretical aspect of science with general beliefs, and the need to combine science with the behavior of people as social beings. In other words, the need for recreating common sense/mind is increasing. Moreover, the common mind forms the basis for impressions and meanings, and otherwise any collectivity is not available (Moscovici, 1981: 185). In this context, SR may be considered as a cognitive phenomenon that gives the opportunity to make sense of the world, and in general, may be evaluated as a common sense theory (Cirhinlioğlu et al., 2006: 164).

SR includes common theories, views, and information that societies produce based on their experiences. Unlike the Social Cognition Theory, in which the cognitive processes of individuals are emphasized, the SR approach assumes the existence of a society's "common" cognition and postulates that the social psychology should consist of "long-term" memory, which covers the history of a society and "short-term" memory of current events. In this framework, the SR, although included in social psychology, envisages the joint work of fields such as linguistics, history, social psychology, social psychiatry, anthropology, sociology, and media studies (Öner, 2002: 29).

There are multiple definitions of SR according to researchers on this topic. SR is mainly defined as "the sum of the thoughts and feelings expressed by the actors in a social group in their verbal and explicit behavior which generate an object for the group" (Wagner et al., 1999: 96), shared perceptions about the nature of the fact and the cause of events. Forming the implicit, widely accepted knowledge and beliefs on which attitudes are based (Dickinson & Dickinson, 2006: 197) and mainly concerned with the analysis of structures and processes that allow knowledge and beliefs to be shared by society. The basic premise of the theory is that the social and the psychological

 DOI: 10.4324/9781003161868-7

are inextricably linked (Cirhinlioğlu et al., 2006: 164). According to the SR approach, the social entity (i.e., society, culture, or group) and the individual are not seen as opposing domains, on the contrary, social phenomenon is argued to be a product of communication and interaction between individuals, as it shapes the content of individual minds. In other words, the "thinking society" is characterized by interpersonal and mediated communication in the arena where reality is constructed and negotiated (Zhou & Ap, 2009: 79; Zhou et al., 2014: 583). In this setting, individuals and groups can be considered as passive receptors that transmit representations autonomously, Aforementioned people make critical assessments in the street, cafe, workplace, hospital, or laboratory; dive into philosophies that can be quite influential in their relationships, preferences, and way of educating their children; andhereby they make up random stories and comments (Moscovici, 1981: 183). In this regard, it is a common view that SR is a social phenomenon that individuals of a community design together during their daily speech and actions. Basically, SRs arise to meet the individual's need to understand the world. The purpose of all representations is to turn the unfamiliar into familiar. Abstract and incomprehensible conceptual expressions become more understandable through SR (Cirhinlioğlu et al., 2006: 163).

Representations provide a basis for the individual in this world with the identities they create. When representations are adopted, they explain the individual's relationship to the world and set them within this world. The symbolic value inherent in representations comes from this dual operation of describing the world and finding a place in it (Duveen, 2013: 93). In this respect, rather than a psychology of cognition about social life, SR theory can be considered as a theory which suggests that psychological activities take place in social life. Social psychology theories focus on attitudes or attributions and rely on relatively narrow definitions of psychological behavior. In such theories, social cognition is viewed as cognitive processes related to social stimuli, but these "social stimuli" are considered innate traits whereas social life itself remains untheorized. Under the influence of this theoretical gap, social cognition is seen as the activity of individual minds facing the social world. In the theory of SR, attitudes and attributions emerge as a result of participation in social life; they form the invisible face of an iceberg; this invisible part contains structures that enable the subject to construct meaningful attitudes and attributions (Duveen & Lloyd, 2013a: 175). In this context, it is possible to consider social representations as the sum of information that emerges during communication, shared by the members of the society, and reinforced by conversations. In other words, SRs are social forms of knowledge (Cirhinlioğlu et al., 2006: 164).

The formation processes of given social forms provide us with explanatory signs on this subject. According to Moscovici (1981: 193–194) these processes can be expressed as anchoring, classification and labelling. Anchoring is the individual's inclusion of unknown, intriguing, and distressing things according to own classification logic. This situation allows the individual to compare with the categories in mind and classify new information accordingly. Everything that is not classified and named is seen as wild, non-existent yet dangerous. By placing the object or person in the favored category and subjecting it to a label belonging to one's own language, the individual has brought it closer to self and removed it from being a foreigner. The individual has taken the opportunity to talk about someone, to make comments, and to see the extraordinary as ordinary. In this regard, individuals create representations by categorizing the non-categorized and naming the unnameable. From this point of view, SR can be considered as the classification and naming process and a method of establishing relationships between categories and tags.

Although the functioning of these processes is similar, the meanings attributed to social representations, group dynamics and the qualities of the object can be determinants in the final form of SRs. Indeed, as stated by Cirhinlioğlu et al. (2006: 166), individuals generalize objects to emphasize the typicality or similarity of them and reduce the difference between a given object and its prototype. On the other hand, they consider the features of that object in detail in order to emphasize the differences. The process type depends on the object concerned, the values of SR, and the purpose of the group.

If SRs are considered more holistically, it is an observable situation that the socialization of individuals as a part of society paves the way for the formation of collective thoughts. On this ground, the individual is teaching while learning. The result is a collective body of knowledge. At this point, it would be useful to check on Durkheim's explanations about collective representations.

Durkheim reviews the mind as a "tabula rosa" that must be filled with representations. This process is not something that can be fulfilled with innate characteristics. Also, if representations were merely functional, there would be no way of constructing truth or reality. The concept of truth would then be null and void, with the exception of individuals finding the same representation by chance. This situation contradicts rational thinking. The idea that collective representations are based on reality enables Durkheim to move forward with a sociology of knowledge based on facts. The point here is that the representations cannot arise by an individual in seclusion with a tabula rosa filled with the power of thought alone. That is, collective representations are transmitted through the process of socialization, in which the individual is taught what to think through the assimilation of collective representations (Pickering, 2002: 101).

Social Representation Framework

The theory of SR has a special place in social psychology with both the problems it poses and the spectrum of phenomena it deals with. This situation brings many criticisms and misunderstandings with it (Moscovici, 1988: 211). According to Potter and Litton (1985: 81) there are four closely related problems in the theory of social representations: (1) the relationship between groups and SRs; (2) consensus issues and the level at which SRs are shared; (3) the functioning of SRs in specific contexts of use; (4) the role of language in SR.

These problems and their criticisms emphasize that there are some aspects of SR theory that need to be revised. According to Moscovici there is much to be learned from criticism, and there appears to be a long way to go before reaching a satisfactory theory of social thought and communication. Such a theory, as currently defined, may not fit well with the social psychology model. However, it should be underlined that the SR approach, unlike the classical understanding of collective representation, answers important social and scientific questions and adopts the constructivist perspective that has become widespread in social psychology from the very beginning (Moscovici, 1988: 211). What is more, according to Moscovici (1981: 191), modern science has broken up with the public, which we can call a fall out with common sense or common mind.

SR can be considered as a theory that can reduce this separation with the paradigm it proposes. This theory has the potential to provide explanations for how various groups of people understand and respond to social phenomena, and it may be appropriate to utilize this theory, especially if the subject of study requires multiple social perspectives or bears potential of volatility and uncertainty (Zhou et al., 2014: 584). Yet the diversity of methodological approaches is a substantial issue. Deeper mental and social processes should be approached in different ways, including linguistic analysis and observing how people think, if empirical methods are to be used to find out how people think (Moscovici, 1988: 211).

Social Representations in Tourism

SRs are belief systems about daily life that consist of shared public communication about broad topics such as sexuality, health, insanity, and tourists. Not being limited to attitudes and values, they are driven by large-scale themes and images, derive their meaning from multiple sources, and organize people's daily spheres of understanding and behavior. SRs are people's everyday theories and networks of knowledge about large parts of the social world (Pearce et al., 1996: 177). What SR theory promises is to illuminate social processes in the world constructed by individuals in everyday life and to show how attitudes, beliefs, and attributions are formed in terms of these socially derived frameworks. SR connects thought, understanding and action to a set of cultural resources shared by large and small social communities (Potter& Litton, 1985: 81). SR turns the unknown to become known, it may vary based on culture, yet there may be common representations in different cultures, it is affected by information systems such as religion and science, but ultimately affects the systems (Pearce, 2005: 181).

Researchers contributing in the field of tourism planning studies have mainly focused on environmental issues, regional patterns, and regional distributions of tourists and tourism facilities. More recently, some have advocated a paradigm shift towards a socially critical agenda emphasis in tourism studies rather than planning and control (Davis, 2001: 125). It is a known fact that tourism is not only an economic phenomenon, but also has cultural, political, and environmental aspects. When examining tourism impacts, it is necessary to deal not only with tangible impacts such as income and foreign currency but also with intangible impacts such as social and cultural impacts (Pizam & Milman, 1986: 29).

The social and cultural effects of tourism are the ways in which tourism contributes to changes in value systems, individual behavior, family relationships, collective lifestyles, moral behavior, creative expression, traditional ceremonies, and community organization – in other words, the effects on local people due to their direct and indirect relations with tourists (Pizam & Milman, 1986: 29). Considering tourism in this context, as a social phenomenon, it offers a wide scope for the formation of SRs. When tourists and local people are taken as a group, it is usual for various social discourses and representations to occur depending on the nature of these groups. The theory of SR provides an effective approach for the in-depth study of the social, in tourism research, where multiple variables are examined.

Social impact studies provide important information to understand the perceptions of local people and tourists about these impacts, especially when planning tourism (Ap, 1990: 611), however, sociocultural impacts of tourism are difficult to measure and often have indirect or even unknown characteristics (Pizam & Milman, 1986: 29). Considered as a theory, SR adopts an emic, contextual, and process-oriented perspective to understand the reality of the social actor; helps define reality; and being a critical component of group and individual identity guides both action and thought. In this respect, SRs may offer potential insight for the research on attitudes of the society towards tourism development and the behaviors or impacts of tourists. The interests, values, qualities, and explanations that individuals perceive about tourism are directly related to the SR framework (Pearce et al., 1996). Accordingly the SR approach has the potential to contribute to the "social" construction of knowledge about tourism and its effects (Moscardo, 2009: 159), especially by offering conceptual framework to explain how local people understand tourism impacts and the behavior of some tourist groups and how they react collectively to them (Monterrubio & Andriotis, 2014: 290).

According to Fredline (2005: 271) the word "Social" in SR theory refers to the fact that these representations are shared by social groups and help facilitate communication. In the tourism context, the theory proposes that local people often have representations of tourism that underpin their perceptions of tourism's impacts, and that these representations are fed by direct experiences, social interaction, and other sources of information such as the media. As Moscardo (2011: 426) states, representations provide groups with the opportunity to construct and share common social realities and guide how they will react to the relevant phenomenon.

To put it more clearly, the reactions to the effects of tourism are shaped by SRs, and these representations are generally fed by the media and social interaction. In addition, it is possible to associate the reactions of local people to tourism with many variables mentioned hereinafter.

Local people's perceptions on social impacts may depend on resources, economic conditions and lifestyle, socio-demographic characteristics, and interaction intensity with tourists (Butler, 1974: 100; Almeida-García et al., 2016; Monterrubio, 2019: 18) or external variables such as the level and intensity of tourism development, type of tourism, seasonality of the destination, and national development level, and internal variables such as economic and/or employment dependence on tourism, community attachment, distance from tourism areas, personal values, social identity, and/or status (Wassler et al., 2019: 2). There are many studies showing that economic reasons cause positive perceptions or moderate thinking about tourism effects (e.g., Perdue et al., 1990; Smith & Krannich, 1998; Sharma & Gursoy, 2015; Almeida-García et al., 2016). Similarly, Suess and Mody, in their study examining the effects of tourism, concluded that the economic dependence on tourism leads to a more positive perception of the economic and social effects of tourism. In the same study, Las Vegas residents indicated that they were willing to pay even higher taxes, regardless of the type of tourism development proposed by the authorities. Such a result shows that rural communities reinforce the "hegemonic" SR of tourism to characterize the capitalist urbanism ethic that dominates the economic development discourse (Suess & Mody, 2016: 20). Similar variables affecting perceptions, in addition to being abundantly in numbers, have the content to offer at least a superficial framework to researchers. The new page that SRs have turned over in tourism research provides space for asking questions of who, how and why and suggests the semantic depth that will be brought by examining the results in more detail.

The perspective of the SR framework may be utilized also to explain the stereotypes of local people about tourists. SR first draws attention to these sharing some common views towards tourists and proceeds by asking how this group has got this view. Another important step for SR theory is to focus on the consequences of having such stereotypes. The important point here is that possible responses to tourists and related tourism developments are part of a system linked to people's everyday understanding and offer a guide for how they will react. According to this view, tourist stereotypes are not only community-driven efforts to have position and identity but also powerful formatives of action and interaction (Pearce, 2005: 20–21).

However, social representation theory is seen as challenging, yet transformative and useful in tourism research (Meliou & Maroudas, 2010: 125). Pearce et al. (1991: 148) have argued that a new model blended with social representation will offer a unified view of equal-social representation to tourism, surpassing stage models (such as the destination life cycle model and the irridex model) in which tourism impacts are measured. In this new approach, the possible reactions of local people to tourism developments will depend on a cost-benefit accounting of the tourism impacts they experience. This accounting, together with the information presented in the media, is likely to prepare various segments of society to share one of the SRs about tourism. For those who are likely to suffer from tourism damage, these representations can be exemplified as tourism – destroying the environment, tourism – only for the rich, and tourism – taking over our city. Where equity estimations are more neutral, tourism would be perceived as dangerous, and it needs to be properly managed (tourism – good if controlled). Finally, when personal benefits outweigh the costs, the representations are tourism – our future, tourism – the savior of the city; tourism – tomorrow's industry.

Given the information aforementioned above, SR offers a useful perspective that can transform well-known models, examine situational and contextual variables, and refer to innovative and comprehensive research in this regard. In addition, thanks to the SR approach, the metaphors formed in the society become clear and offer important clues in terms of examining the social.

Next to these conveniences, there is a point that should be underlined. While SR theory is more attractive than alternatives with the logic of reconciling different levels of interest in tourism among various subgroups of local people and the transfer of representations from person to person; this extra complexity makes testing much more difficult. Considerable progress needs to be made, to confirm the validity and reliability of existing social impact measures, and to more clearly understand the differences within and between various communities (Fredline, 2005: 271–272). For instance, some studies suggest that growing crowds is a major problem for all members of a community but there may be those who enjoy the vibrancy that crowds of people can bring to a community and may not see crowding as a problem. Two different groups of people may have reacted differently based on what they have experienced in the past, but there may be other factors that differentiate them now (Beeton, 2006: 62–63). Similarly, in their study in Hawaii, Pearce et al. (1996) found that while a part of the society sees tourism as the "engine of growth", another part sees it as a "vulture that destroys cultures" (Pearce et al., 1996: 177). As noted earlier, group variables would affect the way reactions and SRs are formed. This is seen as complicating for the empirical task. According to Beeton (2006: 62–63), as a different interpretation to this issue, ongoing studies, for instance, continue to group "age" and "crowd anxiety" together. Whereas, if the research is done from the perspective of SR, the participants are asked to list what are the issues that interest them. If the "crowd" (as a response) appears more than a few times, the research is focused on those who respond in this way and these group members are examined for their personal characteristics. In this type of research, instead of the age variable, it is possible that the members of this group may have lived in the region for similar periods of time, may have come from another place to settle here, have similar education levels, similar occupation types, similar attitudes towards development, etc. Once these similarities are identified, this group now becomes a specific group with a specific SR (or attitude).

In the light of all these reviews, it is clear that investigating the local people within the framework of SR may enable complex, somewhat challenging but in-depth studies. Various studies, albeit few, continue to be conducted on tourists as well, and the SRs they create about local people.

The theory of social representations also promises to be an important conceptual pathway in the construction of a new generation of studies that explore tourists' thoughts on their experiences and post-experience outcomes (Pearce, 2005: 186). According to Monterrubio (2019: 20), the theory of SR remains largely an unexplored option that may be utilized not only to find out local people's perceptions of tourism and tourist (i.e., local gaze) but also to explain tourists' beliefs, attitudes and behaviors (i.e., tourist gaze) towards locals. Within the framework of host-guest interaction, this theory has the potential to go beyond one-sided approach style and gain a more integrated understanding of how both locals and tourists view each other (i.e., mutual view) and interpret the phenomenon of shared tourism.

The metaphors of tourists and their SRs are of the same type. The way tourists communicate with each other, reference groups, or service personnel is also central to SRs (Pearce, 2005: 182). At this point, another strength of the SR approach is that it provides the opportunity to partially explain the effects of travel experience in the context of changes in visitor attitudes and knowledge. SRs can indeed be strengthened and reinforced by relevant experiences. In the studies on attitude, there is evidence that direct contact with objects and environments provides great confidence in participants' beliefs, values, and knowledge systems. Travel may also be influencing SRs by this confirmatory method. The opportunity of empirical investigation about the impact of destination image, attitudes and beliefs that emerge as a result of touristic experience, stems from this broad view of people's information systems (Pearce, 2005: 185). To put a parenthesis on this subject, representations continue to occur even in situations where there is no direct contact or experience is limited. In such settings, groups borrow representations from other sources. In such cases, media political figures, other important people or groups will be used as a reference (Fredline & Faulkner, 2000: 768).

As Pritchard and Morgan (2001: 177) argue, the representations used in destination marketing are not value-free expressions of the identity of a place, but rather are the culmination of historical, social, economic, and political processes and reveal much about the social construction of given place in the context of cultural change, identity, and discourse. Given representations used in marketing campaigns do not only reflect the responses of destination marketers (and advertising agencies) to a dynamic external environment but are also constructed expressions of the cultural and political identities of destinations. It is possible to see how the representations of those places may change depending on those who construct a particular tourism area intersubjectively. Therefore, representations promise to be practical for researchers and stakeholders in destinations to comprehend the complexities of tourism marketing and the power relations underlying them which means to estimate the formation of new power relations emerging from the tourism process.

Conclusion

In this chapter, it is evident that there are conceptually very broad definitions of SRs. From all these definitions one can conclude that SR theory is a relatively new point of view that refers to shared values and collective ideas formed during socialization processes. Keywords such as common knowledge, common cognition, common perceptions, and common mind are used quite frequently, and all these words refer to the field of socially constructed behavior. Although working in this theoretical framework has some difficulties, it is possible to say that it will have functional benefits, especially in the examination of multivariate social relations.

Although the theory has not been studied frequently in the tourism literature, there are a number of guide studies that offer various clues to researchers and open new horizons for researchers aiming to examine multivariate social events. In the field of tourism, it is visible that the reactions of the local people to the tourism effects have been studied so far, apart from that, some studies have focused on the perceptions of the tourists about the local people. Although the variables are multiple and seen as challenging for research, in such studies where the social effects of tourism are examined, the SR approach draws an optimistic picture.

References

Almeida-García, F., Pelaez-Fernandez, M. A., Balbuena-Vazquez, A., & Cortés-Macias, R. (2016). Residents' perceptions of tourism development in Benalmádena (Spain). *Tourism Management, 54*, 259–274.

Andriotis, K., & Vaughan, R. D. (2003). Urban residents' attitudes toward tourism development: The case of Crete. *Journal of Travel Research, 42*(2), 172–185.

Ap, J. (1990). Residents' perceptions research on the social impacts of tourism. *Annals of Tourism Research, 17*(4), 610–616.

Beeton, S. (2006). *Community development through tourism*. Landlinks Press.

Butler, R. W. 1974. The social implications of tourist development. *Annals of Tourism Research, 2*(2), 100–110.

Cirhinlioğlu, F. G., Aktaş, V., & Özkan, B. Ö. (2006). Sosyal temsil kuramına genel bir bakış. *Cumhuriyet Üniversitesi Edebiyat Fakültesi Sosyal Bilimler Dergisi, 30*(2), 163–174.

Davis, J. B. (2001). Commentary: Tourism research and social theory-expanding the focus. *Tourism Geographies, 3*(2), 125–134.

Dickinson, J. E., & Dickinson, J. A. (2006). Local transport and social representations: Challenging the assumptions for sustainable tourism. *Journal of Sustainable Tourism, 14*(2), 192–208.

Duveen, G. (2013). Psychological development as a social process. In Serge, M., Sandra, J. and Brady, W. (Eds.), *Development as a social process contributions of Gerard Duveen* (pp. 90–112), Routledge.

Duveen, G & Lloyd, B. (2013a) Social representation as a genetic theory. In Serge, M., Sandra, J. and Brady, W. (Eds.), *Development as a social process contributions of Gerard Duveen* (pp. 173–181). Routledge.

Duveen, G & Lloyd, B. (2013b) the Significance of Social Identities. In Serge, M., Sandra, J. and Brady, W. (Eds.), *Development as a social process contributions of Gerard Duveen* (pp. 157–172), Routledge.

Fredline, E. (2005). Host and guest relations and sport tourism. *Sport in Society, 8*(2), 263–279.

Fredline, E., & Faulkner, B. (2000). Host community reactions: A cluster analysis. *Annals of Tourism Research, 27*(3), 763–784.

Meliou, E., & Maroudas, L. (2010). Understanding tourism development: A representational approach. *Tourismos, 5*(2), 115–127.

Monterrubio, C. (2019). Hosts and guests' social representations of nudism: A mutual gaze approach. *Annals of Tourism Research, 75*, 18–28.
Monterrubio, J. C., & Andriotis, K. (2014). Social representations and community attitudes towards spring breakers. *Tourism Geographies, 16*(2), 288–302.
Moscardo, G. (2009). Tourism and quality of life: Towards a more critical approach. *Tourism and Hospitality Research, 9*(2), 159–170.
Moscardo, G. (2011). Exploring social representations of tourism planning: Issues for governance, *Journal of Sustainable Tourism, 19*(4–5), 423–436.
Moscovici, S. (1981). On social representations. *Social Cognition: Perspectives on Everyday Understanding, 8*(12), 181–209.
Moscovici, S. (1988). Notes towards a description of social representations. *European Journal of Social Psychology, 18*(3), 211–250.
Öner, B. (2002). Sosyal temsiller. *Kriz Dergisi, 10*(1), 29–35.
Pearce, P. L. (2005). *Tourist behaviour: Themes and conceptual schemes*. Channel View Publications.
Pearce, P. L., Moscardo, G., & Ross, G.F. (1991). Tourism impact and community perception: An equity-social representational perspective. *Australian Psychologist, 26*(3), 147–152.
Pearce, P. L., Moscardo, G., & Ross, G. F. (1996). *Tourism community relationships*. Oxford: Elsevier.
Perdue, R. R., Long, P. T., & Allen, L. (1990). Resident support for tourism development. *Annals of Tourism Research, 17*(4), 586–599.
Pickering, W. S. (2002). What do representations represent? The issue of reality. In Pickering, W. S. (Ed.), *Durkheim and representations* (pp. 114–133). Routledge.
Pizam, A., & Milman, A. (1986). The social impacts of tourism. *Tourism Recreation Research, 11*(1), 29–33.
Potter, J., & Litton, I. (1985). Some problems underlying the theory of social representations. *British Journal of Social Psychology, 24*(2), 81–90.
Pritchard, A., & Morgan, N. J. (2001). Culture, identity and tourism representation: Marketing Cymru or Wales? *Tourism Management, 22*(2), 167–179.
Sharma, B., & Gursoy, D. (2015). An examination of changes in residents' perceptions of tourism impacts over time: The impact of residents' socio-demographic characteristics. *Asia Pacific Journal of Tourism Research, 20*(12), 1332–1352.
Smith, M. D., & Krannich, R. S. (1998). Tourism dependence and resident attitudes. *Annals of Tourism Research, 25*(4), 783–802.
Suess, C., & Mody, M. (2016). Gaming can be sustainable too! Using social representation Theory to examine the moderating effects of tourism diversification on residents' tax paying behavior. *Tourism Management, 56*, 20–39.
Wagner, W., Duveen, G., Farr, R., Jovchelovitch, S., Lorenzi-Cioldi, F., Markova, I., & Rose, D. (1999). Theory and method of social representations. *Asian Journal of Social Psychology, 2*, 95–125.
Wassler, P., Nguyen, T. H. H., & Schuckert, M. (2019). Social representations and resident attitudes: A multiple-mixed-method approach. *Annals of Tourism Research, 78*, 102740.
Zhou, Y., & Ap, J. (2009). Residents' perceptions towards the impacts of the Beijing 2008 Olympic Games. *Journal of Travel Research, 48*(1), 78–91.
Zhou, Y., Lu, T., & Yoo, J. J. E. (2014). Residents' perceived impacts of gaming development in Macau: Social representation perspectives. *Asia Pacific Journal of Tourism Research, 19*(5), 579–599.

8

TRAVEL CAREER PATTERN THEORY OF MOTIVATION

Hera Oktadiana and Manisha Agarwal

Introduction

Understanding why people travel is critical to perceive various facets of tourists' behavior. Motivation studies are of interest to the vast array of group of people, including academia, practitioners, media observers, and communities. Various tourist motivation concepts have been applied in the tourism field such as Plog's allocentric and psychocentric needs, Iso-Ahola's level of arousal model, and Pearce and Lee's travel career pattern. Some argumentations indicate that travel motivation studies do not concern with a theory and behavioral concept coined by social science and psychology scholars (cf. Bowen & Clarke, 2009; Hsu & Huang, 2008). Travel career pattern provide a rich description of travel motivation with its 14 cores factors developed from previous studies. This chapter aims to explicate the development of travel career pattern (TCP) motivation theory and how it has been used in different settings of tourists' behavior.

Travel career pattern supersedes travel career ladder (TCL) approach, which was proposed by Pearce (1988, 1993), as well as the work of Pearce and Caltabiano (1983) and Moscardo and Pearce (1986). TCL, which was partly built on Maslow's needs-hierarchy theory, described travel motivation as having five different levels or hierarchy. The term ladder was later viewed as ascending the steps and received commentaries on the meaning. TCL was then reformulated into a more sophisticated pattern of motives to explore the full complexity of tourist motivation. TCL was changed into travel career pattern that puts more emphasis on the dynamic multi-level motivational structure (cf. Pearce, 2005; Pearce, 2011; Pearce, 2019).

The approach has been continuously reshaped and refined based on empirical data and the new theoretical developments in psychology. Several key points underpin the theory and methodology of travel career pattern. First, following Maslow's work, motivation is dynamic process, thus it can evolve with experience. Second, the work encapsulates multiple factors rather than just using a single trait approach to understand travel behavior. Third, the approach can be used to describe motivation of certain communities. In other words, groups of people may have common vision of travel due to the shared cultural and social views. Fourthly, several methods can be applied to assess motivation. Finally, the approach fits well with the neuroscience groundwork on the motivational states and affect (Pearce, 2019).

The concept of travel career pattern has been used in various studies by a range of authors to understand travel motivation of different tourist groups. Some examples of the studies include Western and Asian tourists (e.g., Li et al., 2015; Pearce & Lee, 2005; Wu et al., 2019), health and spa tourists (e.g., Panchal & Pearce, 2011), backpackers (e.g., Paris & Teye, 2010), repeat

DOI: 10.4324/9781003161868-8

tourists (e.g., Agarwal & Pearce, 2019), Muslim tourists (e.g., Oktadiana et al., 2017), and students (e.g., Song & Bae, 2018). The approach finds constant indication that an individual's travel motives consist of core motives (novelty, escape/relax, and relationship), middle motives (the desire to be in nature, self-development, host-site involvement, and self-actualization), and outer layer motives (nostalgia, status issues and isolation). While the patterns change with more travel experience, all travelers pay attention to the core motives.

The theory of the travel career pattern provides a broad and rich conceptualization of travel motives. It may suit the needs of researchers conducting longitudinal work or case studies of special traveler or tourist markets.

Development and Concept of Travel Career Pattern

The travel career ladder (TCL) model proposed that tourist motivation comprises five levels of needs; relaxation, safety/security, relationship, self-esteem and development, and self-actualization/fulfillment (Pearce, 2005). Pearce argued that tourists can have multiple travel motivations, although there is one dominant level. The structure of TCL was based on Maslow's work and the career concept in leisure and tourism. Following the career concept, people's travel motives change according to their travel experiences and lifespan. The implication is that people shift to a series of motivation levels. As proposed by Pearce (2005, p. 54), TCL denoted "as people accumulate travel experiences, they progress upward through the levels of motivation". The model of TCL (Pearce, 1988) is depicted in Figure 8.1.

According to the model, the direction of change is variable, as stated by Pearce (1988, p. 34): "Some individuals may "ascend" the ladder predominantly on the left-hand side of the system, while others may go through all steps on both the left and right-hand side of the model".

The early application of TCL was to study motivations of tourists visiting Timbertown theme park in Australia. Seven motives were examined in the survey: (1) having a good time eating and drinking, (2) relaxing in a nice setting, (3) Seeing a new, different, and interesting place, (4) enjoying a day out with the family, (5) seeing and doing things with close friends, (6) improving my knowledge of early Australian history, and (7) really feeling a part of the place (Pearce, 1988, p. 79). The study showed the pattern with TCL and accommodation arrangement, repeat versus first-time visitors, and the duration of stay in the theme park. The results indicated that visitors who stayed one or more nights and transited in the area, the first-time visitors, and shorter stay visitors showed interest in the novelty (seeing a new, different, and interesting place), whereas the repeat visitors emphasized self-development/self-esteem, followed by novelty and relaxation (Pearce, 1988).

Travel career ladder (TCL) was later altered to travel career pattern (TCP) to tone down the image of the steps in the physical ladder and hierarchy. The term travel career remains significant as the concept indicates the changing motivational travel patterns of travelers with travel experiences (Pearce, 2005). Pearce specified some important terms in TCP which include travel needs or travel motives, self/others-oriented motives, motivation pattern, travel career ladder, and travel career pattern. Travel motives can be described as forces (biological and socio-cultural) that drive travel behavior. Self/others-oriented motives convey internal motivation and external influences. Motivation pattern denotes multiple travel motives – not merely emerged from a single driver. Travel career implies that an individual's travel motivation is dynamic, depending on the life stage, age, and previous travel experience. The conceptual definition of TCL is defined as "an older theoretical model describing travel motivation through five hierarchical levels of needs/motives in relation to travel career levels", whereas TCP is "conceptually modified TCL with more emphasis on the change of motivation patterns reflecting career levels than on the hierarchical levels" (Pearce, 2005, p. 55).

Self-actualization Behavior motivated by travellers desire to transcend oneself, to feel a part of the whole wold, to experience inner peace and harmony, to develop oneself to one's full potential	
Self-esteem *Other directed* Behavior influenced by external rewards, prestige, glamour of traveling. Some psychological (eating, drinking are cultivate as connoisseur self-esteem needs)	↑ *Self directed* Behavior influenced by internally controled processes; development of skills, special interests, competence, mastery
Love and Belongingness *Other directed* Behavior influenced by desire to be with others, group membership, receiving affection and attention, initating relationship	↑ *Self directed* Behavior influenced by giving love, affection and involving others in the group. Maintaining and strenghtening relationships.
Safety and Security *Otherd irected* Behavior influenced by a concern for safety, welfare of others	↑ *Self directed* Behavior influenced by a concern for one's own safety
Physiological Needs *Externally oriented* Behavior motivated by need for external excitement: novel settings, activites and places. Behavior is stimulus hungry.	↑ *Internally oriented* Survival needs behavior influenced by self-directed need to eat, drink, maintain body systems. Need for relaxation or bodily reconstitution.

Figure 8.1 The Travel Career Ladder.
Source: Adapted/redrawn from Pearce (1988, p. 31)

The first study conducted by Pearce and Lee (2005) used two-steps process in analyzing travel motivation of tourists. The first stage was exploratory interviews, and the second stage was a survey. A group of 12 participants participated in the interviews. They were asked about their previous and future travel motivation. The participants were then a two-paged structured questionnaire on travel motive items. The detailed explanation of the interviews could be found in Lee and Pearce (2002). The findings from this initial research suggested several primary motives which include novelty-seeking, self-development and cultural experiences, relationships, and escape. In relation to the major survey, there were 143 initial motivational motives generated from numerous tourism and leisure literature. After being reviewed, the items were reduced into 74 travel motives. These items were used to construct the questionnaire for the survey.

The analysis from the questionnaires produced 14 motivation factors. The most dominant factors are novelty, escape/relax, and relationship (strengthen). This is followed by autonomy, nature, self-development (host-site involvement), stimulation, self-development (personal development), relationship (security), self-actualization, isolation, nostalgia, romance, and recognition. Further, Pearce examined the links between travel motivation and travel experience. Three basic travel experience features considered domestic and international travel experiences, and age.

The application of 14 motivational factors was started with two studies to investigate travel motives of the Western and Korean tourists. The first study by Pearce and Lee (2005) aimed to investigate pleasure travel motivation in the context of western culture. This study was conducted in Australia. There were 1,012 participants involved in this study (57% Australians, 22% from the United Kingdom, and 21% from other Western countries). The respondents were asked to provide ratings of the 74 motive items using a Likert scale. The second study (Pearce, 2005) also used 74 items and 14 core motive factors to discover travel motivation of 824 South Korean tourists. Data were collected in the three South Korea's cities of Daegu, Daejon, and Seoul. The findings of both studies were similar. Thus, TCP theory can be applied for cross-cultural context even though the focus of travel motivation factors may differ between cultures (Pearce, 2005, 2011). Pearce also argued that the key travel motivations of tourists at both low and high travel career levels were novelty, escape/relax, and kinship. Moreover, regardless of the travel career, travelers' motives for pleasure travel are shaped by the most important and less important motives. The results from the two studies underpin TCP framework as shown in Figure 8.2.

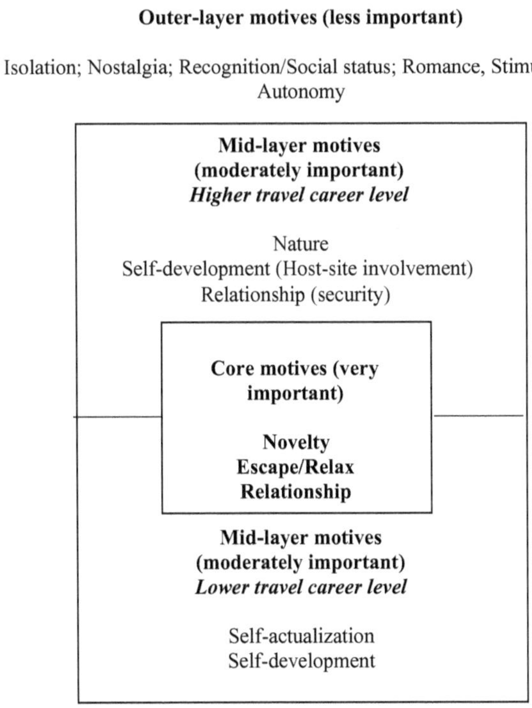

Figure 8.2 The Initial Travel Career Pattern Concept.
Source: Adapted/redrawn from Pearce (2005, p. 79)

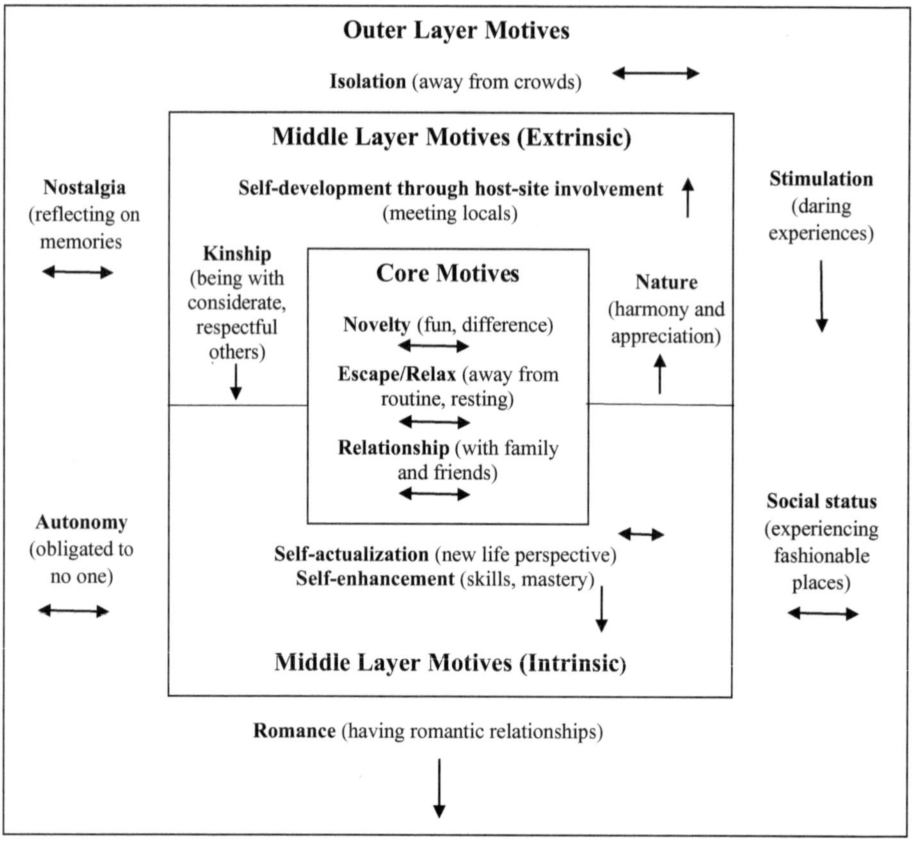

Figure 8.3 The Renewed Travel Career Pattern.
Source: Redrawn from Pearce (2011, p. 62) and Pearce (2019, p. 30)

TCP framework was later updated by Pearce to capture the travel multi-motive of wider market of travelers across countries and time. The refreshed version of TCP model is depicted in Figure 8.3.

Note: The arrows indicate the changing emphasis with travel experience (increasing, decreasing, or neutral)

Pearce (2011, 2019) asserted that the core motives (to experience novelty, to escape or relax, and to build relationships), were viewed as important for the travelers regardless of their travel experiences. Two extrinsic facets of middle layer motives, self-development, and involvement with nature, were seen as important for the experienced travelers, whereas romance and autonomy were considered less salient for the experienced travelers. Those with limited travel experiences generally perceived that all motives are similarly important.

The Use of Travel Career Pattern in Tourism Research

Several tourism studies using TCP approach corroborate that the concept can be applied in various settings (See Table 8.1). Although survey is the primary method to examine tourists' travel motives in TCP framework, secondary data was also utilized (cf. McKercher & Koh, 2018). Further, a qualitative approach using essay was also applicable for TCP framework (cf. Filep & Greenacre, 2007).

Table 8.1 Tourism Studies Using TCP Model

Authors and Publication Year	*Methodology*	*Description*
Agarwal and Pearce (2019)	Data collection: Survey to 492 international repeat tourists visiting India (using 13 motives and 26 items) Data analysis: descriptive statistics, Confirmatory Analysis (CFA), MANOVA	Core motives (novelty, escape/relax, and relationships) were viewed important for the repeat tourists. This was followed by relationship (security), nature, self-development (host-site involvement), self-development (personal development), self-actualization, autonomy, isolation, nostalgia, stimulation, and recognition.
Filep and Greenacre (2007)	Qualitative approach is applied to examine the travel motivations of Australian university students going to Spain by using essay. Data analysis: Conceptual content analysis Quantitative TCP approach (survey) is used for researching travel motivation of those university students and 172 youth backpackers visiting Australia. (using 15 motives and 57 items) Data analysis: Principal component analysis, Cluster analysis, *t*-tests	The main motivation of the students comprise relationship/belonging, curiosity/mental stimulation, self-development and safety/comfort, which echo the TCP motives of kinship, novelty, self-actualization, and escape/relax. The results of quantitatve approach show four most important factors are novelty, stimulation, belonging/immersion, and self-actualization. The least important factors include escape/ social status, romance/friendship, isolation, and nostalgia/comfort.
Li, Pearce, and Zhou (2015)	Data were collected from 640 participants in Hangzhou to assess the motivation of Chinese tourists (using 71 motivation items)	Nature is considered as the most important factor for the Chinese tourists, followed by novelty, self-actualization, self-development (host-site involvement), escape/relax, self-development (personal development), relationship strengthen, stimulation, isolation, nostalgia, relationship safety, autonomy, recognition, and romance.
McKercher and Koh (2017)	Data collection: secondary data from Singapore Tourism Board	"A clear association is observed between the role of specific attractions and the importance placed on different motivation layers identified by Pearce" (p. 669). The core-layer motives of the need to relax and recharge are important. Indonesia: core motive (relationship building, escape from stress and daily routine, recharge, feeling togetherness). China: Middle and outer layer motives (social and status value). Malaysia: Core (escape from daily routine), Middle (host-site involvement such as attending events, educating children about different culture). Australia: Core (escape the daily stress, discover new experiences, bond with the loved one). India: Core (break from stress, recharge, spend time with family and friends). Japan: a mix of core and middle layer (relaxation, novelty, unique experiences).

(Continued)

Authors and Publication Year	*Methodology*	*Description*
		The Philippines: mix of core (escape and relaxation + relationship with the loved ones) and middle layer (sense of achievement). Hong Kong: core (escapism, relaxation, spending time with family) and middle and outer layer (personal enrichment and social status). South Korea: core and middle (escapism, restoration, personal enrichment, personal actualization). Thailand: core (relationship building and novelty).
Oktadiana, Pearce, Pusiran, and Agarwal (2017)	Data collection: Survey with 356 respondents of Indonesian and Malaysian Muslim tourists (using 13 motives and 26 items) Data analysis: Descriptive statistics, Factor analysis	Relationship (strengthen), nature, and novelty are the top three most important travel motivations for the Muslim tourists. The results are rather different with the Western tourists who perceived novelty as the most important factor, followed by escape/relax, nature and relationship. In the context of Muslim tourists, nature which is a middle layer motive in the original study is viewed significant. Other travel motivations include escape/relax, stimulation, relationship (security), self-development (host-site involvement), autonomy, isolation, self-development (personal development), self-actualization, nostalgia, and recognition.
Panchal and Pearce (2011)	Data collection: survey to 319 travellers in Thailand, the Philippines and India (all the 14 motivation factors were applied, added by two motive items: "to maintain my health" and "to improve my health") Data analysis: Descriptive statistics, K-means cluster analysis and Independent *t*-tests	Analyzing the linkage of the health and travel motives of tourists. Escape/relax and novelty are the most two important motives, tailed by nature, self-actualization, health, isolation, self-development (host-site involvement), personal development, stimulation, strengthen relationship, secure relationship, autonomy, nostalgia, romance, and recognition. Health is perceived as moderately important factor. However, it is linked to the core motives of escape/relax and novelty, as well as nature and self-actualization.
Paris and Teye (2010)	Data collection: Survey involving 347 respondents from 30 nationalities Data analysis: Descriptive statistics, Principal component analysis, K-means cluster analysis, Discriminant analysis, and *t*-tests.	Cultural knowledge and relaxation are considered as the important motivation for backpackers. Cultural knowledge includes explore other cultures, increase my knowledge, and interact with the local people, whereas relaxation is related to escape and getting away from routine activities. Four motivational factors (personal/social development, experimental, budget travel, and independence) are dynamic according to backpacker's travel career, previous travel experience, and age.'

Authors and Publication Year	*Methodology*	*Description*
		The more experiences the backpackers have, the less important the personal/social growth become. Limited travel experience backpackers consider all travel motivation are important.
Song and Bae (2018)	Data collection: Survey with 585 respondents of international students living in Korea (using 14 motives and 74 items) Data analysis: Descriptive statistics, Latent Profile Analysis (LPA) or known as a latent class analysis, ANOVA	Examining travel motivation of international students in Korea. Comparing travel careers between Korea as study abroad destination, home country and the third countries (not just a single destination) There are four latent profiles. Core (lowest travel motivation but high in core travel motives – novelty, escape/relax, relationship). Longing (perceive traveling as the object of longing to show off travel experiences – emphasis on the outer layer motives including romance, recognition and luxury). Middle (strong travel motivation at the middle layer, both internal and external) – majority of the respondents are in the twenties. Veteran (strong travel motivation at all levels, particularly personal development in the internal middle layer motivation).
Wu et al. (2019)	Data collection: Survey, involving 21,972 participants from China, Hong Kong, Taiwan, and others countries. Data analysis: ANOVA with post hoc tests and cross-tabulations + chi-square to test the significance.	Linking TLC (travel life cycle) and TCP. Core travel motives are common motivations over the different phases of travel life cycle. Tourists at an early stage of travel life cycle (Stage 1) valued the middle layer motives. Stage 1 of travel life cycle (25–34 years old): tourists like to increase experience and gain knowledge by visiting attraction, shopping, watching shows (novelty, self-development). Stage 2 of travel life cycle (25–44 years old): travelers prefer to do business (strengthen relationship). Stage 3 of travel life cycle (45–55 years old): tourists tend to enjoy special interest tourism, that is, gambling (novelty, escape/relax). Stage 4 of travel life cycle (>55 years old): tourists like to dine (relax, strengthen relationship).

While most studies employed the original 14 travel motives, some research applied 13 motives (e.g., Agarwal & Pearce, 2019; Oktadiana et al., 2017). TCP model with 13 motives is also currently used in a study of coffee tourism to understand the behavior of Indonesian tourists (personal communication with Heri Setiyorini, 23 June 2021). The 13 motives excluded romance as a travel motivation. The rationale to remove romance from the survey is due to a couple of reasons. First, asking about the casual intimate relationship during the holiday was considered inconvenience

Table 8.2 TCP Questionnaire Using 13 Travel Motives and 26 Items

Core Motives	*Middle Layer Motives*	*Outer Layer Motives*
Novelty • Having fun • Experiencing something different	Relationship (security) • Feeling personally safe and secure • Meeting people with similar values/interests	Autonomy • Being independent • Doing things my own way
Escape/relax Resting and relaxing Being away from daily routine	Nature • Viewing the scenery • Getting a better appreciation of nature	Isolation • Experiencing the peace and calm • Being away from the crowds of people
Relationship (strengthen) • Doing something with my family/friend(s) • Strengthening relationships with my family/friend(s)	Self-development (host-site involvement) • Experiencing different culture • Meeting new and varied people	Nostalgia • Thinking about good times I've had in the past • Reflecting on past memories
	Self-development (personal development) • Develop my personal interests • Developing my skills and abilities	Stimulation • Feeling excitement • Having daring/adventuresome experience
	Self-actualization • Understanding more about myself • Working on my personal/ spiritual values	Recognition • Being recognized by other people • Having others know that I have been there

for the Muslim market. Second, some cultures and people were quite disturbed with the romance factors that involve intimacy and sexual relationship. The 13 factors consist of 26 items. These items are derived from the 74 original items with the highest loading factors on the 13 factors. The first attempt using 13 motives with 26 items was successfully implemented (cf. Oktadiana et al., 2017). Therefore, the shorter version of TCP questionnaire is also practicable to study tourists' travel motivation. The 26 items are presented in Table 8.2.

It is interesting to note that although TCP had replaced travel career ladder (TCL), a recent study by Aldao and Mihalic (2020) still utilize TCL concept to develop a framework in assessing travel motivation and social media on tourist behavior in Norwegia.

Conclusion

The first aim of this study is to elaborate travel career pattern theory of motivation. This objective is met by detailed explanation of the development and concept of TCP, starting from the construction of travel career ladder and its application, to the advancement of TCP. The second purpose is to depict the use of travel career pattern in tourism research. Various examples of studies were presented to answer the second aim. Different research settings and context verifies that TCP model is workable to understand tourist behavior.

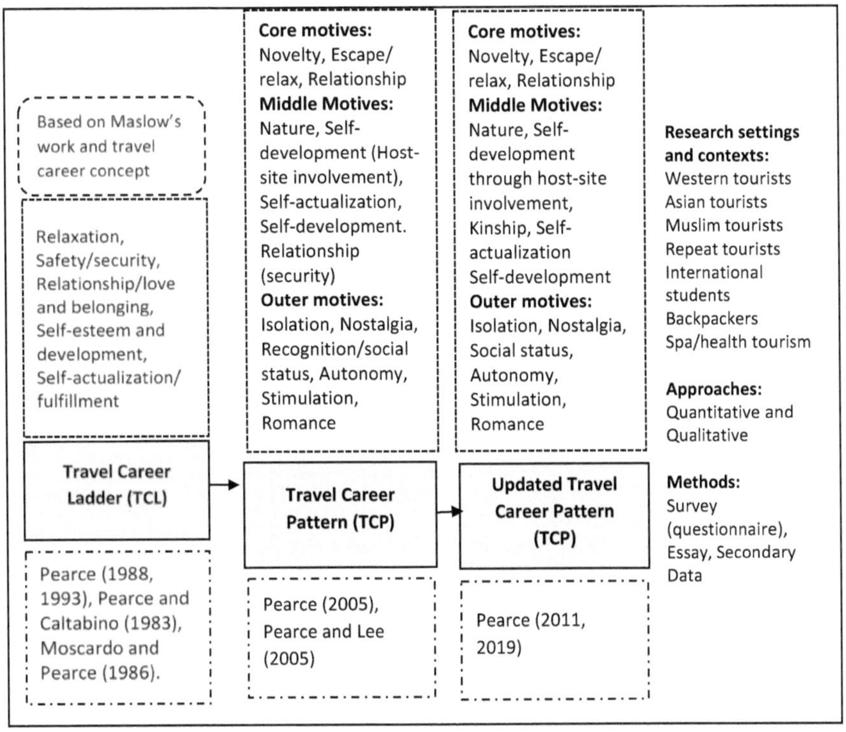

Figure 8.4 The Scheme of TCP and Its Applications.

The diagrammatic of TCP concept, its development and application in tourism researches is shown in Figure 8.4.

Future studies may replicate TCP concept in other cultural context such as African, Middle Eastern, Persian, Western Asian, South American and European. Further, comparison of travel motivation between cultures can be of interest. Other studies may seek travel motivations of particular tourists such as first-time travelers, RV tourists, adventure tourists, and food tourists. TCP may also be further tested qualitatively using for example, interviews and focus group discussions. It is also worth exploring TCP in the current pandemic situation and the post-pandemic.

The contributions of this chapter are in two folds. The first, it offers a meticulous insight into the construction of TCP concept, starting from travel career ladder model. The second, it provides information and study examples for other tourism scholars and researchers regarding the application of TCP to study tourists' travel motivation.

Acknowledgments

This chapter is dedicated to the late Philip L. Pearce, a beloved husband of Hera Oktadiana and a PhD supervisor of Manisha Agarwal.

References

Agarwal, M., & Pearce, P. L. (2019). Back to India again: Motivational insights from the travel career pattern approach. In *Proceedings of the 29th Annual Council for Australasian University Tourism and Hospitality Education Conference*, pp. 204–214. *CAUTHE 2019: 29th Annual Council for Australasian University Tourism and*

Hospitality Education Conference: sustainability of tourism, hospitality & events in a disruptive digital age, 11–14 February 2019, Cairns, QLD, Australia.

Aldao, C., &, Mihalic, T. A. (2020). New frontiers in travel motivation and social media: The case of Longyearbyen, the High Arctic. *Sustainability, 12*, 1–19. Doi:10.3390/su12155905

Bowen, D., & Clarke, J. (2009). *Contemporary tourist behaviour: Yourself and others as tourists.* Wallingford: CABI.

Filep, S., & Greenacre, L. (2007). Evaluating and extending the travel career patterns model. *Tourism, 55*(1), 23–38.

Hsu, C. H. C., & Huang, S. (2008). Travel motivation: A critical review of the concept's development. In A. Woodside& D. Martin (Eds.), *Tourism management analysis, behaviour and strategy* (pp. 14–27). Wallingford: CABI.

Lee, U-I, & Pearce, P. L. (2002). Travel motivation and travel career patterns. In: *Proceedings of first Asia pacific forum for graduate students research in tourism* (pp. 17–35). 22 May, Macao. Hong Kong: The Hong Kong Polytechnic University.

Li, H., Pearce, P. L., & Zhou. L. (2015). Documenting Chinese tourists' motivation patterns [online]. In: E. Wilson, and M. Witsel (Eds.), *CAUTHE 2015: Rising tides and sea changes: Adaptation and innovation in tourism and hospitality* (pp. 235–246). Gold Coast, QLD: School of Business and Tourism, Southern Cross University.

McKercher, B., & Koh, E. (2017). Do attractions "attract" tourists? The case of Singapore. *International Journal of Tourism Research, 19*(6), 661–671.

Moscardo, G. & Pearce, P. L. (1986). Historical theme parks: An Australian experience in authenticity. *Annals of Tourism Research, 13*(3), 467–479.

Oktadiana, H., Pearce, P. L., Pusiran, A. K., & Agarwal, M. (2017). Travel career patterns: The motivations of Indonesian and Malaysian Muslim tourists. *Tourism Culture & Communication, 17*(4), 231–248.

Panchal, J. & Pearce, P. L. (2011). Health motives and the travel career pattern (TCP) model. *Asian Journal of Tourism and Hospitality Research, 5*(1), 32–44.

Paris, C. M., & Teye, V. (2010). Backpacker motivations: A travel career approach. *Journal of Hospitality Marketing & Management, 19*(3), 244–259.

Pearce, P. L. (1988). *The Ulysses factor: Evaluating visitors in tourist settings.* New York: Springer-Verlag.

Pearce, P. L. (1993). Fundamentals of tourist motivation. In D. Pearce & R. Butler (Eds.), *Tourism research: Critiques and challenges* (pp. 85–105). London: Routledge and Kegan Paul.

Pearce, P. L. (2005). *Tourists behaviour: Themes and conceptual schemes.* Clevedon: Channel View Publications.

Pearce, P. L. (2011). *Tourist behaviour and the contemporary world.* Clevedon: Channel View Publications.

Pearce, P. L. (2019). *Tourists behaviour: The essential companion.* Cheltenham and Northampton: Edward Elgar Publishing.

Pearce, P. L. & Caltabiano, M. L. (1983). Inferring travel motivations from travellers' experiences. *Journal of Travel Research, 22*(2), 16–20.

Song, H., & Bae, S. Y. (2018). Understanding the travel motivation and patterns of international students in Korea: using the theory of travel career pattern. *Asia Pacific Journal of Tourism Research, 23*(2), 133–145.

Wu, J., Law, R., D. K. C. Fong, & Liu, J. (2019). Rethinking travel life cycle with travel career patterns. *Tourism Recreation Research, 44*(2), 272–277.

9

SOCIAL COMPARISON THEORY AND TOURISM

Volkan Genç and Seray Gülertekin Genç

Introduction

When people make judgments about their perceived reality with others, two sources of concern may dominate said judgments. The first is the desire to make a correct judgment regarding the truth. The second is wanting to avoid negative thinking about oneself to make a good impression on others (Arkonaç, 1993). According to Gilbert, Price, and Allan (1995), the need to compare oneself with others is found in many contexts and leads to a spirit of competition.

Leon Festinger (1954) argued that people are motivated to assess whether their perception of reality is correct. If we have objective scales to evaluate our judgments, we can use them as a physical benchmark. For example, we can measure how many meters we run per minute by holding a stopwatch. But our beliefs about reality are not always testable with such objective scales. How can we test our belief that consumers, for example, should be protected from social media?

In such cases, the person compares her/his judgment with other people's judgments to assess whether her/his judgment is correct. She/he asks others for their opinion on this matter. If, as a result, her/his judgment matches the views of others, her/his self-confidence increases. By contrast, since the person wants to have a correct perception of the truth, she/he changes her/his judgment in line with the judgment of others; that is, she/he agrees with their opinion (Arkonaç, 1993). In cases where these objective standards do not exist, people obtain the necessary information for evaluation by comparing their opinions and abilities with the opinions and abilities of other people. This is called the social comparison process, and the information obtained is called social comparison information (Goethals, 1986).

Social comparison theory is a highly debated socio-psychological theory that has been developed and constantly renewed by the study of various theorists such as Wheeler (Wheeler & Koestner, 1984; Wheeler et al., 1982); Hakmiller (1966), Wills (1981), Goethals (1986), Alicke (1985), Wood (1989), Gibbons and Gerrad (1989), and Suls (1986) since Festinger first put it forth in the 1950s. However, social comparisons are not always a voluntary and desirable process; people are spontaneously and automatically exposed to various comparison information as a natural result of group life. Experimental and theoretical contributions (Gilbert et al., 1995) are among the most important of these studies, which show that this may lead to some self-worth problems, especially in heterogeneous groups that include people with a wide variety of talents and views.

Individuals who compare themselves with individuals around them tend to categorize themselves in social classes. Thus, social comparison enables the individual to create and maintain a

DOI: 10.4324/9781003161868-9

social identity. Tourist psychology and behavior are among the main areas where individuals make social comparisons about themselves and their environments. Some would say that tourists have a distinct tendency to separate the self from others (Doran et al., 2015). There are also studies on how people perceive themselves and others while on vacation (Doran & Larsen, 2014; Jani & Han, 2014; Siegel & Wang, 2019). Given that biases in social comparison are among the most stable and recurring phenomena in social psychological research (Chambers & Windschitl, 2004), it is surprising that few academics in the field of tourism examine differences in perception of self and others in a natural environment. In this context, social behavior theory is an indispensable social-psychological element for tourism. This study aims to reveal the relationship between social behavior theory and tourism. In this context, first of all, the theory of social behavior, its development and causality, and its relationship with tourism will be explained.

Concept of Social Comparison Theory

As stated in the previous section, the social comparison theory asserts that the individual in question feels the need to evaluate her/his opinions and abilities. Therefore, she/he compares her/his opinions and abilities with other people's opinions and capabilities (Festinger, 1954). According to this theory, people want to have a positive self-image and be positively perceived by other people (Goethals, 1986). Therefore, the operative comparison process will be biased because the person sees herself/himself as being better or correct compared with other group members. According to the social comparison theory, when one judges others in the context of a group discussion, she/he will shift to the extreme of this alternative to differentiate herself/himself from those other members in a more positive way (Forsyth, 2000). What causes polarization is the comparison made between oneself and others. This situation is similar to the unidirectional upward drive that Festinger (1954) suggests in comparing people with themselves to evaluate their abilities (e.g., when you are immersed in study, you compare your knowledge to someone who is a little more hard-working than you, rather than someone who is lazier).

The important factor here is that the person knows the positions of other members regarding social values that are at the forefront of the discussion. Therefore, it may not be necessary to actively discuss this issue or issues with other people to polarize the person or persons who have this knowledge in any other way than group discussion (Forsyth, 2000). Social identity theory emphasized how individuals achieve and/or maintain high self-esteem by making or avoiding comparisons with other groups known as outgroups. The phenomenon of social comparison has started to be understood more by "self-verification" rather than "self-knowledge" – in other words, by self-validation processes rather than self-evaluation processes (Arkonaç, 1993).

According to Festinger (1954), people choose others similar to them to make comparisons while evaluating their opinions and abilities. This claim is closely related to the claim that people try to make accurate and objective assessments as much as possible. Comparisons with the highest diagnostic value, which can serve the purpose of making an accurate and valid evaluation of the individual, are comparisons made with others similar to her/him. Comparisons made by an individual with a very high or very low skill cannot be expected to provide information about what the person's ability level is. Similarly, a person with a very different character from the individual can't provide information about the accuracy of the individual's opinions. The most plausible solution here is to select a benchmark with a high diagnostic level that will allow the individual to make an accurate prediction about his/her belief or ability – that is, to make comparisons with others who are similar to her/him. However, Festinger (1954) also noted that due to the emphasis on individual achievement in Western cultures, when abilities are compared, people tend to compare with others who are slightly better off. For example, it has been observed that people perceive themselves as more altruistic, more moral, more just, and less prejudiced (Doran & Larsen, 2014).

People often practice a certain type of social comparison in which they compare their existence to a certain norm or standard (Alicke & Govorun, 2005).

The desire of a person to perceive what is real and her/his tendency to fit into the group results from the social influence that provides information. The effect here is a real social impact because the person has decided to conform to the opinions of others. In the absence of group pressure, action is taken in line with this decision. In the informative social influence, compliance behavior is motivated by the desire to act correctly. In situations where the perception of physical reality cannot be directly tested with objective scales, the less confident the person is of her/his view and the more uncertainty the stimulus situation carries within her/him, the stronger the influence of the informative social effect will be (Arkonaç, 1993).

Another reason for showing the behavior to conform to the opinions of others is that the person is under the influence of the normative influence. The normative effect is defined as adapting to the positive expectations of another (in- or out-group) person about herself/himself (Turner, 1991). In the normative effect, it is the desire to please other people that motivates compliance behavior. The person adapts out of a desire to be accepted by others, to be approved, and to avoid punishments and rejection. Human beings have a variety of needs, and we look for ways to satisfy them. Satisfying our needs makes us dependent on one another. For this reason, it is important to make good impressions on other people and make them like us. Thus, our disagreement with other people can lead to rejection. On the other hand, the state of the agreement will increase the positive evaluations of others about us; thus, our membership in that group is preserved (Arkonaç, 1993).

Development of Social Comparison Theory

Many authors start the modern theory of social comparison with Festinger's classic 1954 article, "A Theory of Social Comparison Processes" (Goethals et al., 1991; Wheeler, 1991). According to Arrowood (1993), Festinger wanted to imply, by the plural term "processes," at least two basic processes: *evaluating opinions* and *evaluating skills*. Social psychologists have obtained a wide variety of evidence from their work in recent years that social comparison is not a single process but consists of a series of relatively independent processes (Suls & Wills, 1991).

Different researchers base the origin of the theory on Kurk Lewin's group communication and group impact studies (Suls & Wills, 1991), Hyman and Kelley's reference group studies (Goethals et al., 1991), and Muzafer Sherif's autokinetic effect and social impact studies (Gergen & Gergen, 1986). Although there is some truth to all this, we know that Festinger takes the theoretical background of social comparison theory from "informal social communication" studies, as the empirical background and experimental evidence were obtained from "level of aspiration" experiments. Informal social communication theory is a theoretical approach that tries to understand the phenomenon of uniformity pressure within the group regarding opinions and discusses its implications for the individual and the group (Festinger, 1950).

Informal social communication theory has tried to understand how opinions in social groups are affected (opinion influence processes) and shaped (opinion formation). Social comparison theory, on the other hand, was also concerned with abilities (Festinger, 1954). According to Wheeler (1991), Festinger added processes related to abilities as well as views to informal social communication theory and substituted the term "communication" with the term "comparison." According to him, this substitution is necessary because one cannot change their own or someone else's talent by talking (via communication). Still, one can achieve uniformity by competing with that person, cooperating, or ceasing to compare (Wheeler, 1991: 4–5). Another point where social comparison theory differs from informal social communication theory is that although both theories deal with ensuring intra-group homogeneity in terms of opinions, the first emphasizes the group and group locomotion while the second foregrounds the needs of the individual. While the first conceptualizes the phenomenon as a group process, the second treats it as an interpersonal process (Goethals & Darley, 1977).

The social comparison theory put forward by Festinger (1954) has been subjected to many criticisms. Suls (1986), in his study on this subject, stated that Festinger's theory failed to adequately define the parameters of comparison. The same author claims that people are not inclined to compare personal characteristics or temperaments with those of other people in his study. The second criticism of the theory concerns to what extent people compare themselves with others. The third criticism is that the theory focuses only on voluntary and deliberate comparison processes in which people compare themselves with their counterparts (Suls, 1986). The fourth criticism involves the selective use of the information obtained as a result of the comparison. Suls (1986) argues that while Festinger's definition of social comparison focuses on the selective use of knowledge, it does not focus on the selective acquisition of knowledge. The fifth criticism is about the direction of social comparison (Blanton et al., 2001). Festinger (1954) assumed that people have a one-sided urge to compare themselves with people who are more capable than they are. This uncertainty regarding the upward, downward, or even horizontal direction of social comparison confuses the general predictions made by the theory (Taylor & Lobel, 1989).

As a result of these criticisms, researchers have tried to develop the theory put forward by Festinger (1954). This context depends on the social comparison, downward comparison, fear-attachment, social comparison as social cognition, and individual differences in social comparison (Buunk & Gibbons, 2007) and built social comparison (Goethals et al., 1991) paradigms are presented.

Shortly after the publication of Festinger's (1954) paper, the fear-attachment theory came in the form of the second breakthrough in social comparison theory, Schachter's (1959) pioneering work on stress and commitment. In fear attachment theory, Schachter argued that social comparison is the main cause of attachment under stress and is more important than the desire for cognitive clarity regarding the nature of the threat.

Thornton and Arrowood (1966) and Hakmiller (1966) were the first social psychologists to experiment with downward comparison theory. This theory showed that individuals who are threatened to a certain extent prefer to socially compare themselves with others who are considered worse in this dimension. Subsequently, Friend and Gilbert (1973) showed that individuals threatened with failure in a test tend to avoid information about others who are better off after these studies resulted in a considerable turn in social comparison theory that emphasized the role of threatened downward comparisons (Buunk & Gibbons, 2007).

While social comparison theory as social cognition has been associated with past work on social cognition (Gibbons & Gerrard, 1997), it is amazing that until now, social comparison research existed independently from studies on cognitive processes that characterize psychological functioning in any condition. The social cognition approach brings an informed perspective to the social comparison process (Buunk & Gibbons, 2007). Especially, it is assumed that to understand the consequences of social comparisons on self-assessment and self-perception, one should examine what self-knowledge is made available during the comparison and how this information is then used to judge and evaluate oneself. In short, it is necessary to examine the cognitive processes that occur during the comparison (Buunk & Gibbons, 2000).

Individual differences in social comparison are a newer paradigm than studying the effects of social comparison inspired by social cognition, examining the relationship between personality variables and social comparison processes. Even though past studies considered variables such as fear of negative evaluation and self-consistency to be integral to social comparison, the role of personality according to social comparison has only recently been systematically examined (Buunk & Gibbons, 2007; Morse & Gergen, 1970; Wilson & Benner, 1971).

Goethals et al. (1991) first used the term "constructive social comparison" to describe social comparisons made using information that has either nothing to do with reality or partial or incomplete relation to reality, as opposed to actual or realistic social comparison. According to Goethals et al. (1991), realistic social comparison refers to self-evaluations based on the use and analysis of actual information about social reality, while constructive social comparison refers to self-evaluations based on predictions, assumptions, or rational implications of social reality. Constructive comparisons are social comparisons that take place "in one's head," as the authors say, and they are mostly self-serving – that is, motivational – in nature (Goethals & Klein, 2000: 28). Although there is no strict rule of thumb, constructive social comparisons are often biased rather than objective, arising in response to the need to heighten the subjective sense of well-being of the person entering the comparison. People often resort to constructive social comparisons when they do not need real social comparison information if they are dissatisfied with or want to avoid real comparison information (Goethals et al., 1991). Such social comparisons are cognitive constructs of social reality (Goethals et al., 1991) and are produced to perform certain psycho-social functions. This approach has made significant contributions to understanding how reality is interpreted, distorted, and reproduced in the social comparison process.

Causes of Social Comparison

As can be understood from the studies mentioned above, social comparisons play a more central role in our lives than it seems. Social comparisons are "an inevitable way beyond social interaction" (Doran et al., 2015: 556). Buunk and Gibbons (2007) argued that even Festinger did not realize that social comparison processes were such a comprehensive phenomenon. Festinger (1954) thought that individuals resort to social comparisons only to infer "the value of their beliefs and skills" and, again, only when physical standards are not available. Today, however, individuals make social comparisons on subjects that cover much wider and various dimensions (Klein, 1997).

In addition to the variety of topics in social comparisons, the purposes of individuals to enter into comparisons also vary; today, it is widely accepted that the individual can make comparisons with different motivations not only for self-assessment but according to the needs that appear in a particular context. According to Taylor, Wayment, and Carillo (1996), the motivation of individuals to enter into social comparisons can be summarized in four categories: the need for self-evaluation, the need for self-improvement, the need for self-enhancement or see themselves positively, and the need for affiliation (see Figure 9.1).

The Motivation of Self-Evaluation

According to Taylor, Wayment, and Carillo (1996), an individual experiences uncertainty when her/his knowledge of herself/himself falls below a certain level. This provides motivation and mobility to self-evaluate and re-stabilize the self. Thus, the diagnostic value of social comparison information becomes important as the information referred to herein must reduce the uncertainty and allow an accurate assessment of the self. Comparisons with the highest need for self-evaluation are comparisons with similar others. Self-evaluation processes are largely explained by Festinger's (1954) principles of classical social comparison theory.

The Motivation of Self-Improvement

In some cases, people can use social comparison information to improve their skills or do better what they can already do. In this case, they can learn how to improve their abilities from these more successful goals by comparing themselves with people who are better off than

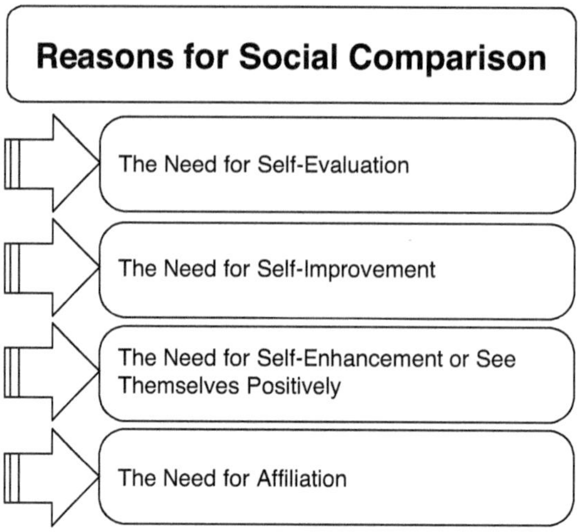

Figure 9.1 Reasons People Make a Social Comparison (Taylor, Wayment, & Carillo, 1996).

them – that is, by making an upward comparison. These upward targets can even be a source of inspiration for them. Collins (1996) gives the example of dieters sometimes hanging pictures of thin people on the refrigerator to inspire and motivate them. According to Collins (1996), objective standards (e.g., body-mass index information) do not provide the information an individual needs to improve herself/himself but only show that she/he is good or bad, whereas knowing where s/he stands concerning a better-off person provides information regarding what she/he can aim for.

Some researchers have argued that downward goals – that is, people who are worse off than themselves – can also help the individual improve herself/himself by showing what s/he should not do (Smith, 2006; Suls, 1977; Taylor & Lobel, 1989). On the other hand, some studies (Blanton et al., 1999; Gerrard et al., 2005; Gibbons et al., 2000) suggest that making upward comparisons helps one to improve her/his skills. Some studies have also shown that upward comparison creates an effect that lowers a person's self-worth (Major et al., 1991; Wood & Taylor, 1991). In such situations, individuals tend toward various defensive strategies.

The Motivation of Increasing Self-Worth

The third stage of social comparison, increasing self-worth, motivates people to feel better about themselves. For example, when people want to do their best in a job, they don't want to think of those with more success than themselves. Instead, they want to compare themselves with those who are weaker (i.e., less successful) than themselves as a way of boosting self-esteem. The result of such social comparison can be either advantageous or disadvantageous. Individuals comparing themselves to those in worse or inferior positions tend to glorify themselves. This situation reveals the concept of downward comparison (Kaya, 2012: 42). All of this leads to an increase in perceptions of self-worth.

The Motivation of Affiliation

The need for affiliation refers to establishing relationships with others, getting into a group, and developing social relationships. This need is satisfied by developing social relationships, emotional attachment, being in a social group, and sharing your feelings with other people.

According to many researchers, how one looks to others plays an important role in self-evaluation (Goethals & Darley, 1977). For this reason, when making comparisons, people prefer others who share common characteristics or the same fate. For example, cancer patients follow an upward and downward comparison strategy because they live the same fear and fate together (Teközel, 2007). This motivational stage is also defined as the "development of a sense of community." Thus, people can use comparisons to understand their situation and abilities, identify common characteristics with others, and create community/relationships with others (Gül, 2016: 29).

Group Dynamics in Social Comparison Theory

People are affected by others in various ways. Social comparison theory is a theory about group dynamics and individuals' perceptions of views and abilities (Forsyth, 2000). The theory stems from the work of small and interacting groups. Festinger (1954) used the comparison principle to explain why groups tend to be homogeneous in attitudes and values. It remains a central principle in studies on the power of reference groups to influence members' attitudes. The comparison principle has also informed researchers and theorists' analysis of group formation, commitment, social identity, majority influence, minority influence, group discussion, polarization, social idleness, brainstorming, and transferring beliefs and values from groups to individuals (Forsyth, 2000).

After studying students' attitudes over four years, Newcomb (1943) concluded that their attitudes changed to match their college classmates' attitudes. Studies by Asch (1955), Milgram (1963), and Moscovici (1994) reached similar conclusions. Thus, the beliefs and self-evaluations of individuals were shaped and reshaped by the groups to which they belong in these studies. But what is the source of one's power over the members of the group? Should members of groups feel intimidated or threatened? Should they be given prizes? This situation is based on social comparison theory. Studies show that group members change their ideas and beliefs not because they are directly under pressure from others but because they identify their strengths, assets, weaknesses, and obligations by comparing them with specific individuals in their group.

Different researchers attribute the origin of the theory to Kurt Lewin's group communication and group impact studies, Hyman and Kelley's reference group studies, Muzafer Sherif's autokinetic effect, and social impact studies (Gergen & Gergen, 1986). These studies have expanded Festinger's perspectives on social comparison theory.

Festinger's first publication, based on research conducted as an undergraduate student, was his master's thesis under the direction of Kurt Lewin at the University of Iowa Child Welfare Station. In it, he discussed the social factors affecting individuals (Hertzman & Festinger, 1940). This research showed that if the subjects find themselves above the group average, they will lower their expectations. Conversely, if they score below the group average, they will increase their expectations. The status of the group also made a difference. Undergraduate participants increased their willingness the most when they scored below high school students and lowered their level of desire the most when they scored above graduate students. The relevance of this research to social comparison theory is clear (Forsyth, 2000).

The second considerable impact was the research on informal communication in small groups that Festinger and colleagues started at the Group Dynamics Research Center founded by Lewin at MIT. This research resulted in the informal theory of social communication (Festinger, 1950), a direct precursor to Festinger's social comparison theory that followed four years later. The earlier theory assumed that people in groups would like to obtain monotonous views, either because group consensus was needed to provide confidence in one's view or because the consensus was needed in coordinating group goals. Various experiments on group communication and rejection of opinion deviations confirmed the elements of the theory (Back, 1951; Festinger & Thibaut, 1951).

Kurt Lewin is considered the pioneer of the concept of group dynamics. Lewin likened groups to living organisms and argued that groups have their ideas. Individuals included in the group think, feel, and behave differently than when they are alone (Craig & Alexander, 1997). The group dynamics studies initiated by Lewin deeply affected the conceptual developments around small groups in social psychology.

Muzafer Sherif (1935, 1936), on the other hand, determined the formation of social norms for the first time with classical research conducted and examined in the laboratory. If a group of people finds itself in an unstructured, uncertain situation where there are no reference points to describe their expectations, perceptions, or activities, they automatically seek information from others in the group. Sherif did not think the group members were reluctantly obeying the judgments of others but instead felt they were using the information contained in others' responses to review their opinions and beliefs (Sherif, 1966). Sherif decided that the autokinetic effect provided an ideal opportunity to study this normative process. This effect occurs when individuals sitting in a completely dark room mistakenly believe that a fixed point of light is moving. Sherif's study confirmed the functioning of social comparison processes under controlled conditions. Furthermore, Sherif found that it documents the development of a norm rather than a momentary change in judgment resulting from group pressure.

Hyman (1980) directly addressed the comparison processes in the analysis of reference groups but focused on how people chose goals for comparison rather than the interpersonal consequences of the comparison process. Although these early researchers did not use the term "social comparison," the processes they studied were dependent on individuals' perceptions of the people's talents, abilities, and attitudes (Forsyth, 2000).

In 1954, Festinger brought together these various theoretical issues and experimental findings in social comparison theory, arguing that individuals seek information about the accuracy of their ideas and the adequacy of their abilities. Thus, they satisfy themselves by comparing themselves with others and have a basic need for accuracy and cognitive clarity.

Social Comparison Theory in Tourism

Social comparisons are considered an important part of psychological functioning that affects people's judgments, experiences, and behavior. Comparisons between individuals and other individuals help people evaluate their performance (Doran et al., 2015). Social comparison theory determines the social positions of relative individuals by comparing the properties of individuals with those of others (Schiffman & Kanuk, 2010). Social comparison theory emphasizes the role of others in a given environment as the source of comparative knowledge (Argo et al., 2006), so socio-environmental comparison covers many areas of tourism.

The argument that people want to perceive themselves as different from mainstream tourist typology has a long history in tourism research (MacCannell, 1976). At the same time, there are different views on the tendency of some tourists to separate the self from others. Previous research shows that people prefer to see themselves as authentic (or individual) travelers rather than stereotypical tourists (Doran et al., 2015). Thus, social comparisons are not a new phenomenon for tourism research. Changing tourism movements and increasing social media activities have also changed the dimensions of social comparison. When the literature is examined, it is seen that many tourism studies have been conducted in the context of social comparison theory.

Jacobsen (2000) investigated the attitudes of charter tourists. Although most of the respondents were found to have positive (9%) or neutral (81%) views about tourists, some reported negative (10%) views about tourists. Jacobsen (2000) concluded that the participants in the second category represented an "anti-tourist" attitude, thereby highlighting the tendency to act distant from those considered to represent typical tourists. Uriely et al. (2002) discovered that backpackers underestimated their visits to popular tourism destinations. Researchers interpreted this as an ideology to maintain the contrast between backpacking and traditional mass tourism.

Prebensen et al. (2003) asked German tourists visiting Norway to indicate whether they saw themselves as a "typical German tourist" or a "non-typical German tourist." They found that almost 90% of their respondents consider themselves non-typical German tourists. McCabe and Stokoe (2004) concluded that people build their tourist experience by comparing their behavior with the behavior of others.

Smith (2006) argues that comparing similarities with others in an environment is especially important in generating the emotions discussed in this study. Further support for social comparison and consumption emotional bonds can be derived from the concept of emotional contagion (Hatfield et al., 1994), which suggests that individuals' emotional expressions in an environment lead others to imitate those emotions (Hatfiel et al., 1994). Stapel and Marx (2007) argue that the similarity aspect is the essence of social comparison processes. It is a criterion for self-evaluation, which has indirect effects on individuals' effects and behavior.

Larsen and Brun (2011) found that risk perception differs according to the tourist typology. Within the scope of social comparison theory, Doran and Larsen (2014) found that tourists perceive themselves as having more environmentally friendly attitudes than other tourists. It shows that there are differences in the perception of self and others regarding social comparisons of environmental sustainability. The findings also show that tourists generally have extremely positive views of themselves on environmental sustainability issues and that their environmental attitudes reflect perceived desired standards.

Jani and Han (2014) found that social comparison significantly affects hotel guests' satisfaction and that this effect plays a critical role in triggering behavioral intentions. Thus, the results confirm that social comparison is included in predicting hotel guest experiences. On the other hand, Doran, Larsen, and Wolff (2015) show that people perceive their travel goals differently from those who perceive them as typical tourists (authentic) and that these tendencies are generalized among people in different types of tourism.

"Keeping up with the Joneses," as Siegel and Wang (2019) put it, means that people socially want to have the same objects and do the same things as their peers to keep step with them. In the study conducted by the authors, the Generation Y cohort determines its distinctive characteristics in social networking habits concerning travel behaviors. Türk (2020) determined that the psychological empowerment perceptions of employees working in accommodation establishments are effective on their identification with the organization and that social comparison has a negative regulatory influence on this effect. These findings suggest that the tendency to separate the self from typical tourists represents a significant part of being a tourist.

Their study on Instagram users, Machado, Santos, and Medeiros (2021), shows that jealousy and social comparisons increase the intention to visit a destination. The strongest effect is related to the social comparison variable. Their study of social comparison on bloggers, Mariani, Styven, and Nataraajan (2021), found that non-professional travel bloggers tended to compare themselves to others on Facebook much more often than professional bloggers.

Conclusion

Wherever there are people, there is comparison. Sometimes, this comparison can be up or down. Comparisons are an indispensable phenomenon of tourism. The social comparison theory, which has been put forward since the 1950s, is of great importance for tourists and employees.

With social comparisons, tourists experience positive and negative emotions by positioning themselves somewhere (downward or upward comparison). Social comparisons are seen as the most important source of emotional contagion. They provide essential roadmaps for tourism marketers and businesses to understand tourists. Studies have emphasized the role of social comparisons in the behavior of individuals who go on vacation. At the same time, social comparison theory has played a great role in the impact of social media platforms on individuals in recent years. Here, tourism managers need to understand better the roles that push individuals to social comparison and develop marketing strategies.

As clearly stated in this chapter, the theory of social comparison should be considered in detail and realistically at every step in tourism. There will inevitably be some important changes in the context of tourism, especially after COVID-19. In this context, there is a need to evaluate the role of social comparison in terms of tourism in multiple ways. As a result, it is expected that tourism businesses that follow the market, pay attention to the social, cultural, and environmental developments and dominant trends, and observe their competitors' situation will have a competitive advantage in their sectors thanks to social comparisons.

References

Alicke, M. D. (1985). Global self-evaluation as determined by the desirability and controllability of trait adjectives. *Journal of Personality and Social Psychology, 49*, 1621–1630.

Alicke, M. D., & Govorun, O. (2005). The better-than-average effect. In M. D. Alicke, D. A. Dunning, & J. I. Krueger (Eds.), *The self in social judgment* (pp. 85–106). New York: Psychology Press.

Argo, J. J., White, K., & Dahl, D. W. (2006). Social comparison theory and deception in the interpersonal exchange of consumption information. *Journal of Consumer Research, 33*, 99–108. doi:10.1086/504140

Arkonaç, S. (1993). *Grup ilişkileri [Group relations]*. Alfa Basım Yayım Dağıtım, İstanbul.

Arrowood, A. J. (1993). Social comparison: Contemporary theory and research (book reviews). *Psychological Record, 43*(2), 331–332.

Asch, S. E. (1955). Opinions and social pressures. *Scientific American, 193*(5), 31–35.

Back, K. W. (1951). Influence through social communication. *Journal of Abnormal and Social Psychology, 46*, 9–23.

Blanton, H., Buunk, B. P., Gibbons, F. X., & Kuyper, H. (1999). When better-than-others compares upward: The independent effects of comparison choice and comparative evaluation on academic performance. *Journal of Personality and Social Psychology, 76*, 420–430.

Blanton, H., Vanden Eijnden, R. J. J. M., Buunk, B. P., Gibbons, F. X., Gerrard, M., & Bakker, A. (2001). Accentuate the negative: Social images in the prediction and promotion of condom use. *Journal of Applied Social Psychology, 31*, 274–295.

Buunk, A. P., & Gibbons, F. X. (2007). Social comparison: The end of a theory and the emergence of a field. *Organizational Behavior and Human Decision Processes, 102*(1), 3–21.

Buunk, B. P., & Gibbons, F. X. (2000). Towards an enlightenment in social comparison theory: Moving beyond classic and renaissance approaches. In J. Suls & L. Wheeler (Eds.), *Handbook of social comparison: Theory and research* (pp. 487–499). New York: Plenum.

Chambers, J.R., & Windschitl, P.D. (2004). Biases in social comparative judgments: The role of nonmotivated factors in above-average and comparative-optimism effects. *Psychological Bulletin, 130*(5), 813–838.

Collins, R. L. (1996). For better or worse: The impact of upward social comparison on self-evaluations. *Psychological Bulletin, 119*, 51–69.

Craig, M., & Alexander, H. (1997). *The message of social psychology* (1st ed.), Cambridge: Blackwell.

Doran, R., & Larsen, S. (2014). Are we all environmental tourists now? The role of biases in social comparison across and within tourists, and their implications. *Journal of Sustainable Tourism, 22*(7), 1023–1036.

Doran, R., Larsen, S., & Wolff, K. (2015). Different but similar: Social comparison of travel motives among tourists. *International Journal of Tourism Research, 17*(6), 555–563.

Festinger, L. (1950). Informal social communication. *Psychological Review, 57*, 271–282.

Festinger, L. (1954). A theory of social comparison processes. *Human Relations, 1*, 117–140.

Festinger, L., & Thibaut, J. (1951). Interpersonal communication in small groups. *Journal of Abnormal and Social Psychology, 46*, 92–99.

Forsyth, D. R. (2000). Social comparison and group processes. In J. Suls & L. Wheeler (Eds.), *Handbook of social comparison: Theory and research* (pp. 81–104). New York: Plenum

Friend, R. M., & Gilbert, J. (1973). Threat and fear of negative evaluation as determinants of locus of social comparison. *Journal of Personality, 41*, 328–340.

Gergen, K. L., & Gergen, M. M. (1986). *Social psychology* (2nd ed.). New York: Springer-Verlag.
Gerrard, M., Gibbons, F., Lane, D., & Stock, M. (2005). Smoking cessation: Social comparison level predicts success for adult smokers. *Health Psychology, 24*, 623–629.
Gibbons, F. X., Blanton, H., Gerrard, M., Buunk, B. P., & Eggleston, T. J. (2000). Does social comparison make a difference? Optimism as a moderator of the impact of comparison level on outcome. *Personality and Social Psychology Bulletin, 26*, 637–648.
Gibbons, F. X., & Gerrard, M. (1989). The effects of upward and downward social comparison on mood states. *Journal of Social and Clinical Psychology*, 8, 14–31.
Gibbons, F. X., & Gerrard, M. (1997). Health images and their effect on health behavior. In B. Buunk & F. X. Gibbons (Eds.), *Health, coping and well-being: Perspectives from social comparison theory* (pp. 63–94). Mahwah, NJ: L. Erlbaum Associates.
Gilbert, D. T., Giesler, R. B., & Morris, K. A. (1995). When comparisons arise. *Journal of Personality and Social Psychology, 69*, 227–236.
Gilbert, P., Price, J., & Allan, S. (1995). Social comparison, social attractiveness and evolution: How might they be related? *New Ideas in Psychology, 13*, 149–165.
Goethals, G. R. (1986). Social comparison theory: Psychology from the lost and found. *Personality and Social Psychology Bulletin, 12*(3), 261–278.
Goethals, G. R., & Darley, J. M. (1977). Social comparison theory: An attributional approach. In J. M. Suls & R. L. Miller (Eds.), *Social comparison processes: Theoretical and empirical perspectives* (pp. 259–278). Washington, DC: Hemisphere.
Goethals, G. R., & Klein, W. M. (2000). Interpreting and inventing social reality. In *Handbook of social comparison* (pp. 23–44). Boston, MA: Springer.
Goethals, G. R., Messick, D. M., & Allison, S. T. (1991). The uniqueness bias: Studies of constructive social comparison. In J. M. Suls & T. A. Wills (Eds.), *Social comparison, contemporary theory and research* (pp. 149–173). Hillsdale, NJ: Lawrence Erlbaum
Gül, E. (2016). Ergenlerde sosyal görünüş kaygısı ve sosyal karşılaştırmanın fonksiyonel olmayan tutum ve bilişsel çarpıtmalarla ilişkisi [English Title: The relationship of social appearance anxiety and social comparison with dysfunctional attitudes and cognitive distortions in adolescents] (Unpublished Master's Thesis), Üsküdar Üniversitesi Sosyal Bilimler Enstitüsü, İstanbul.
Hakmiller, K. L. (1966). Threat as a determinant of downward comparison. *Journal of Experimental Social Psychology* (Suppl. 1), 32–39.
Hatfield, E., Cacioppo, J. T., & Rapson, R. L. (1994). *Emotional contagion*. New York: Cambridge University Press.
Hertzman, M., & Festinger, L. (1940). Shifts in explicit goals in a level of aspiration experiment. *Journal of Experimental Psychology, 27*, 439–452.
Higgins & Kruglanski, A. W. (Eds.), *Social psychology: Handbook of basic principles* (pp. 133–168). New York: Guilford Press.
Hyman, H. H. (1980). The psychology of status. *Archives of Psychology, 38*(269). (Originally published in 1942).
Jacobsen, J. K. S. (2000). Anti-tourist attitudes: Mediterranean charter tourism. *Annals of Tourism Research, 27*(2), 284–300.
Jani, D., & Han, H. (2014). Testing the moderation effect of hotel ambience on the relationships among social comparison, affect, satisfaction, and behavioral intentions. *Journal of Travel & Tourism Marketing, 31*(6), 731–746.
Kaya, A.G. (2012). Sosyal dışlanmaya verilen tepkilerin sosyal karşılaştırma süreçleri açısından incelenmesi, [English Title: Examination of reactions to social exclusion in terms of social comparison processes] (Unpublished Ph.D. Thesis), Hacettepe Üniversitesi, Sosyal Bilimler Enstitüsü, Ankara.
Klein, W. M. (1997). Objective standards are not enough: Affective, self-evaluative, and behavioral responses to social comparison information. *Journal of Personality and Social Psychology, 72*,763–774.
Larsen, S., & Brun, W. (2011). 'I am not at risk-typical tourists are'! Social comparison of risk in tourists. *Perspectives in Public Health, 131*(6), 275–279.
MacCannell, D. (1976).*The tourist: A new theory of the Leisure Class*. New York: Schocken Books, Inc.
Machado, D., Santos, P., & Medeiros, M. (2021). Effects of social comparison, travel envy and self-presentation on the intention to visit tourist destinations. *Brazilian Business Review, 18*(3), 297–316.
Major, B., Testa, M., & Bylsma, W. H. (1991). Responses to upward and downward social comparisons: The impact of esteem-relevance and perceived control. In J. M. Suls & T. Wills (Eds.), *Social comparison: Contemporary theory and research* (pp. 237–260). Hillsdale, NJ: L. Erlbaum Associates.

Mariani, M. M., Styven, M. E., & Nataraajan, R. (2021). Social comparison orientation and frequency: A study on international travel bloggers. *Journal of Business Research, 123*, 232–240.

McCabe, S., & Stokoe, E. H. (2004). Place and identity in Tourists' Ac-counts. *Annals of Tourism Research, 31*(3), 601–622.

Milgram, S. (1963). Behavioral study of obedience. *Journal of Abnormal and Social Psychology, 67*, 371–378.

Morse, S., & Gergen, K. J. (1970). Social comparison, self-consistency and the concept of the self. *Journal of Personality and Social Psychology*, 16(1), 148–156.

Moscovici, S. (1994). Three concepts: Minority, conflict, and behavioral styles. In S. Moscovici, A. Mucchi-Faina, & A. Maass (Eds.), *Minority influence* (pp. 233–251). Chicago, IL: Nelson-Hall.

Newcomb, T. M. (1943). *Personality and social change: Attitude formation in a student community.* New York: Dryden.

Prebensen, N. K., Larsen S, Abelsen B. (2003). I'm not a typical tourist: German tourists' self-perception, activities, and motivations. *Journal of Travel Research, 41*(4), 416–420.

Schachter, S. (1959). *The psychology of affiliation.* Palo Alto, CA: Stanford University Press.

Schiffman, L., & Kanuk, L. (2010). *Consumer behavior* (10th ed.). Upper Saddle River, NJ: Pearson Prentice Hall.

Sherif, M. (1935). A study of some social factors in perception. *Archives of Psychology* (Columbia University), 187, 60.

Sherif, M. (1936). *The psychology of social norms.* Oxford: Harper.

Sherif, M. (1966). *The psychology of social norms.* New York: Harper & Row. (Originally published in 1936).

Siegel, L. A., & Wang, D. (2019). Keeping up with the joneses: Emergence of travel as a form of social comparison among millennials. *Journal of Travel & Tourism Marketing, 36*(2), 159–175.

Smith, R. H. (2006). Assimilative and contrastive emotional reaction to upward and downward social comparison. In J. M. Suls & L. Wheeler (Eds.), *Handbook of social comparison: Theory and research* (pp. 173–200). New York: Kluwer Academic/Plenum Publisher.

Stapel, D.A., & Marx, D.M. (2007). Distinctiveness is key: How different types of self-other similarity moderate social comparison effects. *Personality and Social Psychology Bulletin*, 33(3), 439–448.

Suls, J. (1986). Notes on the occasion of social comparison theory's thirtieth birthday. *Personality and Social Psychology Bulletin, 12*, 289–296.

Suls, J., & Wills, T. A. (Eds.). (1991). *Social comparison: Contemporary theory and research.* Hillsdale, NJ: Lawrence Erlbaum.

Suls, J. M. (1977). Social comparison theory and research: An overview from 1954. In J. M. Suls & R. L. Miller, Journal of Travel & Tourism Marketing (Eds.), *Social comparison process: Theoretical and empirical perspectives* (pp. 1–20). Washington, DC: Hemisphere Publishing Corporation.

Taylor, S. E., & Lobel, M. (1989). Social comparison activity under threat: Downward evaluation and upward contacts. *Psychological Review, 96*(4), 569–575.

Taylor, S. E., Wayment, H. A., & Carrillo, M. (1996). Social comparison, self-regulation, and motivation. In R. M. Sorrentino & E. T. Higgins (Eds.), *Handbook of motivation and cognition* (pp. 3–27). New York: Guilford Press.

Teközel, İ. M. (2007). Gerçekliği inşa etkinliği olarak sosyal karşılaştırma, [English Title: Social comparison as a reality construction activity] (Unpublished Ph.D. Thesis), Ege Üniversitesi, Sosyal Bilimler Enstitüsü, Sosyal Psikoloji Anabilim Dalı, İzmir.

Thornton, D. A., & Arrowood, A. J. (1966). Self-evaluation, self enhancement, and the locus of social comparison. *Journal of Experimental Social Psychology, 2*(Suppl. 1), 40–48.

Türk, O. (2020). Psikolojik güçlendirmenin örgütsel özdeşleme üzerindeki etkisinde sosyal karşılaştırmanın düzenleyici rolü: Konaklama işletmelerinde bir uygulama, [The moderator role of social comparison in psychological empowerment on organizational identification: An application in hospitality businesses] (Unpublished Ph.D. Thesis), Nevşehir Hacı Bektaşı Veli University, Social Science Institute, Nevşehir.

Turner, J. C. (1991). *Social influence.* Pacific Grove, CA: Brooks/Cole.

Uriely, N., Yonay, Y., & Simchai, D. (2002). Backpacking experiences: A type and form analysis. *Annals of Tourism Research, 29*(2), 520–538.

Wheeler, L. (1991). A brief history of social comparison theory. In J. Suls & T. A. Wills (Eds.), *Social comparison: Contemporary theory and research* (pp. 3–21). Hillsdale, NJ: L. Erlbaum Associates.

Wheeler, L., & Koestner, R. (1984). Performance evaluation: On choosing to know the related attributes of others when we know their performance. *Journal of Experimental Social Psychology, 20*, 263–271.

Wheeler, L., Koestner, R., & Driver, R. E. (1982). Related attributes in the choice of comparison others. *Journal of Experimental Social Psychology, 18*, 489–500.

Wills, T. A. (1981). Downward comparison principles in social psychology. *Psychological Bulletin, 90*, 245–271.

Wilson, S. R., & Benner, L. A. (1971). Ability evaluation and self-evaluation as types of social comparison. *Sociometry, 36*, 600–607.
Wood, J. V. (1989). Theory and research concerning social comparisons of personal attributes. *Psychological Bulletin, 106*, 231–248.
Wood, J. V., & Taylor, K. L. (1991). Serving self-relevant goals through social comparison. In J. Suls & T. A. Wills (Eds.), *Social comparison: Contemporary theory and research* (pp. 23–49). Hillsdale, NJ: Lawrence Erlbaum.

10

HOTEL CSR MAY NOT ALWAYS LEAD TO POSITIVE OUTCOMES

The Role of Attributions about Motives Behind CSR Initiatives

Erhan Boğan and Yakup Kemal Özekici

Introduction

Nowadays, environmental and social responsibility practices, which are seen as an important component for companies to maintain their economic existence and survive in competition, are increasingly common in the tourism and hospitality sector (Boğan, 2021). This situation naturally attracts the attention of academicians. Research shows that responsible initiatives are an important tool for companies to gain positive returns from different stakeholder groups. For example, these initiatives result in positive employee responses including job satisfaction (Appiah, 2019), affective organizational commitment (Kim et al., 2017), organizational citizenship behavior (Boğan & Dedeoğlu, 2020), and work performance (Hur et al., 2021). Also, when the reactions of the local community to CSR initiatives are examined, which is a topic that shas attracted the attention of researchers for the last three to four years, research findings show that these initiatives increase the community commitment and support for additional tourism (Gursoy et al., 2019).

In parallel with the increasing interest of business researchers, it is seen that comprehensive and detailed research about CSR in the tourism and hospitality sector is becoming more evident day by day. For example, while the researchers initially examined the direct effect of employee CSR perception on organizational citizenship behavior (Kim et al., 2017), today, they examine variables that play a mediating role in this relationship and the variables that moderate this relationship (Boğan & Dedeoğlu, 2020; Ko et al., 2018). Also, initially the relationship between CSR and employee reactions was explained by social identity theory or social exchange theory; however, nowadays we encounter complex models that are explained using self-determination theory and attribution theory (Boğan, 2020a). This fact shows that corporate social responsibility research has developed significantly and researchers are focusing on the big picture. In this way, new research topics emerge and more specific recommendations are developed for practitioners.

Current research aims to highlight that the reactions of stakeholder groups (employees, customers, and society) to CSR practices will not always be positive. At this point, it is aimed to draw a framework using attribution theory. According to this theory (Kelley & Michela, 1980), people observe the activities done in their environment and make opinions or comments about the possible cause of the activities they observe. In other words, attributions are made about the possible cause of these activities. These attributions shape their subsequent attitudes and behaviors toward these activities or those who carry out these activities. In the current study, it is emphasized that

 DOI: 10.4324/9781003161868-10

the strength or shape of the reactions given by the relevant stakeholder group changes according to the type of attributions they make regarding the activity they perceive.

Corporate Social Responsibility (CSR)

The concept of CSR emphasizes that companies, which are economic organizations, have certain responsibilities toward their stakeholders since their establishment (Post et al., 1999). For example, providing reliable products to consumers and providing full and accurate information about the products can be examples of social responsibility to customers. Also, supplying local products and services and providing employment opportunities to local community can be examples of social responsibility to community. The specified responsibilities are actually the primary responsibility of the company. However, companies can go beyond what is expected and take initiatives or activities that provide benefits for stakeholders (Post et al., 1999), for example, providing financial support to the construction of schools and parks, andproviding information about the nutritional elements of the foods included in the menu items, aiming at preventing obesity and directing consumers to consume healthy food. Carroll (1979) calls philanthropic responsibility when companies carry out responsibility activities that go beyond expectations. Speaking of which, Archie B. Carroll, who played one of the leading roles in the development of the CSR concept, should be mentioned. Carroll (1979, 1991) listed the responsibilities of businesses in four categories: economic, legal, ethical, and philanthropic. Economic responsibility is to make profit as a result of producing and selling goods and services. The business is economically responsible to the owner or shareholders. Legal responsibility, on the other hand, emphasizes that the business must comply with legal regulations while carrying out its activities. Ethical responsibility, on the other hand, states that it is necessary to act in accordance with the norms and moral rules that are not included in the legal regulations but generally accepted by the society. Finally, philantropic responsibility encompasses responsibilities that arise with intrinsic motives often associated with philanthropic activities.

Progress in today's communication and information technology make business activities open to the stakeholder. There exist not only good examples of social responsibility practices but also corporate wrongdoings in the media (Romani et al., 2013). In fact, this situation can be attributed to the increasing awareness level of stakeholder groups in a holistic way. For example, today consumers expect social and environmental responsibility practices in businesses. From a consumer point of view, businesses that meet these expectations satisfy consumers (Su et al., 2017), make them loyal (Martínez & Nishiyama, 2017), and even are willing to pay premium for the products of the company (Xu & Gursoy, 2015). Similarly, millennials perceive companies that implement social and environmental practices as more attractive employers (Boğan & Dedeoğlu, 2019a). Therefore, the positive pressure created by different stakeholder groups on businesses leads companies to exhibit good examples of CSR.

CSR Practices in Hotel Companies

The increasing awareness in the hospitality industry that the concept of CSR is not the same as environmental protection practices has pushed researchers and practitioners to research and practices that will cover different dimensions of CSR beyond the environment (Garay & Font, 2012; Gursoy et al., 2019). Today, many hotel businesses carry out social and environmental practices. However, it is difficult to say that these practices are carried out in all companies, covering the entire sector. It is seen that mostly large and international chain enterprises come to the fore in the sector. Because the social and environmental awareness level of these enterprises is quite high

(McGehee et al., 2009; Nyahunzvi, 2013). However, it should be noted that this should not mean that small businesses do not have social and environmental practices. As a matter of fact, Boğan (2015) emphasizes that business managers who show examples of social responsibility mostly do not have enough awareness that these practices can be used as a strategic tool. In other words, businesses do not find it appropriate to openly carry out voluntary practices that increase the quality of life of people. The factor that is effective in this thought is the dominance of cultural elements in business management. Another reason for the inadequacy of CSR applications in small businesses is that these enterprises do not have sufficient financial resources to allocate to such responsible initiatives. Another reason, perhaps the most important barrier, is that the decision makers in the enterprise have not reached sufficient social and environmental awareness (Boğan, 2015; Jenkins, 2006; Merwe & Wöcke, 2007).

Although CSR activities in small companies are not used enough as a strategic tool, findings of researchers indicate that CSR practices contribute to the financial performance (Kang, Lee & Huh, 2010) and strengthen corporate reputation (Boğan, 2020b) and to gain positive returns from different stakeholder groups (Boğan, 2021). This has enabled CSR practices, especially by large enterprises, to be communicated to stakeholders through different channels (Holcomb et al., 2007). Hotels carry out responsibility activities in a way that covers different areas as well as adopting a specific area of responsibility. For example, Intercontinental Hotels, on the one hand, take initiatives to increase the quality of life of the society, on the other hand, they carry out environmentally friendly practices such as reducing carbon emissions and energy use, creating minimal waste and protecting water resources (IHG, 2021). In addition, the budget that businesses spend on CSR practices may differ from company to company. The factors that are effective here are the profitability of the company, the importance that the owner and shareholders give to CSR, and the positive feedback received from the CSR practices.

Stakeholders' Responses to CSR Practices in Hotel Companies

As mentioned before, the increasing interest in CSR practices in the tourism and hospitality industry is due to the positive reactions of the stakeholder groups to the businesses that carry out these practices. CSR strengthen the corporate reputation of the business among its stakeholders and add strategic value to the business by conducive to obtaining positive reactions from these stakeholders. Among these stakeholder groups, current study focused three stakeholder groups as customers, employees, and the local community.

Customer Responses to CSR

In tourism and hospitality industy, consumers expect environmental, social, and economical responsible initiatives from companies (Srivastava & Singh, 2020). Companies that meet the expectations correctly can gain positive reactions from consumers. Ahn et al. (2020) indicated that social and environmental responsible strategies of hotel companies have positive impact on all components (cognitive, affective, and conative) of customers' brand loyalty. Su, Pan, and Chen (2017) revealed that customer CSR perception has positive impact on corporate reputation and customer satisfaction. Martínez and Nishiyama (2017) indicated that CSR has positive impact on hotel brand equity.

Social identity theory (Ashforth & Mael, 1989) provides an important perspective on why these positive reactions occur. According to this theory, consumers are influenced by the activities of the company from which they purchase products and services. Ethical and social responsible behaviors strengthen company reputation. This reputation reflects positively on the social identity of the consumer group of the business (Bhattacharya & Sen, 2003). To put it more clearly, the consumer gets a share of his social identity from the corporate identity with a strong social responsibility reputation. In return, consumers exhibits a positive attitude and behavior toward the business in question. Indeed, Su et al. (2017) empirically supported that hotel guests' CSR perceptions have positive impact on customer-company identification (CCI) that partially mediates the link between perceived CSR and green consumer behavior. In another study, Srivastava and Singh (2020) indicated that customers CSR perception has positive influence on CCI that partially mediates the relationship between perceived CSR and customer retention.

However, it would not be right to expect CSR practices to have a similar effect for the entire consumer audience. This includes when the responsibility activity is performed (Shin et al., 2021), the type of CSR implementation performed (Kim & Austin, 2020), the perceived CSR commitment of the company (Rim et al., 2020), the demographics of consumers (Hur, Kim & Jang, 2016) and the level of social and environmental awareness of consumers (Jang et al., 2015). For example, Shin et al. (2021) indicated that hotel strategic philanthropic activities during Covid-19 pandemic negatively effect hotel performance and prospective hotel customers' booking behavior. Allowing free rooms to health care staff fighting the pandemic is the examined example of strategic philanthropy. This finding is explained by the potential safety risk perception of consumers due to the allocation of rooms to healthcare personnel struggling with the pandemic.

Employee Responses to CSR

In the hospitality literature, a similar momentum has been gained in the hospitality literature in parallel with the social and environmental responsible practices of relations with consumer reactions in the first periods, and in the following periods, in parallel with the association of researchers in the management literature with the attitudes and behaviors of employees. The study of Lee et al. (2012) should be accepted as one of the pioneering studies in the empirical context. Researchers examined the direct effect of employee CSR perception of franchised foodservice companies on organizational trust and job satisfaction in South Korea. Considered as a multidimensional construct, perceived CSR consists of economic, legal, ethical, and philanthropic dimensions. While the findings revealed the positive effect of economic and philanthropic CSR on organizational trust, it was determined that the CSR dimension that was effective on job satisfaction was only ethical.

In the following periods, the research models became more complicated due to the development of measurements that measure the concept of perceived CSR (Ko et al., 2019; Wong & Kim, 2020), and the increase of researchers interested in the subject in different sub-hospitality sectors (AlSuwaidi et al., 2021; Boğan & Dedeoğlu, 2019b; Ko, Moon & Hur, 2018; Supanti & Butcher, 2019). In fact, the development of different scales at the point of perceived CSR led to new findings that did not overlap with the empirical findings obtained in the past. More specifically, for example, in the direct effect of employees' perception of CSR on organizational citizenship behavior, Kim et al. (2017) reported a positive effect, while Boğan and Dedeoğlu (2020) determined that there was no effect. One of the possible reasons for this situation is that the perceived CSR scales used are different. Because Kim et al. (2017) examined the construct of perceived CSR as multidimensional, while Boğan and Dedeoğlu (2020) examined it as unidimensional. This situation led researchers to discover mediating variables in established direct relationships (AlSuwaidi et al., 2021; Boğan & Dedeoğlu, 2020; Hur et al., 2021). For example, Hur et al. (2021) indicated that job crafting mediates the relationship between CSR and job performance.

However, when evaluated in general, it has been reported that the CSR perception of hotel employees is reflected in some positive employee attitudes and behaviors. To mention a little in this context, it is indicated that employees' CSR perception has positive impact on employee well-being, employee green behavior (AlSuwaidi et al., 2021), employee engagement, pride in organization, voluntary pro-environmental behavior (Raza et al., 2021), employee engagement, compassion and meaningfulness (Nazir & Islam, 2020a), perceived external prestige, pride in organization, OCB (Boğan & Dedeoğlu, 2020), compassion at work, intrinsic motivation and creativity at work (Hur et al., 2018). These findings are instructive in terms of revealing what kind of results the social responsibility activities of the hotels will achieve for the employees.

Local People Responses to CSR

In addition to consumers and employees, for only the last few years, researchers have focused on how social responsibility practices are perceived by local community and what results they have (e.g., Boğan et al., 2020; Gursoy et al., 2019; Pereira & dos Anjos, 2021; Su et al., 2020). In fact, this research topic can be seen as a relatively late research topic considering the importance of local people, one of the important stakeholders to be considered in sustainable destination development. To put it more clearly, the attitudes of the local people toward tourism and their support for tourism are accepted as an important indicator of the success of the destination as a whole. Therefore, it is an extremely important issue how the social responsibility practices are perceived and interpreted by the local people.

In this context, the study of Bohdanowicz and Zientara (2009) should be accepted as a pioneering study. In this study, it has been stated that hotels can contribute to the quality of life of the local people and their employees with their socially and environmentally responsible initiatives. This view is supported by the social responsibility initiative of Scandic, one of the important hotel operators of Scandinavia, known as *Omtanke* in Swedish. However, as mentioned above, it is seen that this topic has attracted the attention of new researchers empirically. Su et al. (2017) indicated that destination social responsibility has a positive effect on community satisfaction, resident trust, resident identification, and destination development economic performance. Gursoy et al. (2019) determined that the social responsibility perception of the local people regarding hotel businesses in Alanya, which is a mature tourism destination of Turkey, has a positive impact on community satisfaction and community commitment and support for additional tourism development.

Empirical studies indicated that social responsibility practices at different levels have positive impact on community satisfaction (Gursoy et al., 2019; Pereira & dos Anjos, 2021; Su et al., 2018; Su et al., 2020), perceived benefits (Lee et al., 2018; Pereira & dos Anjos, 2021), community commitment (Gursoy et al., 2019; Pereira & dos Anjos, 2021), attitude toward tourism (Boğan et al., 2020), community identification (Su, Swanson & He, 2020), support for tourism development (Gursoy et al., 2019; Su, Huang & Huang, 2018; Su, Swanson & He, 2020), tourism impacts (Su et al., 2018; Su, Huang & Pearce, 2018), environmental responsible behavior (Su et al., 2018), and perceived quality of life (Lee et al., 2018; Su et al., 2018).

Attribution Theory

As a human being, people try to understand, think, and question what is going on around them. As a matter of fact, at this point, they search for the causes of the events or phenomena that they come across. In this process, they use some clues in their mind in order to find the right answer in their own way, and form a general opinion depending on the consistency of these clues. This is the essence of attribution theory. That is, people seek a reason to explain the events they encounter or to put them in their minds. The general opinion of the individual is reflected in his attitudes and behaviors toward this event, his perpetrator or his environment (Hewett et al., 2018).

The most important claim of the attribution theory, the foundations of which were laid by Heider (1958), is that the causal relationship perceived by the individual is reflected in his subsequent behaviors and actions. One of the important principles of attribution theory developed by Heider (1958) is the distinction between actions due to personal causes and actions due to environmental causes. In other words, the attributions made by people depend on the factor that caused the behavior or action. This factor can be intrinsically related to the individual's motivation or ability, extrinsic factors, or both intrinsic and extrinsic factors (Hewett et al., 2018).

Kelley (1967), who developed Heider (1958)'s attribution theory, brought an important vision to the attribution theory (Hewett et al., 2018). The basis of Kelley's (1967, 1973) attribution theory is the principle of covariance. Kelley (1967) outlined three types of covariance information that influence an observer's ability to attribute a person's behavior to internal or external causes. The first is distinctiveness. It expresses how much the person exhibits the same behavior in similar situations. If a teacher is constantly irritable both at home and at school (low discrimination), the perceiver will make an internal attribution to his or her irritable behavior. In other words, this teacher is already an irritable person. Observations of different individuals about the person reveal a second information of covariance, which is called consensus. Using the same example, if other teachers at school also agree (high consensus) that the person is irritable, they will make an internal attribution about that person. So that's his character. Third is consistency. It shows the extent to which an individual behaves consistently over time. If the viewers thought that the teacher was irritable in the past, they will make an internal charge, thinking that this is his character. There are different combinations of information that will serve as a reference when making attributions to an individual's behavior (Hewett et al., 2018).

The Role of Attributions about the Motives Behind CSR Practices

Perhaps the most important issue that was overlooked in the relationship between perceived CSR and stakeholder responses in the early periods is the attributions of the stakeholders regarding the basic motivation of these practices (Cha et al., 2016). In other words, stakeholders comments regarding these practices were ignored. This situation attracted the attention of researchers because different results were obtained as a result of testing the same linear relationships in different studies. In other words, the researchers thought that one of the main reasons for these different results could be the attributions of the relevant stakeholders regarding the motivation for these practices. In fact, the opinions we have about why an action we don't know about are often shaped by our attitudes and behaviors toward the action or the party who committed the action (Kelley & Michela, 1980). In fact, according to the attribution theory, people are more concerned with why others do it than with what they do (Gilbert & Malone, 1995). In particular, the inclusion of social and environmental practices, which are considered relatively in the background in terms of their primary responsibilities, in business activities and announcing these initiatives through different channels may cause the stakeholders to be suspicious. The use of these activities as an advertising tool and the expectation of direct or indirect economic returns from these initiatives have made the sincerity of businesses questionable (Zhang et al., 2018). As a reflection of this, the concept of greenwashing and greenhushing, in which intentions are questioned in return for environmental practices in hotel businesses, has recently been researched (Chen et al., 2019; Ettinger et al., 2021; Rahman et al., 2015).

In the literature, the attributions to CSR activities are egoistic, strategic, value-driven, and stakeholder-driven, which are most commonly examined by Ellen, Webb, and Mohr (2006), and intrinsic/extrinsic, which are similar in scope but called by different names (Boğan & Sariisik, 2020; Vlachos et al., 2013) and substantive/symbolic (Donia et al., 2017) attributions. In the study of Ellen et al (2006) attribution framework, egoistic attribution means that the enterprise prioritizes its interests in the CSR activity. In strategic attribution, it is thought that the business focuses on both its own benefit and social benefit in these activities. In stakeholder-driven motives, on the other hand, it is thought that the main factor pushing the business to CSR activities is stakeholder pressure or expectation. Finally, in value-driven motivations, these activities are considered as a factor triggered by the values of the business. In intrinsic and substantive attributions, which are considered to be very close concepts in terms of content, it is essential that the business cares about responsible behaviors in real terms or that responsible behaviors are carried out to solve any problem in a real sense. On the other hand, in symbolic and external attributions, there is a belief that the business uses responsible activities to achieve corporate goals and therefore does not perform these activities with sincere intentions.

Empirical research has revealed that stakeholders' attributions about the motives behind social responsibility practices significantly affect their attitudes and behaviors toward the company. In general, stakeholder groups react positively when these activities are interpreted as intrinsic or substantive; however, when these activities are interpreted as extrinsic or symbolic, it has been determined that some positive attitudes and behaviors are negatively affected by these attributions. It can be clearly said that there are a limited number of studies examining the subject in the context of tourism and hospitality (Boğan & Sariisik, 2020; Su et al., 2020; Su et al., 2020). Boğan and Sariisik (2020) examined the effect of intrinsic and extrinsic attributions about social responsibility practices of hotel employees on affective commitment in Turkey. In addition, the authors examined some organizational factors affecting these attributions. In this context, the variables perceived in-out CSR alignment and behavioral integrity in an organization were expected to affect these attributions. As a result of the analyzes made with the data collected from the employees of the chain hotels operating in the Istanbul region, it was determined that only intrinsic CSR attributions have positive impact on affective commitment. In addition, while perceived in-out CSR alignment and behavioral integrity in the organization have a positive effect on intrinsic attributions, both of these constructs have negative effect on extrinsic attributions.

Within the scope of the factors affecting the stakeholders' attributions of the CSR activities, Shin et al. (2021) carried out one of the pioneering studies. Accordingly, epidemic/disaster periods can be shown as an indicator of the sincerity of company in terms of CSR practices. In other words, continuing these practices during epidemic/disaster periods can be interpreted as these practices are carried out with a substantive motive. It can be accepted as an indicator of sincerity that businesses do not only engage in economic activities during epidemic periods, but also make monetary donations and product donations to those in need. Some hotels even allocate hotel rooms to support healthcare professionals in the fight against the pandemic, which can be considered a clear indication of their sincerity (He & Harris, 2020; Shin et al., 2021).

In another study, Su et al. (2020) examined the effect of tourists' destination social responsibility attributions on destination trust and intention to visit. In addition, the researchers tested the moderate role of destination reputation. The researchers examined the attributions in two dimensions as intrinsic and extrinsic. The analyzes revealed that when tourists make intrinsic attributions related to DSR, destination trust and intention to visit are positively affected by these attributions. In addition, the mediating role of the destination trust in the relationship between DSR attributions and intention to visit is supported. Finally, it is found that destination reputation has a moderate effect in established linear relationships. On the other hand, Su, Gong, and Huang (2020) examined the effect of DSR strategy (proactive and reactive) on attributions (altruism and egoistic) and intention to visit from the perspective of tourists. The findings revealed that tourists have high altruism attribution and destination visit intention when the proactive DSR strategy is applied. In addition, it has been determined that the destination visit intentions of tourists who receive a reactive DSR message from internal sources are higher.

Conclusion

The current research aims to draw a framework for the reactions of stakeholder groups (employees, customers, and society) to CSR practices that will not always be positive based on attribution theory. In recent years, it has mostly encountered that responsible and environmental practices have increased in the tourism and hospitality sector, especially in large-scale chain companies. While some companies attach high importance to these initiatives, some may make these initiatives superficial. However, in the end, these initiatives enter the perception of the stakeholders through different channels and they return to the company directly or indirectly. Therefore, companies should be aware that these initiatives do not always result in positive feedback. Many factors are effective at this critical stage. However, the current study highlights the attributions made by stakeholders based on some clues in their minds as to why these initiatives were made.

Attributions to socially and environmentally responsible initiatives are examined in several different dimensions. Although there is no definite literary agreement about what the attributions are, we think that the intrinsic/extrinsic and substantive/symbolic dimensions, which are quite close to each other, are more specific and instructive. In intrinsic and substantive attributions, the individual has the conviction that the business is sincere in these initiatives and that its real goal is to benefit society. However, in extrinsic and symbolic attributions, the individual has a belief that the enterprise symbolic lays these practices and the company has some direct or indirect expectation from these practices. In fact, extrinsic attributions can have negative consequences for stakeholder groups (e.g., society) who do not directly rely on the economic performance of the business. However, extrinsic attributions have the potential to create positive effects for employees who are directly affected by the economic performance of the business. Because with these initiatives, they will gain benefits as the business wins. In fact, all these cases show that social and environmental practices have a sensitive nature. However, it should be noted that when the business convinces stakeholder groups that it is sincere in these initiatives, it is clear that positive feedback is reinforced. On the other hand, if there is a belief that it is carried out with symbolic or egoistic motives, negative results may occur, aside from the expected positive returns.

References

Ahn, J., Wong, M. L., & Kwon, J. (2020). Different role of hotel CSR activities in the formation of customers' brand loyalty. *International Journal of Quality and Service Sciences, 12*(3), 337–353.

AlSuwaidi, M., Eid, R., & Agag, G. (2021). Understanding the link between CSR and employee green behaviour. *Journal of Hospitality and Tourism Management, 46*, 50–61.

Appiah, J. K. (2019). Community-based corporate social responsibility activities and employee job satisfaction in the US hotel industry: An explanatory study. *Journal of Hospitality and Tourism Management, 38*, 140–148.

Ashforth, B. E., & Mael, F. (1989). Social identity theory and the organization. *Academy of Management Review, 14*(1), 20–39.

Bhattacharya, C. B., & Sen, S. (2003). Consumer–company identification: A framework for understanding consumers' relationships with companies. *Journal of Marketing, 67*(2), 76–88.

Boğan, E. (2015). The impact of employee perception of corporate social responsibility practices on employee trust in organization: An application of four and five-star hotels in Alanya. Master Thesis. Akdeniz University, Antalya/Turkey.

Boğan, E. (2020a). Turizm ve ağırlama sektöründe çalışanların kurumsal sosyal sorumluluk faaliyetlerine yönelik tepkileri üzerine bibliyometrik bir çalışma. *Seyahat ve Otel İşletmeciliği Dergisi, 17*(1), 87–102.

Boğan, E. (2020b). The effect of hotel employees' corporate social responsibility perception on affective commitment and employer attractiveness: the mediating role of corporate reputation. *Alanya Academic Review, 4*(2), 381–398.

Boğan, E. (2021). A review of prominent theories in perceived CSR-employee outcomes link in hospitality literature. *Journal of Multidisciplinary Academic Tourism, 6*(2), 99–105.

Boğan, E., & Dedeoğlu, B. B. (2019a). The influence of corporate social responsibility in hospitality establishments on students' level of commitment and intention to recommend. *Journal of Hospitality, Leisure, Sport & Tourism Education, 25*, 100205.

Boğan, E., & Dedeoğlu, B. B. (2019b). The effects of hotel employees' CSR perceptions on trust in organization: Moderating role of employees' self-experienced CSR perceptions. *Journal of Hospitality and Tourism Insights, 2*(4), 391–408.

Boğan, E., & Dedeoğlu, B. B. (2020). Hotel employees' corporate social responsibility perception and organizational citizenship behavior: Perceived external prestige and pride in organization as serial mediators. *Corporate Social Responsibility and Environmental Management, 27*(5), 2342–2353.

Boğan, E., Dedeoğlu, B. B., & Balıkçıoğlu Dedeoğlu, S. (2020). The effect of residents' perception of hotel social responsibility on overall attitude toward tourism. *Tourism Review*. https://doi.org/10.1108/TR-08-2019-0353

Boğan, E., & Sariisik, M. (2020). Organization-related determinants of employees' CSR motive attributions and affective commitment in hospitality companies. *Journal of Hospitality and Tourism Management, 45*, 58–66.

Bohdanowicz, P., & Zientara, P. (2009). Hotel companies' contribution to improving the quality of life of local communities and the well-being of their employees. *Tourism and Hospitality Research, 9*(2), 147–158.

Carroll, A. B. (1979). A three-dimensional conceptual model of corporate performance. *Academy of Management Review, 4*(4), 497–505.

Carroll, A. B. (1991). The pyramid of corporate social responsibility: Toward the moral management of organizational stakeholders. *Business Horizons, 34*(4), 39–48.

Cha, M. K., Yi, Y., & Bagozzi, R. P. (2016). Effects of customer participation in corporate social responsibility (CSR) programs on the CSR-brand fit and brand loyalty. *Cornell Hospitality Quarterly, 57*(3), 235–249.

Chen, H., Bernard, S., & Rahman, I. (2019). Greenwashing in hotels: A structural model of trust and behavioral intentions. *Journal of Cleaner Production, 206*, 326–335.

Donia, M. B., Tetrault Sirsly, C. A., & Ronen, S. (2017). Employee Attributions of Corporate Social Responsibility as Substantive or Symbolic: Validation of a Measure. *Applied Psychology, 66*(1), 103–142.

Ellen, P. S., Webb, D. J., & Mohr, L. A. (2006). Building corporate associations: Consumer attributions for corporate socially responsible programs. *Journal of the Academy of Marketing Science, 34*(2), 147–157.

Ettinger, A., Grabner-Kräuter, S., Okazaki, S., & Terlutter, R. (2021). The desirability of CSR communication versus greenhushing in the hospitality industry: The customers' perspective. *Journal of Travel Research, 60*(3), 618–638.

Garay, L., & Font, X. (2012). Doing good to do well? Corporate social responsibility reasons, practices and impacts in small and medium accommodation enterprises. *International Journal of Hospitality Management, 31*(2), 329–337.

Gilbert, D. T., & Malone, P. S. (1995). The correspondence bias. *Psychological Bulletin, 117*(1), 21–38.

Gursoy, D., Boğan, E., Dedeoğlu, B. B., & Çalışkan, C. (2019). Residents' perceptions of hotels' corporate social responsibility initiatives and its impact on residents' sentiments to community and support for additional tourism development. *Journal of Hospitality and Tourism Management, 39*, 117–128.

He, H., & Harris, L. (2020). The impact of Covid-19 pandemic on corporate social responsibility and marketing philosophy. *Journal of Business Research, 116*, 176–182.

Heider, F. (1958). *The psychology of interpersonal relations*. New York: Wiley

Hewett, R., Shantz, A., Mundy, J., & Alfes, K. (2018). Attribution theories in human resource management research: A review and research agenda. *The International Journal of Human Resource Management, 29*(1), 87–126.

Holcomb, J. L., Upchurch, R. S., & Okumus, F. (2007). Corporate social responsibility: what are top hotel companies reporting?. *International Journal of Contemporary Hospitality Management, 19*(6), 461–475.

Hur, W. M., Kim, H., & Jang, J. H. (2016). The role of gender differences in the impact of CSR perceptions on corporate marketing outcomes. *Corporate Social Responsibility and Environmental Management, 23*(6), 345–357.

Hur, W. M., Moon, T. W., & Choi, W. H. (2019). The role of job crafting and perceived organizational support in the link between employees' CSR perceptions and job performance: A moderated mediation model. *Current Psychology*, 1–15.

Hur, W. M., Moon, T. W., & Choi, W. H. (2021). The role of job crafting and perceived organizational support in the link between employees' CSR perceptions and job performance: A moderated mediation model. *Current Psychology*, 40(7), 3151–3165.

Hur, W. M., Moon, T. W., & Ko, S. H. (2018). How employees' perceptions of CSR increase employee creativity: Mediating mechanisms of compassion at work and intrinsic motivation. *Journal of Business Ethics, 153*(3), 629–644.

IHG (2021). Introducing our 10-year responsible business-plan. Retrieved from https://www.ihgplc.com/responsible-business

Jang, Y. J., Kim, W. G., & Lee, H. Y. (2015). Coffee shop consumers' emotional attachment and loyalty to green stores: The moderating role of green consciousness. *International Journal of Hospitality Management, 44*, 146–156.

Jenkins, H. (2006). Small business champions for corporate social responsibility. *Journal of business ethics, 67*(-3), 241–256.

Kang, K. H., Lee, S., & Huh, C. (2010). Impacts of positive and negative corporate social responsibility activities on company performance in the hospitality industry. *International Journal of Hospitality Management, 29*(1), 72–82.

Kelley, H. H. (1967). Attribution theory in social psychology. *Nebraska Symposium on Motivation*, 15, 192–238.

Kelley, H. H. (1973). The processes of causal attribution. *American Psychologist*, 28(2), 107–128.

Kelley, H. H., & Michela, J. L. (1980). Attribution theory and research. *Annual Review of Psychology, 31*(1), 457–501.

Kim, H. L., Rhou, Y., Uysal, M., & Kwon, N. (2017). An examination of the links between corporate social responsibility (CSR) and its internal consequences. *International Journal of Hospitality Management, 61*, 26–34.

Kim, S., & Austin, L. (2020). Effects of CSR initiatives on company perceptions among Millennial and Gen Z consumers. *Corporate Communications: An International Journal, 25*(2), 299–317.

Ko, A., Chan, A., & Wong, S. C. (2019). A scale development study of CSR: hotel employees' perceptions. *International Journal of Contemporary Hospitality Management, 31*(4), 1857–1884.

Ko, S. H., Moon, T. W., & Hur, W. M. (2018). Bridging service employees' perceptions of CSR and organizational citizenship behavior: The moderated mediation effects of personal traits. *Current Psychology, 37*(4), 816–831.

Lee, C. K., Kim, J. S., & Kim, J. S. (2018). Impact of a gaming company's CSR on residents' perceived benefits, quality of life, and support. *Tourism Management, 64*, 281–290.

Lee, Y. K., Lee, K. H., & Li, D. X. (2012). The impact of CSR on relationship quality and relationship outcomes: A perspective of service employees. *International Journal of Hospitality Management, 31*(3), 745–756.

Martínez, P., & Nishiyama, N. (2017). Enhancing customer-based brand equity through CSR in the hospitality sector. *International Journal of Hospitality & Tourism Administration, 20*(3), 329–353.

McGehee, N. G., Wattanakamolchai, S., Perdue, R. R., & Calvert, E. O. (2009). Corporate social responsibility within the US lodging industry: An exploratory study. *Journal of Hospitality & Tourism Research, 33*(3), 417–437.

Merwe, M., & Wöcke, A. (2007). An investigation into responsible tourism practices in the South African hotel industry. *South African Journal of Business Management, 38*(2): 1–15.

Nazir, O., & Islam, J. U. (2020a). Effect of CSR activities on meaningfulness, compassion, and employee engagement: A sense-making theoretical approach. *International Journal of Hospitality Management, 90*, 102630.

Nyahunzvi, D. K. (2013). CSR reporting among Zimbabwe's hotel groups: a content analysis. *International Journal of Contemporary Hospitality Management, 25*(4), 595–613.

Pereira, T., & Gadotti dos Anjos, S. J. (2021). Corporate social responsibility as resource for tourism development support. *Tourism Planning & Development*, 1–21. doi.org/10.1080/21568316.2021.1873834

Post, J. E., Lawrence, A. T., & Weber, J. (1999). *Business and society: Corporate strategy, public policy, ethics* (9th ed.), Boston: Irwin McGraw-Hill.

Rahman, I., Park, J., & Chi, C. G. Q. (2015). Consequences of "greenwashing": Consumers' reactions to hotels' green initiatives. *International Journal of Contemporary Hospitality Management*, 27(6), 1054–1081.

Raza, A., Farrukh, M., Iqbal, M. K., Farhan, M., & Wu, Y. (2021). Corporate social responsibility and employees' voluntary pro-environmental behavior: The role of organizational pride and employee engagement. *Corporate Social Responsibility and Environmental Management*, 28(3), 1104–1116.

Rim, H., Park, Y. E., & Song, D. (2020). Watch out when expectancy is violated: An experiment of inconsistent CSR message cueing. *Journal of Marketing Communications, 26*(4), 343–361.

Romani, S., Grappi, S., & Bagozzi, R. P. (2013). My anger is your gain, my contempt your loss: Explaining consumer responses to corporate wrongdoing. *Psychology & Marketing, 30*(12), 1029–1042.

Shin, H., Sharma, A., Nicolau, J. L., & Kang, J. (2021). The impact of hotel CSR for strategic philanthropy on booking behavior and hotel performance during the COVID-19 pandemic. *Tourism Management, 85*, 104322.

Srivastava, S., & Singh, N. (2020). Do Corporate Social Responsibility (CSR) initiatives boost customer retention in the hotel industry? A moderation-mediation approach. *Journal of Hospitality Marketing & Management, 3*(4), 459–485.

Su, L., Gong, Q., & Huang, Y. (2020). How do destination social responsibility strategies affect tourists' intention to visit? An attribution theory perspective. *Journal of Retailing and Consumer Services, 54*, 102023.

Su, L., Huang, S., & Huang, J. (2018). Effects of destination social responsibility and tourism impacts on residents' support for tourism and perceived quality of life. *Journal of Hospitality & Tourism Research, 42*(7), 1039–1057.

Su, L., Huang, S. S., & Pearce, J. (2018). How does destination social responsibility contribute to environmentally responsible behaviour? A destination resident perspective. *Journal of Business Research, 86*, 179–189.

Su, L., Lian, Q., & Huang, Y. (2020). How do tourists' attribution of destination social responsibility motives impact trust and intention to visit? The moderating role of destination reputation. *Tourism Management, 77*, 103970.

Su, L., Pan, Y., & Chen, X. (2017). Corporate social responsibility: Findings from the Chinese hospitality industry. *Journal of Retailing and Consumer Services, 34*, 240–247.

Su, L., Swanson, S. R., & He, X. (2020). A scale to measure residents perceptions of destination social responsibility. *Journal of Sustainable Tourism, 28*(6), 873–897.

Supanti, D., & Butcher, K. (2019). Is corporate social responsibility (CSR) participation the pathway to foster meaningful work and helping behavior for millennials?. *International Journal of Hospitality Management, 77*, 8–18.

Vlachos, P. A., Panagopoulos, N. G., & Rapp, A. A. (2013). Feeling good by doing good: Employee CSR-induced attributions, job satisfaction, and the role of charismatic leadership. *Journal of Business Ethics, 118*(3), 577–588.

Wong, A. K. F., & Kim, S. S. (2020). Development and validation of standard hotel corporate social responsibility (CSR) scale from the employee perspective. *International Journal of Hospitality Management, 87*, 102507.

Xu, X., & Gursoy, D. (2015). Influence of sustainable hospitality supply chain management on customers' attitudes and behaviors. *International Journal of Hospitality Management, 49*, 105–116.

Zhang, L., Yang, W., & Zheng, X. (2018). Corporate social responsibility: the effect of need-for-status and fluency on consumers' attitudes. *International Journal of Contemporary Hospitality Management, 30*(3), 1492–1507.

11

ATTITUDES IN TOURISM AND TRAVELING AS A TOOL/INSTRUMENT FOR ATTITUDE CHANGE

Nisan Yozukmaz and Burhan Kiliç

Introduction

Attitude is a central concept that has existed for a long time (Allport, 1935) and remains important in the fields of social psychology (Bohner & Dickel, 2010) and consumer behavior literature as it is accepted that attitude predicts behavior even though the situation and behavior consistency differ in various situations (Tussyadiah et al., 2018). In recent years, there has been an increase in studies on automatic or implicit features of attitudinal processes. Basic research in social psychology on these topics has encouraged applied research (Bohner & Dickel, 2010) in, for example, employee psychology (e.g., Johnson et al., 2010) and consumer psychology (e.g., Gibson, 2008).

There are many studies on tourist attitudes towards specific tourism types, host cultures, environments, or other tourists. Also in recent years, residents' and locals' attitudes have begun to be a focus of tourism research. In terms of tourism, the impacts of traveling on tourists' attitudes and perceptions are one of the most important subjects that has been extensively studied since the 1980s. However, it has been generally approached with a marketing perspective. From the social psychological perspective, the issue of attitude change in tourism literature has been examined with a focus on the question of whether tourism causes attitude change. In other words, previous studies have dealt with the issue from the point of whether travel could be a mediator of attitude change in tourists. There are prominent studies in which contact theory is used within the literature. For example, whether Israeli tourists' attitudes have changed after their trip to Egypt (Amir & Ben-Ari, 1985), whether Greek tourists' attitudes towards Turkish people have changed (Anastasopoulos, 1992), whether prejudice, ethnocentrism, and stereotyping have turned into tolerance, compassion, goodwill, justice, and respect by traveling (Kelly, 2003), whether tourism could be a tool of maintaining peace between ethnic groups (Etter, 2007), whether the attitudes of domestic tourists in Turkey towards South Easterners have changed (Çelik, 2019), whether intercultural interaction has caused post-travel attitude change (Fisher & Price, 1991; Nyaupane et al., 2008) have been examined in tourism literature. Apart from such ethnic or political perspectives, attitude change has also been investigated on tourism students' attitudes towards guests with disabilities (Bizjak et al., 2011) and has started to be analyzed about technological developments (Tussyadiah et al., 2018).

In line with the previous research, this chapter focuses on the subject of attitude change in tourism literature and discusses whether traveling may be an instrument in the process of attitude change. For this purpose, first the concept of attitude is examined in this section. And then how

DOI: 10.4324/9781003161868-11

attitudes change through tourism is explained in detail with examples from previous studies and the chapter provides a conclusion with some suggestions for future research.

The Concept of Attitude

An attitude is the evaluation of an object of thought. Objects of attitude can consist of anything that an individual holds in mind, including people, groups, things, and ideas, in other words, from material things to abstract notions. Most attitude researchers agree on these basic definitions, but more detailed models of the concept of attitude are quite diverse. The definitions vary considerably from the fact that attitudes are fixed entities stored in memory to the fact that they are temporary judgments produced from the information available at that moment (Bohner & Dickel, 2010; Gawronski, 2007).

In general, attitudes are considered to be a part of the process of socialization (Daruwalla & Darcy, 2005) or a kind of social knowledge consisting of feelings, experiences, and beliefs that emerge with objects of attitude (Bizjak et al., 2011). According to researchers like Zanna and Rempel (2008), Fazio and Petty (2008) and Fishbein (2008), attitudes are evaluations of people, objects, and ideas which means they have an attitude object and carry and evaluation for it (Arkonaç, 2015). They are also defined as evaluations involving positive or negative responses to anything (Aronson et al., 1998; Bizjak et al., 2020; Fishbein & Ajzen, 2003). Fazio et al. (2003) suggest that attitudes are functional structures that carry many things out for people. Attitudes constitute the social universe of the individual and while doing this, they facilitate decision-making process and allows the individual to move between people and objects they encounter on a daily basis. Smith (1964) argues that an attitude towards an object saves the individual from wasting energy or generally from the tough process of understanding how to relate themselves to that object, thus enabling the individual to move forward more easily in everyday life (cf. Bizjak et al., 2011).

Attitude Formation

According to Baysal (1981), the factors in attitude formation can be listed as genetic factors; physiological conditions (maturation, illness, drug abuse, etc.); direct experience with the subject of attitude; personality, socialization process (social adaptation); group membership; and social class.

According to a more common definition, attitudes involve a response to an object with either two (Subramaniam & Silverman, 2007) or three components (İnceoğlu, 1993): cognitive, affective, and behavioral. The first two are basic components in providing explanations of how attitudes are constituted (Subramaniam & Silverman, 2007). If an individual's knowledge about a subject requires them to view it positively (cognitive element), the individual is positive about that subject (emotional element). They show this through words or actions (behavioral element) (İnceoğlu, 1993). The last component (behavior) is also known as the three-component attitude model (Bizjak et al., 2010).

Cognitively based Attitudes

Attitudes are constituted over an evaluation or consideration of people's beliefs about the attitude object (Aronson et al., 1998; Fazio & Petty, 2008; Subramaniam & Silverman, 2007). The function of cognitively based attitudes is to predict the value of an object (Aronson et al., 1998), that is, people sometimes construct attitudes based on comparative logic, taking into account the value they identify with the characteristics that define an object (Fazio & Petty, 2008).

Cognitive component is the rational element of attitudes consisting of ideas, knowledge and beliefs and is related to understanding, evaluation, planning, decision making, and thinking. The primary function of cognitive systems is to understand and interpret important aspects of personal experiences (Koç, 2019).

Affectively based Attitudes

Affectively based attitudes are based on emotions. The emotion evoked by an object is a strong determinant of individual and societal attitudes (Fazio & Petty, 2008) as attitudes measure the degree of feelings or emotional attraction towards the object of attitude (Subramaniam & Silverman, 2007). But where do these attitudes come from? Aronson et al. (1998) argue that attitudes originate from different sources such as people's values, sensual responses, or aesthetic responses. Moreover, as Zajonc (2008) points out, emotional responses are often fundamental, and the initial level of reaction to the environment is emotional. Others result from operant or classical conditioning.

> In the classical conditioning, the initially neutral stimulus is repeatedly paired with another stimulus that elicits an emotional response until the initially neutral stimulus alone elicits the response. In the case of operant conditioning, learning occurs when the responses increase or decrease in frequency because they have positive or negative consequences.
>
> *(Bizjak et al., 2011: 845)*

Attitude and then behavioral change can be targeted by dealing with emotional themes such as romance. In addition, the use of cute puppies, babies, and handsome or beautiful models in advertisements can be given as an example of attitude change efforts by addressing emotional attributes. In research on consumer behavior, it has been observed that emotional marketing messages are more effective on consumers who have low awareness of their needs (i.e., those who do not know their own needs very well) compared to cognitively based marketing messages (Koç, 2019).

Behaviorally based Attitudes

These attitudes are based on observations that show how individuals behave towards an object of attitude (Aronson et al., 1998). Unless the behavior is based on an external force, the attitude of an individual directly derives from the behavior (Fazio & Petty, 2008), of course only under specific circumstances: if their first attitudes are weak or changeable and unless secondary attitudes have plausible explanation of the behavior (Aronson et al., 1998; Bizjak et al., 2010).

Attitude Change

The conceptualizations of attitude vary. Attitudes are thought to be stored in memory or formed at the moment. At the same time, attitudes can be measured using explicit individual report tools or implicit responsive time-based measures. Different views on attitude definitions and measurements are related to the theoretical understanding of attitude change (Bohner & Dickel, 2010). The attitude measured in a certain time may change or be changed after a while with various effects. In other words, the phenomenon of attitude is not static, but dynamic (Baysal, 1981). According to Arkonaç (2015), advertisements, conversations in television programs, health pages in newspapers, healthy life programs on television, intellectuals, and many others call people every day for everything from what kind of lifestyle people should have in the name of modern life to what people think and decide how they will be considered a modern person. The purpose of these publications and non-governmental organizations is to influence their behavior, or change or strengthen their existing attitudes. And for this, there are two processes in explaining attitude change: (a) cognitive consistency and (b) cognitive dissonance approaches.

According to Bizjak et al. (2010: 845), "when attitudes change, they do so in response to social influence". Every attitude can be affected by what other individuals think, whether it is towards a political candidate or towards a car brand (Aronson et al., 1998). Many events aiming to change individual attitudes in the society happen all the time. When any social action is desired to be influenced, the starting point is the attitudes of the individuals who make up that society. Educators, revolutionaries, politicians, leaders of minority groups, businessmen, managers, and trade unionists are closely concerned with how to change existing attitudes on the subject to innovate, and how to create new attitudes if there is no previous attitude on it (Baysal, 1981). Daruwalla and Darcy (2005) claim that attitude change will help individuals act more effectively in their daily lives.

According to Cooper and Croyle (1984), attitude change can be analyzed with three contexts: (1) the person, because attitudes change depending on goals, values, emotions, language, and human development, (2) social relationships linking attitude change to influential messages, social media, and culture, (3) sociohistorical processes highlighting the impact of events including sociopolitical, economic, and climatic incidents. According to Koç (2019) there are three strategies for attitude change: (1) adding a new salient belief, (2) increasing the power of an existing positive belief, and (3) improving the evaluation of a strong belief.

Many theories about how attitudes are formed and how they change have been put forward. Four different theoretical approaches have been used in various studies on attitudes, especially on the problem of attitude change. These are cognitive theories including "learning", "social judgment", "consistency" or "balance", and "functional" theories (Kağıtçıbaşı, 1999):

Learning theories: These theories have been applied to attitude change in a general way. Accordingly, by considering attitude change as a learning process, effective communication studies as well as classical conditioning and attitude development experiments were carried out. As in classical conditioning, objects associated with pleasant experiences are evaluated positively, while those associated with unpleasant experiences are evaluated negatively (Tavşancıl, 2002).

Social judgment theories: To like or dislike something, to like it or not, requires having a judgment about that thing. According to these theories, a strongly attached attitude is more likely to reject views different from itself than to accept it. Conversely, attitudes that are not too strongly attached are more likely to accept different views than to reject them. Here, too, by using the analogy mechanism, the probability of seeing and accepting those views more similar to theirs than they actually are increases. Although these theories provide supporting findings, it is a basic framework for understanding attitude change rather than making specific measurable predictions about attitude change (Kağıtçıbaşı, 1999).

Consistency theories: Consistency theories focus on the human effort to provide consistency between different attitudes. Balance theory, developed by Heider (1958) is a theory whose main idea is that an unbalanced system can change towards balance. According to this theory, unbalance causes pressure on the person to change towards a stable state. Because attitude structures that are not in balance are uncomfortable and unpleasant. The basic principle of balance theory is that people tend to maintain balance in their attitude structure. However, the balance theory does not foresee that the imbalance situation will always be tried to be eliminated, it suggests that there is such a tendency (Kağıtçıbaşı, 1999).

Functional theories: Katz (1960) suggested that attitudes have four personality functions: (a) utilitarian function, (b) knowledge function, (c) ego-defensive function, and (d) value-expressive function. For an attitude change to occur, there must be a mismatch between the needs met by the attitude and the attitude. Attitude change can be achieved by understanding the function of attitude for the individual and creating strategies to produce a mismatch between attitude and one or more of the attitude functions.

Attitudes and Attitude Change in Tourism

The link between attitudes towards a tourism destination and behavioral intention to visit that destination and participate in tourist activities is a well-researched topic in the literature. For Example, Huang and Hsu (2009) who studied the intention of Beijing tourists to revisit Hong Kong, revealed the great impact of attitude on intentions. Phillips, Asperin and Wolfe (2013) revealed the effect of the attitude towards Korean cuisine consumption on the intention to visit Korea and try Korean cuisine.

The best examples in tourism according to Bizjak et al. (2010), appear in the attitudes of tourism employees towards different tourist groups. Tourists from Russia in one season may not be welcome, but within a week they can become the most eminent guests in a tourist destination. Bizjak et al. (2011) examined whether tourism students' attitudes and perceptions towards people with disabilities who were also tourists might have changed. It was observed in the study that education about people with disabilities could cause attitude changes in students.

Daruwalla and Darcy (2005) carried out studies supporting that the attitudes of employees in the tourism industry and tourism students towards people with disabilities could change with an intervention program. Daruwalla and Darcy (2005) claim that this change would be more effective and last longer if subjects spent time with people with disabilities in a controlled environment.

Tussyadiah et al. (2018) examined attitude change through virtual reality in tourism. Two studies were conducted. One study was conducted in Hong Kong and the other in the UK. In the study, it was observed that the sense of presence in VR experiences had positive results. First, the feeling of being in the virtual space enhanced the pleasure from VR experiences. The feeling of being there increased the likelihood of choosing and liking the destination. Positive attitude change increased the level of visit intention.

Traveling as a Tool of Attitude Change

The common idea that emerges from existing studies in tourism literature is that attitude changes are related to some factors such as social distance, prior expectations, and travel experience (Nyaupane et al., 2008). These concepts are discussed under different perspectives as (1) contact theory, (2) social distance theory, and (3) expectancy-value theory for better understanding the mediating role of traveling in attitude change.

Contact Theory

Contact theory suggests that attitudes can change when people come together, and behaviors can also change when attitudes change (Emerson et al., 2002). This theory has generally been applied in studies on tourism and peace. In terms of tourism, there is an idea that the attitudes and behaviors of individuals and groups can change through intercultural communication and interactions which is called in literature as contact theory (Allport, 1954). D'Amore (1988) claims that tourism may act as a bridge between psychological and cultural gaps between people and that tourism may contribute to the recognition of diversity in the world. The essential idea here is that many everyday interactions that occur between tourists and local people can lead to the awareness and understanding required to foster global relationships between people, communities, and even nations (D'Amore, 1988). Studies show that the nature of the experiences resulting from interactions between tourists and host communities has a powerful impact on change in tourists' attitudes (Nyaupane et al., 2008).

To explain the relationship between tourism and peace, it is required to look at people's intercultural interactions, particularly in tourist-host roles. From theoretical perspective, tourism can offer opportunities for social and cultural awareness, understanding, and acceptance, as it brings people from different backgrounds together. Thus, prejudices, conflicts, and tensions between individuals can be reduced, which is reflected in the interaction between hosts and guests (Nyaupane et al., 2008). However, previous research contains different results.

Carlson and Widaman (1988) demonstrated that the level of international tolerance of the participants increased and a more positive attitude could be formed after the trip. However, Krippendorf (1982) stated that tourism could have the opposite impact. Pizam et al. (1991) could not prove that the attitude of tourists changed after they visited the host destination. Fisher and Price (1991) developed a model of the relationship between international tourism and post-vacation attitudes of visitors towards the culture of the host community and the destination. It was observed that the results did not support the direct relationship between travel motivations and post-vacation attitude change, only the educational motivation was significantly related.

In a study on understanding the mechanics of peace through tourism, Etter (2017) evaluated the relevance of the conditions defined in contact theory in explaining the attitudes of tourists towards cultural groups in a destination. The following situational conditions for a positive attitude change were used in the study: (1) perception of equal status among individuals, (2) existence of common goals, (3) sanction of authorities (laws, customs, or local atmosphere), (4) voluntary and intimate interaction, (5) lack of negative personality traits. As a result, it turned out that all these factors were significant for positive attitude change, but common goals, voluntary and intimate interaction, and lack of negative personality traits were more important.

Amir and Ben-Ari (1985) evaluated a cognitive intervention to improve relations between Israelis and Egyptians. A specially designed booklet introducing Egypt to Israeli tourists was used as an intervention tool. Findings show that intercultural interaction due to tourism movements does not guarantee positive attitude change. Researchers argued that the interaction should be supported by individual and situational factors for intergroup relations to improve. According to the results, tourism could have a specific impact on attitude change because tourists interact voluntarily with the host population and therefore have the motivation to gain new knowledge. The effects of tourism, in this context, can be seen as an ice-breaking stage. Anastasopoulos (1992) examined the change in Greek tourists' attitudes towards the Turkish people before and after their visit to Turkey. As a result, visiting Turkey had a negative influence on the perceptions of Greek tourists towards the host community.

Many studies have examined the changes in the attitudes of students who have studied abroad for a specific period. According to Nyaupane et al. (2008), students in study abroad programs experience a higher level of interaction with host communities than "institutionalized mass tourists" (Cohen, 1972). Pizam et al. (1991) examined the changes in the attitudes of American students who had visited the former Soviet Union with a pre/post-trip survey. The results indicated that when students decided to visit the destination, their attitude towards host communities improved. Carlson and Widaman (1988) investigated in one of their studies the effect of one-year study abroad in a university in Europe on attitudes towards different cultures. The study supported the claim that the international understanding of students increased in terms of international political and intercultural interest and cultural sophistication. It was observed that the group in the study abroad program had more positive but some critical attitudes towards the United States. There were many crucial differences that showed the aspects influencing students' attitudes. For example, the students who had prior experience such as living or traveling in Europe before their first year in a university showed greater political and intercultural interest, and traveling for studying abroad acted as an "equalizer". Other factors such as gender and university major also had an impact on the results. For instance, humanities majors and female students had higher intercultural interest even before and after studying in a different country. The general results of the study show that studying in a different country contributes to the increase in international consciousness and endorses attitudes and behaviors required for a higher level of international understanding. Litvin (2003) studied the impact of students' travels on perceptions and attitudes towards host communities. Participants were 50 students who traveled from Singapore to Israel and Egypt on a 12-day trip. According to the results of pre-trip and post-trip surveys, the answers of 32 of the 62 attitude statements differed significantly between the pre and post-tests. Attitudes towards Egyptians and Egypt changed in negative direction, but attitudes towards Israelis and Israel changed in positive direction. As a result of the study, it was deduced that the direction of attitude change altered with regard to the country visited and the origin.

Does Tourism Reduce Social Distance?

Some studies have examined the change in attitudes that occur through interactions between tourists and host communities. As pointed by Riordan (1978), equal status between locals and tourists is one of the important criteria for a positive result in the communication process between two different cultures. This criterion can be better explained by social distance theory in intercultural studies and sociology. Social distance is defined as "cultural differences between two groups" (Poole, 1926), and since it was brought to light by sociologists such as George Simmel and Robert Park in the late nineteenth century, this theory has been applied in studies of ethnicity, gender, class, religion, peace, conflict, and other types of social relations. Although social distance is a mental state function (Giddings, 1895), people maintain social distance through spatial distinctions such as choosing a place of residence, work, and leisure (Ethington, 1997; Shibutani, 1955). Social distance ranges from differences between sisters or brothers to differences between different ethnicities and races. The social distance between cultures has been measured in some studies in terms of nationality, as individuals living in the same country often share a fixed and dominant cultural character (Nyaupane et al., 2008; Reisinger & Turner, 2002; Thyne et al., 2006).

According to social distance theory, local people in host destinations are more tolerant and more accepting towards people who are more culturally and socially identical with them (Thyne et al., 2006). The results of a study conducted by Thyne et al. (2006) which was on the significance of the nationality of tourists in terms of residents' acceptance and attitude towards tourists support social distance theory. In the study, it was found that while Australians are probably the nation the most similar to New Zealanders socially, they have not been the most preferred nation. This may be due to different reasons, such as the influence of American culture, economic gain from American tourists, and the competition between New Zealand and Australia, which is evident in sports such as rugby (Thyne et al., 2006).

Martin et al. (1995) suggested that social distance helped build expectation. In their study of Americans visiting the UK, the researchers observed that Americans were not at all satisfied with their travel to the UK, since they had similar expectations to those in their own country, because they had common cultural norms and language. The study showed that social distance was not the only thing that had an impact on post-trip attitudes. The expectations of how these experiences actually occurred also had impact.

Expectancy-Value Theory

For visitors, pre-trip expectations play a significant part in determining the direction of attitude change. Marion (1980) argues that students with prejudiced or established ideas about the host destination may be less positive after their visit because their unrealistic expectations are too high to be met. Weissman and Furnham (1987) reached similar results in their study of the experiences and expectations of visitors. This may be explained by expectancy- value theory. This theory claims that all unmet expectations will result negatively (Feather, 1982). In another study by Rogers and Ward (1993) on the differences in expectation-experience, the psychological adaptations of second-year students returning to New Zealand during intercultural re-entry were examined.

The study findings showed that expectations had a predictive power when considered in terms of difference from actual experience.

In the study conducted by Nyaupane et al. (2008), two dominant psychological and social theories were used to examine attitude change and tourism. First, pre-trip and post-trip attitudes of American students who visited Australia, Fiji, Austria, and the Netherlands for summer school were compared. The primary purpose was to examine the role and relevance of social distance theory and expectancy theory. Based on social distance theory, both pre-trip and post-trip attitudes were expected to be higher for Australia which would be followed by Austria and the Netherlands, then Fiji. Expectancy theory was tested by comparing the differences between pre-trip expectations and attitudes and post-trip expectations and attitudes. According to the expectancy theory, students with higher expectations about the destination were expected to have higher level of pre-trip attitude scores. This would lead to negative or fewer changes in attitudes. In addition, the study examined the role of experiences related to tourism or not in the formation of post-trip attitudes.

Conclusion

An attitude is a tendency that is attributed to an individual and that affects their, feelings, thoughts, and behaviors regarding a psychological object. What an attitude generates is not just a tendency to behavior or just an emotion; it is an integration of behavioral tendency. Researchers generally agree that attitudes consist of three main components: (1) the cognitive component is based on one's beliefs, opinions and views about an attitude object, (2) the emotional element is based on one's emotions and emotional reactions to an attitude object, and (3) the behavioral element is based on the past behavior and behavioral intentions of the person towards an attitude object. Attitudes change but it is not something that happens easily and quickly.

There are many theories on how attitudes change dealing with the issue from different perspectives such as learning, social judgment, balance, and functionality (Kağıtçıbaşı, 1999). However, with a tourism research perspective, the concept of attitude change has mostly dealt with theories of social contact, social distance, and expectancy-value. The main discussion is the mediating role of traveling in attitude change in tourists towards host communities or vice versa. Some studies suggest that traveling to different places can create a positive change in attitudes through interactions with people. Some studies could not produce evidence for this idea. However, it can be claimed after a detailed literature review that tourists can create *communitas* (Turner, 1973) with other tourists or local communities and this feeling of *communitas* may contribute to a positive attitude change because this feeling is above all notions such as race, ethnicity, nationality, and other societal norms.

The peace in the world is also a subject of tourism research in terms of positive attitude changes in tourists or locals towards each other. The aim is to find evidence for the idea that traveling to a different place and interacting with other cultures and other people there can change the views, attitudes, and opinions of people positively and there may be a common point beyond all political or racial issues. This is actually a noble viewpoint and therefore the number of studies on this topic should increase.

As another suggestion, especially in the field of consumer behavior, the impact of the Covid-19 pandemic on attitude change can be a focus of future research. Empirical studies on this subject need to be carried out in the future. This issue should also be evaluated within the scope of tourists' and locals' psychology.

References

Allport, G. W. (1935). Attitudes. In C. Murchinson (Ed.), *Handbook of social psychology* (pp. 798–844). Worcester, MA: Clark University Press.

Allport, G. W. (1954). *The nature of prejudice.* Cambridge: Addison Wesley.

Amir, Y. & Ben-Ari, R. (1985). International tourism, ethnic contact, and attitude change. *Journal of Social Issues, 41*(3), 105–115. https://doi.org/10.1111/j.1540-4560.1985.tb01131.x

Anastasopoulos, P. G. (1992). Tourism and attitude change: Greek tourists visiting Turkey. *Annals of Tourism Research, 19*(4), 629–642. https://doi.org/10.1016/0160-7383(92)90058-

Arkonaç, S. A. (2015). *Sosyal Psikolojiye Giriş I.* İstanbul Üniversitesi Açık ve Uzaktan Eğitim Fakültesi Ders Notları. http://auzefkitap.istanbul.edu.tr/kitap/sosyalhizmetler_ao/sosyalpsiko.pdf

Aronson, E., Wilson, T. D. & Akert, R. M. (1998). *Social psychology* (3rd. ed.). New York: Longman.

Baysal, A. C. (1981). Sosyal psikolojide tutumlar. *İstanbul Üniversitesi İşletme Fakültesi Dergisi, 10*(1), 121–138.

Bizjak, B., Knežević, M. & Cvetrežnik, S. (2011). Attitude change towards guests with disabilities: Reflections from tourism students. *Annals of Tourism Research, 38*(3), 842–857. https://doi.org/10.1016/j.annals.2010.11.017

Bohner, G. & Dickel, N. (2010). Attitudes and attitude change. *Annual Review of Psychology, 62,* 391–417. https://doi.org/10.1146/annurev.psych.121208.131609

Carlson, J. & Widaman, K. (1988). The effects of study abroad during college on attitudes toward other cultures. *International Journal of Intercultural Relations, 12,* 1–17. doi:10.1515/jsarp-2013-0001

Cohen, E. (1972). Toward a sociology of international tourism. *Social Research, 39*(1), 164–189.

Cooper, J. & Croyle, R. T. (1984). Attitudes and attitude change. *Annual Review of Psychology, 35*(1), 395–426. DOI: 10.1146/annurev.ps.35.020184.002143

Çelik, S. (2019). Does tourism change tourist attitudes (prejudice and stereotype) towards local people? *Journal of Tourism and Services, 10*(18): 35–46. DOI:10.29036/JOTS.V10I18.89

D'Amore, L. (1988). Tourism–a vital force for peace. *Tourism Management, 9(*2):151–154. https://doi.org/10.1016/0261-5177(88)90025-8

Daruwalla, P. & Darcy, S. (2005). Personal and societal attitudes to disability. *Annals of Tourism Research, 32*(3), 549–570. https://doi.org/10.1016/j.annals.2004.10.008

Emerson, M. O., Kimbro, R. T. & Yancey, G. (2002). Contact theory extended: The effects of prior racial contact on current social ties. *Social Science Quarterly, 83*(3), 745–761. https://doi.org/10.1111/1540-6237.00112

Ethington, P.J. (1997). The intellectual construction of "Social Distance": Toward a recovery of Georg Simmel's social geometry. Cybergo, epistemologie, histoire, didactique. https://doi.org/10.4000/cybergeo.227

Etter, D. (2007). Situational conditions of attitude change within tourism settings: understanding the mechanics of peace through tourism. http://www.iipt.org/educators/OccPap11.pdf

Fazio, R. H., Blascovich, J. & Driscoll, D. M. (2003). On the functional value of attitudes: The influence of accessible attitudes on the ease and quality of decision making. In M. A. Hogg (Ed.), *Social psychology* (pp. 301–346). London, Thousand Oaks, CA and New Delhi: Sage Publications.

Fazio, R. H. & Petty, R. E. (2008). *Attitudes their structure, function, and consequences.* New York: Psychology Press.

Feather, N. (1982). *Expectations and actions: Expectancy-value models in psychology.* New Jersey: Lawrence Erlbaum Associates, Inc.

Fishbein, M. (2008). An investigation of the relationship between beliefs about an object and the attitude toward that object. In R. H. Fazio & R. E. Petty (Eds.), *Attitudes: Their structure, function and consequences* (pp. 137–143). New York: Psychology Press.

Fishbein, M., & Ajzen, I. (2003). In M. A. Hogg (Ed.). *Social psychology* (Vol. 1, pp. 325–347). Berlin: Springer.

Fisher, R. J. & Price, L. L. (1991). International pleasure travel motivations and post-vacation cultural attitude change. *Journal of Leisure Research, 23*(3), 193–208. https://doi.org/10.1080/00222216.1991.11969853

Gawronski, B. (2007). Editorial: Attitudes can be measured! But what is an attitude? *Social Cognitive, 25,* 573–581. DOI:10.1521/SOCO.2007.25.5.573

Gibson, B. (2008). Can evaluative conditioning change attitudes toward mature brands? New evidence from the Implicit Association Test. *Journal of Consumer Research, 35,* 178–188. DOI: 10.1086/527341

Giddings, F. (1895). Sociology and the abstract sciences. The origin of the social feelings. *Annals of the American Academy of Political and Social Science, 5,* 94–101. https://doi.org/10.1177/000271629500500506

Heider, F. (1958). *The psychology of interpersonal relations.* London: Lawrence Erbaum Associates, Publishers.

İnceoğlu, M. (1993). *Tutum, algı, iletişim.* Ankara: Verso.

Huang, S. & Hsu, C. H. C. (2009). Effects of travel motivation, past experience, perceived constraint, and attitude on revisit intention. *Journal of Travel Research, 48*(1), 29–44. https://doi.org/10.1177/0047287508328793

Johnson, R. E., Tolentino, A. L., Rodopman, O. B. & Cho, E. (2010). We (sometimes) know not how we feel: predicting job performance with an implicit measure of trait affectivity. *Personnel Psychology, 63,* 197–219 https://doi.org/10.1111/j.1744-6570.2009.01166.x

Kağıtçıbaşı, Ç. (1999). *Yeni İnsan ve İnsanlar (10. Baskı).* Sosyal Psikoloji Dizisi:1, İstanbul: Evrim.

Katz, D. (1960). The functional approach to the study of attitudes. *Public Opinion Quarterly, 24*(2), 163–204. https://doi.org/10.1086/266945

Kelly, I. (2003). The peace proposition: Tourism as a tool for attitude change (online). In: Braithwaite, R. L. & Braithwaite, R. W. (Eds.) *CAUTHE 2003: Riding the wave of tourism and hospitality research.* Lismore: N.S.W.: Southern Cross University: 618–630. https://citeseerx.ist.psu.edu/viewdoc/download?doi=10.1.1.618.3123&rep=rep1&type=pdf

Koç, E. (2019). *Tüketici Davranışı ve Pazarlama Stratejileri.* (8. Baskı). Ankara: Seçkin.

Krippendorf, J. (1982). Toward new tourism policies: The importance of environmental and sociocultural factors. *Tourism Management 3*(3), 135–148. https://doi.org/10.1016/0261-5177(82)90063-2

Litvin, S. (2003). Tourism and understanding: The MBA study mission. *Annals of Tourism Research, 30*(1), 77–93. https://doi.org/10.1016/S0160-7383(02)00048-8

Marion, P. (1980). Relationships of student characteristics and experiences with attitude changes in a program of study abroad. *Journal of College Student Personnel 21*, 58–64.

Martin, J., Bradford, L. & Rohrlich, B. (1995). Comparing pre-departure expectations and post-sojourn reports: A longitudinal study of U.S. students abroad. *International Journal of Intercultural Relations, 19*(1), 87–110. https://doi.org/10.1016/0147-1767(94)00026-T

Nyaupane, G. P., Teye, V. & Paris, C. (2008). Innocents abroad: Attitude change toward hosts. *Annals of Tourism Research, 35*(3), 650–667. https://doi.org/10.1016/j.annals.2008.03.002

Phillips, W. J., Asperin, A. & Wolfe, K. (2013). Investigating the effect of country image and subjective knowledge on attitudes and behaviors: U.S. upper Mid- westerners' intentions to consume Korean Food and visit Korea. *International Journal of Hospitality Management, 32*, 49–58. https://doi.org/10.1016/j.ijhm.2012.04.003

Pizam, A., Jafari, J. & Milman, A. (1991). Influence of tourism on attitudes: US students visiting USSR. *Tourism Management, 12*(1), 47–54. https://doi.org/10.1016/0261-5177(91)90028-R

Poole, W. (1926). Social distance and personal distance. *Journal of Applied Sociology, 11*, 114–120.

Reisinger, Y. & Turner, L. (2002). Cultural differences between Asian Tourist Markets and Australian hosts: Part 1. *Journal of Travel Research 40*(3), 295–315. https://doi.org/10.1177/004728750204000308

Riordan, C. (1978). Equal-status interracial contact: A review and revision of the concept. *International Journal of Intercultural Relations, 2*(2), 161–185. https://doi.org/10.1016/0147-1767(78)90004-4

Rogers, J. & Ward, C. (1993). Expectation-experience discrepancies and psychological adjustment during cross-cultural reentry. *International Journal of Intercultural Relations,* 17, 185–196. https://doi.org/10.1016/0147-1767(93)90024-3

Shibutani, G. (1955). Reference groups as perspectives. *The American Journal of Sociology, 60*(6), 562–569.

Smith, F. (1964). Prospective teachers' attitudes toward arithmetic. *The Arithmetic Teacher, 11*(7), 474–477.

Subramaniam, P. R. & Silverman, S. (2007). Middle school students' attitudes toward physical education. *Teaching and Teacher Education, 23*(5), 602–611. https://doi.org/10.1016/j.tate.2007.02.003

Tavşancıl, E. (2002). *Tutumların Ölçülmesi ve SPSS ile Veri Analizi.* Ankara: Nobel Yayınları.

Thyne, M., Lawson, R. & Todd, S. (2006). The use of conjoint analysis to assess the impact of the cross-cultural exchange between hosts and guests. *Tourism Management, 27(2)*, 201–213. https://doi.org/10.1016/j.tourman.2004.09.003

Turner, V. (1973). The center out there: Pilgrim's goal. *History of Religions, 12*(3), 191–230.

Tussyadiah, I. P., Wang, D., Jung, T. H. & tom Dieck, M. C. (2018). Virtual reality, presence, and attitude change: Empirical evidence from tourism. *Tourism Management, 66*, 140–154. https://doi.org/10.1016/j.tourman.2017.12.003

Weissman, D. & A. Furnham (1987). The expectations and experiences of a sojourning temporary resident abroad: A preliminary study. *Human Relations, 40*(5), 313–326. https://doi.org/10.1177/001872678704000505

Zajonc, R. (2008). Feeling and thinking: Preferences need no inferences. In R. H. Fazio & R. E. Petty (Eds.), *Attitudes their structure, function, and consequences* (pp. 143–168). New York, Hove: Psychology Press.

Zanna, M. P. & Rempel, J. K. (2008). Attitudes: A new look at an old concept. In R. H. Fazio & R. E. Petty (Eds.), *Attitudes: Their structure, function, and consequences* (pp. 7–17). New York: Psychology Press.

12

EXPLAINING INTERGROUP AND INTRAGROUP DYNAMICS IN TOURISM

A Social Identity Approach

P. Monica Chien and Wanting Sun

Introduction

Tourism, especially international travel, involves the meeting of people from different countries and cultures. Tourists often must interact with other tourists and residents within a destination in order to co-create the tourism experience (Yang, 2015). From a service encounter perspective, residents represent the cultural agents who provide services to tourists and create a friendly and welcoming atmosphere (Chien et al., 2012; Fang et al., 2021). Indeed, intergroup contact forms an integral part of tourism experience and represents an essential foundation to the conviviality and resilience promoted by tourism. Therefore, social interaction among the stakeholders in the tourism ecosystem can significantly shape tourists' satisfaction with the destination (Pizam et al., 1997; Tsang et al., 2016) and residents' support of tourism development (Lai & Hitchcock, 2017). Importantly, tourism is said to contribute to peace through travel-induced contact, as exchange and interactions between residents and tourists can improve perceptions among people, cultivate cross-cultural understanding, and enhance human relations (Farmaki, 2017). At a time of rising racial, social, and religious conflicts, promoting an understanding among different social groups would be an important social and strategic priority for building a coherent and peaceful society at every tourism destination.

Fundamental to the success of sustainable tourism lies in the balanced and harmonious relationship between tourists and the local people they encounter at the destination (Sharpley, 2014). Several studies have examined how residents' interactions with tourists influence their attitudes toward tourism. According to Bimonte and Punzo (2016), since social exchange in tourism involves a meeting of two distinct populations (i.e., the hosts and guests), the quality and nature of their interactions could affect residents' perceptions of tourism and tourists' willingness to pay. Assessing residents' preference for tourists, Thyne et al. (2006) revealed that residents appeared to be more comfortable with tourists who were "familiar", implying that stereotypical perceptions could inhibit interactions and cross-cultural understanding. Similarly, Sinkovics and Penz (2009) showed that Austrians were more reluctant to engage with Japanese tourists than German tourists, possibly due to cultural differences which pose problems for developing socially rewarding and economically viable relationships. The authors suggested that cultural distance may increase hostility, and the removal of prejudice is vital to improve the economic outcomes of the exchange.

Consequently, the organized social context within which residents and tourists interact can determine interpersonal relations, in addition to individuals' psychological functioning (Chien & Ritchie, 2018). Although the issues related to intergroup dynamics have been extensively

DOI: 10.4324/9781003161868-12

examined in the fields of organizational (e.g., Hogg & Terry, 2000; Korschun, 2015), consumer (e.g., White & Dahl, 2007; Yi et al., 2013), and sport (e.g., Chien et al., 2016; Sun et al., 2021) studies, they have not been explicitly addressed in tourism research. Our understanding of the nature and process of these intergroup interactions in tourism is still limited. With the growing concern of tourism-induced disruptions and sustainable tourism recovery post-COVID, a better understanding of mechanisms underlying intergroup relations in the tourism context is of paramount importance.

Since social exchange in tourism involves a meeting of distinct populations and the values and emotions inherent to these groups, the process individuals use to construe intergroup relations could be rather complex. The *social identity theory* (Tajfel & Turner, 1986) provides a useful theoretical framework to understand the intergroup dynamics and predict behavioral outcomes in the tourism ecosystem. This chapter is set out to discuss how phenomena observed in tourism, such as intergroup relations between residents and tourists, residents' portrayal and biased evaluation of tourists (and vice versa), the construal of tourist (mis)behavior, and intergroup conflicts, can be explained by the social identity approach and related concepts. The discussion will identify avenues for future research.

Social Identity Theory and the Construal of Ingroup-Outgroup in Tourism

The intergroup relations between residents and tourists can be explained by the social identity theory (Tajfel & Turner, 1986), referring to the construal of self through the lens of group membership. Individuals derive a sense of self-worth and give meaning to their self-concept by means of their affiliation with social groups, together with associated values and emotional significance (Ellemers et al., 2002). People will categorize themselves and others based on their salient group membership in order to make sense of the social environment, leading to a division of the world into "us" and "them" (Perdue et al., 1990). A group with which an individual identifies is an *ingroup* while a group with which an individual does not identify is called an *outgroup* (Hewstone, 1990). An individual's ingroup and outgroups are presumed to have little in common, with each trying to achieve positive distinctiveness (Aberson et al., 2000).

Tourism, especially international travel, involves the encounter of people who come from different nations and possess varied cultural values (Lai & Hitchcock, 2017). Therefore, an individual's country of origin can be highly salient indicator for social affiliation in the tourism context. From the perspective of a destination's residents, other locals who live in the community and even domestic tourists are viewed as ingroup members given the shared nationality, whereas foreign tourists are seen as outgroup members due to differences in nationality and/or culture. Similarly, tourists may view other tourists from their home country (i.e., their compatriots) as the ingroup, whereas tourists from other countries and residents of the destination are considered as outgroups. The process of self-categorization results in people likely to accentuate the perceived similarities between the self and the other ingroup members on relevant ingroup prototype but exaggerate the differences between the self and outgroups (Hogg & Terry, 2000). When social identities are salient, the norms of relevant ingroups become a crucial source of information about appropriate ways to think, feel, and act, and people rely on the ingroup as a guide for their own thoughts and behavior (Hornsey, 2008). This suggests that residents and tourists may have different views toward social norms and what is considered appropriate tourist behavior at a destination.

People's sense of social identity constitutes a "social cure" which promotes positive effects such as collective self-esteem, group-based pride, and citizenship behavior (Jetten et al., 2017). Studies in psychology have shown that the sense of social identity afforded by group membership delivers well-being benefits, as it provides people with psychological resources to confront and overcome adversity (e.g., Greenaway et al., 2015). In organizational behavior literature, identification with the organization has been found to fulfill individuals' need for self-enhancement, uncertainty reduction, and interpersonal attachment (Hogg & Terry, 2000). When individuals identify with the organization, the salience of other identities recedes, and the sense of oneness or sameness with the organization motivates prosocial or extra-role behaviors that contribute to the collective success of the organization and lead to desirable outcomes (Ashforth & Mael, 1989; Podsakoff et al., 2000). These include helping behavior with fellow employees and developing ties with other internal stakeholders (Podsakoff et al., 2000).

Returning to the context of tourism, extant literature suggests that individuals who integrate their sense of self with sense of the group (e.g., nation, culture) tend to internalize prototypical in-group norms and carry out behaviors that are beneficial to the group. For example, as residents of a destination interact with tourists, they may feel like taking the role of a "destination ambassador" and act as an exemplar of the nation/culture by displaying a hospitable and welcoming manner. They are also more likely to participate in citizenship behaviors, such as helping lost tourists, volunteering, and showcasing civic virtue, since they view the successes and failures of the destination's tourism development as their own. Likewise, when traveling internationally, tourists may be considered as "non-state diplomats" or "goodwill messengers" and expected to speak and act in ways that are consistent with the identity of their home countries, so as not to taint the collective image. Zhang et al. (2019) demonstrate that Chinese tourists tend to promote civilized behaviors when they feel embarrassed about fellow tourists' unruly behaviors overseas and are concerned about "losing collective face". This implies that prosocial or citizenship behavior can be promoted by activating people's national or cultural social identity.

Paradoxical Effect of Ingroup Identification

While identification with one's group can have important implications for intergroup behavior (Tajfel & Turner, 1986), not every member is affected equally by the group membership (Doosje, Branscombe, Spears, & Manstead, 1998). Identification with the ingroup represents a depersonalization process, and the degree to which one uses a group to define oneself varies (Roccas et al., 2006). Social psychology literature suggests that the level of identification with ingroup is a key factor driving individuals' reactions to ingroup-outgroup interactions (e.g., Ellemers et al., 2002; Iyer et al., 2012; Stenstrom et al., 2008). It is, therefore, necessary to distinguish between high and low identifiers.

For individuals who highly identify with the ingroup (e.g., their city or country), they are more invested in the group's positive image and perceive the ingroup more self-conceptually important than low identifiers, and as such, they have stronger needs for maintaining a positive group identity (Hutchison & Abrams, 2003). Those high in identification have a greater tendency to see and think of themselves as ingroup members, display a higher level of self-stereotyping, incline to feel that their group is good and moral, and be concerned about how their group is treated relative to other groups (Ellemers et al., 2002). The study conducted by Roccas et al. (2006) confirms that people who view their national ingroup as superior to other nations attribute high importance to their national identity and strive to contribute to its welfare. Inherent to this *ingroup glorification* phenomenon is the belief that the ingroup is more worthy than other groups, that the group's acts can be justified, and that any criticism toward the group should be rejected or discredited (Roccas et al., 2006). High glorifiers in particular show a stronger belief that group members should adhere to the group norms and would feel insulted if others challenge the group's symbols or authority (Roccas et al., 2006). Applying to the tourism context, residents who highly identity with their City/country, for example, may be more likely to expect tourists to show appreciation of the destination's heritage or protect its natural resources, compared to low-identifying residents. As such, they may feel offended if tourists ignore local customs or mock their cultures, because these individuals are motivated to protect the ingroup's welfare and maintain its image.

It follows that the difference between high and low identifiers can be amplified when the image of their group is threatened, with high identifiers more likely to remain committed to the ingroup in the face of threat (Doosje et al., 1998; Stenstrom et al., 2008). For example, sport fans who identify highly with a team react differently from low identifying fans when their beloved team perform badly (e.g., Wann & Branscombe, 1990). While they are more likely to increase their association with the team's success (i.e., BIRGing or basking-in-reflected-glory) compared to low identifying fans, they also show decreased tendencies to distance themselves from team failure (i.e., CORFing or cutting-off-reflected-failure) (Wann & Branscombe, 1990). When people attach greater significance to their group such as in high-identifying sports fans, they appear to be more resistant to evaluatively threatening information about the team, and use ingroup favoritism as a means to protect the group (Abrams et al., 2013).

Similarly, in the organizational behavior literature, an employee's level of identification with the organization is suggested to affect the way the individual views external stakeholders (Cornwell et al., 2018; Conroy et al., 2017), which can be adversarial at times (Korschun, 2015). Anecdotal evidence also reveals that some of the highly devoted employees show the tendency to belittle customers, while viewing themselves as superior to employees of other companies (Korschun, 2015).

Taken together, the concept of ingroup identification can be used to explain the intergroup dynamics between residents and tourists. These include discontent of some residents toward tourists (i.e., the outgroups) visiting their communities, whose behaviors might be viewed as intimidating or undermining the ingroup identity (Chen et al., 2018). Residents' interaction with tourists is likely influenced by the affinity felt toward the destination. Specifically, high-identifying residents may perceive certain tourist behaviors, such as taking photos of residents' properties or disrupting the quiet atmosphere of the community, as disrespectful and insulting, due to the tourists' disregard of the symbols, rules, or regulations central to the group. Tourists' unruly behaviors may be considered as a pervasive threat to the ingroup, thereby causing distress among the residents. As highlighted by Chen et al. (2018), on one hand, ingroup glorification may bestow residents a sense of superiority, and their unfavorable attitudes or contemptuous social judgments toward tourists reflect their perceived national superiority; on the other hand, residents may experience a feeling of deprivation given the competition and despoliation of valuable resources (Chen et al., 2018). Likewise, tourists' ingroup glorification may manifest when visiting certain destinations, such as in the postcolonial context (Bandyopadhyay, 2019), which may lead to non-observance of local norms or rules.

Ingroup Bias and Intergroup Attribution

Because identification with a group and enhancement of that group identity can increase self-esteem and self-image, people are intrinsically motivated to preserve and maintain a positive identity of the group to which they belong (Tajfel & Turner, 1986). One way to achieve this is by bolstering evaluations of one's own group in relation to others in perceptual, attitudinal, or behavioral domains (Aberson et al., 2000). Judgment of ingroup and outgroup members are based on norms and stereotypes (Ellemers et al., 2002). This appraisal often occurs in a biased manner with people subjectively evaluating their ingroup more favorably than outgroups and giving preferential treatment to ingroup members (i.e., *ingroup bias*; Aberson et al., 2000), whereas outgroups are given unjustly negative evaluations in order to emphasize superiority of the ingroup (i.e., *outgroup derogation*; Ellemers et al., 2002). Research has demonstrated, for example, that Hong Kong residents hold negative stereotypes against Mainland Chinese tourists (Tung et al., 2020). Ingroup identification has also been associated with prejudice toward outgroup members (Ariyanto et al., 2009). The process of *intergroup attribution* represents the ways members of different social groups explain the behavior and its consequences of their own ingroup members as well as members of other social groups, and such evaluation is colored by the characteristics associated with the groups involved (Islam & Hewstone, 1993).

Intergroup attribution bias has been observed in contexts where the intergroup relations are characterized by rivalry (e.g., competition between sport teams), and where there is an intense intergroup conflict (e.g., confrontation between Muslims and Christians) (Ariyanto et al., 2009; Doosje & Branscombe, 2003). Tourism represents such setting because residents and tourists compete for limited resources, such as public transport, leisure facilities, and space. In particular, high identification with one's ingroup can lead to legitimization of the group's action (Abrams et al., 2013), resulting in the rejection that the group has committed immoral deeds (Roccas et al., 2006). High-identifiers require more evidence to be convinced of the ingroup wrongdoing (Miron et al., 2010), are motivated to improve the ingroup's status position (Ellemers et al., 2002), and experience less guilt for ingroup misconduct (Iyer et al., 2012). As high identifiers seek conformity, they may – directly or indirectly – facilitate unethical behavior of other ingroup members, cover it up, or validate the behavior by changing the content of ingroup and outgroup stereotypes (Conroy et al., 2017; Roccas et al., 2006). This psychological process may underpin intergroup exchange in tourism, a context where intergroup emotions are visceral and the level of mistrust between the groups can be high (Ariyanto et al., 2009). For example, hotel employees may attribute withholding of benefits or information to tourists by their co-workers as a way to protect the organization/nation's rights, instead of admitting that their ingroup members have carried out an unethical behavior. Given demand or mistreatment from tourists can be considered as a threat to the ingroup, hotel employees are likely to condone the behavior of their co-workers and justify it as moral.

Studies in sports have found that when exposed to news about player transgressions, fans often maintain favorable attitudes toward the team they are affiliated with despite negative publicity and the number of team member implicated in a sport scandal, while rival fans hold negative attitude (e.g., Chien et al., 2016). A sport scandal can be considered a group failure experience, where an ingroup member engages in acts destructive to the group. Given the importance of the team to the fans' social identity, it would be unthinkable to abandon the team in the face of such a situation, or to switch loyalties to another team (i.e., an outgroup) (Sun et al., 2021). Instead, fans may be more tolerant of their own group's transgressions than they would be of similar outgroup transgressions, especially when doubt is cast on the case against the suspected perpetrator (Miron et al., 2010). Evidence of this moral hypocrisy arises in intergroup settings, whereby a transgression enacted by an ingroup member is perceived to be more acceptable than the identical behavior enacted by an outgroup member (Valdesolo & DeSteno, 2008). Because the team stands as an important source of fans' self-definition, preserving its legitimacy appears to trump the use of more objective moral principles (Valdesolo & DeSteno, 2008).

In line with findings that people are likely to attribute negative behavior of the ingroup member to temporary situational factors but attribute negative behavior of the outgroup member to stable characteristics of the outgroup (*ultimate attribution error*; Hewstone, 1990), it can be argued that individuals may downplay the implications of a misbehavior (e.g., queue jumping) committed by their compatriot tourists, attribute the misbehavior to the situational factors rather than the nature of their ingroup (e.g., the tourists are in a hurry), and exculpate the perpetrators through ingroup favoritism (e.g., these people are on holiday and are not familiar with the local customs), so that positive group distinction is maintained. Indeed, Zhang et al. (2019) found that individuals tend to rationalize the misconduct of compatriot tourists via cultural difference and even outgroup discrimination. They are, however, likely to respond to the same misconduct by tourists from other countries differently. In other words, intergroup attributions are ethnocentric, in that people tend to attribute positive ingroup behaviors to internal causes and negative behavior to external causes (Islam & Hewstone, 1993). Conversely, attributions are disproportionally more internal for negative outgroup behaviors than positive outgroup behaviors (Islam & Hewstone, 1993), showing outgroup derogation. It would be important for future research to better understand the process underlying intergroup attribution in tourism and to provide validation across interaction contexts. Subsequent empirical investigations could be conducted to understand how residents' or tourists' level of identification with their nation, community, or culture influences their evaluation of, and behavioral intention toward, the outgroups when the two social groups mingle.

The Black Sheep Effect and Intragroup Differentiation

The tendency to favor ingroup members over outgroup members is pervasive but not a forgone conclusion in intergroup relations (Hutchison et al., 2008). Although in general, individuals are prone to give favorable treatment and find excuses for ingroup members who misbehave or deviate from the group norms, in some circumstances, people may perceive these deviants negatively, even more so than an outgroup perpetrator (Iyer et al., 2007). Referring to the *black sheep effect*, individuals may judge ingroup rule-breakers more harshly than outgroup perpetrators for a similar transgression if they believe that the transgression has reflected negatively on the ingroup as a whole (Marques et al., 2001). The ingroup deviant, or the "black sheep", is seen as an atypical ingroup member and are evaluated negatively because their behavior departs from the group's salient norms and values, which undermines distinctiveness of the group in intergroup comparison (Jetten & Hornsey, 2014). The concept can be illustrated by a recent case in Venice, where four Japanese tourists were charged more than €1,000 for their meals (Henley, 2018). The incident infuriated some local residents and businesses, as it not only damaged the restaurant's reputation but also the image of Venice as a desirable tourist destination. Consequently, the Venetia Hoteliers Association offered the tourists a free stay in a luxury hotel as a compensatory gesture, and the restaurant also faced a hefty fine.

According to Marques et al. (2001), ingroup favoritism (over-rating salient ingroup members) and intra-group differentiation (under-rating salient ingroup members) can co-exist so as to preserve the perceived ingroup positivity. High-identifying members of the group tend to report less tolerance toward ingroup deviants, and they are more likely to derogate and punish ingroup members who engage in rule-breaking behavior to a greater extent than low identifying members (Iyer et al., 2012). Excluding undesirable members from the ingroup thus serves the important function of protecting the ingroup stereotype while reinforcing important group norms (Hutchison et al., 2008), as shared group membership can increase one's embarrassment or guilt by association due to vicarious shame (Doosje et al., 1998). Downgrading and derogating undesirable ingroup members can be a cognitive strategy used to legitimize ingroup positivity and restore debilitated group cohesion (Jetten & Hornsey, 2014).

Support for these ideas come from several studies showing that deviant ingroup members are evaluated more negatively than outgroup members who engage in the same behavior. When examining vegetarians' evaluations of other vegetarians who eat meat occasionally, Hornsey and Jetten (2003) found that people who highly identified as vegetarians devaluated the rule-breakers strongly and displayed more negative affect toward them, comparing to participants who were non-vegetarians. Likewise, studies investigating sport fans' evaluations of transgressions committed by sport teams showed that ingroup fans (i.e., fans affiliated with the implicated team) tend to isolate the responsibility to the perpetrators instead of devaluating the team as a whole (e.g., Kelly et al., 2018). In tourism, although not examining the black sheep effect per se, Zhang et al. (2019) provided some evidences indicating the existence of the black sheep effect when investigating Chinese people's reflections on Chinese tourists' uncivilized behaviors. People distanced the misbehaving Chinese tourists from the normative Chinese tourists, portraying the misbehaving tourists as older, under-educated, low income, and package tour consumers (Zhang et al., 2019).

In light of the above results, individuals are referred to as deviant tourists who possess qualities or attitudes that differ from the prototype (e.g., rowdy tourists), who display socially undesirable or antisocial behaviors (e.g., urinating in street), or who engage in transgressions (e.g., damaging heritage sites) that violate expectancies. These tourists not only cause disruptions to the local community but also negatively influence the collective image of the their home country through the process of (self)stereotyping (Tung et al., 2020), resulting in certain undesirable features becoming associated with the nation and its people (e.g., all American tourists are rude) (Zhang et al., 2019). Black sheep effect may arise when people witness a rule-breaking behavior from their compatriot tourists, especially if the behavior reflects badly on the ingroup, since people generally expect ingroup members to be more virtuous than outgroup members (Abrams et al., 2013; Marques et al., 2001). As deviant compatriot tourists create uncertainty about the image of the ingroup as a whole, isolating the responsibility to the misbehaving tourists, exclude them from the ingroup representation, and punish them become an important strategy to restore a positive group identity (Jetten & Hornsey, 2014). This explains why many people took to social media to share their disgust toward compatriot tourists' misconducts while traveling overseas, as derogation of deviant ingroup members help restore threatened group positivity and cohesion.

The Dark Side of Intergroup Interaction

While existing studies have mainly emphasized tourism as a prosocial phenomenon and conducive for cross-cultural understanding and peace-making, tourism-induced disruptions have started to gain research attention in recent years (e.g., Chien & Ritchie, 2018; Cheung & Li, 2019). One of the hallmarks of relations between members of different groups is the tendency for conflict (Stenstrom et al., 2008) due to differences in culture, value and communication, which can provoke negative sentiments and responses that hinder tourism's contribution to peace (Chien & Ritchie, 2018). In the cases of popular tourist destinations such as Byron Bay (Australia) and New Orleans (the USA), such conflicts appear to be aggravated by the growth of home-sharing economy such as Airbnb (Bainbridge & Armitage, 2017). Residential areas have been increasingly used for Airbnb style lodgings, and the coexistence of tourists and the resident population have engendered tension (Gutiérrez et al., 2017). In particular, tourists may feel they have gained a "license to sin" (De Witt Huberts et al., 2012) when on holiday, thus knowingly or unknowingly engage in rule-breaking behavior, such as violating local norms of conduct, which can lead to intergroup conflict.

Intergroup conflicts might be seen as a potential threat to the ingroup that undermines positive social identity and inflict harms (e.g., stress, anxiety) on the ingroup members (Stenstrom et al., 2008). Consequently, the perceived threat from outgroups generates intense emotional reactions and may invoke aggression or even warlike responses (Lickel et al., 2006). In line with the social identity theory, high-identifying members of the group (e.g., residents) not only respond more strongly to the outgroup compared to low-identifying members but are also motivated to seek revenge of a fellow ingroup member who becomes entwined in an intergroup conflict (Spanovic et al., 2010). Importantly, because other ingroup members embody the prototype of the ingroup, an attack on a fellow ingroup member may be perceived as an attacked on the ingroup as a whole (Miron et al., 2010). Empathy toward the ingroup member, the need to protect and enhance group pride, and compliance with the normative influence exerted by the ingroup underlies vicarious retribution (Lickel et al., 2006).

This psychological process is termed *vicarious retribution*, where neither the agent of retaliation nor the target of retribution is directly involved in or precipitate the intergroup conflict (Lickel et al., 2006). Ingroup identification biases or distorts how the outgroup is perceived, such as the degree of social interdependence among outgroup members and their blameworthiness in the intergroup conflict, which triggers outgroup-directed retaliation (Stenstrom et al., 2008). The aggression is deemed morally legitimate as the individuals met out punishment to those who have wronged them to gain a sense of justice (Spanovic et al., 2010). Applying this line of reasoning to the tourism context, residents who highly identify with their city and/or country are more likely to engage in retributive behavior toward tourists, if they feel that their fellow residents or the community have been harmed (e.g., inappropriate tourist behavior toward the neighbors or their properties). Under this circumstance, the residents' appraisal of the tourists and their behaviors may evoke strong emotions, and even aggression or intention to retaliate other tourists, although they are not part of the initial conflict – a manifestation of the need of restoring threatened in-group positivity and maintaining intergroup distinction.

While intergroup conflict is expected to widen the gulf, intergroup contact may close the gap between tourists and residents. Researchers have shown that past positive contact with the outgroup – characterized by pleasant, cooperative intergroup interactions – typically produces more positive intergroup attitudes, reduces anxiety about interacting with outgroup members, and buffers against some of the harmful effects of current negative contact (Hayward et al., 2018; Islam & Hewstone, 1993). Negative contact, in contrast, makes group membership more salient and therefore serves as a stronger predictor of prejudice and avoidance than positive contact (Hayward, Tropp, Hornsey, & Barlow, 2017; Paolini, Harwood, & Rubin, 2010). Specifically, while negative contact can engender negative emotions such as anger and anxiety, frequent positive contact can exert strong influence on positive affective outcomes such as empathy toward the outgroup, resulting in the intergroup incident being viewed in less conflictual ways (Hayward et al., 2017).

Extending these findings to tourism, prior positive contact between residents and tourists may be particularly important in mediating subsequent negative contact between the groups and change the dynamics during an intergroup conflict. Yet, we know little about what constitutes positive and negative contact between residents and tourists in real-world settings, as contact can be experienced myriad ways (Hayward et al., 2018) such as a short conversation in passing or hosting of outgroup members in private residence. Future studies could provide an audit of how intergroup contact is truly experienced in tourism, as well as the perceived intensity or quality of both positive and negative intergroup contact experience.

Conclusion

Emerging research hints at the importance of understanding intergroup interactions between residents and tourists, suggesting that such intergroup dynamics open pathways to better understand the economic and social impact of tourism development (Cheung & Li, 2019; Chien & Ritchie, 2018; Fang et al., 2021). Social identity theory presents a useful framework to understand intergroup relations in tourism and predict stakeholder responses (see Figure 12.1 for a schematic presentation of the psychological process discussed in this chapter). Individuals' social identity, and

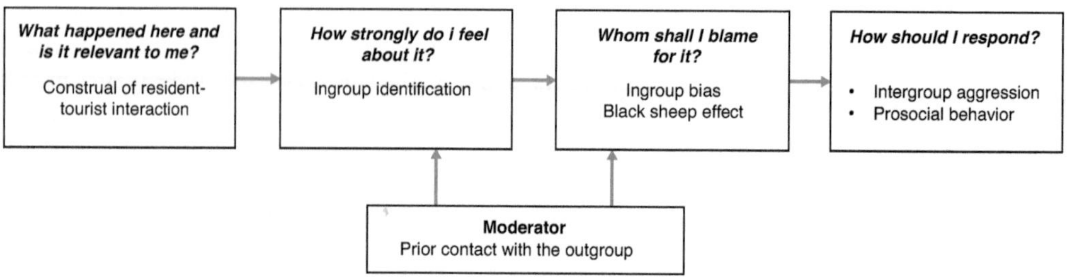

Figure 12.1 A Framework for Understanding Intergroup Dynamics in Tourism.
Source: Adapted from Lickel et al., 2006

the values and norms that come with it, can influence their appraisal of the intergroup exchange, including construal of the behavior of both ingroup and outgroup members. The interesting tourist-resident dynamics following the exposure of intergroup contact sheds light on the impact of intergroup interactions on tourist experience and residents' quality of life, thus contributing to the evolving fields of tourist/resident behavior and sustainable tourism research agenda.

Because research on the intergroup relations in tourism is still evolving, a systematic approach is needed to understand how the nature of interaction, and the motivation of individuals involved, influence responses. Identification with the ingroup is conceptualized in this chapter as identification with the resident's or tourist's nation. The concept of social identity complexity (Roccas & Brewer, 2002), however, suggests that most people are simultaneously members of a range of different, cross-cutting social groups, including those derived from highly meaningful and clearly delineated groups (e.g., nation, religion) as well as those referring to more abstract and perhaps ambiguous social categories (e.g., tour group) (Ellemers et al., 2002). There may be situations in which multiple social identities are relevant and may become equally salient, due to the perceived threat directed at both domains of the individual's ingroups (Roccas & Brewer, 2002). In other words, individual's self-categorization may be context dependent, and the ingroup-outgroup categorization will likely be calibrated based on the combination of different social categories, where behavior would reflect the important social norms associated with that ingroup (Brewer & Pierce, 2005; Roccas & Brewer, 2002). Future research should consider how residents and tourists construe their social identities and their activation in different social settings to assess the intergroup outcomes. Similarly, the nature and range of intergroup contact between residents and tourists can be diverse, and a truly comprehensive account of intergroup interaction must solicit both resident and tourist perspectives. An enhanced recognition of resident-tourist intergroup relations – both positive and negative – is therefore needed.

The downstream consequences of intergroup interaction remain unclear. Recent research has begun to uncover some promising findings. For example, after exposure to the negative image projected on them owing to tourist misbehavior of their compatriots, tourists expressed the willingness to carry out prosocial behaviors such as being careful with their own behaviors, reminding travel companions to behave properly, and stopping inappropriate behaviors of their compatriots when traveling abroad (Zhang et al., 2019). Individuals may be willing to engage in prosocial behaviors when they perceive an opportunity to improve their group stereotypes (Hopkins et al., 2007). Other researchers have implied that individuals' negative emotions derived from previous negative intergroup contact may also spill over to tourism. A study conducted by Luo and Zhai (2017) of Weibo users' discussions of Occupy Central, a political protest calling for the general election of Hong Kong's chief executive, revealed an intention to boycott a destination, when negative emotions and group confrontations between people from Hong Kong and people from Mainland China were provoked by an issue unrelated to tourism (Luo & Zhai, 2017). Further exploration of whether different prosocial/antagonistic behaviors are associated with exposure to different types of resident-tourist interactions is warranted.

Questions remain as to whether negative contact with one group of tourists (e.g., Chinese) will spill over to other groups of tourists (e.g., Korean). Other dimensions of intergroup interactions may be considered by future research, such as the duration of the contact (repetitive versus one-off), involvement of a misconduct by either tourists or residents, and intentionality of the behavior.

Methodologically, extant work on intergroup interactions in tourism is mainly qualitative and correlational, making it difficult to conclude that positive(negative) contact causally decreases (increases) bias or negative perceptions. Further extension of this work can experimentally simulate resident-tourist interaction, such as an intergroup conflict, to examine how social identity of the individuals involved interacts with contextual factors to influence their construal of the event as well as evaluations of the ingroup and outgroup members. It is also crucial to examine the long-term effects of intergroup interaction on a range of outcomes such as residents' well-being and the destination's restorative qualities.

References

Aberson, C. L., Healy, M., & Romero, V. (2000). Ingroup bias and self-esteem: A meta-analysis. *Personality and Social Psychology Review, 4*(2), 157–173.

Abrams, D., Randsley de Moura, G., & Travaglino, G. A. (2013). A double standard when group members behave badly: Transgression credit to ingroup leaders. *Journal of Personality and Social Psychology, 105*(5), 799.

Ariyanto, A., Hornsey, M. J., & Gallois, C. (2009). Intergroup attribution bias in the context of extreme intergroup conflict. *Asian Journal of Social Psychology, 12*(4), 293–299.

Ashforth, B. E., & Mael, F. (1989). Social identity theory and the organization. *Academy of Management Review, 14*(1), 20–39.

Bainbridge, A., & Armitage, R. (2017, June 18). Airbnb in Australia shows the sharing economy has a 'dark side'. *ABC News*. Retrieved from: http://www.abc.net.au/news/2017-06-18/airbnb-in-australia-the-sharing-economy-has-a-dark-side/8624122

Bandyopadhyay, R. (2019). Volunteer tourism and "The White Man's Burden": Globalization of suffering, white savior complex, religion and modernity. *Journal of Sustainable Tourism, 27*(3), 327–343.

Bimonte, S., & Punzo, L. F. (2016). Tourist development and host–guest interaction: An economic exchange theory. *Annals of Tourism Research, 58*, 128–139.

Brewer, M. B., & Pierce, K. P. (2005). Social identity complexity and outgroup tolerance. *Personality and Social Psychology Bulletin, 31*(3), 428–437.

Chen, N., Hsu, C. H., & Li, X. R. (2018). Feeling superior or deprived? Attitudes and underlying mentalities of residents towards Mainland Chinese tourists. *Tourism Management, 66*, 94–107.

Cheung, K. S., & Li, L. H. (2019). Understanding visitor–resident relations in overtourism: Developing resilience for sustainable tourism. *Journal of Sustainable Tourism, 27*(8), 1197–1216.

Chien, P. M., Kelly, S. J., & Weeks, C. S. (2016). Sport scandal and sponsorship decisions: Team identification matters. *Journal of Sport Management, 30*(5), 490–505.

Chien, P. M., & Ritchie, B. W. (2018). Understanding intergroup conflicts in tourism. *Annals of Tourism Research, 72*(C), 177–179.

Chien, P. M., Ritchie, B. W., Shipway, R., & Henderson, H. (2012). I am having a dilemma: Factors affecting resident support of event development in the community. *Journal of Travel Research, 51*(4), 451–463.

Conroy, S., Henle, C. A., Shore, L., & Stelman, S. (2017). Where there is light, there is dark: A review of the detrimental outcomes of high organizational identification. *Journal of Organizational Behavior, 38*(2), 184–203.

Cornwell, T. B., Howard-Grenville, J., & Hampel, C. E. (2018). The company you keep: How an organization's horizontal partnerships affect employee organizational identification. *Academy of Management Review, 43*(4), 772–791.

De Witt Huberts, J. C., Evers, C., & De Ridder, D. T. (2012). License to sin: Self-licensing as a mechanism underlying hedonic consumption. *European Journal of Social Psychology, 42*(4), 490–496.

Doosje, B., & Branscombe, N. R. (2003). Attributions for the negative historical actions of a group. *European Journal of Social Psychology, 33*(2), 235–248.

Doosje, B., Branscombe, N. R., Spears, R., & Manstead, A. S. (1998). Guilty by association: When one's group has a negative history. *Journal of Personality and Social Psychology, 75*(4), 872.

Ellemers, N., Spears, R., & Doosje, B. (2002). Self and social identity. *Annual Review of Psychology, 53*(1), 161–186.

Fang, Y., Chien, P. M., & Walters, G. (2021). Understanding the emerging Chinese working holiday market. *Journal of Vacation Marketing, 27*(1), 3–16.

Farmaki, A. (2017). The tourism and peace nexus. *Tourism Management, 59*, 528–540.

Greenaway, K. H., Haslam, S. A., Cruwys, T., Branscombe, N. R., Ysseldyk, R., & Heldreth, C. (2015). From "we" to "me": Group identification enhances perceived personal control with consequences for health and well-being. *Journal of Personality and Social Psychology, 109*(1), 53.

Gutiérrez, J., García-Palomares, J. C., Romanillos, G., & Salas-Olmedo, M. H. (2017). The eruption of Airbnb in tourist cities: Comparing spatial patterns of hotels and peer-to-peer accommodation in Barcelona. *Tourism Management, 62*, 278–291.

Hayward, L. E., Tropp, L. R., Hornsey, M. J., & Barlow, F. K. (2017). Toward a comprehensive understanding of intergroup contact: Descriptions and mediators of positive and negative contact among majority and minority groups. *Personality and Social Psychology Bulletin, 43*(3), 347–364.

Hayward, L. E., Tropp, L. R., Hornsey, M. J., & Barlow, F. K. (2018). How negative contact and positive contact with Whites predict collective action among racial and ethnic minorities. *British Journal of Social Psychology, 57*(1), 1–20.

Henley, J. (2018, January 26). Venice restaurant that hit tourists with £1,000 bill faces £17,000 fines. *The Guardian*. Retrieved from: https://www.theguardian.com/world/2018/jan/25/venice-restaurant-overcharged-tourists-fines-osteria-da-luca

Hewstone, M. (1990). The 'ultimate attribution error'? A review of the literature on intergroup causal attribution. *European Journal of Social Psychology, 20*(4), 311–335.

Hogg, M. A., & Terry, D. I. (2000). Social identity and self-categorization processes in organizational contexts. *Academy of Management Review, 25*(1), 121–140.

Hopkins, N., Reicher, S., Harrison, K., Cassidy, C., Bull, R., & Levine, M. (2007). Helping to improve the group stereotype: On the strategic dimension of prosocial behavior. *Personality and Social Psychology Bulletin, 33*(6), 776–788.

Hornsey, M. J. (2008). Social identity theory and self-categorization theory: A historical review. *Social and Personality Psychology Compass, 2*(1), 204–222.

Hornsey, M. J., & Jetten, J. (2003). Not being what you claim to be: Impostors as sources of group threat. *European Journal of Social Psychology, 33*(5), 639–657.

Hutchison, P., & Abrams, D. (2003). Ingroup identification moderates stereotype change in reaction to ingroup deviance. *European Journal of Social Psychology, 33*(4), 497–506.

Hutchison, P., Abrams, D., Gutierrez, R., & Viki, G. T. (2008). Getting rid of the bad ones: The relationship between group identification, deviant derogation, and identity maintenance. *Journal of Experimental Social Psychology, 44*(3), 874–881.

Islam, M. R., & Hewstone, M. (1993). Dimensions of contact as predictors of intergroup anxiety, perceived out-group variability, and out-group attitude: An integrative model. *Personality and Social Psychology Bulletin, 19*(6), 700–710.

Iyer, A., Jetten, J., & Haslam, S. A. (2012). Sugaring o'er the devil: Moral superiority and group identification help individuals downplay the implications of ingroup rule-breaking. *European Journal of Social Psychology, 42*(2), 141–149.

Iyer, A., Schmader, T., & Lickel, B. (2007). Why individuals protest the perceived transgressions of their country: The role of anger, shame, and guilt. *Personality and Psychology Bulletin, 33*(4), 572–587.

Jetten, J., Haslam, S. A., Cruwys, T., Greenaway, K. H., Haslam, C., & Steffens, N. K. (2017). Advancing the social identity approach to health and well-being: Progressing the social cure research agenda. *European Journal of Social Psychology, 47*(7), 789–802.

Jetten, J., & Hornsey, M. J. (2014). Deviance and dissent in groups. *Annual Review of Psychology, 65*, 461–485.

Kelly, S. J., Weeks, C. S., & Chien, P. M. (2018). There goes my hero again: Sport scandal frequency and social identity driven response. *Journal of Strategic Marketing, 26*(1), 56–70.

Korschun, D. (2015). Boundary-spanning employees and relationships with external stakeholders: A social identity approach. *Academy of Management Review, 40*(4), 611–629.

Lai, I. K. W., & Hitchcock, M. (2017). Local reactions to mass tourism and community tourism development in Macau. *Journal of Sustainable Tourism, 25*(4), 451–470.

Lickel, B., Miller, N., Stenstrom, D. M., Denson, T. F., & Schmader, T. (2006). Vicarious retribution: The role of collective blame in intergroup aggression. *Personality and Social Psychology Review, 10*(4), 372–390.

Luo, Q., & Zhai, X. (2017). "I will never go to Hong Kong again!" How the secondary crisis communication of "Occupy Central" on Weibo shifted to a tourism boycott. *Tourism Management, 62*, 159–172.

Marques, J., Abrams, D., & Serôdio, R. G. (2001). Being better by being right: Subjective group dynamics and derogation of in-group deviants when generic norms are undermined. *Journal of Personality and Social Psychology, 81*(3), 436–447.

Miron, A. M., Branscombe, N. R., & Biernat, M. (2010). Motivated shifting of justice standards. *Personality and Social Psychology Bulletin, 36*(6), 768–779.

Paolini, S., Harwood, J., & Rubin, M. (2010). Negative intergroup contact makes group memberships salient: Explaining why intergroup conflict endures. *Personality and Social Psychology Bulletin, 36*(12), 1723–1738.

Perdue, C. W., Dovidio, J. F., Gurtman, M. B., & Tyler, R. B. (1990). Us and them: Social categorization and the process of intergroup bias. *Journal of Personality and Social Psychology, 59*(3), 475.

Pizam, A., Pine, R., Mok, C., & Shin, J. Y. (1997). Nationality vs industry cultures: Which has a greater effect on managerial behavior? *International Journal of Hospitality Management, 16*(2), 127–145.

Podsakoff, P. M., MacKenzie, S. B., Paine, J. B., & Bachrach, D. G. (2000). Organizational citizenship behaviors: A critical review of the theoretical and empirical literature and suggestions for future research. *Journal of Management, 26*(3), 513–563.

Roccas, S., & Brewer, M. B. (2002). Social identity complexity. *Personality and Social Psychology Review, 6*(2), 88–106.

Roccas, S., Klar, Y., & Liviatan, I. (2006). The paradox of group-based guilt: Modes of national identification, conflict vehemence, and reactions to the in-group's moral violations. *Journal of Personality and Social Psychology, 91*(4), 698.

Sharpley, R. (2014). Host perceptions of tourism: A review of the research. *Tourism Management, 42*, 37–49.

Sinkovics, R. R., & Penz, E. (2009). Social distance between residents and international tourists – Implications for international business. *International Business Review, 18*(5), 457–469.

Spanovic, M., Lickel, B., Denson, T. F., & Petrovic, N. (2010). Fear and anger as predictors of motivation for intergroup aggression: Evidence from Serbia and Republika Srpska. *Group Processes & Intergroup Relations, 13*(6), 725–739.

Stenstrom, D. M., Lickel, B., Denson, T. F., & Miller, N. (2008). The roles of ingroup identification and outgroup entitativity in intergroup retribution. *Personality and Social Psychology Bulletin, 34*(11), 1570–1582.

Sun, W., Chien, P. M., & Weeks, C. S. (2021). Sport scandal and fan response: The importance of ambi-fans. *European Sport Management Quarterly*, doi: 10.1080/16184742.2021.1912131

Tajfel, H., & Turner, J. C. (1986). The social identity theory of intergroup behavior. In S. Worchel & W. G. Austin (Eds.), *Psychology of intergroup relations*. Chicago: Nelson Hall.

Thyne, M., Lawson, R., & Todd, S. (2006). The use of conjoint analysis to assess the impact of the cross-cultural exchange between hosts and guests. *Tourism Management, 27*(2), 201–213.

Tsang, N. K., Prideaux, B., & Lee, L. (2016). Attribution of inappropriate visitor behavior in a theme park setting-a conceptual model. *Journal of Travel & Tourism Marketing, 33*(8), 1088–1105.

Tung, V. W. S., King, B. E. M., & Tse, S. (2020). The tourist stereotype model: Positive and negative dimensions. *Journal of Travel Research, 59*(1), 37–51.

Valdesolo, P., & DeSteno, D. (2008). The duality of virtue: Deconstructing the moral hypocrite. *Journal of Experimental Social Psychology, 44*(5), 1334–1338.

Wann, D. L., & Branscombe, N. R. (1990). Die-hard and fair-weather fans: Effects of identification on BIRGing and CORFing tendencies. *Journal of Sport and Social Issues, 14*(2), 103–117.

White, K., & Dahl, D. W. (2007). Are all out-groups created equal? Consumer identity and dissociative influence. *Journal of Consumer Research, 34*(4), 525–536.

Yang, F. (2015). Tourist co-created destination image. *Journal of Travel & Tourism Marketing, 33*(4), 1–15.

Yi, Y., Gong, T., & Lee, H. (2013). The impact of other customers on customer citizenship behavior. *Psychology & Marketing, 30*(4), 341–356.

Zhang, C. X., Pearce, P., & Chen, G. (2019). Not losing our collective face: Social identity and Chinese tourists' reflections on uncivilised behaviour. *Tourism Management, 73*, 71–82.

13

TRAVEL AND TRANSFORMATION

An Examination of Tourists' Attitude Changes

Jessica Mei Pung

Defining Tourist Transformation

Gnoth (1997) built on tourist attitude as a construct to discuss tourists' behavior and to better understand their underlying motivations. He described emotions (i.e., drives, feelings, instincts) and cognitions (i.e., knowledge, beliefs) as antecedent influences that shape tourists' attitudes towards their experiences. Based on Fishbein and Ajzen (1975) and Lutz (1981)'s theories, attitudes consist of affect, cognition and conation (Cocolas et al., 2020; Gnoth, 1994), with the latter component being a behavioral response which can translate into actual behavior (Da Silva & Alwi, 2006). This conceptualization has provided the basis for understanding shifts in tourists' attitudes at the destination. In more recent tourism research (Lee, 2009; Li et al., 2016; Li & Wang, 2020), tourists' attitude at the destination is studied and measured to predict tourist behavioral intention, which means whether tourists intend to re-visit a destination and/or to recommend it. These studies have been found to mostly use the theory of planned behavior and the contact model as conceptual backgrounds (Hadinejad et al., 2019). Beyond this research strand, the extent to which tourism can change tourists' attitude in their life domains, is being gradually investigated by some scholars (Li & Wang, 2020). In particular, there has been an increased focus on understanding which factors contribute to changing the attitudes of on-site tourists, towards the destination and towards local residents (e.g., Çelik, 2019; Paris et al., 2014).

Cohen and Cohen (2019) suggest that studies in tourism are increasingly adopting social and psychological perspectives, to go beyond tourism as an economic activity, and to explore, for example, the important role of emotions and other socio-psychological components of tourist experiences. In order to deeply understand the different experience components and situations that can impact tourists' change in attitudes, and to analyze attitude change as an outcome of travel, this chapter draws upon transformative tourism research. Transformative travel consists in tourism experiences that lead to positive changes in values and attitudes of those participating (Christie & Mason, 2003). Furthermore, the transformation of a tourist is defined as a process that is "facilitated by contextual stimuli which strike the tourists and lead to reflecting and integrating new knowledge, skills and beliefs", increasing "tourists' existential authenticity [...] "cross-cultural understanding and pro-environmental awareness, with potential consequences on long-term behaviour" (Pung et al., 2020a, p. 2). The conceptualization of transformative travel draws on numerous disciplines, such as social psychology, geography, and philosophy, and describes tourists as having an embodied experience at the destination, consisting in inward interpretation of social structures and cultural encounters, and outward performativity, which could

DOI: 10.4324/9781003161868-13

translate into developing habits or into life-changing decisions. For the purpose of the chapter, these concepts and related processes are further examined in the following sections.

The Tourist Transformation Model

A tourist transformation model has been proposed by Pung et al. (2020a), to identify and conceptualize common processes that tourists experience when feeling transformed. Tourist transformation is first facilitated by specific factors at the destination, namely feeling like being in an extraordinary environment that is removed from their daily life and structures (Kirillova et al., 2017b), being struck by cultural differences with the destination residents, and facing challenging problems during the trip (Brown, 2009). These factors elicit stimuli that lead to experiencing peak episodes, disorienting dilemmas or a physical performance, causing a sense of fragmentation in the tourists, who reflect on their individual role in the world (Pung et al., 2020a). Through interpreting and recalling aspects of the experience, tourists integrate new values and knowledge about society, in their renewed perspective (Reisinger, 2013). Such existential and educational self-change experience can change travelers' attitudes and long-term behavior (Lean, 2012). As such, Pung et al. (2020a) argue that tourist transformation can consist in existential transformation, transformative learning, and/or behavioral change, according to the specific processes experienced during the trip.

Common dimensions of transformative travel experiences have also been revealed and defined by Soulard et al. (2020), in their development of a transformative travel experience scale. In different contexts, transformative experiences through tourism are characterized by four factors, named: local residents and culture, self-assurance, disorienting dilemma, and joy (Soulard et al., 2020). As such, tourists mainly experience transformation through interactions with local residents at the destination, and by learning about their perspectives and cultural heritage, which are critical for travelers to develop an integrated and expanded worldview (Decrop et al., 2018; Robledo & Batle, 2017). Joy prevails as a positive emotion that is felt by tourists experiencing transformation at the destination, especially when discovering a new culture and being immersed in a novel environment. By self-assurance, Soulard et al. (2020) indicate that transformed tourists report a sense of empowerment and self-growth, especially in feeling truer to themselves and more confident (Kirillova et al., 2017c; Laing & Frost, 2017; Wolf et al., 2017). A disorienting dilemma represents a common process which, according to Soulard et al. (2020), affects travelers only after they return to their home, who have difficulty in readapting to their previous daily life. Tourism dimensions and factors leading to tourists' attitude change, and drawing from both Pung et al. (2020a) and Soulard et al. (2020) studies, they are illustrated in detail in this chapter. Following Pung et al. (2020a)'s approach, as well as studies on attitude as a tripartite construct (Fishbein & Ajzen, 1975; Gnoth, 1994), transformative processes and outcomes of tourism experiences are hereby differentiated into emotional/existential, cognitive, and behavioral changes.

Emotional and Existential Changes through Tourism

Existential Transformation and Liminality

Within tourism, existential experiences are characterized by "personal or intersubjective feelings activated by the liminal process of tourist activities" (Wang, 1999, p. 351). Such condition is especially provided by travel as it consists in being in a physically distant and novel environment, which temporarily suspends societal constraints and norms (Brown, 2013; Graburn, 1983).

Liminality thus allows individuals to experience their authentic selves, and to feel free to engage in self-expression (Kirillova et al., 2017b; Turner, 1969; Wang, 1999). This results in tourists performing novel behaviors, roles and identities that are different from those embodied in their daily life, and which are more accepted in the liminoid state of tourism (Pung et al., 2020b; Wu et al., 2020). Such acceptance also provides opportunities for interactions and meaningful connections, as it allows to establish a sense of community (Kirillova et al., 2017b; Kontogeorgopoulos, 2017). Experiences of communitas have especially been described in relation to backpackers and volunteer tourists, who often connect with other travelers by engaging in the same activities, and by having flexible arrangements in the same extra-ordinary dimension (Bui & Wilkins, 2018; Kontogeorgopoulos, 2017). As such, the transformative force of liminality lies in the opportunities for tourists to encounter the Other, and thus experience moral and emotional changes, in their self-concept and in reading their past (Park, 2016). In the context of festivals, Wu et al. (2020) also argue that the liminal experience can contribute to long-lasting transformation, involving the development of alternative wider social connections, life-changing relationships, and emotional attachments.

Peak Episodes and the Role of Emotions

Existential transformation has been defined as a type of transformation through tourism that involves pre-travel expectations and post-travel meaning making (Kirillova et al., 2017b). According to Kirillova et al. (2017b), existential transformation is particularly facilitated by a triggering (or "peak") episode that functions as a surprising moment, and consists in recognizing the meaningfulness of their travel experience, and in feeling connected with other individuals. These peak experiences are also exemplified as reaching the top of a mountain, witnessing a striking local performance (Pung et al., 2020a), experiencing wildlife or impressive sceneries (Kirillova et al., 2017c). Peak episodes involve heightened cognition, as they are vividly recollected by transformed tourists, and often involve the realization of life transiency and connection to something (Kirillova et al., 2017c). These moments have a specific timing, often occurring towards the end of the trip. They have a strong emotional valence, whether positive (e.g., joy) or negative (e.g., horror, discomfort) (Arnould & Price, 1993; Rickly-Boyd & Metro-Roland, 2010; Saunders et al., 2017; Walker & Manyamba, 2020). Joy has been specifically identified as a factor of transformative tourism experiences, as an elation feeling from being immersed at the destination (Soulard et al., 2020), interacting with local residents and accessing fascinating cultures (Decrop et al., 2018). Emotions felt during the transformative experience are intense. Their recollection has an essential role in making sense of the triggering episode, pushing tourists to realize something new and to let go of reality as it was viewed before their trip (Kirillova et al., 2017c).

Another existential component of transformative tourism can be flow experiences, as they involve mindful travel and self-directed activities (Gnoth & Matteucci, 2014; Sheldon, 2020). Tourists can engage in non-reflective, but rewarding and skillful activities, and thus be "in the flow" (Graburn, 1983). This state involves practicing skills, a loss of sense of time, and being less conscious about the self (Csikszentmihalyi, 1990). The feeling of immersion and challenge in flow activities resembles rituals, which lead to reflecting on values of everyday life (Graburn, 1983). These flow experiences (e.g., dancing, rafting, mountaineering, scuba diving) can be transformational (Cater et al., 2020; Feder Mayer et al., 2019; Pomfret & Bramwell, 2016; Wu & Liang, 2011), in providing tourists with an opportunity to slow down and to nurture their traits (Sheldon, 2020). Compared to being measured through tourist satisfaction or attachment towards the destination, outdoor tourists' emotional responses have been recently argued to be best captured by flow experience. As such, not only do flow activities provide a sense of well-being and restoration

(Tsaur et al., 2013; Wöran & Arnberger, 2012), but they are also a means to understand tourists' transformed behavior (Kim & Thapa, 2018).

Existential Authenticity and Self-Awareness

Liminality, peak experiences and flow at the destination can lead tourists to obtain a sense of existential authenticity (Kirillova et al., 2017b), which consists in "a potential existential state of Being that is to activated by tourist experiences" (Wang, 1999, p. 352). Existential authenticity consists in the realization and negotiation of existential concerns (on life meaning, alienation, freedom) that are made more apparent in tourism contexts. Peak episodes at the destination can lead tourists to realizing the importance of meaningful relationships, or to have a clearer idea of their identity and of their desired path in life (Kirillova et al., 2017b). Thus, traveling can become a spiritual journey that helps individuals understand who they are, and make sense of their values (Kirillova et al., 2017b). Self-discovery can be a result of experiencing togetherness in family holidays, intimacy with others in recreation tourism or pilgrimage tourism, as well as of facing challenges in outdoor tourism (Steiner & Reisinger, 2006). According to Kirillova et al. (2017a), it has a greater effect when tourists engage with cultural heritage, sightseeing, a theme park, and in backpacking or volunteering, compared to the seaside or sunbathing. As such, tourism can be considered a catalyst for change when it allows to evaluate life and to explore the self (Brown, 2009, 2013). Self-exploration and better understanding of one's skills and socio-cultural identity during the experience are thus conducive to enhance travelers' sense of existential authenticity, self-awareness, and to redirect their life goals and priorities (Pung et al., 2020b).

Cognitive Changes through Tourism

Transformative Learning and Applications to Tourism

Transformation through tourism also involves cognition, as travelers mentally process and compare their observations to their past experiences and reference points (Teoh et al., 2021). Studies investigating transformation through tourism also use transformative learning theory, which was proposed in the discipline of education and adult learning, and draws on psychological and sociological perspectives (Mezirow, 1978, 1991). This theorized process of individuals acquiring new or transformed schemes or perspectives has been described to occur through ten phases. Individuals first experience a disorienting dilemma, and as a result undergo self-examination and engage in a critical assessment of previous assumptions, sharing similar processes of self-change with others, and exploring the adoption of new roles, relationships, and actions (Mezirow, 1991). These steps are then followed by planning a course of action, acquiring knowledge and skills for their plan, experimenting new roles, and building competence and self-confidence in performing new roles and relationships. As a final phase, transformed individuals reintegrate the new acquired perspective into their life (Mezirow, 1991). Transformative learning theory highlights the rational and cognitive aspects of learning, as it relates to the change of meaning schemes in relation to understanding the world and others (Pung et al., 2020a; Walter, 2016). Compared to the constructs and emotional perspectives surrounding the definition of existential transformation, the transformative learning process seems to consolidate external values towards objective facts, enhancing the cognitive component of tourists' changed attitudes (Gnoth, 1994).

Transformative learning theory has been employed in tourism research as a conceptual framework to explain how travelers make meaning of a disorienting dilemma encountered during the trip, and interpret their experience by engaging in reflection and consciousness about their pre-trip assumptions and worldviews (Reisinger, 2013). Transformative learning has been strongly linked to educational travel, specifically study abroad experiences, as overseas students absorb content that challenges their pre-assumptions (Stone & Duffy, 2015). The use of Mezirow's theory also revealed how volunteer tourists solve and overcome problems in an unfamiliar situation, develop relationships with individuals of different cultures, consequently feeling more resilient and enriched with new life perspectives (Coghlan & Weiler, 2018). Upon the travelers' return home, transformative experiences of volunteer tourism lead to reconcile and build learning skills, knowledge, and confidence to change individual behavior (Coghlan & Gooch, 2011). Phases of transformative learning have also been drawn and proposed by Wolf et al. (2017), as features of a transformation process to be implemented and experienced in protected areas and parks, for visitors to ultimately achieve personal benefits, such as competence and environmental appreciation.

Disorienting Dilemmas: Culture Shock, Challenges

When using Mezirow (1991)'s theory, experiencing a disorienting dilemma has been especially examined, to describe its triggering role in initiating tourists' self-change (Robledo & Batle, 2017; Walter, 2016). Disorienting dilemmas in tourism experiences are mainly initiated by experiencing culture shock and encountering challenges during the trip (Pung et al., 2020a). When visiting a novel and distant environment, tourists interact with individuals that are different from them: they are exposed to diversity, to different cultural beliefs, customs, ways of relating, and behaving (Brown, 2009; Lean, 2012; Taylor, 1994). A disorienting dilemma may arise with travelers starting to question their own frame of reference as a result (Robledo & Batle, 2017). As such, the culture shock deriving from experiencing an unfamiliar cultural and social environment can facilitate the occurring of a disorienting dilemma during the trip (Coghlan & Gooch, 2011). Social exchanges between local hosts and tourists, and intergroup social contact, provide transformative opportunities to learn about the visited region, to discover other cultures, to increase mutual understanding and empathy towards outgroups, and to reduce stereotypes or negative feelings towards perceived differences (Decrop et al., 2018; Li & Wang, 2020; Soulard et al., 2020).

Adversities and uncertainty faced while traveling can also contribute to self-change. Travelers can incur in an unexpected event or situation to be overcome (e.g., being away from home, living with strangers, working in alien social and cultural environments, managing travel plan disruptions, fear of being attacked, managing road accidents, facing prejudice, shyness, engaging in extreme sports) (Hischorn & Hefferon, 2013; Kontogeorgopoulos, 2017; Pung et al., 2020b; Robledo & Batle, 2017). These challenges, which can be mental, emotional, physical, or environmental, are perceived to occur outside of the travelers' control, and are addressed and solved successfully with courage (Hischorn & Hefferon, 2013; Wolf et al., 2017). Tourists that persevere and overcome challenges during their experience are transformed as they obtain a sense of achievement, as well as mastery of specific skills (Arnould & Price, 1993; Hischorn & Hefferon, 2013; Kontogeorgopoulos, 2017).

The Role of Reflection and Consciousness

As Walker and Manyamba (2020) note, it is how travelers respond to disorienting dilemmas that really determines whether a transformative learning process will occur or not. It is important for travelers to respond to disorienting dilemmas, facilitated by culture shock and challenges, with critical reflection of high cognitive level and rational discourse, as travelers make meaning of the salient events and emotions experienced during the trip (Cavender et al., 2020). Reflection represents a process and space for travelers to interpret the experience during and after the trip (Robledo & Batle, 2017). This facilitates self-exploration, and the contemplation of life choices and values, of tourists' past and future (Kontogeorgopoulos, 2017). More comprehensively, Coghlan and Weiler

(2018) define tourists' transformation as involving content reflection (of what they think and perceive), process reflection (of how they perform their functions), premise reflection (understanding the motives for their actions, and detecting change), and relational reflection (noting behavioral change in relating with others). Pung et al. (2020a) provide another distinction of transformative reflection processes, based on their "direction": the interpretation of the experience may be introspective and directed to the traveler's self, and leading to integrating values and changing life meaning. Conversely, transformative learning involves a reflection about the world and an interpretation of social interactions with others, and is therefore oriented towards the traveler's outside world (Pung et al., 2020a).

Existing research in transformative tourism has also discussed whether travelers' consciousness is mainly heightened at the destination, or after their trip (e.g., Pung et al., 2020a). Several scholars (e.g., Coghlan & Weiler, 2018; Kirillova et al., 2017b; Soulard et al., 2020) argue that, while first presented during the trip, the powerful effects of disorienting dilemmas are felt after travelers return to their everyday life, now met with alienation, and perceived as inauthentic and based on materialistic principles (Soulard et al., 2020). After the trip, they make sense of their predicaments and detect undergoing transformations (Kirillova et al., 2017b). In Pung et al. (2020b)'s study, reflection was found to occur during the tourism experience, and was connected to the need of space and time for travelers during the trip, to question and better understand themselves. As such, critical reflection has been addressed in tourism research, and different techniques can be cultivated during the travel experience, such as journaling, blogging, discussion, and dialogue, as well as storytelling, video journaling, poetry, games, silent reflection, or non-written communication (Knollenberg et al., 2014; Stone & Duffy, 2015; Teoh et al., 2021). Among transformative experiences allowing deep engagement with the place, Wolf et al. (2017) suggests meditative walks, involving stations presenting interpretative content with religious or cultural elements, or silent walks, focusing on introspective thought. Guiding can also promote reflection through dialogue, hands-on and art-based activities, as well as through allocating time for inactivity and deep introspection (Walter, 2016).

Cross-cultural Awareness, Self-efficacy and Empowerment

Among changes connected to the cognitive and learning sphere, exposure to the Other during travel leads to question subjective beliefs, stereotypes and prejudice, to then expand the travelers' worldview and frame of reference (Taylor, 1994). Open-mindedness about other cultures means that foreign beliefs and behaviors are better understood and tolerated through inter-cultural interactions in tourism (Brown, 2009; Taylor, 1994). Travelers can develop a sense of global citizenship and cross-cultural awareness, consisting in improved understanding and acceptance of culturally distant others, and an acquired international perspective (Bell et al., 2016; Reisinger, 2013). This also involves changed attitudes towards the hosts, and the development of qualities and skills in negotiating cross-cultural encounters (Paris et al., 2014; Reisinger, 2013). Increased cross-cultural awareness has been found in the context of studying abroad and volunteer tourism, and more generally when tourists feel challenged by cultural differences (Pung et al., 2020b). Challenges (exemplified above) being solved during the trip can also help travelers master specific skills, as well as reveal increased independence (Pung et al., 2020b). By overcoming unexpected issues and situations at the destination, tourists realize that they can take control and improve their functioning, boosting self-efficacy (Kakoudakis et al., 2017). Being empowered has been associated with women independent travel and the negotiation of constraints and social structures (Myers, 2017), and has been recently extended to represent an additional outcome of transformative learning, for both females and males who become more confident through tourism (Pung et al., 2020b; Soulard et al., 2020).

Behavioral Changes through Tourism

Transformative Tourism Consequences: Attitude Change

The previous sections discussed how contextual stimuli (e.g., challenges, cultural shock, and liminality) at the destination facilitate the occurring of peak episodes or disorienting dilemmas, which are reflected upon and interpreted. Such interpretation leads to emotional and existential changes in the travelers' self, as well as cognitive changes, as tourists may integrate new knowledge about themselves, about others, and the outside world. According to the theory of planned behavior, when new values, knowledge, and beliefs about an object are formed, they strongly influence and determine changes in attitude towards a behavior (Ajzen & Dasgupta, 2015; Gnoth, 1997; Katz, 1960). Changes in attitudes are a function of acquired behavioral beliefs and evaluative responses to these beliefs (Ajzen, 1991). The acquisition of environmental information leads to the emergence of social cognitions (values and attitudes), with the aim of optimal functioning (Homer & Kahle, 1988). In social adaptation theory, values are thus described as abstract social cognitions that reflect adaptation and form attitudes (Homer & Kahle, 1988). This also occurs in tourism contexts, where tourists' novel needs and a modified value system change their attitude towards an object (e.g., destination) and towards a behavior (Gnoth, 1997; Li et al., 2016). Through a reinforced transformative process, the temporary acquisition of habits and attitude change can lead to influence travelers' behavior, directed to address an important and newly configured perspective, meaning, or sense of self (Pung et al., 2020a).

Long-Term Behavioral Changes

With these premises, once travelers return to their home and everyday life, they may still be immersed in "relationships, conversations, roles, routines, sanctions, objects and symbols" that characterized their trip (Lean, 2012, p. 167). Individuals are informed by their heightened introspection at the destination, and continue to interpret their travel experience (Lean, 2012; Pung et al., 2020b). This allows them to consciously adopt new behaviors, or modify their lifestyle, in the long term, by establishing new routines or changing pre-existing performances (Lean, 2012). As exemplified by Brown (2013), individuals may attempt to live a more authentic life through small changes, such as following a healthier lifestyle, taking up a hobby, or using technological devices less. Together with improved cross-cultural awareness and open-mindedness from interacting with destination residents and different social groups, travelers can also show increased patience and maturity (Noy, 2004), a shift towards cultural cosmopolitanism (Cohen, 2010) and towards civic attitudes (Teoh et al., 2021). Interpersonal dynamics are also negotiated, with travelers attaching increased importance to meaningful relationships (Kirillova et al., 2017b), or rearranging pre-existing personal relationships (Lean, 2012; Pung et al., 2020b). Transformed tourists may also develop friendships with other travelers that reveal to be long lasting. Moreover, Müller et al. (2020) analyzed how individuals engaging in volunteer tourism broadened their perspectives and re-evaluated career choices, with travel experiences providing guidance on directing their professional paths. The development of job-market skills, such as resource management, self-evaluation, interpersonal social skills, or problem-solving (Pearce & Foster, 2007) as a result of transformative tourism, can also enhance employability (Brown, 2009; Inkson & Myers, 2003; Müller et al., 2020; O'Reilly, 2006).

Enhancing Sustainable Practices

Experiencing harmony with nature is a key dimension of tourism extraordinary experiences (Arnould & Price, 1993). According to Weaver (2005), comprehensive ecotourism can involve a combination of an emphasis of sustainability on a global level, donations, and volunteer activity engagement, thus providing opportunities for reclaiming and rehabilitating human-altered habitat. Ecotourism not only improves natural settings and enhances their protection but also has the potential to provide a space promoting natural values and physical health (Cheng et al., 2014). Through these cues and by abiding to the eco-site rules, tourists can develop a greater appreciation for natural settings, and better understand the environmental impact of their behavior, thus experiencing changes in their attitude towards the environment (Cheng et al., 2014; Puhakka, 2011; Pung et al., 2020b). More comprehensively, tourists participating in community-based ecotourism experiences can learn about the environment as well as be educated on preserving local cultures. This type of experience, along with educators and guides acting as role models, can ultimately foster tourists' sustainable practices (Walter, 2016). Environmental responsible behavior would then involve adopting actions such as increased energy management, green consumption, waste recycling, and political and community activism (Iwata, 2001; Thapa, 2010). This behavior is especially shaped during and after the (eco-)tourism experience, by the tourists' perceived benefit from the activities characterizing the experience, and degree of involvement and satisfaction in these, so programs and activities need to be designed to involve active engagement and interpretation (Chiu et al., 2014; Moscardo, 2017).

Discussion and Conclusion

Attitude change is conveyed by specific socio-psychological aspects of tourism experiences, with emotional, existential, and cognitive connotations. This chapter attempted to link the construct of attitude, discussed in social psychology, to individual behavior and how it can be transformed through travel experiences (Li et al., 2016). Emotional and existential changes are facilitated by the liminoid dimension of tourism, which allows for self-exploration outside of daily social structures, and the development of communitas with fellow travelers. Cathartic episodes and intensified emotions during the trip are also catalysts for subverting tourists' existential values and life meaning. These factors, together with flow experiences of being immersed in rewarding activities and practicing skills at the destination, all contribute to increase the travelers' self-awareness, and to enhance their sense of existential authenticity. Tourists' changes in attitudes are also conveyed through overcoming challenges and experiencing culture shock, which can represent a disorienting dilemma faced by travelers at the destination, and lead to the acquisition of new knowledge and an integrated perspective (i.e., cognitive changes). The tourists' process of reflection was highlighted in this chapter as having an essential role, in both allowing travelers to be mindful of changing meanings during the trip, and in interpreting the whole experience post-travel and thus realizing self-change. Developing self-efficacy, cross-cultural understanding, and empowerment are outcomes of these processes, which, together with existential changes, can influence and transform travelers' attitudes. Changes in behavior take place when travelers return home, and they modify their social relationships, career paths, and adopt environmentally friendly practices. The conceptualization of these processes allowed for a critical examination of the interplay between travel experiences and attitude change.

The full account of these constructs allows to draw conclusions about the extended potential of transformative tourism experiences, which lead to the examined changes in travelers. Among identified self-change processes, intergroup social exchanges were discussed to play an important role in changing tourists' attitudes, in favor of a greater empathy, mutual understanding, and a reduction of negative emotions and stereotypes being linked to different cultural groups (Li & Wang, 2020). Based on this rationale, transformative experiences contribute to the hopeful tourism paradigm, further promoting reciprocity and ethical principles in the travel and tourism industry. Such enriched framework of tourist transformation also provides a step forward towards travelers' awareness on aspects of social and environmental justice, and towards better understanding sustainability opportunities in tourism (Stone & Duffy, 2015). From a practical perspective, this chapter also provides insights on the variety of factors to be focusing on when designing tourist experiences that provide self-change. Emotions should be intensified, for example, through the conscious creation of places and routes with unique natural features and inspiring aesthetics (Sheldon, 2020). Challenging experiences, such as adventure activities and flow practices, that engage tourists' presence and whole self, can also generate transformative moments by conveying a personal sense of achievement (Cheers et al., 2017). These and other types of experiences have a pivotal role in generating changes in attitudes of tourists, whose transformation can lead to better host-guest relationships, and in the longer term have positive impacts on society.

References

Ajzen, I. (1991). The theory of planned behavior. *Organizational Behavior and Human Decision Processes, 50,* 179–211.

Ajzen, I., & Dasgupta, N. (2015). Explicit and implicit beliefs, attitudes, and intentions: The role of conscious and unconscious processes in human behavior. In P. Haggard & B. Eitam (Eds.), *The sense of agency* (pp. 115–144). Oxford University Press.

Arnould, E. J., & Price, L. L. (1993). River magic: extraordinary experience and the extended service encounter. *Journal of Consumer Research, 20,* 24–45.

Bell, H. L., Gibson, H. J., Tarrant, M. A., Perry, L. G., & Stoner, L. (2016). Transformational learning through study abroad: US students' reflections on learning about sustainability in the South Pacific. *Leisure Studies, 35*(4), 389–405.

Brown, L. (2009). The transformative power of the international sojourn: An ethnographic study of the international student experience. *Annals of Tourism Research, 36*(3), 502–521.

Brown, L. (2013). Tourism: A catalyst for tourism authenticity. *Annals of Tourism Research, 40,* 176–190.

Bui, H. T., & Wilkins, H. C. (2018). Social interactions among Asian backpackers: Scale development and validation. *Current Issues in Tourism, 21*(10), 1097–1114.

Cater, C., Albayrak, T., Caber, M., & Taylor, S. (2020). Flow, satisfaction, and storytelling: A causal relationship? Evidence from scuba diving in Turkey. *Current Issues in Tourism,* DOI: 10.1080/13683500.2020.1803221

Cavender, R., Swanson, J. R., & Wright, K. (2020). Transformative travel: Transformative learning through education abroad in a niche tourism destination. *Journal of Hospitality, Leisure, Sport & Tourism Education, 27,* Article 100245. https://doi.org/10.1016/j.jhlste.2020.100245

Çelik, S. (2019). Does tourism reduce social distance? A study on domestic tourists in Turkey. *Anatolia, 30*(1), 115–126.

Cheers, J. M., Belhassen, Y., & Kujawa, J. J. (2017). Spiritual tourism. *Tourism Management Perspectives, 24,* 186–187.

Cheng, M., Jin, X., & Wong, I. A. (2014). Ecotourism site in relation to tourist attitude and further behavioural changes. *Current Issues in Tourism, 17*(4), 303–311.

Chiu, Y. H., Lee, W.-I., & Chen, T.-H. (2014). Environmentally responsible behavior in ecotourism: Antecedents and implications. *Tourism Management, 40,* 321–329.

Christie, M. F., & Mason, P. A. (2003). Transformative tour guiding: Training tour guides to be critically reflective practitioners. *Journal of Ecotourism, 2*(1), 1–16.

Cocolas, N., Walters, G., Ruhanen, L., & Higham, J. (2020). Air travel attitude functions. *Journal of Sustainable Tourism, 28*(2), 319–336.

Coghlan, A., & Gooch, M. (2011). Applying a transformative learning framework to volunteer tourism. *Journal of Sustainable Tourism, 19*(6), 713–728.

Coghlan, A., & Weiler, B. (2018). Examining transformative processes in volunteer tourism. *Current Issues in Tourism, 21,* 567–582.

Cohen, S. A. (2010). Chasing a myth? Searching for 'self' through lifestyle travel. *Tourist Studies, 10*(2), 117–133.

Cohen, S. A., & Cohen, E. (2019). New directions in the sociology of tourism. *Current Issues in Tourism, 22*(2), 153–172.

Csikszentmihalyi, M. (1990). *Flow: The psychology of optimal experience.* New York: Harper & Row.

Da Silva, R. V., & Alwi, S. F. S. (2006). Cognitive, affective attributes and conative, behavioural responses in retail corporate branding. *Journal of Product & Brand Management, 15*(5), 293–305.

Decrop, A., Del Chiappa, G., Mallargé, J., & Zidda, P. (2018). "Couchsurfing has made me a better person and the world a better place": The transformative power of collaborative tourism experiences. *Journal of Travel & Tourism Marketing, 35*(1), 57–72.

Feder Mayer, V., dos Santos Machado, J., Marques, O., & Gonçalves Nunes, J. M. (2019). Mixed feelings? Fluctuations in well-being during tourist travels. *The Service Industries Journal*. https://doi.org/10.1080/02642069.2019.1600671.

Fishbein, M., & Ajzen, I. (1975). *Belief. Attitude, intention and behaviour: An introduction to theory and research*. Addison Wesley.

Gnoth, J. (1994). *Expectations and satisfaction in tourism: An exploratory study into measuring satisfaction* [Doctoral Dissertation, University of Otago]. University of Otago Library. http://hdl.handle.net/10523/3617

Gnoth, J. (1997). Tourism motivation and expectation formation. *Annals of Tourism Research, 24*(2), 283–304.

Gnoth, J., & Matteucci, X. (2014). A phenomenological view of the behavioral tourism research literature. *International Journal of Culture, Tourism and Hospitality Research, 8*(1), 3–21.

Graburn, N. H. H. (1983). The anthropology of tourism. *Annals of Tourism Research, 10*(1), 9–33.

Hadinejad, A., Noghan, N., Moyle, B. D., Scott, N., & Kralj, A. (2019). Future research on visitors' attitudes to tourism destinations. *Tourism Management, 83*, 104215.

Hischorn, S., & Hefferon, K. (2013). Leaving it all behind to travel: Venturing uncertainty as a means to personal growth and authenticity. *Journal of Humanistic Psychology, 53*(3), 283–306.

Homer, P., & Kahle, L. (1988). A structural equation test of the value-attitude-behavior hierarchy. *Journal of Personality and Social Psychology, 54*(4), 638–646.

Inkson, K., & Myers, B. (2003). "The big OE": Self-directed travel and career development. *Career Development International, 8*(4), 170–181.

Iwata, O. (2001). Attitudinal determinants of environmentally responsible behaviour. *Social Behavior and Personality, 29*(2), 183–190.

Kakoudakis, K. I., McCabe, S., & Story, V. (2017). Social tourism and self-efficacy: Exploring links between tourism participation, job-seeking and unemployment. *Annals of Tourism Research, 65*, 108–121.

Katz, D. (1960). The functional approach to the study of attitudes. *Public Opinion Quarterly, 24*(2), 163–204.

Kim, M., & Thapa, B. (2018). Perceived value and flow experience: Application in a nature-based tourism context. *Journal of Destination Marketing & Management, 8*, 373–384.

Kirillova, K., Lehto, X. & Cai, L. (2017a). Existential authenticity and anxiety as outcomes: The tourist in the experience economy. *International Journal of Tourism Research, 19*(1), pp. 13–26.

Kirillova, K., Lehto, X., & Cai, L. (2017b). Tourism and existential transformation: An empirical investigation. *Journal of Travel Research, 56*(5), 638–650.

Kirillova, K., Lehto, X., & Cai, L. (2017c). What triggers transformative tourism experiences? *Tourism Recreation Research, 42*(4), 498–511.

Knollenberg, W., McGehee, N. G., Bynum Boley, B., & Clemmons, D. (2014). Motivation-based transformative learning and potential volunteer tourists: Facilitating more sustainable outcomes. *Journal of Sustainable Tourism, 22*(6), 922–941.

Kontogeorgopoulos, N. (2017). Finding oneself while discovering others: An existential perspective on volunteer tourism in Thailand. *Annals of Tourism Research, 65*, 1–12.

Laing, J. H., & Frost, W. (2017). Journeys of well-being: Women's travel narratives of transformation and self-discovery in Italy. *Tourism Management, 62*, 110–119.

Lean, G. L. (2012). Transformative travel: A mobilities perspective. *Tourist Studies, 12*(2), 151–172.

Lee, T. H. (2009). A structural model to examine how destination image, attitude, and motivation affect the future behavior of tourists. *Leisure Sciences, 31*(3), 215–236.

Li, F., & Wang, B. (2020). Social contact theory and attitude change through tourism: Researching Chinese visitors to North Korea. *Tourism Management Perspectives, 36*, Article 100743. https://doi.org/10.1016/j.tmp.2020.100743

Li, M., Cai, L. A., & Qiu, S. (2016). A value, affective attitude, and tourist behavioural intention model. *Journal of China Tourism Research, 12*(2), 179–195.

Lutz, R. J. (1981). The role of attitudes in the theory of marketing. In H. H. Kassarjan & T. S. Robertson (Eds.), *Perspectives in Consumer Behavior* (3rd ed., pp. 233–249). Scott, Foresman.

Mezirow, J. (1978). Perspective transformation. *Adult Education, 28*(2), 100–110.

Mezirow, J. (1991). *Transformative dimensions of adult learning*. Jossey-Bass.

Moscardo, G. (2017). Critical reflections on the role of interpretation in visitor management. In J. N. Albrecht (Ed.), *Visitor management in tourism destinations* (pp. 170–187). CABI.

Müller, C. V., Scheffer, A. B. B., & Closs, L. Q. (2020). Volunteer tourism, transformative learning and its impacts on careers: The case of Brazilian volunteers. *International Journal of Tourism Research, 22*, 726–738.
Myers, L. M. (2017). Independent women travelers' experiences and identity development through multi-sensual experiences in New Zealand. In C. Khoo-Lattimore & E. Wilson (Eds.), *Women and travel: Historical and contemporary perspectives*. Apple Academic Press.
Noy, C. (2004). This trip really changed me: Backpackers' narratives of self-change. *Annals of Tourism Research, 31*(1), 78–102.
O'Reilly, C. (2006). From drifter to gap year tourist: Mainstreaming backpacker travel. *Annals of Tourism Research, 33*(4), 998–1017.
Paris, C. M., Nyaupane, G. P., & Teye, V. (2014). Expectations, outcomes and attitude change of study abroad students. *Annals of Tourism Research, 48*, 275–277.
Park, H. (2016). Tourism as reflexive reconstructions of colonial past. *Annals of Tourism Research, 58*, 114–127.
Pearce, P., & Foster, F. (2007). A "University of Travel": Backpacker learning. *Tourism Management, 28*(5), 1285–1298.
Pomfret, G., & Bramwell, B. (2016). The characteristics and motivational decisions of outdoor adventure tourists: A review and analysis. *Current Issues in Tourism, 19*(14), 1447–1478.
Puhakka, R. (2011). Environmental concern and responsibility among nature tourists in Oulanka Pan Park, Finland. *Scandinavian Journal of Hospitality and Tourism, 11*(1), 76–96.
Pung, J. M., Gnoth, J., & Del Chiappa, G. (2020a). Tourist transformation: Towards a conceptual model. *Annals of Tourism Research, 81*, Article 102885. https://doi.org/10.1016/j.annals.2020.102885
Pung, J. M., Yung, J. M., Khoo-Lattimore, C., & Del Chiappa, G. (2020b). Transformative travel experiences and gender: A double duoethnography approach. *Current Issues in Tourism, 23*(5), 538–558.
Reisinger, Y. (2013). *Transformational tourism: Tourist perspectives*. CABI.
Rickly-Boyd, J. M., & Metro-Roland, M. M. (2010). Background to the fore: The prosaic in tourist places. *Annals of Tourism Research, 37*(4), 1164–1180.
Robledo, M. A., & Batle, J. (2017). Transformational tourism as a hero's journey. *Current Issues in Tourism, 20*(16), 1736–1748.
Saunders, R., Weiler, B., & Laing, J. (2017). Transformative guiding and long-distance walking. In S. Filep, J. Laing & M. Csikszentmihalyi (Eds.), *Positive tourism* (pp. 167–184). Routledge.
Sheldon, P. J. (2020). Designing tourism experiences for inner transformation. *Annals of Tourism Research, 83*, Article 102935. https://doi.org/10.1016/j.annals.2020.102935
Soulard, J., McGehee, N., & Knollenberg, W. (2020). Developing and testing the Transformative Travel Experience Scale (TTES). *Journal of Travel Research, 60*(5), 923–946.
Steiner, C. J., & Reisinger, Y. (2006). Understanding existential authenticity. *Annals of Tourism Research, 33*(2), 299–318.
Stone, G. A., & Duffy, L. N. (2015). Transformative learning theory: A systematic review of travel and tourism scholarship. *Journal of Teaching in Travel & Tourism, 15*(3), 204–224.
Taylor, E. W. (1994). Intercultural competency: A transformative learning process. *Adult Education Quarterly, 44*(3), 154–174.
Teoh, M. W., Wang, Y., & Kwek, A. (2021). Conceptualising co-created transformative tourism experiences: A systematic narrative review. *Journal of Hospitality and Tourism Management, 47*, 176–189.
Thapa, B. (2010). The mediation effect of outdoor recreation participation on environmental attitude-behavior correspondence. *The Journal of Environmental Education, 41*(3), 133–150.
Tsaur, S. H., Yen, C. H., & Hsiao, S. L. (2013). Transcendent experience, flow and happiness for mountain climbers. *International Journal of Tourism Research, 15*(4), 360–374.
Turner, V. (1969). *The ritual process: structure and anti-structure*. Aldine.
Walker, J., & Manyamba, V. N. (2020). Towards an emotion-focused, discomfort-embracing transformative tourism education. *Journal of Hospitality, Leisure, Sport & Tourism Education, 26*, Article 100213. https://doi.org/10.1016/j.jhlste.2019.100213
Walter, P. G. (2016). Catalysts for transformative learning in community-based ecotourism. *Current Issues in Tourism, 19*(13), 1356–1371.
Wang, N. (1999). Rethinking authenticity in tourism experience. *Annals of Tourism Research, 26*(2), 349–370.
Weaver, D. B. (2005). Comprehensive and minimalist dimensions of ecotourism. *Annals of Tourism Research, 32*(2), 439–455.
Wolf, I. D., Ainsworth, G. B., & Crowley, J. (2017). Transformative travel as a sustainable market niche for protected areas: A new development, marketing and conservation model. *Journal of Sustainable Tourism, 25*(11), 1650–1673.

Wöran, B., & Arnberger, A. (2012). Exploring relationships between recreation specialization, restorative environments and mountain hikers' flow experience. *Leisure Sciences, 34*(2), 95–114.
Wu, C. H. J., & Liang, R. D. (2011). The relationship between white-water rafting experience formation and customer reaction: A flow theory perspective. *Tourism Management, 32*(2), 317–325.
Wu, S., Li, Y., Wood, E. H., Senaux, B., & Dai, G. (2020). Liminality and festivals – insights from the East. *Annals of Tourism Research, 80*, Article 102810. https://doi.org/10.1016/j.annals.2019.102810

14

DOES TOURISM IMPACT ON PREJUDICE, DISCRIMINATION, ASSIMILATION, GENOCIDE, SEGREGATION, INTEGRATION?

Buket Buluk Eşitti and Erol Duran

The Integrated Threat Theory

The integrated threat theory is designed to determine the dimensions of prejudices and attitudes of individuals toward all outgroups, including gender, race, sexual orientation, national origin, and disability (Stephan et al., 2000a, p. 64). The integrated threat theory in its most general expression deals with the situation where one group sees the existence of another as a threat to itself. The theory offers a new perspective to examine biased attitudes that feed from intergroup relations and threat definitions by synthesizing various theoretical perspectives on the relationships between threat perceptions and intergroup attitudes. Integrated threat theory is important for intercultural research, as it helps to explore social issues and to understand the cognitive processes behind people's attitudes, beliefs, and behaviors. Some of the applicable research topics of the integrated threat theory are expressed by Redmond (p. 3), "religious intolerance", "public attitudes towards immigration", "racial profiling and stereotyping", "public attitudes toward same gender relationships", "support for feminist movements", "diversity" and "national identity", and "different motives in the workplace".

The integrated threat theory suggests that many variables are effective on threats, and thus attitudes also affect attitudes indirectly. In cases where intergroup solidarity emerges strongly, threat perception toward the outgroup may be higher. Negative contacts with outgroup members again increase the level of perceived threat. However, the scarcity of in-group members' knowledge of the outgroup, in other words, knowing little about the outgroup members, is another factor that causes the perceived threat to rise (Stephan & Stephan, 2017).

Cross-cultural differences have been the main determinants of the categories humanity has built throughout history and the resulting ways of treating the other (Hanel et al., 2018). While power and ability were the determinants of differences between groups in the primitive ages, culture became the main qualifier and descriptor of this difference in time. Culture is not a reference that humanity only uses when defining the other. Culture has also come to be seen as a basic reference source that should be preserved and transferred between generations because of the belief that it contains the codes of being different from the "opposite" – "other" and naturally being similar within itself and providing ontological continuity. Thus, the habit of groups with different cultural references to perceive other groups as a threat to their own culture has emerged (Stephan & Stephan, 2017). This habit, in which humanity is still in the grove today, can be counted as one of the main reasons for the prejudices against the alien-different/not described in "us" to remain alive.

DOI: 10.4324/9781003161868-14

Within "the context of the integrated threat theory, it is assumed that there are four basic threat factors. These threats are "realistic threat", "symbolic threat", "negative stereotype" and "intergroup anxiety" (Colombo et al., 2012, p. 135).

Realistic threats include perceived threats to the physical and material well-being of a group and its members (Stephan et al., 1999). Realistic threats are mainly economic, physical, and political (Stephan & Stephan, 1993). According to González et al. (2008), realistic threats arises because of competition over scarce resources such as houses and jobs or when the ingroup feels their resources are being threatened by the outgroup. The desire of the ingroup to protect their resources becomes the motivating factor behind prejudice, negative attitudes, and discrimination toward members of the outgroup. This often occurs when groups living together in a shared context compete for scarce resources and develop conflicting goals (Curseu et al., 2007).

Symbolic threats include threats to the world view of a group and are related to values, norms, beliefs, and attitudes originating from different cultures (Ward & Berno, 2011, p. 1559). It is also possible that symbolic threats lead to the most vicious behavioral responses to outgroups such as genocide, segregation, integration, torture, and mutilation. In the context of immigration policy, symbolic threats would be expected to be linked to a preference for the assimilation of outgroups.

Negative stereotypes create threats by negatively creating the behavior and expectations of outgroup members (Riek et al., 2006, p. 338). When individuals have stereotypes of an outgroup, they expect the group to behave in a certain manner. Negative outgroup stereotypes are related to feelings of threat and fear (Craig & Richeson, 2012).

Intergroup anxiety, which is the last threat element that constitutes the integrated threat theory, includes the fear of rejection or the feeling of threat in relation to the interactions between people, the individual's perception of himself as inadequacy in active interaction with the members of the group, and the resulting behaviors resulting from being mocked or embarrassed (Ward & Berno, 2011, p. 1559). Gudykunst (1993) explained how individuals take multiple steps to reduce such anxiety in his anxiety/uncertainty management theory. Research showed that prejudice increases as intergroup anxiety increases (Islam & Hewstone, 1993). Stephan and Renfro (2002) asserted that a distinction should be made between individual and group-level processes and whether threats are made toward individuals or groups. Intergroup anxiety is a feeling of being personally threatened during interaction(s) with outgroup members, while realistic threats, symbolic threats, and stereotypes are directed mainly at the in-group. This distinction between groups versus individual-level threats has been supported (Tam et al., 2007).

In the light of all the mentioned above, the theoretical basis of this chapter was formed within the scope of the concepts of integrated threat theory, prejudice, discrimination, assimilation, genocide, segregation, and integration.

The Impact of Tourism on Prejudice, Discrimination, Assimilation, Genocide, Segregation, Integration

Prejudice

Stereotypes and prejudices regarding different cultures begin to create obstacles to positive dialogue and cooperation with the intensification of intercultural relations. Generally, stereotypes and prejudices are two different concepts. Stereotypes are based on cognitive components in our perceptions of group members (Stangor, 2014). However, prejudices arise before the truth is known (Gürses, 2005). Although stereotypes are generally equated with the concept of prejudice in a narrower sense, prejudices can arise not only against countries or cultures but also against sub-segments and different layers of the same group (Marshall, 2003).

In addition to our stereotypes, we may also develop prejudice as an unjustifiable negative attitude toward and outgroup or its members (Stangor, 2014). being able to express prejudice as a judgment or concept based on insufficient evidence about a subject or a society in advance, without thinking about it and researching (Avcıkurt, 2015), it is only a matter of concern to a certain group members. It is an image that is formed and spread in the society and emphasizes selectivity in the perception of reality (Mendras, 2009).

Stating that the basic issues that need to be examined in order to understand the causes of prejudice can be examined the relationship between stereotypes and prejudices, Stephan and Stephan (1996, p. 409) found a consistent and significant relationship between stereotypes and prejudices in a study they conducted. The authors also emphasized that fear or various threats can play a prejudiced role and can be a recurring problem in the intergroup relations literature. Prejudices arise from groups with different cultures, evaluating other groups according to the social characteristics of their own groups. When it comes to the interaction of culturally different groups, the dominant group often reveals prejudices when one group tries to overcome the other in different ways. On the other hand, the group that cannot be dominant can have hostile feelings toward the dominant group (Gürses, 2005, p. 144).

Lippmann (1998, p. 99) emphasized that another consequence of inadequate communication, which emerges as a different problem in intergroup relations, is being misunderstood and other misunderstandings. The author underlined that information that will eliminate being misunderstood and other misunderstandings can be realized through an effective communication process, away from the negative effects of stereotypes and prejudices. Generally, demographic variables such as age, education, socioeconomic status, gender, and ethnicity are associated with prejudice, but these relationships provide relatively little information about the origin of bias (Stephan & Stephan, 1996, p. 410). For example, Brigham (1971) found that the emergence of stereotypes between Africans and Americans is significantly related to their attitudes toward each other, as well as individual characteristics.

Due to people have stereotypes or prejudices toward those who are not like them, it becomes difficult for societies to get closer and to get to know each other. This situation causes societies to keep their distance from each other and creates a feeling of hostility between them (Choi et al., 2017). One of the most important ways of this prevention situation is to increase contact between communities and allow them to get to know each other. One of the most important tools to achieve this theme is the tourism sector, which creates opportunities to travel and get to know other peoples and cultures (Zaei & Zaei, 2013).

Tourism "is not the only sector that has an economic effect, but it also is a tool that procures interaction between societies. Tourism is a mind-opening experience, which teaches people that the world is not made up of a single model of living and other models of living exist. Many studies have demonstrated that this experience changes the attitudes of people (Wintersteiner & Wohlmuther, 2014).

The change of prejudices can be achieved through interaction and communication. In this case, the tourism sector can be an important tool for this. As a matter of fact, through tourism, different tourist groups and societies know and interact with each other, reducing prejudices between societies and breaking down stereotypes. In its simplest form, changes in the attitudes of tourism workers toward different groups are a good example.

Fisher and Price (1991) determined in their research that holiday satisfaction, and intercultural interaction are effective in changing attitudes after vacation. The researchers also determined that travel motivation indirectly influences the change of attitudes. Anastasopoulos (1992) determined in his study that the attitudes of Greek tourists who visited Turkey changed negatively. The author explained the reason for this with the influence of "quality of life", "institutions", and "culture" factors in Turkey. Sirakaya-Turk et al. (2014) revealed in their research that the prejudicial attitudes of German tourists who visited Turkey through package tours toward Turks changed negatively after the trip. It was determined that the reason for this was the "dissatisfaction with the tour guide" and "the shopping experiences".

In their study Scott-Thomas et al. (2014) have stated that individually visiting tourists may be more effective in decreasing prejudices. Çelik (2019)'s study aimed to determine the pre-post-travel attitudes of local tourists who visited Şırnak, located in the Southeastern Anatolia Region of Turkey, regarding Şırnak. As a result of the study, it was concluded that the prejudicial attitudes of most of the tourists toward Şırnak before the trip are neutral, they get negative reactions from their surroundings when they say that they will visit Şırnak, their attitudes before the trip change positively after the trip, they are satisfied with the tour and it is effective in changing the biased attitudes of tourism.

Lastly, Çelik (2019b) in his study aimed to determine whether attitudes (prejudices and stereotypes) of domestic tourists in Turkey toward South-easterners change after their travel experiences. Additionally, data was gathered during the process of study from people who have never visited the Southeast to create a comparison group. As a result of the study, it was seen that the prejudices of participants who have never come to the Southeast have been higher. Also, positive changes were determined in the attitudes (prejudices, positive attitudes) of people who have undergone experiences of travel.

Consequently, it can be stated that as a result of the tourists developing their attitudes by ignoring the cultural differences of each other during the interaction processes, the perception of threat is realized on the sides. It was stated that perceived threats consist of four types of threats, including realistic threats, symbolic threats, negative stereotypes, and intergroup anxiety, which include the integrated threat theory. It is stated that negative thoughts and prejudices emerge as a result of the perception of integrated threats. It is observed that the prejudice of the tourists to each other in the mutual communication process causes tensions and discriminatory behaviors between the groups.

Discrimination

"Discrimination" comes from the Latin word *discriminare* and is defined as the transformation of prejudices against individuals into behavior due to their characteristics such as gender, race, religion, language and color (Wood et al., 2013). In other words, discrimination can be expressed as treating people belonging to one group differently than people belonging to another group (Lahey, 2005). In this context, this concept, which is expressed as discriminatory attitudes and behaviors toward people with certain characteristics, has a generally negative meaning (Metin, 2018, p. 73).

In the relevant literature, discrimination is discussed in three ways: directly, indirectly, and positively (Cankurtaran & Beydili, 2016). Direct discrimination means the application of an administrative act or a legal regulation against discrimination and individuals taking different actions based on the prohibition of discrimination against individuals or groups of different languages, religions, races, ages, and sexual orientations (Manav, 2013). Indirect discrimination points to the situation where the results of an application that is apparently equally valid for everybody and which is not discriminatory affect the individuals included in a certain group negatively (Arısoy & Demir, 2007). Positive discrimination is expressed as the name given to the practices developed in favor of the groups who are exposed to discrimination in society and consequently have limited or no access to certain rights. These practices generally occur in the fields of health, education, and employment (Metin, 2018).

In addition to the above, everyday discrimination is a subset of general discrimination because it reflects thoughts and beliefs about experiencing discrimination that are chronic or episodic but generally minor (Williams & Mohammed, 2009). Everyday discrimination is not restricted to tourists, and even individuals living within the same community (i.e., residents) could experience discrimination. For instance, residents could be discriminated against when they interacted with others on a daily basis in various contexts based on their perceived social categorical membership (e.g., Asian Americans and African Americans). The existence of social categorization differentiates residents into" "us" and "them", which could stimulate discriminatory actions. Tourists, due to more obvious social differences, could experience more discrimination compared to residents. The resident-to-resident everyday discrimination (i.e., by Americans against their fellow residents) reflects residents' social dynamics and could serve as a reference on social group relations. It is possible that residents could project such discriminatory dynamics onto tourists for the sake of assimilation needs against an outside reference group. Such actions could affect host-guest relations that are important for a sustainable tourism development (Tse & Tung, 2020).

Reports of discrimination have further exacerbated and discrimination against Chinese tourists have spread quickly with the COVID-19 pandemic (Devakumar et al., 2020). In South Korea, some business owners was barred Chinese from private establishments (Fottrell, 2020). In Japan, residents was avoid contact with Chinese (Ipsos MORI, 2020). In England, abuses against Chinese in public was reported (Preston-Ellis, 2020). These actions have sparked a deluge of concerns about host-guest relations and the broader challenges of managing discrimination between social groups.

Though some research has focused on the extent of host contact enjoyed by international sojourners, there has been little research into the incidence and impact of discrimination against international students, either racial or religious, which is described by Pai (2006) as a hidden problem. One of these research, Brown and Jones (2013)'s article makes a contribution to the existing and extensive literature on the international student experience by reporting on the incidence of racism and religious incidents experienced by international students at a university in the south of England. It was seen that out of a survey of 153 international postgraduate students, 49 had experienced some form of abuse. In most cases, this took the form of verbal abuse though racism manifested physically for nine students. In addition, strong emotional reactions were reported, including sadness, disappointment, homesickness and anger. It was also seen that there was a consequent reluctance to return to the UK as a leisure tourist or to offer positive word of mouth to future students. This research was offered a portrait of the reception offered to international students against a backdrop of increased racism in the UK. A link was thus made between the micro experience and macro forces. Implications of racist abuse for student satisfaction and future international student recruitment were drawn.

Lee and Scott (2017) investigated African Americans' travel behavior using Bourdieu's concept of habitus and vignette technique. In-depth, face-to-face interviewees were conducted with 13 middle class African Americans. Five salient themes were identified: (1) racial discrimination during traveling, (2) indirect experience of racism, (3) fear of racism (4) Black travel habitus: racism-related travel choice, and (5) accommodating park officials. The findings showed that informants' strong fear of racism is manifested in their distinctive travel behavior. They affirmed that African Americans' travel pattern needs to be conceived as a defensive mechanism against potential racial discrimination.

Philipp (1994) explored racial differences in reported tourism preferences. In his study, respondents were tested using statements related to four basic travel preference dichotomies: dependence vs. autonomy, activity vs. relaxation, order vs. disorder, and familiarity vs. novelty. Data from 213 randomly selected black (n = 96) and white (n = 117) respondents in a Southeastern US metropolitan area provided evidence that some preferences can be significantly associated with race. In addition to the usual theoretical explanations for differences in black/white leisure behavior, marginality, and ethnicity, it appeared an understanding of racial prejudice and discrimination may help explain some differences in travel preferences."

On the other hand, "residents' discrimination against tourists has long been a major problem. Previous studies have documented discrimination in products and pricing strategies against tourists (Sharifi-Tehrani et al., 2013; Chiaravutthi, 2019). Tourists have also been harassed by locals when they are seen as loud and unreasonable. Discrimination against tourists through low levels of threats (e.g., anti-tourist messages) and extreme violence (e.g., physical attacks) have been reported (Cheung & Li, 2019).

Tse and Tung (2020)'s research aimed to uncovers the relationships between (1) everyday discrimination perceived by residents, (2) the types of harmful actions they may perform on tourists, and (3) their support for further discriminatory responses against tourists in the context of a major societal event (i.e., in this case, COVID-19). The research showed that initial evidence that residents who perceive everyday discrimination are more likely to adopt harmful and discriminatory responses against tourists. For instance, it was seen that the residents who reported everyday discrimination (e.g., been called names or insulted) were more likely to have performed harmful actions against tourists (e.g., be unfriendly or mock a tourist) and support further discriminatory responses amidst COVID-19 (e.g., exclude tourists from public and private spaces). It was also seen that the residents were not more sympathetic to discrimination after having perceived discrimination themselves; instead, it was seen that there was a significant relationship from everyday discrimination to supporting harmful and discriminatory responses against tourists.

Consequently, it was stated that intergroup anxiety, which is among the threat variables in the integrated threat theory, is the anxiety individuals experience when they come into contact with or think that they will come into contact with people from different cultures or groups. As a result of the studies conducted in this sense, it is noteworthy that people think that they will experience feelings such as embarrassment, tension, and anxiety as a result of contact with the outgroup, they will be physically harmed, discriminated against, and negatively evaluated by the outgroup.

Assimilation

Assimilation has been the hegemonic theory of ethnic group relations in sociology, during the past century and refers to the social, economic, and political integration of an ethnic minority group into mainstream society which means also finding ways to live cooperatively, playing by common rules that define the parameters of intergroup conflict (Kivisto, 2004). In other words, individuals go through cultural learning and behavioral adaptation in order to overcome cultural differences when they encounter a new culture at a destination. The level of assimilation varies, depending on the individual and/or cultural distance. However, Gordon (1964) distinguished the assimilation process into seven steps: (1). Behavioral assimilation; (2). Structural assimilation; (3). Marital assimilation (amalgamation); (4). Identificational assimilation; (5). Attitude receptional assimilation; (6). Behavior receptional assimilation; (7). Civic assimilation."

The "process of assimilation is an inevitable phenomenon in the tourist experience. Within a tourism destination, tourists require a certain motivation to participate in the cultural assimilation process in order to connect with a non-native culture (the destination culture). However, the level of cultural assimilation may differ from Gordon (1964)'s steps, which were developed based on an immigrant context.

While traveling, all tourists become aware of the distance of various dimensions toward a destination. Choi, Lee, and Noh (2016)'s study was aimed to examine how tourist's psychological distance influences cultural assimilation in a tourism destination. This study attempted to verify the relationship of psychological distance and cultural assimilation, and the several study results were found. First and most importantly, the study results found out the cause-and-effect in the relationship between the psychological distance that tourists feel and the level of their cultural assimilation. It was also found that the psychological distance that the tourists feel has been found to serve as the positive factor toward the cultural experience that the tourists feel at the destination. The authors stated that both geographic and economic distances have a direct positive effect on cultural similarity awareness. In addition, the linguistic and political distances were found to be the influence factors of acculturation intention."

As "it is stated above, tourism can also lead to acculturation and assimilation. One of the strongest indicators of such impacts is loss of native language as a result of an of an influx of tourist languages (White, 1974). Anaya (1989) studied Indo-Hispano communities of the Southwest and warned that residents must hold on to their history and traditional values in order to stave off assimilation, to avoid being doomed to exist as a tourist commodity, admired for its quaint folkways but not taken seriously.

Zhang et al.'s (2018) studies was aimed to identify the relationship among the following factors: cross-cultural awareness, tourist experience, authenticity, tourist satisfaction, acculturation, and assimilation. The study was also aimed to determine what role that tourist activities play in acculturation. Furthermore, the study looked to provide a feasibility plan for the effective management, protection, and sustainable development of World Cultural Heritage Sites. The authors chose Chinese in Korea (immigrants, workers, and international students) who visited the historic villages of Korea (Hahoe and Yangdong) as the research object, and used 430 questionnaires for analysis. The results showed that (1) Chinese in Korea, who have higher cultural awareness, had more interests in objective authenticity (e.g., historical traditions, cultural heritage, and architecture) of world heritage sites; (2) Chinese in Korea could feel and appreciate the true value of traditional culture through tourist experience; (3) The objective authenticity and existential authenticity have a positive effect on tourist satisfaction; and (4) Higher tourist satisfaction could effectively promote cultural integration and assimilation, and prevent cultural separation and marginalization. Consequently, the authors stated that it is possible to say tourist satisfaction may promote cultural integration and assimilation. They also stated that thus, tourism can be an effective solution to human rights violations, cultural contradictions, and other problems.

Poudel (2014) was aimed to study the sociocultural impact in Tharus' culture, customs, tradition, and lifestyle by tourism. As a result of the study, it was seen that local people have changed their life style, their traditional values, cultural aspects and are following the borrowed values and aspects in the name of modernization. In addition that, author stated that the same phenomenon has occurred in Sauraha, where the host culture has been highly influenced and affected by the guest culture. Consequently, author stated that cultural assimilation occurs when two cultures come in contact and the more technologically influential group gains supremacy while the latter society readily adopts the former.

In addition to the above, it is possible to say that dramatic changes have been taken place due to the modernization, urbanization, and tourist activities. It is seen that the cultural conflicts have been taking place when gross disparity between the affluence of tourists and poverty-ridden people of the host region comes together. Thus, it can be said that in the process of development of tourism, weak people of host region sometimes lose their moral values.

Genocide is a very specific term that refers to violent crimes committed against groups with the aim of eliminating the existence of a group (May, 2010). The word "genocide" is a compound word formed by the combination of the word "genos", which means race, nation or ancestry in Greek, and the suffix "cide", which means killing in Latin, and is made with words taken from two different languages.

The pairing of the words "genocide" and "tourism" may seem unlikely. Indeed, the lay person may be surprised that such an activity exists. Nevertheless, to give some idea of the potential scale of genocide tourism, it is worth noting that, by the early 1990s, some half a million tourists per year were visiting the site of the Auschwitz concentration camp and that, nowadays, the level of visitation exceeds a million visitors per year (Beech, 2009). Rwanda, conversely, now attracts just 30,000 international tourists a year, yet not only is this a significant increase after almost a decade of negligible tourist arrivals, but also many of these tourists visit memorials to the 1994 genocide, in particular, the Kigali Genocide Memorial and the Ntarama Church Memorial. Thus, this arguably most extreme form of dark tourism is a significant area of tourism, and one which deserves serious and sensitive study. While the Holocaust may immediately spring to mind as the archetype of genocide, the definition of "genocide" itself is problematic and it is, therefore, necessary to consider first what qualifies as genocide and what falls outside a definition of genocide (Melson, 1992).

Dark tourism, which can be given as an example of one of the alternative tourism types that are of different and special interest, is a witness to wars, genocides, political structure of countries, all kinds of disasters they have experienced in their past, natural disasters, poverty, technological disasters, torture or deaths of celebrities that affect societies and world history (Lennon, 2017). It is described as a type of travel made to see the areas that lead. With the increasing power of the media, this type of tourism, which includes events such as war, genocide, death, natural disaster and poverty, has shaped people's interest in travel by further developing. For this reason, the destinations of dark tourism are known by people all over the world due to the media (Kozak & Kama, 2015, pp. 3–6).

According to Aristotle, the concept of "catharsis", which points to the result of an experience together with sadness tourism, means purification and cleansing of the soul. The underlying reason for people's interest in sadness tourism and the motivation that guides them consists of the desire to see places where mass deaths occurred, genocides, places left with deep traces due to past wars, war zones, and their desire to feel events again (Varol, 2015). The places where there are mass murders and monuments in Bosnia, the killing fields that emerged in the 1970s and the Landmine Museum are given as examples.

Tanaś (2013) examined the influence of acts of terror and war crimes on both the development of tourism, or a lack of it in his study. At the same time, the author aimed to determine the new tourism products which are based on death as an element of cultural and historical heritage, set in death spaces, meet the expectations of a variety of groups, determining tourism potential in some areas. The author also defended that terrorist attacks, the victims of which are tourists as well as local citizens, constrain or completely prevent tourism development. The author stated that terrorism is then a barrier to the development of the tourism economy. According to the author, with time, however, memory of the crime and its victims, due to documentation and commemoration, may become an impulse to organize spaces which will be included in tourism.

Vaes and Wicklund (2002) stated that the quality of cognitive, emotional, and behavioral responses to threats may depend on the nature of perceived threats being symbolic or realistic. The authors also stated that symbolic threats can lead to dehumanization, deprivation, moral exclusion of the outgroup and less empathy toward the outgroup rather than realistic threats. In addition, it was stated that symbolic threats should result in increased compliance with the norms and values of the inner group in particular. It was also stated that symbolic threats could also cause the most vicious behavioral reactions to outgroups such as genocide and torture, and in the context of immigration policy, symbolic threats are expected to be linked to the assimilation preferences of outgroups.

Segregation is defined as the policy of keeping one group of people apart from another and treating them differently, especially because of race, sex, or religion (Massey et al., 2009). Conceptually, one must be careful not to necessarily equate segregation with inequality. Segregation is made up of two dimensions: vertical segregation and horizontal segregation. The phenomenon of occupational sex segregation can be used to explain each: pay differentials between men and women across occupations within a given labor force characterize vertical segregation, while horizontal segregation illustrates the separation of various individuals in terms of the concentration of the sexes in different types of occupations – but does not necessarily indicate discrimination or inequality (neither does it show the absence of discrimination or inequality). Theoretically, then, it is possible for individuals to be completely segregated horizontally without any vertical dimension, or vice versa. A given labor market, however, is more often segregated to different extents along both vertical and horizontal lines (Blackburn et al., 2001).

Sex segregation occurs both horizontally and vertically (Charles & Bradley, 2002). In horizontal segregation, while women and men have the same position and official role in the organization, it is possible that they assume different duties and responsibilities. This is because the same job does not contain equal responsibility, challenge, and opportunity. In vertical segregation, women will have fewer opportunities to grow and be promoted professionally as they work in lower positions and with less responsibility. Therefore, men occupy higher positions in organizations and women at lower levels.

While the tourism sector is considered important in terms of creating job opportunities for the unemployed and especially women unemployed in developing countries (Lee & Kang, 1998), it is emphasized that employment in tourism is sexed and strengthens segregation based on sex (Hemmati, 2000).

In many studies on women's employment in the tourism sector, it is seen that women have limited access to full-time, well-paid, qualified, and managerial jobs compared to the male workforce, where they have the same qualifications in education, job placement, and taking advantage of career opportunities (Iverson & Deery, 1997; Long & Kindon, 2005; García-Pozo et al., 2012). In many studies, it was alleged that women were subjected to segregationary practices considering sex differences while forming management staff (Li & Leung, 2001; Manwa & Black, 2002; Skalpe, 2007; Thrane, 2008). These studies showed that men earn more income due to higher status jobs, women earn less, and wage inequality is a form of sex segregation (Riley & Szivas, 2003; Thrane, 2008).

Empirical studies also confirm the existence of significant sex-based differences in the income within the tourism sector. Burgess (2000) explained that male managers working in the finance department of hotels in England receive higher wages than female managers. Skalpe (2007) found that there are around 20% of women CEOs in Norwegian tourism businesses, and women have been employed in smaller firms offering less wages and are subject to wage segregation. Thrane (2008) proved that while socio-demographic variables such as education, work experience, parenthood, and marriage were constant in tourism businesses in Norway during 1994–2002, male employees received approximately 20% higher wages than female employees. Costa et al. (2012) revealed in their studies that although there are more women among tourism graduates in Portugal, men receive higher salaries and fill the top positions in the tourism sector.

Consequently, it is seen that the widespread literature around the world shows that women are exposed to sexist segregation in terms of recruitment, remuneration, and promotion.

Religious segregation is the separation of people according to their religion (Knox, 1973). When the subject is considered in terms of tourism, it is seen that especially Muslim countries such as Malaysia, Turkey, and Egypt, as well as in countries such as Australia, Singapore, France, Japan, Philippines, New Zealand, and Brazil, suitable accommodation services are provided for Muslim tourists with high religious sensitivity (Mohsin et al., 2015). In other words, Muslim-friendly destinations offer plenty of "Halal" services (such as Halal holiday, Halal food and beverages, Halal health care facilities and services, Halal cruise, sex-segregated swimming pools) along with comfortable places for Muslims to perform their daily prayers.

Consequently, it was stated that the integrated threat theory, four threats were differentiated with a focus on realistic and symbolic threats (Stephan et al., 2009). It was also stated that integrated threat theory has both negative (Riek et al., 2006; Stephan et al., 2005) and positive (Li & Zhao, 2012) effects of integrated threat on segregation as stated above.

Integration is generally being disclosed as the action or process of successfully joining/mixing with a different group of people (Taskin, 2019). Tourists are the building blocks of the tourism industry. Tourism is also the situation where the tourist goes to a different destination outside the cultural framework they are constantly living and used to and experiences different cultural structures by spending a certain period of time (at least 24 hours) in that destination. Therefore, the phenomenon of tourism actually makes it possible for people from different cultural backgrounds to meet and get to know each other. The perceptions, attitudes, and behaviors of people involved in tourism activities toward people with different cultural structures outside their cultural structure can turn into understanding and sympathy over time. In short, it can be said that tourism brings along cultural diversity and increases tolerance toward difference (Chow, 2005).

On the other hand, cultural diversity is perceived positively in destinations and countries that send tourists to tourism activities effectively, and people who come to these countries as immigrants do not experience discrimination and more negative behavior (Moufakkir, 2014). Therefore, it can be said that immigrants are more prone to integrating into host countries more comfortably and contributing to social tolerance than the countries facing discrimination. From this point of view, it can be said that tourism has a very constructive and positive contribution especially to the immigration problem and integration problems of our age, as well as many positive sociological benefits.

Kum (2020) stated that the readiness of the host communities to welcome refugees requires a distinct but gradual process of integration comprising interrelated legal, economic, social, and cultural dimensions, all of which are important for refugees' ability to integrate successfully as fully included members of the host society. In addition, the author stated that the integrated threat theory is considered most appropriate method to apply in this context because of the integrational focus of most policy contexts when it comes to addressing the challenges of refugees.

Conclusion

Tourism has always been a way of life. As long as our civilization has existed, people have travelled. They have moved from one place to another for leisure, recreational, trade, or family purposes. According to the definition of the World Tourism Organization, tourism is defined as the activities of a person who goes to a place other than his/her usual surroundings to stay for less than a specified period and whose main purpose is to travel outside of trying an activity that earns money in the place they visit (Doğanay & Zaman, 2019).

Tourism is the temporary movement of people to the destinations outside their normal places of work and residence, the activities undertaken during their stay in those destinations, and the facilities created to cater to their needs (Mathieson & Wall, 1982, p. 1). During their stay in the destination, tourists interact with local residents and the outcome of their relationship changes the host individuals' and host community's quality of life, value systems, labor division, family relationships, attitudes, behavioral patterns, ceremonies and creative expressions (Pizam & Milman, 1984). Travelling/tourism is a genuinely powerful and unique force for change in community (Kunwar, 2006). It has several impacts in society as well as culture of the host country. These impacts, along with other impacts, can be listed as influences that affect the perceptions, attitudes, and behaviors of host communities of prejudice, discrimination, assimilation, genocide, discrimination, and integration.

This chapter has sought to impact of ***tourism*** on prejudice, discrimination, assimilation, genocide, segregation, integration and examined it in relation to integrated threat theory. In general, the integrated threat theory focuses on the conditions caused by intergroup contact and changing intergroup relations. The integrated threat theory is designed to determine the dimensions of prejudices and attitudes of individuals toward all outgroups, including gender, race, sexual orientation, national origin and disability (Stephan et al., 2000, p. 64).

People who live in groups may have to struggle with their surroundings in order to survive and if necessary, they may have to fight when necessary. Despite the fact that human rights have reached a universal dimension, in some parts of the world, problems of intergroup discrimination and pressures still continue, and when there is any interaction or communication between people with different cultures and perspectives, conflicts may inevitably arise (Ruggiero, 2013, p. 143). Classical social psychological theories emphasize the important role of "threat" and "competition" in determining the attitudes of groups toward each other in intergroup relations. Intergroup threats can pave the way for conflicts as they can affect people's perceptions, feelings, and behaviors. Threat assessment can also provoke negative emotions among groups, such as fear, anger, harassment, anger, resentment, disappointment, contempt, and insecurity. In addition, threat perception can also reduce emotional empathy toward outgroup members, and perception of threat types can also cause bias, regardless of whether the threat is real or not (Avcıkurt, 2015). Stephan and Renfro (2002, p. 267) determined that intergroup threat leaders depend on factors such as intergroup contact, in-group identity, and status inequalities.

As a result, it can be stated that the threat in intergroup relations through *tourism* can be perceived and interpreted in various ways. Threats can be at the social, economic, or political level and generally include the opportunity to compete on limited resources (Ward & Berno, 2011, p. 1558). It can be stated that people living in different societies and having different cultures should have the same rights, in other words, the right to free consumption in access to consumer goods, services, and cultural products from different societies (Tuna & Özbek, 2012, p. 89). Therefore, it can be said that the existence of scarce resources arises as a result of factors such as hostile attitudes and discriminatory behavior through *tourism,* and perceived competition between groups. It can be stated that another negativity in intercultural relations that emerges through *tourism* is the belief in perceiving other cultures as a threat to their own culture. Consequently, this causes to positive or negative perceptions, attitudes, and behaviors in host individuals related to prejudice, discrimination, assimilation, genocide, discrimination, and integration.

References

Anastasopoulos, P. G. (1992). Tourism and attitude change: Greek tourists visiting Turkey. *Annals of Tourism Research*, 19(4), 629–642.

Anaya, R. (1989). Aztln: A homeland without boundaries. In R. Anaya & F. Lomelí (Eds.), *Aztlán: Essays on the Chicano Homeland* (pp. 230–241). El Norte Publications.

Arısoy, İ., & Demir, N. (2007). Gender equality in EU social law in the context of fight against discrimination. *Ege Academic Review*, 7(2), 707–725.

Avcıkurt, C. (2015). *Turizm sosyolojisi: Genel ve yapısal yaklaşım*. Detay Yayıncılık.

Beech, J. (2009). *Genocide tourism*. Channel View Publications.

Blackburn, R. M., Brooks, B., & Jarman, J. (2001). The vertical dimension of occupational segregation. *Work, Employment and Society*, 15(3), 511–538.

Brigham, J. C. (1971). Ethnic stereotypes. *Psychological Bulletin*, 76, 15–38.

Brown, L., & Jones, I. (2013). Encounters with racism and the international student experience. *Studies in Higher education*, 38(7), 1004–1019.

Burgess, C. (2000). Hotel accounts-do men get the best jobs? *International Journal of Hospitality Management*, 19(4), 345–352.

Cankurtaran, Ö., & Beydili, E. (2016). Ayrımcılık karşıtı sosyal hizmet uygulamasının gerekliliği üzerine. *Toplum ve Sosyal Hizmet*, 27(1), 145–160.

Charles, M., & Bradley, K. (2002). Equal but separate? A cross-national study of sex segregation in higher education. *American Sociological Review*, 573–599.

Cheung, K. S., & Li, L. H. (2019). Understanding visitor-resident relations in overtourism: Developing resilience for sustainable tourism. *Journal of Sustainable Tourism*, 27(8), 1197–1216.

Chiaravutthi, Y. (2019). Price discrimination against tourists: Is it ethical? *Development Economic Review*, 13(1), 8–27.

Choi, J. W., Lee, C. J., & Noh, E. J. (2016). The influence of psychological distance to cultural assimilation on tourism destination. *Indian Journal of Science and Technology*, 9(35), 1–9.

Choi, O. J., Lee, K. S., Lee, K. T., & Kim, J. H. (2017). Influences of stereotype and social distance on prejudice toward African Americans. *Journal of Psychology in Africa*, 27(1), 13–17.

Chow, C. S. (2005). Cultural diversity and tourism development in Yunnan Province, China. *Geography*, 90(3), 294–303.

Colombo, M., Cherubini, P., Montali, L., & Marando, L. (2012). There's foreigner and foreigner: xenophobic reasoning and anti-immigrant discourse. *Global Journal of Community Psychology Practice*, 3(4), 135–145.

Costa, C., Carvalho, I., Caçador, S., & Breda, Z. (2012). Future higher education in tourism studies and the labor market: Gender perspectives on expectations and experiences. *Journal of Teaching in Travel & Tourism*, 12(1), 70–90.

Craig, M. A., & Richeson, J. A. (2012). Coalition or derogation? How perceived discrimination influences into minority intergroup relations. *Journal of Personality and Social Psychology*, 102, 759–777.

Curseu, P. L., Stoop, R., & Schalk, R. (2007). Prejudice toward immigrant workers among Dutch employees: Integrated theory revisited. *European Journal of Psychology*, 37, 125–140.

Çelik, S. (2019). Factors affecting the impact of tourism on attitude change: A qualitative research. In T. J. Da Silva, Z. Breda, F. Carbone (Eds), *Role and Impact of Tourism in Peacebuilding and Conflict Transformation* (pp. 238–254). USA: IGI Global

Çelik, S. (2019b). Does tourism change tourist attitudes (prejudice and stereotype) towards local people? *Journal of Tourism and Services*, 10(18), 35–46.

Devakumar, D., Shannon, G., Bhopal, S. S., & Abubakar, I. (2020). Racism and discrimination in COVID-19 responses. *The Lancet*, 395(10231), 1194.

Doğanay, H., & Zaman, S. (2019). *Türkiye turizm coğrafyası*. Pegem Akademi Yayıncılık.

Fisher, R. J., & Price, L. L. (1991). International pleasure travel motivations and post-vacation cultural attitude change. *Journal of Leisure Research*, 23(3), 193–208.

Fottrell, Q. (2020, 04 May). *"No Chinese allowed": Racism and fear are now spreading along with the coronavirus*. https://www.marketwatch.com/story/no-chinese-allowed-racism-and-fear-are-now-spreading-along-with-the-coronavirus-2020-01-29.

García-Pozo, A., Campos-Soria, J. A., Sánchez-Ollero, J. L., & Marchante-Lara, M. (2012). The regional wage gap in the Spanish hospitality sector based on a gender perspective. *International Journal of Hospitality Management*, 31(1), 266–275.

González, K. V., Verkuyten, M., Weesie, J., & Poppe, E. (2008). Prejudice towards Muslims in the Netherlands: Testing integrated threat theory. *British Journal of Social Psychology*, 47, 667–685.

Gordon, M. M. (1964). *Assimilation in American life*. Oxford University Press.

Gudykunst, W. B. (1993). Toward a theory of effective interpersonal and intergroup communication: an anxiety/uncertainty management (AUM) perspective. In R. L. Wiseman & J. Koester (Eds.), *International and Intercultural Communication Annual, Intercultural Communication Competence* (pp. 33–71). Sage Publications.

Gürses, İ. (2005). Önyargının nedenleri. *Uludağ Üniversitesi İlâhiyat Fakültesi Dergisi*, 14(1), 143–160.

Hanel, P. H., Maio, G. R., Soares, A. K., Vione, K. C., de Holanda Coelho, G. L., Gouveia, V. V.,... & Manstead, A. S. (2018). Cross-cultural differences and similarities in human value instantiation. *Frontiers in Psychology*, 9, 849.

Hemmati, M. (2000). Women's employment and participation in tourism. *Sustainable Travel&Tourism*, 5(1), 17–20.

Ipsos MORI (2020, 05 May). *COVID-19 – One in seven people would avoid people of Chinese origin or appearance*. https://www.ipsos.com/ipsos-mori/en-uk/covid-19-one-seven-people-would-avoid-people-chinese-origin-or-appearance.

Islam, M. R., & Hewstone, M. (1993). Dimensions of contact as predictors of intergroup anxiety, perceived out-group variability, and out-group attitude: An integrative model. *Personality and Social Psychology Bulletin*, 19(6), 700–710.

Iverson, R. D., & Deery, M. (1997). Turnover culture in the hospitality industry. *Human Resource Management Journal*, 7(4), 71–82.

Kivisto, P. (2004). What is the canonical theory of assimilation? *Journal of the History of the Behavioral Sciences*, 40(2), 149–163, Spring.

Kozak, A. M., & Kama, S. (2015, 29–30 April). Dark (hüzün) turizmi olarak Çanakkale. VIII. Graduate Student Research Congress, Nevşehir, Turkey.

Kum, H. A. (2020). Refugees as troubled learners in UK schools: An evaluation and reconceptualization of the education for diversity in UK schools. The integrated threat theory approach. *International Journal for Innovation Education and Research*, 8(10), 126–144.

Kunwar, R. R. (2006). *Tourists and tourism: Science and industry interface*. Kathmandu: International School of Tourism and Hotel Management Publish.

Knox, H. M. (1973). Religious segregation in the schools of Northern Ireland. *British Journal of Educational Studies*, 21(3), 307–312.

Lahey, J. N. (2005). *Do older workers face discrimination?* Center for Retirement Research at Boston College.
Lee, C. K., & Kang, S. (1998). Measuring earnings inequality and median earnings in the tourism industry. *Tourism Management*, 19(4), 341–348.
Lee, K. J., & Scott, D. (2017). Racial discrimination and African Americans' travel behavior: The utility of habitus and vignette technique. *Journal of Travel Research*, 56(3), 381–392.
Lennon, J. (2017). *Dark tourism*. Oxford University Press.
Li, L., & Leung, R. W. (2001). Female managers in Asian hotels: Profile and career challenges. *International Journal of Contemporary Hospitality Management*, 13(4/5), 189–196.
Li, T., & Zhao, Y. (2012). Help less or help more-perceived intergroup threat and out-group helping. *International Journal of Psychological Studies*, 4(4), 90–98.
Lippmann, W. (1998). *Public opinion: With a new introduction by Michael Curtis*. Transaction Publishers.
Long, V. H., & Kindon, S. L. (2005). *Gender and tourism development in Balinese villages*. Routledge.
Manav, E. (2013). 2000/43, 2000/78, 2006/54 Sayılı AB direktifleri çerçevesinde iş hukukunda ayrımcılıkla mücadele ve Türkiye'deki uygulamalar. *Dokuz Eylül Üniversitesi Hukuk Fakültesi Dergisi*, 15 (Special Issue), 731–779.
Manwa, H., & Black, N. (2002). Influence of organizational culture on female and male upward mobility into middle and senior managerial positions: Zimbabwean banks and hotels. *International Journal of Cross Cultural Management*, 2(3), 357–373.
Marshall, G. (2003). *A dictionary of sociology*. USA.
Massey, D. S., Rothwell, J., & Domina, T. (2009). The changing bases of segregation in the United States. *The Annals of the American Academy of Political and Social Science*, 626(1), 74–90.
Mathieson, A., & Wall, G. (1982). *Tourism: Economic, physical and social impacts*.Longman Group Limited
May, L. (2010). *Genocide: A normative account*. Cambridge University Press.
Melson, R. (1992). *Revolution and genocide: On the origins of the Armenian genocide and the Holocaust*. University of Chicago Press.
Mendras, H. (2009). *The principles of sociology*. D. Appleton and Company.
Metin, B. (2018). İşgücü piyasasında yaşa dayalı ayrımcılık sorunu ve yaşa dayalı ayrımcılıkla mücadele: ABD ve AB uygulamaları çerçevesinde Türkiye için bir değerlendirme. *Akademik Araştırmalar ve Çalışmalar Dergisi (AKAD)*, 10(18), 72–89.
Mohsin, A., Ramli, N., & Alkhulayfi, B. A. (2015). Halal tourism: Emerging opportunities. *Tourism Management Perspectives*, 19, 137–143.
Moufakkir, O. (2014). What's immigration got to do with it? Immigrant animosity and its effects on tourism. *Annals of Tourism Research*, 49, 108–121.
Pai, H. (2006, 29 August). *Overseas aid The Guardian*. https://www.theguardian.com/education/students/overseasstudents.
Philipp, S. F. (1994). Race and tourism choice: A legacy of discrimination? *Annals of Tourism Research*, 21(3), 479–488.
Pizam, A., & Milman, A. (1984). The social impacts of tourism. *Industry and Environment*, 7(1), 11–14.
Poudel, J. (2014). Socio-cultural impact in tourism: A case study of Sauraha, Nepal. *Journal of Advanced Academic Research*, 1(2), 47–55.
Preston-Ellis, R (2020, 04 May). *Asian teens punched, kicked and spat at in three separate racist coronavirus attacks in Exeter in just one day*. https://www.devonlive.com/news/devon-news/asian-teens-punched-kicked-spat-3922546.
Riek, B. M., Mania, E. W., & Gaertner, S. L. (2006). Intergroup threat and outgroup attitudes: A meta-analytic review. *Personality and Social Psychology Review*, 10, 336–353.
Riley, M., & Szivas, E. (2003). Pay determination: A socioeconomic framework. *Annals of Tourism Research*, 30(2), 446–464.
Ruggiero, V. R. (2013). A guide to sociological thinking. Sage.
Sharifi-Tehrani M., Verbič M. & Chung J.Y. (2013). An analysis of adopting dual pricing for museums: The case of the national museum of Iran. *Annals of Tourism Research*, 43, 58–80.
Sirakaya-Turk, E., Nyaupane, G., & Uysal, M. (2014). Guests and hosts revisited: Prejudicial attitudes of guests toward the host population. *Journal of Travel Research*, 53(3), 336–352.
Skalpe, O. (2007). The CEO gender pay gap in the tourism industry-Evidence from Norway. *Tourism Management*, 28(3), 845–853.
Stangor, C. (2014). *Principles of Social Psychology*, V.1.0. BCcampus.
Stephan, C. W., Stephan, W. G., Demitrakis, K. M., Yamada, A. M., & Clason, D. L. (2000a). Women's attitudes toward men: An integrated threat theory approach. *Psychology of Women Quarterly*, 24, 63–75.

Stephan, W. G., Diaz-Loving, R., & Duran, A. (2000b). Integrated threat theory and intercultural attitudes: Mexico and the United States. *Journal of Cross-Cultural Psychology*, 31(2), 240–249.

Stephan, W. G., & Renfro, C. L. (2002). *The role of threat in intergroup relations*. Blackwell Publishing.

Stephan, W. G., Renfro, C. L., Esses, V. M., Stephan, C. W., & Martin, T. (2005). The effects of feeling threatened on attitudes toward immigrants. *International Journal of Intercultural Relations*, 29(1), 1–19.

Stephan, W. G., & Stephan, C. W. (1993). Cognition and affect in stereotyping: Parallel interactive networks. In D. M. Mackie & D. L. Hamilton (Eds.). *Affect, cognition, and stereotyping: Interactive processes in group perception*, Academic Press.

Stephan, W. G., & Stephan, C. W. (1996). Predicting prejudice. *International Journal of Intercultural Relations*, 20, 409–426.

Stephan, W. G., & Stephan, C. W. (2017). Intergroup threat theory. *The International Encyclopedia of Intercultural Communication*, 1–12.

Stephan, W. G., Ybarra, O., & Bachman, G. (1999). Prejudice toward immigrants. *Journal of Applied Social Psychology*, 29, 2221–2237.

Stephan, W. G., Ybarra, O., & Morrison, K. R. (2009). Intergroup Threat Theory. T. D. Nelson (Ed.), In *Handbook of prejudice, stereotyping, and discrimination*. Psychology Press.

Tam, T., Hewstone, M., Cairns, E., Tausch, N., Maio, G., & Kenworthy, J. (2007). The impact of intergroup emotions on forgiveness in Northern Ireland. *Group Processes & Intergroup Relations*, 10(1), 119–136.

Tanaś, S. (2013). The meaning of genocide and terror in cognitive tourism. *Turyzm*, 23(1), 7–15.

Taskin, E. (2019). Analysis of projects related to the integration of migrants. *Horizon Insights*, 2(2), 25–32.

Thrane, C. (2008). Earnings differentiation in the tourism industry: Gender, human capital and socio-demographic effects. *Tourism Management*, 29(3), 514–524.

Tse, S., & Tung, V. W. S. (2020). Residents' discrimination against tourists. *Annals of Tourism Research*. 103060. https://doi.org/10.1016/j.annals.2020.103060.

Tuna, M., & Özbek, Ç. (2012). *Yerlileşen yabancılar, Güney Ege Bölgesi'nde göç, yurttaşlık ve kimliğin dönüşümü*. Detay Yayıncılık.

Ward, C., & Berno, T. (2011). Beyond social exchange theory: Attitudes toward tourists. *Annals of tourism research*, 38(4), 1556–1569.

White, P. (1974). The social impacts of tourism on host communities: A study of language change in Switzerland. *Research Paper*. Oxford University.

Williams, D. R., & Mohammed, S. A. (2009). Discrimination and racial disparities in health: Evidence and needed research. *Journal of Behavioral Medicine*, 32(1), 20–47.

Wintersteiner, W., & Wohlmuther, C. (2014). Peace sensitive tourism: How tourism can contribute to peace. In Wohlmuther, C. and Wintersteiner, W. (Eds.), *International Handbook on Tourism and Peace*, (pp. 31–61). Ausria: Drava Print GmbH.

Wood, S., Braeken, J., & Niven, K. (2013). Discrimination and well-being in organizations: Testing the differential power and organizational justice theories of workplace aggression. *Journal of Business Ethics*, 115(3), 617–634.

Vaes, J., & Wicklund, R. A. (2002). General threat leading to defensive reactions: A field experiment on linguistic features. *British Journal of Social Psychology*, 41, 271–280.

Varol, F. (2015, 28–30 May). Hüzün turizminin Türkiye'de var olan potansiyeli üzerine kuramsal bir araştırma. I. Eurasia International Tourism Congress: Current Issues, Trends, and Indicators, Konya, Turkey.

Yıldız, S. (2006). Türk ve Alman toplumlarında kültürel ilişkiler, imgeler ve medya. *Milli Folklor*, 18(72), 37–46.

Zaei, M. E., & Zaei, M. E. (2013). The impacts of tourism industry on host community. *European Journal of Tourism Hospitality and Research*, 1, 12–21.

Zhang, H., Cho, T., Wang, H., & Ge, Q. (2018). The influence of cross-cultural awareness and tourist experience on authenticity, tourist satisfaction and acculturation in World Cultural Heritage Sites of Korea. *Sustainability*, 10(4), 927.

15

RE-EXAMINING THE TOURISM AND PEACE NEXUS

A Social Network Theory Perspective

Gaunette Sinclair-Maragh

Introduction

Tourism as a social force has the propensity to create peace and understanding (Khamouna & Zeiger, 1995). Several other authors are in support of this claim (e.g. Caneday, 1991; Cho, 2007; Holland, 1991; Knopf, 1991; Var et al., 1989). Caneday (1991) further points out that tourism can bring people and nations together for mutual understanding and respect. In support, Knopf (1991) purports that tourism is able to dissipate barriers among people and is therefore presented as the fundamental key to world peace.

The relationship between tourism and peace could consequently be perceived as linear. Nonetheless, some scholars (e.g., Farmaki, 2017; Litvin, 1998; Pratt & Liu, 2016) question this linear relationship between tourism and peace. Farmaki (2017) asks if tourism creates peace or is it that peace generates tourism. Pratt and Liu (2016) find that tourism is not the grounds for peace but the beneficiary of peace. Litvin (1998) purports that most research has not suggested that tourism leads to peace as they tend to be absolute and conjectured. He points out that despite the uncertainty of the relationship between tourism and peace, be it co-relational or causal, tourism does flourish in an environment where there is peace. There could also be an indirect relationship since tourism not only flourishes in peaceful environments but may also contribute to the achievement of peace where lives, communities, nations, and international relations are transformed (Lankford et al., 2008).

This debate calls for further examination of the association between tourism and peace. This is supported by Farmaki (2017) who points to the inconclusive findings of empirical studies on the nexus between both phenomena. Empirical research to confirm the association between both is also scant (Becken & Carmignani, 2016). Furthermore, previous studies relied on case studies to explain the relationship between tourism and peace rather than theories, and hence the need for a consolidation of theory to explain this association (Farmaki, 2017).

Based on the discourse, it is reasonable to continue with the exploration of the association between tourism and peace from a deductive reasoning perspective to reach a logical conclusion, particularly because empirical research to confirm the association between both phenomena is scant (Becken & Carmignani, 2016). Therefore this chapter presents a re-examination of the association between tourism and peace. It specifically explores the notion of tourism as a creator of peace as well as being a beneficiary of peace in terms of peace being a generator of tourism. The chapter also examines the relationship between tourism and peace within the contexts of peace tourism which has not been fully explored in the literature (Van den Dungen, 2014), and

DOI: 10.4324/9781003161868-15

sustainable development which is a current global thrust to create a better future for humanity (Gore, 2015).

The novel of this inquiry is that the social network theory is used to provide underpinning theoretical support to better understand the likely relationship between tourism and peace. This theory is plausible since it establishes the basis for understanding human social organization and characterizes social structure at the individual and population levels (Krause et al., 2007). The social network theory is among the few social science theories that are not reductionist (Kadushin, 2004). This is because it can be applied to small groups as well as the entire global systems; hence it applies to tourism which occurs at all levels: local, national, regional, and global.

Despite the social network theory not being fully developed in tourism studies (Viren et al., 2015), it is plausible within this context as it suggests the use of nodes and ties that connect people and according to Aliperti et al. (2019), tourism involves the interactions of organizations, people, and events. On this basis, nodes and ties will be of relevance to any interaction between tourism stakeholders. Seemingly, social networks are important to the tourism and peace nexus and this chapter presents respective tourism nodes and tourism ties to further explain the relationship between tourism and peace.

Notably, past literature tends to refer to peace within the context of travel or the traveler and not within the setting of tourist or tourism. This, according to Litvin (1998), can create major chasms and could be a semantic issue. For the purpose of this chapter, both terms will be used in the discourse as according to Wohlmuther and Wintersteiner (2014), peace is the cornerstone of travel and tourism.

The Tourism and Peace Nexus

Is tourism a creator of peace or a beneficiary of peace? Litvin (1998) posits that although tourism is a major beneficiary of peace, it is not in itself a contributor of peace. Haessly (2010) argues that tourism contributes to the goal of a more harmonious world while Litvin (1998) contends that there is a correlation between successful tourism and the absence of war, terrorism, and internal strife. Upadhayaya et al. (2011) conclude that tourism is both sensitive to conflict and responsive to peace. This section will evaluate the debate regarding the tourism and peace nexus in addition to examining their relationship within the contexts of peace tourism and sustainable development. The section will also analyze how the social network theory can explain the likely tourism-peace relationship.

Understanding the Concept of Peace

Peace is oftentimes defined in a negative way as the absence of war (D'Amore, 1988). A positive connotation is that it is the state of calmness that is achieved when people of diverse cultures, perceptions, and beliefs understand as well as respect the diversity of other people (Jamgade, 2021). Haessly (2010) contends that peace is more than the absence of war and (Galtung, 1969) claims that it can either be positive or negative.

Positive peace denotes the absence of structural violence (Galtung, 1964). Structural violence is usually a consequence of corporate- or state-sponsored systems that result in inequality caused from the unequal distribution of resources for social, economic, and environmental betterment. The outcome of this includes poverty, hunger, homelessness, and environmental pollution. Positive peace results in positive outcomes such as the resolution of relationships and creation of social systems (Haessly, 2010). On the other hand, negative peace occurs in a non-war environment; it is the absence of personal or physical violence. This type of violence includes physical fights and military warfare (Galtung, 1969) as well as when a country is in a state of readiness for war even though it is not involved in the conflict (Haessly, 2010). Negative peace is demonstrated in the absence of war or armed conflict and the enactment of ceasefire.

Tourism as a Creator of Peace

Tourism has been recognized as having an integral role in the achievement of world peace (Cho, 2007). This is because it plays a vital part in world diplomacy (D'Amore, 1988) by creating geopolitical stability between and among countries and promoting understanding, cooperation, and fraternity across the globe (Caneday, 1991; Holland, 1991; Knoff, 1991). This suggests that tourism is able to connect people and places across the globe and can be a powerful vehicle for peace tourism (Young, 2019). This makes global tourism the world's largest business for peace because people are able to recognize and understand the differences and sameness among each other (Malley, 2002). In support of this discourse, the IIPT (2020) envisions tourism as the "World's First Global Peace Industry" with people acting as ambassadors for peace. They posit the "Credo for the Peaceful Traveller" as they believe that peace begins with the individual.

The experience while being a tourist helps people to understand the meaning of the "The Global Village". Tourism improves relationships among nations and people of the world (D'Amore, 1988). This global phenomenon is known to connect people and unite humanity, thus creating harmony and peace across the world. This is underscored by McIntosh et al. (1995), who declared that having an understanding and appreciation of peoples' way of life, culture, mores, and language make one more of a part of a world community. Tourism promotes peaceful relations through visits to other countries where tourists gain knowledge of their cultures and develop an appreciation of their environment. In doing so cross-cultural understanding is fostered and new friendships established (Raymond & Hall, 2008). This demonstrates that when tourists show respect to the residents and their cultures as well as heritage sites and take care of their biodiversity, they are contributing positively to the lack of conflict.

Several conferences have been hosted to highlight the role of tourism in creating peace, for example, the "International Tourism Passport to Peace" held in 1997, "Peace through Tourism" in 1990, and "Tourism: A Vital Force for Peace" in 1988 (Khamouna & Zeiger, 1995). The latter was held in Canada with approximately 1,000 delegates. This number signifies the importance of tourism in creating peace even three decades ago and demonstrates the interest in tourism as a mechanism to create peace at all levels. In addition, the conversation regarding the role of tourism in creating peace has led to the establishment of several non-governmental/international organizations such as the Institute for Peace through Tourism, Tourism, and Peace and the International Centre for Peace through Tourism (Haessly, 2010).

Tourism is likely to contribute to world peace through proper management of globalization, migration, conflicts, prejudices, and conflicts (Moufakkir & Kelly, 2010). Peace is also created through tourism policies (Haessly, 2010). It can be promoted by way of tourism education by infusing peace-making in the tourism curriculum to focus on values and conflict resolution and through students' educational experiences to foster attitudinal and behavioral changes (Haessly, 2010). Through trans-perceptual learning, tourism education can play a major role in encouraging critical thinking and the understanding of attitudes and finding solutions for the various situations that cause disharmony. Trans-perceptual learning occurs when an individual learns from recognizing reality from the perspective of others (Crews, 1989).

Tourism can enable societies to recover from civil war as well as prevent or reduce the structural and cultural tensions that generate large-scale violence in many conflicts. This is confirmed by Fernando et al. (2014), who pointed out that tourism is able to play a vital role in post-war development strategy as in the case of Sri Lanka. Therefore, as an industry, tourism not only generates economic benefits but is also a conveyor of promoting positive peace.

Tourism as a Beneficiary of Peace/Peace as a Generator of Tourism

Successful tourism requires civil peace (Becken & Carmignani, 2016). For instance, the three decades of separationist war in Sri Lanka which ended in 2009 has impacted tourism during that period (Fernando et al., 2013). It was the view at the end of the war that a long-term reconciliation process between North and South Sri Lanka could be beneficial for the country. This decision was based on the premise that the peace initiative would attract more tourists to visit the destination. This reconciliation process was based on the positive peace concept (Chandrasiri, 2019). In this case, tourism is a beneficiary of peace (Litvin, 1998). Peace has allowed tourism to happen. This is why it is imperative to understand the language of nonviolence to augur knowledge and understanding of comparative values and beliefs (Blanchard & Higgins-Desbiolles, 2013).

Tourism development and activities have definitely emerged post battles and according to Lloyd (1998), these are on a large scale. Battlefield sites are being used for tourism in addition to military grounds and memorials associated with warfare which have become tourist attractions (Baldwin & Sharpley, 2009). One example is battlefield tourism in Flanders, Belgium which resulted from World War I where there was great devastation ranging from the loss of lives to farmlands and other resources (Vanneste & Foote, 2013). This Great War of 1914–1918 resulted in the creation of memorials to commemorate the event. In this case, tourism is used to create and perpetuate the memory of the Great War (Winter, 2009).

The Battle of Waterloo (Seaton, 1999) and the Korean War (Suntikul, 2013) left a legacy for tourism development in those countries. Another post-war opportunity is the emergence of dark tourism from the Massacre of Glencoe in Scotland (Fyall et al., 2006). Peace also brought Hilton to Northern Ireland (Selwitz, 1994). These examples show that tourism has the potential to contribute to economic growth and development in post-conflict countries. Post the civil wars and 1994 genocide in Rwanda, tourism positively impacted the economic and social conditions of the country through ecotourism, pro-poor tourism initiatives, and a community-based tourism model (Alluri, 2009).

There are situations where tourists are given travel advisories by their respective government regarding traveling to destinations where there may be disruptions of peace resulting from social or political instability, criminal activities, or terrorism. Individuals want to be able to travel across the globe in a hassle-free way, without travel restrictions and conflicts (Jamgade, 2021) as well as visit a destination that is peaceful, safe, and secure (Cavlek, 2006). Tourism is, therefore, viewed as a beneficiary of peace as it will not be successful in the absence of peace (Litvin, 1998).

Peace Tourism

This form of tourism "involves visits to places, at home and abroad, which are significant because of their association with such notions as peace-making, peaceful conflict resolution, prevention of war, resistance to war, protesting war, nonviolence and reconciliation" (Van den Dungen, 2014, p. 62). Shin (2005) confirms that tourists, both international and domestic, favor a peaceful environment for visits. Despite the many deliberations regarding the role of tourism in creating peace, the practice of peace tourism has not come to the fore. Peace tourism is almost unknown unlike battlefield tourism or war tourism. There are more destinations for war tourists to visit when compared to those for peace tourists (Van den Dungen, 2014).

Peace tourists travel to places that epitomize a peaceful environment or provide artifacts, monuments, and/or information pertaining to peace. They visit attractions to include peace cities, peace museums, and city peace trails. Peace tourists are interested in local peace history which includes literature, documents, places, actions, and campaigns that promote a culture of peace (Van den Dungen, 2014). Peace is also manifested in buildings, parks, museums, and other cultural landscapes and these are premium attractions for peace tourists.

Destinations can change the perceived image of a battle site or war zone through positive communications to potential markets. For example, although Hiroshima was a battlefield, it is promoted as a city of peace with the Hiroshima Peace Memorial Museum being a major attraction for peace tourists.

Tourism and Peace as Contributors of Sustainable Development

Sustainable development is intended to transform the world by 2030, with a focus on five critical areas: people, planet, prosperity, peace, and partnership (United Nations Department of Economic and Social Affairs/UNDESA, 2015). As per Table 15.1, the aim of the 17 Sustainable Development Goals (SDGs) established by the United Nations General Assembly in 2015 is to end global poverty, protect the planet and for people to enjoy peace and prosperity.

Tourism is identified as a key contributor to this global sustainable development thrust (Stojanovska-Stefanova & Atanasoski, 2017). It protects the environment and supports sustainable development (United Nations World Tourism Organization/UNWTO and Organization of American States/OAS, 2018). In addition, tourism plays a very important role in protecting the planet and in connecting people across the world (Taylor, 2019). World Tourism Day is annually celebrated to increase awareness of the importance of tourism and its social, cultural, political, and economic value (UNWTO, 2015).

Peace is likewise identified as a contributor to global sustainable development (Krieger, 2005). It enhances the sustainable use of the planet (Cairns, 2000) and plays an important role in social growth and development (Wohlmuther & Wintersteiner, 2014). SDG 16 aims at attaining a peaceful, just, and inclusive society. It is intended to advance conflict prevention, build peace and enable prosperity for people and the planet. The annual celebration of "International Day of Peace" further signifies the ideals of peace across the globe.

Melotti et al. (2018) deduced that to promote sustainable development, tourism must be a peace-making factor. This occurs by providing opportunities for residents, preserving the cultural and natural environment and involving them in decisions. Musavengane (2020) agrees that if peace is promoted through tourism this will facilitate the attainment of SDG 16. Villiers (2014) proposes that tourism, peace, and sustainable development are the cornerstones for a better world. Mishra and Verna (2017) add their support to this view, noting that tourism is an engine of peace-oriented sustainable economic development. Farmaki (2017) concludes that tourism is perceived as a force for peace, it is a primary component of sustainable development.

Table 15.2 shows an at a glance illustration of tourism and peace as contributors to sustainable development based on the five critical areas identified by the UN. The indicators suggest that there are similarities between tourism and peace in contributing to global sustainable development. This makes them mutually in agreement in the thrust of attaining sustainable development in all spheres.

Table 15.1 Critical Areas for Sustainable Development

Areas	*Aims*
People	End poverty and hunger
Planet	Protect the planet from degradation by way of sustainable practices
Prosperity	Enable all human beings to enjoy prosperous and fulfilling lives
Peace	Foster peace, just and inclusive societies
Partnership	Achieve global solidarity and interlinkages across all stakeholders, people, and countries

Table 15.2 Contributions of Tourism and Peace to Sustainable Development

Critical Areas for Sustainable Development	*Tourism*	*Peace*
People	Provides economic benefits to improve the standard of living for people (Andereck & Jurowski, 2006)	Fosters economic growth in developing countries (Santhirasegaram, 2008)
Planet	Protects the environment (UNWTO & OAS, 2018)	Enhances sustainable use of the planet (Cairns, 2000)
Prosperity	Enables all human beings to enjoy prosperous and fulfilling lives (Uysal et al., 2016)	Plays an important role in social growth and development (Wohlmuther & Wintersteiner, 2014)
Peace	Is a powerful force for peace (Farmaki, 2017)	Peace and sustainable development will rise and fall together (Krieger, 2005)
Partnership	Legitimate partnership in tourism planning is required for sustainability (Nunkoo, 2017)	Partnership is required for development and peace (Wolfensohn, 2002)

Understanding the Relationship between Tourism and Peace through the Social Network Theory

Social Network Theory

The social network theory is used to understand human social organization and characterizes social structure at the individual and population levels (Krause et al., 2007). The theory emerged from social network analysis (SNA) where social psychologists in the 1940s and 1950s used matrix algebra and graph theory to establish network terms such as groups and social circles which were applied to the field of social psychology (Borgatti & Ofem, 2010). The use of social network analysis was further applied in studying "the effects of different communication structures on the ability of groups to solve problems" (Borgatti and Ofem, 2010, p. 17). It then became very useful in other disciplines such as anthropology, political science, sociology, and economics. SNA became established in the social sciences where it was used to analyze organizational performance, examining the characteristics of the organization as well as the relationship they have with other organizations (Borgatti & Ofem, 2010).

The main component of SNA is the networking characteristic. A network, as described by Borgatti and Ofem (2010, p. 19), is "a set of nodes or actors, along with a set of ties of a single type that connect the nodes". The word node emerged from being a mathematical term that is now being used in the realm of social theory (Kadushin, 2004). Nodes can be people, departments or units, organizations, and industries. Nodes span across friendships, between individuals, communication patterns or conflict between nation-states (Borgatti & Ofem, 2010).

A network is a set of relationships and in the past was addressed in the field of mathematics and engineering but is now widely used in the social sciences (Kadushin, 2004). Social scientists have identified three types of networking: egocentric, socio-centric, and open systems. Egocentric is a network that is connected with a single node or person, for example, someone's good friend or the organization that an entity does business with. It is based primarily on the gathering of information. Socio-centric networks connect a particular set of people, for example, employees in an organization and are viewed as closed system networks. The open system network is one where the connection between the nodes is open without clear boundaries, for example, connections between organizations (Kadushin, 2004).

The set of relationships vary. Relations can be based on similarities, for example, one's race or class which can facilitate or inhibit social ties, and social relations where there are ongoing ties such as kinship. Social relations can also be institutionalized. There are also mental as well as interaction and flow relations. Mental relations refer to an individual's perception of and attitude toward another such as liking or disliking someone. Interaction relation explains how one relates to another, for example, giving advice or information. Flow relations can either be tangible or intangible and are transmitted during interactions such as ideas, material resources, or even viruses.

Application of the Social Network Theory to the Tourism and Peace Nexus

The social network theory is fitting to the dynamics of tourism. This is because tourism itself involves the interactions of organizations, people, and events (Aliperti et al., 2019) and network approaches allow for the identification of cohesive groups of actors who engage and interact with each other (Frank, 1995) or blocks of actors who are structurally connected in similar patterns of interaction (Borgatti & Everett, 1992). Additionally, tourism provides social and economic opportunities (Sinclair-Maragh et al., 2015). The network view indicates that there is a web of relationships in which actors are embedded in providing these opportunities. However, on the other hand, they can constrain such opportunities (Borgatti & Ofem, 2010). The nodes or actors identified by Borgatti and Ofem (2010) are representative of the tourism stakeholders. Peace requires the participation of all stakeholders (Moufakkir & Kelly, 2010).

The set of ties that connect the nodes are the elements and factors required for attaining peace and operating in a peaceful tourism environment. Garnered from the literature, Table 15.3 provides a list of the tourism nodes (stakeholders) and sets of tourism ties that are required in attaining and maintaining peace in the industry at all levels: local, national, regional, and global. This list is not exhaustive as Richmond (2008) suggests that emotive, aesthetic, and linguistic skills are necessary to resolve conflicts and attain peace.

The following explains the application of the social network theory to the tourism and peace nexus within the three types of network as identified by Scott (2000).

Table 15.3 List of Tourism Nodes and Sets of Ties

Tourism Nodes	*Tourism Ties*
Government	Respect
Tourists	Soft Skills
Visitors	Friendship
Local Residents	Communication
Communities	Empathy
Tourism Sectors	Understanding
Tourism Supporting Bodies	Cultural Immersion
Investors	Ecotourism Activities
Multinational Corporations	Trust
Destination Management Organization	Problem-solving Skills
Tourism Industry and Sectors	Critical thinking Skills
Units/Departments	Collaborative Dialogue
	Collective Leadership
	Empowerment
	Knowledge
	Involvement

i Egocentric Network

Being connected with a single node, the egocentric network is demonstrated by each stakeholder's relationship with another. Cases in point: visitor/tourist and residents, government, and investors, or residents, and tourism sectors and government, or its agencies as well as the relationship between other stakeholders.

This type of network will allow for the resolving of stakeholder conflict which as proposed by Janssen (2020) can be addressed through empathy, problem-solving, and critical thinking. Of importance is the dialogue between and among tourism stakeholders. Collaborative dialogue is presented as a tourism tie since it is likely to contribute to meaningful debates among tourism stakeholders. Collective leadership can likewise be a significant contributor to stakeholder dialogue (Gauthier, 2006). When stakeholder tensions are resolved then peace through tourism can be achieved. Successful tourism requires civil peace (Becken & Carmignani, 2016).

Tourism facilitates the exploration of new cultures (Becken & Carmignan, 2016) and the understanding and respect of the living style of residents (Jamgade, 2021). Through community-based tourism, local communities market their cultural and natural assets to tourists as in the case of the Countrystyle Community Tourism Network in Jamaica which models the 'Village as Business' program to sustain and enhance community-based tourism (IIPT, n.d.). In doing so, residents execute their offerings through social and cultural immersion. Along with their demonstration of soft skills and respect among other tourism ties, they will be able to sustain a relationship with tourists.

The cross-cultural understanding developed through these interactions successively breaks down political and ideological barriers (Kim et al., 2007). In addition, the economic support given to local communities reduces poverty (Sinclair-Maragh, 2018). Realizing income and job opportunities from tourism activities increase the satisfaction level of residents (Gajdošík et al., 2019) and this is known to promote peace (Haessly, 2010). In essence, by way of community-based tourism offerings, tourists can build relationships, be it social, mental, interaction, or flow, with residents to achieve sustainability within the environment as well as peace. This is because tourism broadens the mind through the visits and subsequent experiences gained to include the culture and diversity of the destination. Henceforth, intellectual, moral, and spiritual understanding and the broad-mindedness to accept a human being created. This consequently results in the generation of peace (Jamgade, 2021).

To further support the perspective that tourism creates peace, tourists are viewed as playing an important role in the promotion of peace and are oftentimes referred to as peace messengers. Promoting peace through tourists is more impactful than through other channels such as meetings/conferences (Gajdošík et al., 2019).

Although peace begins with the individual, Smith (2004) purports that it should be supported by broad-based strategies. Thus, the importance of the government's role at all levels to ensure that this is incorporated into the tourism plans. With respect to the relationship between government and residents, the latter stakeholder wants to feel empowered, knowledgeable, and involved in tourism. This has been the findings of researchers such as Nunkoo (2015) and Fong and Lo (2015). For this reason, Murphy (2013) recommends a community approach to development and planning. This is to encourage local benefit and a tourism product that is in harmony with the local environment and its people. With the communication and respect among other ties are applicable to the relationship between government and residents to eliminate conflicts.

Trust as a social tie is included in the list in Table 15.3 as it is imperative to the relationship between the tourism nodes. Residents want to be able to trust their government in terms of them making tourism development decisions on their behalf and they want to be involved in the process (Sinclair-Maragh & Gursoy, 2016). Additionally, international travel promotes trust among people from all backgrounds thus, creating peace through tourism (Jimenez & te Kloeze, 2014). Trust is also an important tie in the relationship between countries, in this case, tourism destinations and source markets. It is imperative to strengthen trust in the travel industry to encourage global peace and sustainable tourism. Overall, this will grow tourism businesses (Yin & Zhao, 2006).

The egocentric network can be further explained within the notion of sustainable tourism. The involvement of all the tourism nodes (stakeholders) is ostensibly paramount to sustainable tourism. Ties such as ecotourism activities, empathy, and respect would therefore be critical to the achievement of sustainable tourism. Having these nodes and ties connected in a relationship can result in a peaceful environment. Peace is one of the five indicators of sustainability (Buckley, 2012). Wohlmuther and Wintersteiner (2014) point out that peace and sustainability cannot be separated.

ii Socio-centric Networks

Socio-centric networks connect individuals within their particular settings. The nodes within this type of network include residents in a community tourism area as well as employees in any of the tourism sector, government tourism ministry or agencies, and destination management organizations. Because this network comprises people from varying backgrounds who are in a common space, be it community or place of work, there is the propensity for conflicts. Societal diversity in terms of differences in ethnicity, religion, and political ideology can cause conflicts (Schlee, 2008).

There must be peace and harmony in communities among residents to enable tourism activities and development. The tourism ties in Table 15.3 would be applicable in generating and sustaining this inner-community network relationship. An understanding of one's differences within this closed network system can reduce the possibility of conflicts and disharmony. In expounding on the catalytic effect of tourism in peace building, Jamgade (2021) referred to peace as a state of calmness which occurs when people from varied cultures, perceptions, and beliefs understand and respect each other's diversity in their surroundings.

Tourism ties such as understanding, respect, communication, trust, collaborative dialogue, and collective leadership are likewise applicable to the relationship between and among employees in the tourism sectors and respective entities. These sectors and entities comprise a pool of individuals with varying social, economic, and political backgrounds, and beliefs. Employing the tourism ties in their everyday interactions will promote a peaceful environment. In addition,

Haessly (2010) proposes six principles that are relevant to employees in the tourism industry. They are "Attention to just relationships, respect for human rights, care for the common good, protection of global security, engagement in just and, transforming actions that promote, protect, preserve and sustain a culture of peace".

iii Open System Network

This network suggests that there are no clear boundaries in the relationship as the connection between the nodes is open. It is demonstrated by tourists who visit a destination, locale, or site on their own without a tour operator or tour guide. Seemingly, this type of network suggests individualism and the character of an adventurous person. Nonetheless, this type of tourist is required to have an appreciation of and respect for the local community population, heritage, and cultural assets as well as the natural surroundings. This is supported by Carter and Bramley (2002) who from a heritage values perspective point out the role of resource users in adhering to permissible activities involving valued resources.

Through responsible tourism, tourists can enable sustainable tourism and contribute positively to communities and the environment. This can be achieved by conserving the natural environment and purchasing sustainable products (Levy & Hawkins, 2009). In doing so they are engendering peace. Peace can also be linked to sustainability in tourism through collective leadership, as well as empathy (Janssen, 2020). Communication is likewise an essential tourism tie in this dynamics Boulding (1992) posits that communication articulates one's vision to other stakeholders Even on an individual basis, being mindful and respectful of the traditions, and conserving and protecting the natural environment of the destination and local communities are very important in building and maintaining a relationship and creating peace.

Conclusion

This chapter seeks to re-examine the relationship between tourism and peace. The debate presents tourism as a mechanism to create peace as well as peace is a mechanism to generate tourism. From a deductive reasoning approach, the chapter concludes that both arguments are plausibly discussed in the literature. Overall, tourism, being a global activity, can generate peace. Wars and conflicts which result from the absence of peace can impact tourism. Likewise, peace is a catalyst for tourism and its development as people tend to travel to and in peaceful environments. The literature also speaks of peace tourism, a form of tourism that attracts tourists to places that denote peace through post-war initiatives, peace museums, and other activities that promote a culture of peace.

In addition, the chapter provides an analysis of the contribution of tourism and peace to sustainable development. The deduction is that both constructs have vital roles to play in the global mission of transforming the world by focusing on five critical areas: people, planet, prosperity, peace, and partnership. To advance the literature, the chapter further analyzed the tourism-peace nexus within the context of the social network theory. The uniqueness about this theory is its networking characteristics with the composition of nodes and ties. From an exploration of the literature, tourism nodes and tourism ties were identified and discussed within the tourism and peace nexus. The tourism nodes are the actors or stakeholders to include local residents, tourists, government, and investors. The tourism ties are the elements that create a connection between the stakeholders. These include respect, communication, empathy, understanding, collative dialogue and involvement. The chapter deduces that tourism and peace are not mutually exclusive; they have a mutual relationship for global, regional, national, and local sustainable development. This major contribution to the literature can be useful to future scholars in expounding on the relationship between tourism and peace.

The study will be of importance to tourism stakeholders: governments, destination planners, industry players, and other actors in tourism planning, development, and execution. Future research can quantitatively analyze the relationship between tourism and peace at the local, national, regional, or global level. Although the study of tourism and peace is not new in the literature, it has not adequately addressed the tourism and peace nexus within the context of sustainable tourism development and this presents the opportunity for future studies in this area.

There is also the need to examine the concepts of tourism, peace, and/or peace tourism within the period of the coronavirus (COVID-19) pandemic, the impact of which is unprecedented.

References

Aliperti, G., Sandholz, S., Hagenlocher, M., Rizzi, F., Frey, M., & Garschagen, M. (2019). Tourism, crisis, disaster: An interdisciplinary approach. *Annals of Tourism Research, 79*, 102808.

Alluri, R. M. (2009). The role of tourism in post-conflict peacebuilding in Rwanda. swisspeace Working Paper, 2.

Andereck, K., & Jurowski, C. (2006). Tourism and quality of life. In G. Jennings & E. Nickerson, (Eds.), *Quality tourism experiences* (1st ed., pp. 136–154). Routledge.

Baldwin, F., & Sharpley, R. (2009). Battlefield tourism: Bringing organised violence back to life. In R. Sharpley, & P.R. Stone, P. R. (Eds.), The *darker side of travel* (pp. 186–206). Channel View Publications.
Becken, S., & Carmignani, F. (2016). Does tourism lead to peace? *Annals of Tourism Research, 61*, 63–79.
Blanchard, L. A., & Higgins-Desbiolles, F. (2013). *Peace through tourism: Promoting huma security through international citizenship*. Routledge.
Borgatti, S. P., & Everett, M. G. (1992). Notions of position in social network analysis. *Sociological methodology*, 22, 1–35.
Borgatti, S. P., & Ofem, B. (2010). Social network theory and analysis. In A. J. Daly (Ed.), *The ties of change: Social network theory and application in education* (pp. 17–30). Cambridge: Harvard Press.
Boulding, E. (1992). *New agendas for peace research: Conflict and security re-examined* (1st ed.), Lynne Rienner Publishers.
Buckley, R. (2012). Sustainable tourism: Research and reality. *Annals of Tourism Research, 39*(2), 528–546.
Cairns J, Jr. (2000). World peace and global sustainability. *The International Journal of Sustainable Development & World Ecology*, 7(1), 1–11.
Caneday, L. (1991). Tourism: Recreation of the elite. In J. Zeiger and L. Caneday (Eds.), *Tourism and leisure. dynamics and diversity* (pp. 83–91). Alexandria, VA: National Recreation and Parks Association.
Carter, R. W., & Bramley, R. (2002). Defining heritage values and significance for improved resource management: an application to Australian tourism. *International Journal of Heritage Studies, 8*(3), 175–199.
Cavlek, N. (2006). *Tour operators and destination safety* (pp. 338–355). Routledge.
Chandrasiri, P. R. (2019). Problems and prospects of peace tourism in post-war Sri Lanka. *E Journal of Tourism, 6*(1), 28–42.
Cho, M. (2007). A re-examination of tourism and peace: The case of the Mt. Gumgang tourism development on the Korean Peninsula. *Tourism Management, 28*(2), 556–569.
Crews, R. J. (1989). A values-based approach to peace studies. In D. C. Thomas & M. T. Klare (Eds.), *Peace and world order studies: A curriculum guide* (5th ed., pp. 28–37). Boulder, CO: Westview Press.
D'Amore, L. J. (1988). Tourism—A vital force for peace. *Tourism Management, 9*(2), 151–154.
De Villiers, D. A. W. I. D. (2014). Cornerstones for a better world: Peace, tourism and sustainable development. In *International handbook on tourism and peace* (pp. 78–86).
Farmaki, A. (2017). The tourism and peace nexus. *Tourism Management, 59*, 528–540.
Fernando, S., Bandara, J., Liyanaarachchi, S., & Smith, C. (2014). Managing the tourism-led development strategy in post-war Sri Lanka. *Sri Lankan Journal of Business Economics, Department of Business Economics, 5*(1).
Fernando, S., Bandara, J. S., & Smith, C. (2013). Regaining missed opportunities: the role of tourism in post war development in Sri Lanka. *Asia Pacific Journal of Tourism Research, 18*(7), 685–711.
Frank, K. A. (1995). Identifying cohesive subgroups. *Social Networks, 17*(1), 27–56.
Fong, S. F., & Lo, M. C. (2015). Community involvement and sustainable rural tourism development: Perspectives from the local communities. *European Journal of Tourism Research, 11*, 125–146.
Fyall, A., Prideaux, B., & Timothy, D. J. (2006). War and tourism. *International Journal of Tourism Research, 8*(3), 153–246.
Gajdošík, T., Sokolová, J., Gajdošíková, Z., & Pompurová, K. (2019). Peace promotion through volunteer tourism. In A. M. Nedelea & Nedelea, M. O. (Eds.), *Marketing peace for social transformation and global prosperity*. IGI Global.
Galtung, J. (1964). A structural theory of aggression. *Journal of Peace Research, 1*(2), 95–119.
Gauthier–September, A. (2006). Developing collective leadership: Partnering in multi-stakeholder contexts. *Leadership is global: Bridging sectors and communities*, 1–20
Gore, C. (2015). The post-2015 moment: Towards Sustainable Development Goals and a new global development paradigm. *Journal of International Development, 27*(6), 717–732.
Haessly, J. (2010). Tourism and a culture of peace. In O. Moufakkir & I. Kelly (Eds.), *Tourism, progress and peace* (pp. 1–16). Wallingford: CAB International.
Holland, S. (1991). "Recreation and Tourism: Evolution of the Social Mission." In J. Zeiger and L. Caneday (Eds.), *Tourism and leisure: dynamics and diversity* (pp. 66–81). Alexandria, VA: National Recreation and Parks Association.
International Institute for Peace through Tourism/IIPT (n.d.). IIPT Caribbean, https://peacetourism.org/car/
International Institute for Peace through Tourism/IIPT (2020). About the International Institute for Peace through Tourism (IIPT), https://peacetourism.org/about-iipt/
Jamgade, S. (2021). Catalytic effect of tourism in peace building: sustainability and peace through tourism, In J.T. da Silva, Z. Breda & F. Carbone, F. (Eds.), *Role and impact of tourism in peace building and conflict transformation* (pp. 29–45). IGI Global.

Janssen, M. (2020) *Linking together peace and sustainability in tourism-developing sustainable destinations personal reflection on the future*, https://peacetourism.org/linking-together-peace-and sustainability-in-tourism/
Jimenez, C., & Kloeze, J. T. (2014). Analyzing the peace through tourism concept: The challenge for educators. *Sociology and Anthropology, 2*(3), 63–70.
Kadushin, C. (2004). *Introduction to social network theory*. Boston, MA. Retrieved from https://www.academia.edu/43497590/Introduction_to_Social_Network_Theory_Chapter_2_Some_Basic_Network_Concepts_and_Propositions_Basic_Network_Cocepts
Khamouna, M., & Zeiger, Z. B. (1995). Peace through tourism. *Parks & Recreation (Arlington), 30*(9), 80–86.
Kim, S., Prideaux, B. & Prideaux, J. (2007). Using tourism to promote peace on the Korea Peninsula. *Annals of Tourism Research, 34*(2), 291–309.
Knopf, R. (1991). "Harmony and Convergence between Recreation and Tourism." In J. Zeiger & L. Caneday (Eds.) *Tourism and leisure: dynamics and diversity* (pp. 53–66). Alexandria, VA: National Recreation and Parks Association.
Krause, J., Croft, D. P., & James, R. (2007). Social network theory in the behavioural sciences: Potential applications. *Behavioral Ecology and Sociobiology, 62*(1), 15–27.
Krieger, D. (2005). Peace and sustainable development will rise or fall together. *International Journal of Humanities and Peace, 21*(1), 33.
for achieving common goals – a practical guide for change agents from public sector, private sector and civil society (1st ed., pp. 25–32). Potsdam: Collective Leadership Institute.
Lankford, S. V., Grybovych, O., & Lankford, J. K. (2008). Context for peace and community tourism: A worldview framework. *World Leisure Journal, 50*(3), 199–208.
Levy, S. & Hawkins, D. (2009). Peace through tourism: Commerce-based principles and practices. *Journal of Business Ethics, 89*(1), 569–585.
Litvin, S. (1998). Tourism: the world's peace industry? *Journal of Travel Research* 37(1), 63–66.
Lloyd, D.W. (1998) *Battlefield tourism*. Oxford: Berg.
Malley, M. (2002). Leven's words about tourism, peace still ring true. *Hotel & Motel Management, 217*(19), 8 8.
McIntosh, R. W., Goeldner, C. R., & Ritchie, J. B. (1995). *Tourism: Principles, practices, philosophies* (7th ed.). John Wiley and Sons.
Melotti, M., Ruspini, E., & Marra, E. (2018). Migration, tourism, and peace: Lampedusa as a social laboratory. *Anatolia, 29*(2), 215–224.
Mishra, P. K., & Verma, J. K. (2017). Tourism and peace in economic development perspective of India. *Journal of Environmental Management & Tourism, 8*(4 (20)), 927–934.
Moufakkir, O. & Kelly, I. (2010). *Tourism, progress, and peace* (1st ed.). Wallingford: CAB International.
Murphy, P. (2013). *Tourism: A community approach (RLE Tourism)*. Routledge.
Musavengane, R. (2020). Election risk and urban tourism in sub-Saharan African cities: Exploring peace through tourism in Harare, Zimbabwe. In *Sustainable Urban Tourism in Sub-Saharan Africa* (pp. 92–106). Routledge.
Nunkoo, R. (2015). Tourism development and trust in local government. *Tourism Management, 46*, 623–634.
Nunkoo, R. (2017). Governance and sustainable tourism: What is the role of trust, power and social capital? *Journal of Destination Marketing & Management, 6*(4), 277–285.
Pratt, S., & Liu, A. (2016). Does tourism really lead to peace? A global view. *International Journal of Tourism Research, 18*(1), 82–90.
Raymond, E. M., & Hall, C. M. (2008). The development of cross-cultural (mis) understanding through volunteer tourism. *Journal of Sustainable Tourism, 16*(5), 530–543.
Richmond, O. (2008). Reclaiming peace in international relations. *Millennium: Journal of International Studies, 36*(3), 439–470.
Santhirasegaram, S. (2008). Peace and economic growth in developing countries: Pooled data cross-country empirical study. In *International conference on applied economics ICOAE 2008*, 807–814.
Schlee, G. (2008). *How enemies are made: towards a theory of ethnic and religious conflicts* (1st ed.). Berghahn Books.
Seaton, A. V. (1999). War and thanatourism: Waterloo 1815–1914. *Annals of Tourism Research, 26*(1), 130–158.
Scott, J. (2000). *Social network analysis: A handbook*. Thousand Oaks, CA: SAGE Publications, 23.
Selwitz, R. (1994). Peace brings Hilton to Northern Ireland. *Hotel & Motel Management, 209*(19), 90-90.
Shin, Y. S. (2005). Safety, security and peace tourism: The case of the DMZ area. *Asia Pacific Journal of Tourism Research, 10*(4), 411–426.
Sinclair-Maragh, G. (2018). Responsible tourism and poverty: A case of Sandals Resorts International. In S.A. Hipsher. (Eds.), *Examining the Private Sector's Role in Wealth Creation and Poverty Reduction* (pp. 51–66). IGI Global.

Sinclair-Maragh, G., & Gursoy, D. (2016). A conceptual model of residents' support for tourism development in developing countries. *Tourism Planning & Development, 13*(1), 1–22.
Sinclair-Maragh, G., Gursoy, D., & Vieregge, M. (2015). Residents' perceptions toward tourism development: A factor-cluster approach. *Journal of Destination Marketing & Management, 4*(1), 36–45.
Smith, D. (2004), *Towards a strategic framework for peacebuilding: Getting their act together*, Overview report of the joint Utstein study of peacebuilding, International Peace Research Institute, Oslo, pp. 10–13.
Stojanovska-Stefanova, A., & Atanasoski, D. (2017). UN goal: Sustainable tourism as a key contributor for sustainable development in developing countries. *Yearbook-Faculty of Tourism and Business Logistics, 3*(2), 152–161.
Suntikul, W. (2013). Thai tourism and the legacy of the Vietnam War. In R. Butler & W. Suntikul (Eds.), *Tourism and war* (2nd ed., pp. 105–118). Routledge.
United Nations Department of Economic and Social Affairs/UNDESA (2015). Transforming our world: The 2030 Agenda for sustainable development.sdgs.un.org/2030 Agenda
United Nations World Tourism Organization/UNWTO (2015). World Tourism Day 2015. https://www.unwto.org/world-tourism-day-2015
United Nations World Tourism Organization/UNWTO and Organization of American States/OAS (2018). Tourism and the sustainable development goals: Good practices in the Americas, https://www.oas.org/en/sedi/desd/CT/Documents/OAS_UNWTO_9789284419685.pdf
Upadhayaya, P. K., Müller-Böker, U., & Sharma, S. R. (2011). Tourism amidst armed conflict: Consequences, copings, and creativity for peace-building through tourism in Nepal. *The Journal of Tourism and Peace Research, 1*(2), 22–40.
Uysal, M., Sirgy, M. J., Woo, E., & Kim, H. L. (2016). Quality of life (QOL) and well-being research in tourism. *Tourism Management, 53*, 244–261.
Van den Dungen, P. (2014). Peace tourism. *International handbook on tourism and peace*, 62–77.Vanneste, D., & Foote, K. (2013). 20 War, heritage, tourism, and the centenary of the Great War in Flanders and Belgium. *Tourism and War, 34*, 254.
Var, T., Brayley, R., & Korsay, M. (1989). Tourism and world peace: Case of Turkey. *Annals of Tourism Research, 16*(2), 282–286.
Viren, P. P., Vogt, C. A., Kline, C., Rummel, A. M., & Tsao, J. (2015). Social network participation and coverage by tourism industry sector, *Journal of Destination Marketing & Management, 4*(2), 110–119.
Winter, C. (2009). Tourism, social memory and the Great War. *Annals of Tourism Research, 36*(4), 607–626.
Wohlmuther, C. & Wintersteiner, W. (2014). *International handbook on tourism and peace* (1st ed.), Drava Verlag.
Wolfensohn, J. D. (2002). A partnership for development and peace. Keynote Address delivered at the Woodrow Wilson International Center, Washington, DC, March 6.
Yin, M., & Zhao, S. Z. (2006). Research on a dynamic model of trust building within regional tourism alliances: Evidence from China. *Chinese Economy, 39*(6), 5–18.
Young, T. (2019), *International Day of Peace 2019 – Tourism as a vehicle of peace*. International Institute for Peace through Tourism/IIPT, https://peacetourism.org/international day of-peace 2019.

16

WHAT INFLUENCES ATTITUDE CHANGE?

Tourist Satisfaction, Motivation, Personality, Tolerance Level, Contact Situation (Level, Type, Frequency)

Bekir Eşitti

Introduction

Due to globalization, many sectors were not limited to the borders of a country, and sectoral changes and developments were experienced worldwide. Providing foreign currency to countries, the tourism sector has become one of the key sectors with the effect of globalization. The development of the tourism sector in a particular destination depends on the tourist, who is the main element of the sector. For this reason, tourist behavior has become one of the subjects researched in the relevant literature (Lai, 2018). The desired and expected behavioral state in this sector is the satisfaction of tourists which demonstrates their perceptions and intentions toward tourism activities. Satisfaction, in general, creates positive feelings toward any brand, product, or service and as a result, this positive feeling leads to the brand, product, or service being repurchased or recommended to others whereas dissatisfaction can cause negative attitudes, negative announcements, and this leads to not being preferred. Satisfaction also motivates tourists by directing them to voice positive outcomes, such as repurchase or word-of-mouth marketing of the destination. Motivation is a driving and attractive force behind all behavior of individuals and is accepted as an important variable that explains tourist behavior (Moutinho, 2000).

Motivational research questions the causes of human behavior, how it occurs, and how these behaviors can be directed. It is difficult to answer the question of why tourists visit so that people's motivations can be learned. The differences between the motivations created by the personalities of the tourists cause the attitude change and the difference in the level of satisfaction with the service they receive.

Differences arising from the personalities of tourists can cause an increase or decrease in tourists' attitude, satisfaction, and motivation level. Another theme that affects the satisfaction level and motivation of tourists can be the frequency and duration of their contact with the local community. One of the determinants of tourist behavior is undoubtedly the social bond established by the tourist and the local community. In tourism research, tourist-host social contact situation has been widely believed to play an important role in understanding intergroup relations, tourists' behaviors, and tourists' attitudes toward a destination (Fan et al., 2017). In this regard, this study examines the tourists personality traits, motivations, social contact situation (its level, type, and

DOI: 10.4324/9781003161868-16

frequency), and tolerance level on their attitude change. In this context, this chapter starts with the tourist satisfaction, which is an expected and a very important issue for the tourism sector.

Tourist Satisfaction

Tourism has become one of the biggest industries of today by gaining great momentum around the world due to the increase in leisure time, the increase in living standards, the rapid development of fast and comfortable transportation technologies, the desire of people to see and explore new places, and many other reasons. Providing service that will meet the expectations of customers is one of the common goals of all businesses, as well as businesses that offer tourism services. Researches show that customers can switch from one brand to another just because of better customer service and are ready to pay more to get better customer service. This shows that providing customer satisfaction is the key to the success of companies. Homburg et al. (2005) measure satisfaction through evaluating the experience of consuming products by consumers. Satisfaction in general is the evaluation of all the experiences of a customer in a certain tourism company by the same customer.

In the academic studies on the tourism industry, the question of why people visit has been frequently tried to be answered. While considering the importance of understanding touristic travel motivation, tourist satisfaction is the key in providing competitive advantage and strategy which also plays an important role in decision-making. According to Zeithaml and Bitner (2000), customer satisfaction is the emotional evaluation of the completed purchasing process that occurs after purchasing a good or service. In the study of LeBlanc (1992), satisfaction is considered as the function of the discrepancy arising from the difference between the customer's expectations before purchasing a good or service and the perceived performance after purchasing and using that good or service. In general, customer satisfaction emerges when the benefits of the purchased goods or services match the expectations/desires of the customer. Correia et al. (2009: p. 42) describe satisfaction as "a comparison between expectations and performance of a good or service". The performance of a good or service is indicated by the benefits produced by the good or service when consumed; if the benefits produced meet or exceed the expectations of customer, the customer would be satisfied.

Various basic theories are emphasized in the literature to determine the satisfaction levels of customers by looking at the compatibility between the customer's expectations and the benefit obtained as a result of consumption after using a certain good or service. "Affinity and Contrast Theory", which is among the theories developed for customer satisfaction, argues that if the performance remains within the acceptance zone of a customer, even if it falls behind general expectations, the inconsistency will be ignored, the affinity process will work and the performance will be found acceptable. According to the "Affinity Theory" (Festinger, 1957), consumers assimilate performance against their previous expectations. If the performance is in the rejection zone, the contrast will stand out, the difference will be exaggerated, and the product will be deemed unacceptable, no matter how close to expectations. According to the "Contrast Theory" (Cardozo, 1965), any difference between expectations and experience will be exaggerated in the direction of the difference. Consumers overestimate the perceived level of performance. Thus, the performance level that exceeds the expectations emerges at a higher level than the reality (Altıntaş, 2000; Kılıç & Pelit, 2004). According to this theory, people tend to make changes to justify stressful behavior by adding new parts to cognition that cause psychological dissonance (rationalization) or by avoiding conflicting information and conditions likely to increase the magnitude of cognitive dissonance (confirmation bias). When this theory is considered in terms of tourist satisfaction; the theory indicates that a customer who is faced with a lower performance than he expected with the touristic product will try to minimize the contradiction in his mind if he has made a psychological investment in the product in question. This happens either by lowering the level of expectation or by a more positive perception of performance. Dealing with this kind of nuances of conflicting ideas or experiences is mentally stressful. Experiencing this stress frequently, especially over touristic products or services, will negatively affect customer satisfaction. In the tourism environment, the tendency to demand again becomes more effective, especially today, where the consumer is more conscious and examines all the options in detail before the re-purchase (Peyton et al., 2003).

Another notable theory in the context of customer satisfaction is the "Negativity Theory". The theory developed by Aronson and Carlsmith (1963: p. 584), and asserts that any conflict between expectations and performance will disturb the customer. This theory, just like the others mentioned above, is also based on the disconfirmation process. Dissatisfaction will happen if the perceived performance falls beneath expectations, or if the perceived performance goes beyond the expectations (Isac & Rusu, 2014).

The last theory covered in this study is Deighton's (1984) hypothesis testing theory, which suggests a two-step model for satisfaction generation. First, Deighton hypothesizes, prepurchase information (advertising) plays a substantial role in building up expectations. Customers use their experience with product/service to test their expectations. Second, Deighton believes, customers will tend to attempt to confirm rather than disconfirm their expectations. The theory suggests that customers are biased to positively confirm their product/service experiences. It is an optimistic view, but it turns the management of evidence into a very powerful marketing tool (Isac & Rusu, 2014: p. 84).

Businesses that develop various strategies to ensure customer satisfaction in the tourism sector can achieve their goals. When developing tourist satisfaction strategies, businesses should consider the following suggestions (İnce & Doğantan, 2020: pp. 13–16).

- Tourism businesses should strive to allow customers to express their problems with sincere interest and to find solutions to their problems.
- Customer demands and expectations differ day by day in line with changing technological conditions and developing communication opportunities, and responding to changing expectations requires uninterrupted communication.
- User-friendly website and social media pages should be designed.
- Communication with customers should be maintained through social media or e-mails, and various campaigns and competitions should be organized using this kind of tools.
- Employees who act as a bridge between customers and the business should be satisfied.
- Customer satisfaction should be the focal point in the organizational structure and the organization should be carefully shaped.
- When the degree of response of the offered goods and services to the expectations is measured, the perception of the activities carried out by the enterprises will be easier and it will be possible to determine whether the actual performance is close to the desired level.
- By determining the qualities of the business that provide or hinder satisfaction in the eyes of customers, the factors that provide satisfaction should be developed and the factors that cause dissatisfaction should be eliminated or changed.

A satisfied customer will have the intention to re-purchase continuously from the business. Otherwise, the customer will purchase from the business only once and will make the second and subsequent purchases from the rival businesses. The main strategy to prevent the loss of customers is to ensure customer satisfaction. In order to ensure customer satisfaction, businesses must first know their customers. Attitudes to be adopted to know the customer will return as more sales in the future, and each customer's feeling of loyalty to the business in a way that will affect their environment will provide the opportunity to reach more customers.

Motivation

Motivation is a psychological term representing the driving force that impels individuals to act (Schiffman & Kanuk, 2009). In short, it can be said that motivation is the driving force that leads an individual to act. The concept of tourist motivation is defined in different ways in each of the behavioral sciences, organizational behavior, and consumer behavior fields. Within the scope of tourist behavior, motivation includes a set of factors predisposing a tourist toward a particular activity (Pizam et al., 1979). Within this scope tourist motivation is defined by Swarbrooke and Horner (2007: p. 413) as follows: "motivation is a set of factors that make a consumer want to buy a particular good or service". In terms of consumer behavior, it is not possible to say that all stimulated needs, in other words, motives, will result in purchasing action. If a tourist has an incentive to buy any touristic product or service, the tourist can show three alternative behaviors. It is possible to sort these alternative behaviors as no reaction (no purchase), automatic reaction (unplanned purchase), and entering the decision-making process (planned purchase) (Odabaşı & Gülfidan, 2010: p. 104).

In addition to the general service and product purchasing motivations of tourists, there is also travel motivation. Travel motivation of tourists is defined as the sum of the motives that lead the tourist to travel and to choose a particular touristic destination for their trip (Crompton, 1979: p. 409).

Dann (1981) points out that two reasons make it difficult to study travel motivations. The first of these is that there are individual differences among tourists and the value judgments of tourists differ according to the society they live in. Second, when the tourist is asked what his/her travel motivations are, it is possible to find needs that are in the depths of the tourist's mind and that they are not aware of, apart from the travel motivations that the tourist directly expresses.

Studies carried out to estimate the motivation of tourists is essential especially for marketers. By these studies, they can characterize the strategic tools for performing marketing segmentation, competitive analysis, and market positioning (Guttentag et al., 2018; Pearl, Ok & Au, 2021). It can be said that there are four basic characteristics of motivations that affect tourists purchasing behavior. These four key features of consumer motivations are summarized below (Koç, 2007: p. 135).

- Needs and desires are the basis of the power that moves an individual. A tourist acts according to his needs and desires as an individual.
- Motivations give direction to action and movement; motivations determine when, how, and in what form actions will occur to meet the needs.
- Motivation reduces the feeling of tension in the individual; meeting the needs and desires as a result of the behavior reduces the tension in the individual.
- Motivations are formed within a certain environment; the social environment in which the individual is located affects the formation of motivation.

Various motivational theories have been developed to explain why a tourist engages in certain purchasing behaviors. Although motivation theories are grouped in different ways, the main purpose of motivation theories is to try to clarify the formation of motivation by systematically examining events and phenomena related to motivation. Maslow's hierarchy of needs, Herzberg's motivational and conditional factors theory, achievement needs theory, Lawler and Porter's expectation theory, Edwin Locker's goal theory, equity theory, and behavior correction and empowerment theories are evaluated in this context (Yousaf et al., 2018: pp. 204–205).

Solomon et al. (2009: p. 92) examined motivation theories in terms of consumer behavior under the titles of needs theory, environment theory, interaction theory, and expectation theory.

Some of these theories focus on internal factors while examining motivation, while others focus on external factors. Dann's (1977, 1981) two-dimensional push-pull motivation framework is the most extensively applied motivation framework in the tourism literature (Pearl et al., 2021). In this framework, push factors refer to the internal drives that inspire a tourist to engage in a particular behavior from a tourist's side, whereas pull factors refer to the external factors that cause a tourist to value a specific behavior over another from the supplier's side (Klenosky, 2002). Inner driver is a power potentially found in every human being that, when activated, enables us to move forward irresistibly through concentration.

Swarbrooke and Horner (2007: p. 53) divided the tourists' travel motivations into two main groups as motivations that lead the tourist to go on vacation and motivations that lead the tourist to spend their vacation in a certain touristic destination at a certain time. According to the authors, the travel motivations of tourists can be listed under six headings (Swarbrooke & Horner, 2007: p. 53).

- Physical factors: relaxation, exercise, sun, and health
- Personal development: desire to increase one's knowledge of the world or to gain new knowledge and skills
- Gaining status: traveling with status, suitable destination selection and desire to gain prestige
- Cultural factors: desire to travel, see and get to know new cultures and cultural elements
- Emotional factors: romance, spiritual renewal, nostalgia, adventure, escape and fantasy
- Personal factors: visiting friends or relatives, making new friends and willingness to make others happy

Personality

Personality is the concept that expresses the human being, who is effective in social and working life with his/her abilities and characteristics, as a sole and unique being. It is derived from the Latin word "persona", which means "mask" that theater actors wear on their faces in accordance with their roles.

One aspect of personality is the attitude a person takes in his relations with other people, the behavior he displays and the mask he wears. Man is in constant contact with his environment and often tries to give his feelings, thoughts, attitudes, and behaviors a different form than they are. In some people, this situation is permanent; others want to look different depending on their location. Thus, people try to show themselves as they want or wanted, by hiding behind a mask that is worn time to time or constantly. Therefore, the concept of personality includes the individual's reaction in relations with others and the way he shows himself. Although personality includes some innate tendencies, it is mainly shaped by the interactions the person has with the people around her/his from infancy and the experiences she/he experiences as a result of these interactions.

The concept of personality is defined in different ways. One of the most known of these definitions was made by Weinstein, Capitanio, and Gosling (2008: p. 330): Personality describes and explains the emotions, thoughts, and behaviors of individuals in a stable pattern. Hogan (2009), on the other hand, sees personality as an important factor that reveals interpersonal difference and states that genetic and environmental factors are effective in the formation of personality. According to Hogan, the success or failure of all institutions in a wide range starting from the family to multinational companies and modern nation-states depends to a great extent on the personalities of the people involved in the operation of the institutions. Individuals' working in jobs suitable for their personality structures will bring positive expansions in organizational and individual terms such as job satisfaction and work efficiency and effectiveness. Since consistency, adaptation to the environment, and individual reactions are the basis of personality traits, individual reactions resulting from interaction with other people can also be considered as personality (Şimşek et al., 2011: p. 97). Considering the common aspects of all these definitions of personality, the following items can be reached. (İnanç & Yerlikaya, 2009: pp. 3–4):

- Containing features that make an individual different from others,
- Being consistent, not changing the behavior of the person in similar situations over time,
- The features that distinguish the individual from other individuals are a structured whole, the personality is a system consisting of many features, and these features that make up the system form a pattern that is related to each other and unique to the individual.

Early theories of what personality is varied, including Allport (1960)'s list of 4,000 personality traits, Cattell (1956)'s personality factors, and Eysenck (1952)'s three-factor theory. In order to examine personality, Eysenck first used extraversion and neuroticism with factor analysis, but then added the dimension of openness to experience and reached the three-factor model. After these studies, the five-factor theory emerged to describe key traits that serve as the building blocks of personality (McCrae & Costa, 1987; Somer et al., 2002; Thurstone, 1951). In this framework, personality traits five factors are extraversion, agreeableness, conscientiousness, neuroticism (sometimes named by its polar opposite, emotional stability), and openness to experience (sometimes named intellect) (John et al., 2008; McCrae & Costa, 1987).

Emotional balance/instability-neuroticism: It measures a person's irritability, self-confidence, optimism, shyness, emotionality and whether he is anxious, in short, what kind of emotional structure the person has. Neurotic individuals are characterized by being more prone to experiencing anxiety. On the contrary, emotionally stable individuals are relaxed and calm (Çivitçi & Arıcıoğlu, 2012). Especially those working in a labor-intensive service sector such as tourism are expected to be calm, cold-blooded, and relaxed individuals.

Extraversion: This personality dimension represents the individual's more sociability and friendliness level. Key traits of Extraversion are warmth, sociability, assertiveness, excitement seeking, and a greater tendency to experience positive emotions.Non-extroverted individuals (introverts) are characterized by rather quiet and reserved behaviors (Chamorro-Premuzic, 2007). People with high extrovert personality traits are energetic, friendly and open to establishing new social relationships. In short, extroverted individuals are fun-loving, talkative, humorous, social and affectionate. Introverts, on the other hand, are quiet, distant, shy, passive and love solitude. Tourism sector workers and tourists are expected to be extroverts, and this can ensure a positive relationship between the tourist-worker and local people.

Agreeableness: This personality trait is more related to interpersonal relationships. It refers to the individual's participation in interpersonal cooperation and the degree of approval of this cooperation. Mild people tend to be friendly, close, warm, sociable, and trustworthy toward other individuals, while ungentlemanly people are less agreeable to others, argumentative, uncooperative, and harsh (Glass et al., 2013).

Conscientiousness: It is related to how people control, regulate, and direct their impulses. High conscientiousness is often perceived as being stubborn and focused. Low conscientiousness is associated with flexibility and spontaneity, but can also appear as sloppiness and lack of reliability. Individuals who have low conscientiousness are those who tend to engage in impulsive behaviors, are disorganized, and tend to postpone their tasks (John et al., 2008).

Openness to Experience: People who are open to experience are intellectually curious, open to emotion, sensitive to beauty and willing to try new things. People with high openness to experience are generally non-conservative (Brown et al., 2002).

It is seen that the personality traits of the tourists have an important effect on the quality of the service they receive and the determination of their attitudes and behaviors toward local stakeholders and tourism workers in the destination they are gone. Therefore, it can be said that the attitude of tourists can change according to their dominant personality traits.

Tolerance Level

In terms of tourist behavior, the tolerance level is used to determine the upper limit of how much of something can be tolerated by a tourist. The positive and negative dimensions of the concept of tolerance in tourist behavior, such as "to be tolerant" or "be intolerant", as well as the limits of tolerance and the interaction structures that affect it, have made it one of the important issues to be considered within the scope of differences in today's tourism environment. Because the concept of tolerance simply does not define the attitude and action of "Indulgence" as one of the criteria of the attitude that the tourist takes in his social relations while living with the local people for a temporary time in the society where he is a guest.

It is seen that the concept of tolerance level is closely related to many fields, from attitudes and behaviors in inter-individual relations to the functioning of social organization, from religious attitudes to moral values (Sönmez & Aksan, 2019). There are also visible aspects of tolerance in religious, ethnic, and class relations. Due to the diversity, social exchange theory can be used to explain both the level of tolerance of the tourist toward the local people and the level of tolerance of the local people toward the tourist. The classic social exchange theory associates tourists' attitudes toward a destination with the perceived positive and negative impacts, social-cultural, environmental, and economic impacts. During the social exchange process, tourists and local residents trade resources with each other and make decisions based on their exchange. When applied to tourism research, the theory assumes that residents get economic benefits from tourism development while sharing and trading their social and environmental resources (Harrill, 2004).

The premise of the theory is that people form their attitude of support or opposition based on trading off the perceived personal benefits and costs from tourism development. There are significant results that indicate factors such as financial benefit increase level of support (Williams & Lawson, 2001). However, for the local people, the economic return of tourism alone does not provide sufficient tolerance. In addition to the economic contribution, tourism activities and tourists should also act with respect and understanding to the local social, cultural and environmental values. Likewise, local people should not see the tourist only as an economic commodity. The socialization, environmental values, ethnic and cultural infrastructure that the tourist values are also required by the local community to act with respect and understanding. If this mutual understanding is provided during the tourism activities, the tourist's acceptance, tolerance, and satisfaction of the environment can be ensured.

Witenberg (2000) states that tolerance, which emerges in interpersonal interaction and affects society, is a moral issue in this respect and should be considered together with concepts such as respect, equality, and freedom. Tourists with a high level of tolerance state that they have respect and a positive worldview "no matter who they are" from local people or tourism workers. These tourists like to hear different opinions, and feel strong with others, but an exclusionary lifestyle is not for them. From this point of view, to increase the tolerance level, empathy toward tourists should be increased in social, environmental, and economic terms, especially the individual values of tourists must be considered.

Contact Situation (Level, Type, and Frequency)

Another important point that determines the tolerance is the level and type of contact with tourists. The tourism sector exhibits a labor-intensive feature due to its structure, therefore "human", which is at the center of the contact process, also draws attention as an important resource (Erkuş & Günlü, 2009: p. 7). The contacts of tourists differ according to their level of interest in tourism. Goeldner and Ritchie (2012) list the stakeholders in contact with tourists as destination local residents, local/regional/national public regulators, environmental groups, domestic and foreign tourists, daily visitors, sector representatives, destination management organizations, cultural heritage groups, and social/health/education groups. These groups can be classified as those who make face-to-face contact with the tourist and those who do not and they can establish primary or secondary level relations with tourists according to their contact situation. Especially tourism employees, sector representatives, cultural heritage groups, and local people can communicate with tourists at a primary level.

While tourism employees contact directly and frequently with tourists, regulators indirectly contact with tourists through the legal regulations they put in place. The tourism employees who contact frequently with tourists at the direct level are customs officers, transfer officers, hotel attendants, sellers, guides, agents, animators, and indirect workers such as kitchen staff (Duran & Aslan, 2016). Other stakeholders, on the other hand, contact with tourists infrequently, and mostly at the indirect level.

For the tourist in particular, meeting a local is by chance, while for locals who specializes in tourists, it is an important part of their life. Therefore, the situation of providing mutual benefit may be different for the tourist and also for the locals, and a complex contact situation may arise in which different motivations such as sincerity, discovery, and profit are intertwined. These contact situations are (Duran & Aslan, 2016):

Private contact channels: This contact level based on similarity of personality, concerns or interests. For example, a local with an extrovert personality type who likes to establish relationships and get to know people and a tourist will be mutually satisfied with the relationship they will establish. For example, a tourist participating in sports tourism meets a local in the area he is interested in, he will be happy and then mutual satisfaction will be provided.

The contact that occurs in the primary and secondary level: Such contacts take place on the basis of holistic relationships or in a pragmatist (utilitarian) way. Primary contact groups establish face-to-face relations and friendship ties are high, and trusting relations are established over time. When evaluated in terms of tourist-local contact levels, in primary clusters, the tourist is perceived by the locals as a foreigner and sometimes as a threat. For this reason, the level of tolerance is also low at the same rate. However, it cannot be said that such an understanding toward tourists is adopted in every society. Secondary contact groups are larger and better-organized clusters than primary ones. However, relations develop from a more utilitarian perspective.

The tourist-domestic contact situation that occurs in the so-called front and rear regions: It is shaped by naturalness (authenticity) and artificiality (falsity). In the front areas, the locals treat the tourists as the tourists want to see them, or in other words, artificially, while in the back regions, the locals behave as they are, naturally. What is meant by natural (authentic) is more ethnically "pure", culturally "traditional" and "closed" places unaffected by westernization or other postmodern dominant cultures.

One of the crucial things is to create a positive interaction between host and guest and the positive attitude that is the outcome of this interaction. Findings show that host-guest contact level is related to interaction frequency and intensity (Yilmaz & Tasci, 2015: pp. 127–128). The frequency and quality of interaction between tourists and residents contribute both to the tourist experience and perception of the destination visited, and to the acceptance and tolerance of tourists by residents (Armenski et al., 2011: p. 109). The more often people have the opportunity to interact and communicate, the more harmonious they will be. In this regard, the frequency and intensity of tourists' visits and their contact with tourism stakeholders will positively affect the attitudes tourists toward the tourism businesses. The same is true for tourism stakeholders. The more frequent and intense interaction is established with tourists, the more positive their attitudes toward the business will be. The long-term continuation of the established interaction is also a very important issue.

Briefly, it can be said that contact frequencies, situations, and level of contacts can be determined on attitudes according to different tourist and local resident typologies. The important point here is that the tourist and the local have close sociocultural characteristics or positive relations that arise as a result of common interests.

Conclusion

The main element of the tourism sector is the tourists and the contact network established around them. Therefore, predicting the tourist's attitude or acting according to the tourist attitude are the questions that the stakeholders of the sector seek to answer the most. Considering the literature, it is observed that the personality of tourists, their satisfaction levels, their motivations, and the frequency of contact with both tourism employees and other tourism stakeholders are the main factors that affect the sector.

The factor that ultimately shapes the attitude of tourists is undoubtedly their satisfaction. Tourist satisfaction emerges when the benefits of the purchased touristic goods or services match the expectations/desires of the customer. A tourist who is satisfied with the product or service he receives will have the intention of repurchasing from the tourism business and will turn into a loyal customer over time. dissatisfaction will alienate the customer from the business and will break the contact bond established between the business and the tourist over time. As a result, tourists will make their next purchases from elsewhere.

Keeping in touch with tourists and following their attitude should be the primary goal of businesses. There are also various motivational tools that affect and direct the attitudes of tourists. The environment in which tourism activities take place, the needs and desires of tourists, their actions and feelings direct their motivation. One of the main indicators that shape the attitudes, motivations, and satisfaction of tourists is their personality traits. The concept of personality includes the individual's reaction in contact with others and the way he shows himself. Looking at the literature, it is seen that personality traits are shaped in five basic structures. As stated above, these are extraversion, agreeableness, conscientiousness, neuroticism, and openness to experience. These personality traits do not show dominant characteristics in a person alone. However, having knowledge about personality traits will give practitioners information about the general tendency of the person in terms of his satisfaction, motivation, and attitude toward preferences about close or distant contact. While addressing the subjects of satisfaction, motivation, and attitude, an issue that should be emphasized is the tolerance levels of tourists. It should be taken into account that tolerance level is closely related to attitudes and behaviors. How long can tourists tolerate situations that they encounter and that they do not normally consider before dissatisfaction? In order to find the answer to this question, it is necessary to consider the personality traits, contact preferences, and attitudinal tendencies of the tourists.

References

Allport, G. W. (1960). The open system in personality theory. *The Journal of Abnormal and Social Psychology, 61*(3), 301.

Altıntaş, M. K. (2000). *Tüketici davranışları; müşteri tatmininden müşteri değerine.* İstanbul: Alfa Yayıncılık.

Armenski, T., Dragičević, V., Pejović, L., Lukić, T., & Djurdjev, B. (2011). Interaction between tourists and residents: Influence on tourism development. *Polish Sociological Review*, 173, 107–118.
Aronson, E., & Carlsmith, J. M. (1963). Effect of the severity of threat on the devaluation of forbidden behavior. *Journal of Abnormal and Social Psychology*, *66*(6), 584–588.
Brown, T. J., Mowen, J. C., Donavan, D. T., & Licata, J. W. (2002). The customer orientation of service workers: Personality trait effects on self and supervisor performance ratings. *Journal of Marketing Research*, *39*, 110–119.
Cardozo, R. N. (1965). An experimental study of customer effort, expectation, and satisfaction. *Journal of Marketing Research*, *2*(3), 244–249.
Cattell, R. B. (1956). Second-order personality factors in the questionnaire realm. *Journal of Consulting Psychology*, *20*(6), 411–418.
Chamorro-Premuzic, T. (2007). *Personality and individual differences*. Oxford: Wiley Blackwell.
Çivitçi, N., & Arıcıoğlu, A. (2012). Beş faktör kuramına dayalı kişilik özellikleri, *Mehmet Akif Ersoy Üniversitesi Eğitim Fakültesi Dergisi, 12*(23), 78–96.
Correia, A., Moital, M., Oliveira, N., & Costa, C. F. (2009). Multidimensional segmentation of gastronomic tourists based on motivation and satisfaction. *International Journal of Tourism Policy*, *2*(1/2), 58–71.
Crompton, J. (1979). Motivations for pleasure vacation. *Annals of Tourism Research*, *6*(4), 408–424.
Dann, G.M.S. (1977). Anomie, ego-enhancement and tourism. *Annals of Tourism Research*, *4*(4), 184–194.
Dann, G.M.S., (1981). Tourist motivation: An appraisal. *Annals of Tourism Research*, *8*(2), 187–219.
Deighton, J. (1984). The Interaction of Advertising and Evidence, *Journal of Consumer Research*, 11 (December), 763–770.
Duran, E., & Aslan, C. (2016). *Turizmin sosyal psikolojik dinamikleri*. İstanbul: Paradigma Akademi.
Erkuş, A., & Günlü, E. (2009). İletişim tarzının ve sözsüz iletişim düzeyinin çalışanların iş performansına etkisi: Beş yıldızlı otel işletmelerinde bir araştırma. *Anatolia: Turizm Araştırmaları Dergisi*, *20*(1), 7–24.
Eysenck, H.J. (1952). *The scientific study of personality*. London: Routledge & Kegan Paul.
Fan, D. X., Zhang, H. Q., Jenkins, C. L., & Lin, P. M. (2017). Does tourist–host social contact reduce perceived cultural distance? *Journal of Travel Research*, *56*(8), 998–1010.
Festinger, L. (1957). *A theory of cognitive dissonance*. Stanford, CA: Stanford University Press.
Glass, R., Prichard, J., Lafortune, A., & Schwab, N. (2013). The influence of personality and Facebook use on student academic performance. *Issues in Information Systems*, *14*(2), 119–126.
Goeldner, C., & Ritchie, B. (2012). *Tourism: Practices, principles, philosophies*. New Jersey: John Wiley & Sons, Inc.
Guttentag, D., Smith, S., Potwarka, L., & Havitz, M. (2018). Why tourists choose Airbnb: A motivation-based segmentation study. *Journal of Travel Research*, *57*(3), 342–359.
Harrill, R. (2004). Residents' attitudes toward tourism development: A literature review with implications for tourism planning. *Journal of Planning Literature*, *18*(3), 251–266.
Hogan, R. (2009). *Kişilik ve kurumların kaderi*. Trans. Selen Y. Kölay. İstanbul: Remzi Kitabevi.
Homburg, C., Koschate, N., & Hoyer, W.D. (2005). Do satisfied customers really pay more? A study of the relationship between customer satisfaction and willingness to pay. *Journal of Marketing*, *69*(2), 84–96.
İnanç, B. Y., & Yerlikaya, E. E. (2009). *Kişilik kuramları* (2nd ed.). Ankara: Pegem Akademi.
İnce, İ., & Doğantan, E. (2020). Otel yöneticileri perspektifinden dijital pazarlama. *Anadolu Üniversitesi İşletme Fakültesi Dergisi, 2*(1), 13–26.
Isac, F. L., & Rusu, S. (2014). Theories of consumer's satisfaction and the operationalization of the expectation disconfirmation paradigm. *Economy Series*, *2*(2), 82–88.
John, O. P., Naumann, L. P., & Soto, C. J. (2008). Paradigm shift to the integrative Big Five trait taxonomy: History, measurement, and conceptual issues. In Oliver P. John, Richard W. Robins, and Lawrence A. Pervin (Eds.), *Handbook of personality: Theory and research* (3rd ed., pp.114–158). New York: Guilford.
Klenosky, D. B. (2002). The "pull" of tourism destinations: A means-end investigation. *Journal of Travel Research*, 40(4), 396–403.
Kılıç, İ., & Pelit, E. (2004). Yerli turistlerin memnuniyet düzeyleri üzerine bir araştırma. *Anatolia: Turizm Araştırmaları Dergisi*, *15*(2), 113–124.
Koç, B. (2007). *Tüketici davranışı ve pazarlama stratejileri: Global ve yerel yaklaşım*. Ankara: Seçkin Yayıncılık.
Lai, K. (2018). Influence of event image on destination image: The case of the 2008 Beijing Olympic Games. *Journal of Destination Marketing & Management*, 7, 153–163.
LeBlanc, G. (1992). Factors effecting customer evaluation of service quality of travel agencies: An invastigation of customer perceptions. *Journal of Travel Research*, *30*(4), 10–16.
McCrae, R, R., & Costa, P.T, (1987), Validation of a five-factor model of personality across instruments and observers. *Journal of Personality and Social Psychology*, *52*, 81–90.

Moutinho, L. (2000). Trends in tourism. In L. Moutinho (Ed.), *Strategic management in tourism* (pp. 3–16). Wallingford: CAB International.

Odabaşı Y., & Gülfidan, B. (2010). *Tüketici davranışı*. Istanbul: Mediacat Kitapları.

Pearl, M. C., Ok, C. M., & Au, W. C. (2021). Peer-to-peer dining: A motivation study. *Journal of Hospitality and Tourism Research*. DOI: 10.1177/1096348021990709.

Peyton, R. M., Pitts, S., & Kamery, R. H. (2003). Consumer satisfaction/dissatisfaction (CS/D): A review of the literature prior to the 1990s. In *Allied academies international conference. Academy of organizational culture, communications and conflict. proceedings* (Vol. 8, No. 2, p. 41). Las Vegas: Jordan Whitney Enterprises, Inc.

Pizam, A., Neuman, Y., & Reichel, A. (1979). Tourist satisfaction uses and misuses. *Annals of Tourism Research, 6*(2), 195–197.

Schiffman, L.G., & Kanuk, L.L., (2009). *Consumer behavior.* Harlow: Prentice Hall.

Şimşek, M. Ş., Akgemci, T., & Çelik, A. (2011). *Davranış bilimlerine giriş ve örgütlerde davranış* (7th ed.). Ankara: Gazi Kitabevi.

Solomon, M., Bamossy, G., Askegaard, S., & Hogg, M. (2009). *Consumer behaviour: A European perspective.* London and New York: Pearson Higher Education.

Somer, O., Korkmaz, M., & Tatar, A. (2002). Beş faktor kişilik envanterinin geliştirilmesi-I: Ölçek ve alt ölçeklerinin oluşturulması. *Türk Psikoloji Dergisi, 17*(49), 21–33.

Sönmez, Ö. A., & Aksan, G. (2019). Üniversite öğrencilerinin tolerans düzeylerinin farklı değişkenlerle ilişkisi. *Selçuk Üniversitesi Sosyal Bilimler Enstitüsü Dergisi*, (41), 302–316.

Swarbrooke, J., & Horner, S. (2007). *Consumer behaviour in tourism.* Oxford: ButterworthHeinemann.

Thurstone, L. L. (1951). The dimensions of temperament. *Psychometrika, 16*(1), 11–20.

Weinstein, T. A. R., Capitanio, J. P., & Gosling, S. D. (2008). Personality in animals. In O. P. John, R. W. Robins & L. A. Pervin (Eds.), *Handbook of personality: Theory and research* (ss. 328–350). New York: The Guilford Press.

Williams, J., & Lawson, R, (2001). Community issues and resident opinions of tourism. *Annals of Tourism Research, 28*(2), 269–290.

Witenberg, R. (2000). Do unto others: Toward understanding racial tolerance and acceptance. *Journal of College and Character, 1*(5), 1–8.

Yilmaz, S. S., & Tasci, A. D. (2015). Circumstantial impact of contact on social distance. *Journal of Tourism and Cultural Change, 13*(2), 115–131.

Yousaf, A., Amin, I., & C Santos, J. A. (2018). Tourist's motivations to travel: A theoretical perspective on the existing literature. *Tourism and Hospitality Management, 24*(1), 197–211.

Zeithaml, V. A., & Bitner, M. J. (2000). *Services marketing: Integrating Customer focus across the firm.* Boston, MA: McGraw-Hill.

17
SOCIAL/CULTURAL DISTANCE AND ITS REFLECTIONS ON TOURISM

Aysen Ercan Iştin

Introduction

Developments in the tourism industry on a global scale increase the flow of people with different social and cultural structures to tourism destinations. In tourism destinations, different groups are formed as domestic – foreign tourists, local people – resident foreigners, tourism workers, and tourism enterprises. Thus, there can be positive and negative social and cultural interactions among these groups in customs and traditions, moral values, cultural values, attitudes, behaviors, and beliefs. Social and cultural differences must be managed through a diversity of organizational and individual strategies in terms of eliminating the negative effects of these interactions that harm the development and sustainability of tourism and creating a positive image of tourism destinations in the public. It is very difficult to draw the boundaries of multidimensional and wide-ranging social and cultural interactions that occur through tourism. In this respect, social and cultural distance is important in the examination of the residents' attitudes towards tourism and the attitudes of socially/culturally different nations towards the residents in the destination.

Differences in terms of culture in dimensions such as individualism-collectivism can display both aspects of the scale of social distance developed by Bogardus (1925) in order to measure racist and national attitudes. Relatedly, there is a prospect that the measure and meaning of social distance can differ among cultures. For instance, in a more society of individualistic, when individuals see themselves as more independent and make decisions on important issues, they prioritize themselves and give importance to individual success. In this case, there can be many social distance degrees to hold the grander diversity of social relationships in an individualistic society. On the contrary, in a more collectivist society, individuals see themselves as a part of the society they care about like family, look after the interests of the community rather than individual interests, and take care to behave as expected of them. In this case, there can be less degrees of social distance to hold a grander diversity of social relationships in a collectivist society.

Bogardus (1940) defines social distance as the degree of understanding that exists among individuals, among groups, and among a person and groups, while Hofstede (1980) defines cultural distance as the degree to that values and norms in a country differ from the norms in another country. Groups mentioned in the definition of social distancing and cultural distancing may refer to a range of entities. The most important of these are nations and organizations. In this respect, social and cultural distance is examined based on Hofstede's theory relating cultural dimensions that is one of the most widely used classifications in the related literature and the scale of social distance developed by Bogardus. In this chapter, it is aimed to reveal the reflections of social and

DOI: 10.4324/9781003161868-17

cultural distance in tourism and to offer suggestions to tourism destinations, enterprises and organizations related to tourism in order to eliminate the negative aspects of these reflections.

Social Distance

According to the TDK (2021) dictionary, the word meaning of "distance" is interim, interval and distance. The word distance (TDK, 2021), which is formally expressed as the state of not being very sincere in metaphorical relationships, was defined as "the degree of sympathetic understanding between person and person, person and group and groups" in social psychology by Bogardus (1926) in the 1920s, and pioneered a series of studies (Mather et al., 2017: 2). Bogardus (1947) states that the focus of attention in social distance studies is to feel the reactions of people towards other people and groups of people. According to this approach towards interpersonal and personal group relationships, the main emphasis entirely is human responses, driven by the emotional aspects of personality. Social distance is generally understood as a sociological concept that cannot be reduced to spatial or biological distance. In particular, studies conducted in recent years have shown that people can experience social distance and intimacy without "being together" in the same physical place (Karakayali, 2009: 540). From this point of view, it can be said that social distance cannot be considered as physical coexistence. Therefore, the use of the concept of social distance as a rule in order to preclude from the spread of the disease to more people in the Covid-19 pandemic, which affected the whole world in the last months of 2019, will inevitably cause conceptual complexity in the future. In this case, it can be said that it is a more accurate expression to use the concept of physical distance as a rule for precaution in the Covid-19 pandemic.

Social distance means to the degree to which individuals experience a emotion of familiarity or unfamiliarity among themselves and individuals who are different from their own in terms of social, ethnic, occupational and religious groups. (Hodgetts & Stolte, 2014: 1776). In other words, social distance is the degree of acceptance or rejection of social relations among individuals belonging to different ethnic, racial, or social groups (Sharlamanov & Jovanoski, 2013: 34). In this case, it is important to address the emotional, normative, interactive, and cultural aspects of social distance in order to be defined and used by social scientists (Karakayali, 2009: 540). A common view of social distancing is emotionality. Therefore, in order to analyze the phenomenon of social distance, the concept of emotional social distance, which focuses on the emotional personal relationship of sympathy or antipathy towards the members of the ethnic groups remaining in the society, has been developed in sociology (Sharlamanov & Jovanoski, 2013: 34). Bogardus (1941) defined emotional social distance as the willingness of participants from a social group to establish social contact with members from a particular group of "others" (Leino & Himmelroos, 2019: 1891). According to Bogardus (1941), *mutual sympathy and affection* are essential elements of emotional social distance (Leino & Himmelroos, 2019: 1892). Normative social distancing means to broadly admitted and frequently intentionally articulated norms about who should be taken into "insider" and who should be "outsider/foreigner". Interactive social distance means to the density and frequency of interactions between two groups, the more members of the two groups interact, the more socially they claim to be close. This feature is similar to sociological network theory in which the frequency of interaction between two parties is used as a measure of the power of social vineyards among parties. Finally, cultural distance is affected by the "capital" people have (Sorensen et al., 2021: 8–9).

Focusing on the race problem as one of the most important problems facing the US society, Bogardus (1925) developed a scale to measure social distance (Parrillo & Donoghue, 2005: 257). The scale developed by Bogardus was used for the first time in 1926 on 1725 European-American students from 24 universities (Çöllü & Öztürk, 2006: 385). The scale of social distance is a technique used to measure attitudes belonging to racial and ethnic groups. The fundamental idea behind the scale of social distance is that the more biased a person is towards a specific group, the less that person will want to interact with members of the group. Therefore, the items that create the scale of social distance define the relationships that the participant may wish to go into with a member of the determined cultural group (Geisinger, 2010). The items that create the scale of social distance, which is called the scale of social distance, are given below (Çöllü & Öztürk, 2006: 385).

"According to my initial emotional reactions, let us accept one or more of the various relationship groups presented by members of the ethnic communities listed below."

1 To close kinship gained by marriage (Social distance)
2 To my society as a personal friendship (Social distance)
3 To my street as neighbors
4 To employment in my profession
5 To citizenship in my country
6 Visitors from my country only
7 Since it will be excluded from my country

The most important criticism on the Bogardus scale is that it is not one-dimensional. The fact that an individual is in a social distance with some ethnic groups, that is, living separately with them, as a positive or negative attitude towards that group, makes the scale two-dimensional. Wanting or not wanting to live with ethnic groups is the first dimension, positive or negative attitude towards them is the second dimension. A second criticism about the scale is that it does not have validity and reliability because it is not one-dimensional or homogeneous. Another criticism is that the choice of scale items has no objective basis. Therefore, it is unclear whether the items belong to the same continuum. However, it should not be forgotten that such a scale can be used appropriately when necessary corrections and standardizations are made (Hoşgörür, 1997: 348).

Cultural Distance

The word culture derives from the Latin words *Cultura* or *Colere*. In Latin, these words mean "to care" or "to cultivate". The first meaning of the word *Cultura*, which is used as cultivating, plowing, and growing crops in the agricultural sense, as human experience and human life style emerged after the developments in Germany after 1750 (Usal & Kuşluvan, 1998: 105).

Culture is a conceptual word that has been discussed by anthropologists, sociologists, historians, and philosophers for thousands of years (Mobley et al., 2005: 12). Culture occurs mainly of stereotyped ways of reacting and thought, gained and transferred through symbols, which constitute the discriminating achievements belonging to human groups. In addition, the fundamental essence of culture occurs of traditional thoughts and particularly the values connected to them (Kluckhohn, 1951a: 86). Hofstede (1980: 24) defined culture "as the collective programming of the human mind that distinguishes members of one human group from another". In this sense, culture is a framework of values held collectively.

In the literature, it is seen that many models related to the culture have been put forward. Harrison (1972: 121) classified cultures as power orientation, role orientation, task orientation and person orientation in terms of organizational ideology. These ideologies are rarely found in organizations as pure types, but most organizations tend to focus on one or the other. De Vries and Miller (1986) classified culture as; paranoid culture, avoidance culture, charismatic culture, bureaucratic culture and politicized. On the other hand, Cameron and Quinn's (1992) typologies of hierarchy, clan, market, and adhocracy cultures have an important place in the literature. The model developed by Hofstede (1980) and used in the measurement of cultural values is accepted as the most ambitious theoretical model that guides the intercultural comparison of organizational culture, as well as being the most common model used in the literature (Öncül et al., 2016: 259). In this chapter, the concept of cultural distance and its reflections on tourism are examined based on Hofstede's (1980) theory.

Hofstede's theory has been applied throughout the world for many years in various fields, academia and industry. Hofstede conducted research focusing on the culture of 14,000 IBM employees worldwide. Hofstede's main hypothesis was that "culture influences human attitudes and behaviors". This assumption has been studied and tested on a wide spectrum. The theory has also been used to examine how cultural dimensions affect customer behavior in the marketing business discipline in every industry, including the tourism industry (Buafai & Khunon, 2016: 2984). Hofstede (1980) concluded that culture has four dimensions. These are power distance, masculinity/femininity, individualism/collectivism, and uncertainty avoidance (Manrai & Manrai, 2011). Subsequently, these dimensions increased to six. Hofstede and Bond (1988) added the long-term orientation dimension as they see it as one of the strongest driving forces in the world (Jung et al., 2020: 242). Minkov (2007), on the other hand, added the sixth dimension, which he called indulgence-constraint, in his study (Türker & Karadağ, 2019: 276).

Power Distance

According to Dahl (1957), who first defined the concept of power, power is a relationship among people in which some have more power than others (Dahl, 1957: 201). Power is one of the most central and problematic concepts in sociological theory. In sociological theory, power forms the cornerstone of the conflict model about society. Conflict theorists see power as a barrier to the destruction of society by humans in the competition to satisfy individual needs and desires. Therefore, power remains a slippery and problematic concept (Martin, 1971: 240). In the cultural sense, power shapes the power distance of the culture. One dimension of social culture studies is expressed as power distance and the power distance of societies is divided into two as high power distance and low power distance (Kemikkıran, 2015: 318).

Power distance is described as "the level at which less powerful members of institutions and organizations in a country expect and accept the unequal distribution of power" (Hofstede, 1994: 28). In cultures where the power distance is small (e.g. Australia, Austria, Denmark, Ireland, New Zealand), people hope and admit power relations that are more democratic and relate more equally to one another in any event of their official position. In cultures where the power distance is a great (for instance, Malaysia, Guatemala, Panama, Philippines, Mexico, Venezuela, China) less powerful people admit power relations that are autocratic or patriarchal (Abdullah et al., 2014: 120).

Consistent with the definition of asymmetric control over valuable resources commonly used in social psychology and other disciplines, power is a common feature of social relationships, especially in situations with limited resources. In this case, it is clear that a low-powered individual in a dual power relationship is more dependent on a high-powered individual for desired results (Magee & Smith, 2013: 159).

Masculinity/Femininity

Societies that are disposed to display a thrusting or strong nature are called masculine, while societies that are disposed to display a more affectionate or mild character are called feminine. IBM studies found that women's values diverge less than men's across societies, that men's values are very assertive and competitive from one country to another, and that men's values diverge in maximum level from women's, but that men's values are humble and helpful and equal to women's (Hofstede, 2011: 12–13).

According to the masculinity and femininity dimension, men should be pretentious, harsh, and focused on financial achievement, and women should be humble, sensitive, and care about quality. Masculine individuals prefer success, heroism, and financial success, unlike feminine persons who specified success in terms of close human relations and life quality. Japan, Mexico, Italy, the opposite pole are masculine countries, while Scandinavian countries are feminine countries (Podrug et al., 2006: 4).

Individualism/Collectivism

Collectivism and individualism are used to explain the dominance of individual or social values in people. In the case of collectivism, the feelings of solidarity, cohesion, and loyalty dominate among individuals. Collectivism focuses on others and puts the interests of society first. While the subject that is tried to be agreed upon in collectivism-dominated societies is to maintain healthy communication among individuals, in individualism-dominated societies it is on the establishment of justice. Individualism refers to the degree of people's integration into groups (Uzuntarla et al., 2016: 207). In an individualistic culture, people lose their individual vineyards or bonds and their personal interests dominate due to the great amount of independence a society gives to individuals (Hofstede, 2011). Individualistic societies emphasize self-awareness, autarchy, emotional independence, looking for pleasure, and universality. Moreover, collectivist societies emphasize "us" consciousness, collective identity, group cooperation, and sharing (Kim & Lee, 2000: 153).

Avoidance of Uncertainty

According to Hofstede (2001), "uncertainty avoidance is the degree to which people feel threatened by unknown or indefinity situations". In other words, uncertainty avoidance is tolerance of society for uncertainty, which indicates the extent to which members of a society cope with a difficult situation by minimizing uncertainty. It has been determined that when investors make their investment decisions, uncertainty avoidance is their first choice and uncertainty avoidance has a significant effect on their investment decisions (Hofstede, 2001). Uncertainty avoidance indicates the extent to which a culture programs its members to sense comfortable or uncomfortable in unstructured situations. Unstructured situations are new, surprising, unknown and different from the normal. Uncertainty avoidant cultures try to minimize the possibility of such situations with strict codes of conduct, rules and laws, disapproval of deviant views, and belief in certain truth (Hofstede, 2011: 10).

Long-term Orientation

Long- and short-term orientation refers to "the degree to which individuals in a society are concerned about their future" (Kumar & Dhir, 2020: 5). In the long-term orientation, where the view that the most important events in life will happen in the future, a good individual is adaptable to circumstances and is shaped by what is good or bad. Traditions can be adapted to changing conditions. The individual is open to learning from other countries. Important goals are savings and perseverance. In the short-term orientation, there is the perception that the most important events in life have happened in the past or are happening in the present. A good person is always good because there is personal stability. There are universal rules about what is good or bad. Traditions are seen as sacred. An individual should be proud of his country (Hofstede, 2011: 15).

Indulgence/Constraint

The constraint versus indulgence was first received from the World Values Study and utilized by Hofstede in his cultural dimensions model. The constraint versus indulgence projects the grade to which a society answers to human fundamental needs. The constraint against a high indulgence score indicates a relatively weak constraint of primitive human needs and emotions related to enjoying and enjoying life (entertainment, consumption, spending, and sex). The constraint versus a low indulgence score means relatively stronger constraint of these needs thanks to harsh social norms. The constraint versus indulgence is a large cultural dimension that encompasses various aspects of cultural phenomena and shows intertemporal stability (Guo et al., 2018: 2). The focus of these societies is freedom, happiness, and quality of life. Having control over human impulses and natural desires is a choice. More constraints societies are disposed to value more self-control and protection from natural human impetus (Wallace et al., 2019: 21).

Reflections of Social Distance on Tourism

Sin et al. (1999) argue that cross-cultural studies are often exposed to violent ethnocentrism (Meng, 2010: 342). The concept of ethnocentrism, which was first used by William Graham Sumner in the field of sociology in 1906, refers to seeing the ethnic group as the center of the universe, evaluating other social groups from the point of view of their group, and rejecting people from different ethnic groups by blindly accepting people who are culturally similar to themselves (Öztürk, 2020: 305). Individuals with ethnocentric tendencies also accept cultures similar to their own and reject different cultures (Yıldırım & Gültaş, 2016: 20).

Ethnocentric attitudes, which are thought to be effective in the purchase of touristic products in terms of tourism, may differ according to touristic products, as well as the characteristics of the sample (education level, income level, intellectual structure, etc.) and the characteristics of the country (development level, international relations, economic level, etc.) may also differ. The situation where the development levels of the countries are different, the situation where ethnic origins have culturally different values, and the situation in which the citizens of the country have different perspectives in terms of religion, education, and social values make the ethnocentric attitudes of each country and even the groups within the countries more likely to be different (Altınay Özdemir & Kızılırmak, 2019: 178). An example of this situation can be given as the result of Çelik's (2019) research on tourists visiting Gaziantep and Şanlıurfa cities in Turkey, that the profile of people coming from the Southeast is effective in the formation of social distance.

Research revealing the host society's attitudes towards tourism has increased in recent years. Krippendorf (1987) emphasized that the sociocultural impacts of tourism should be investigated. Parallel to this, there appears to be a major change to put sociocultural influences at the fore of sustainable tourism management planning. Tourism planners, tourism experts, tourism developers, local authorities, and operators concur that host community support is an important constituent of tourism delivery, and that continuous monitoring of community attitudes towards tourism development and visiting tourists is vital (Thyne et al., 2020: 2–3).

Residents' attitudes (positive and negative), who living in a destination, provide a meaningful understanding of their social distance towards tourists. Residents with positive attitudes towards tourism interact closely with tourists and the perceived closeness among them increases. This ultimately reduces the differences in judgment about themselves and others. Conversely, residents who have more negative attitudes towards tourism interact more distantly with tourists (Thyne et al., 2020: 14–15).

Previous research on race and travel has indicated that the ideology of racism has restricted African Americans' tourism mobility and access to many tourism destinations for centuries. In this case, their travel rights are based on a system of inequality. Even today, African Americans continue to be discriminated against when traveling in the United States. Lee and Scott (2017) revealed in their research that African American travelers are frequently subjected to racial maltreatment from service providers and other tourism-related businesses (Hudson et al., 2018: 2–3).

The Civil Rights Act of 1964 outlawed discrimination in public places such as gas stations, restaurants, hotels, and various public entertainment venues, thus removing many of the barriers to African American leisure travel. However, despite the fact that the law has been in effect for nearly six decades, Blacks, in general, remain anxious travelers. African Americans are more likely to travel in large groups than White Americans, and they limit their visits to destinations recommended to them by family members, friends, and acquaintances. Black travelers tend to stick to their vacation itineraries. Their lack of spontaneity and adventurousness usually manifests as a tendency to avoid unknown food and accommodation options and to search for the familiar hotel and restaurant chains (Carter, 2008: 265–266). In this case, racialized experiences are likely to affect people's perceptions and impressions of certain destinations and overshadow positive experiences. Nevertheless, considering the structural conditions of racism evident in daily life, the problems experienced by individuals during their travels may force them to think about their own roles, status, and identities (Stephenson & Hughes, 2005: 154). Therefore, it can be said that racism affects the choice of destination and even causes tourists to limit their destination choices to a certain area where they can feel safe.

It can be said that religion is another factor that is effective in the social distance in tourism activities. Impacts of religious tourism not only directly but also indirectly, as in pilgrimages, because tourists are often exposed to religion on their travels, even when the travel is not started for religious reasons. Religious belief is a cultural attribute that forms tourists' destination perceptions. Religion does not appear to be a clear factor in a tourist's decision-making process, but it can be an important determinant of tourists' preferences (but implicit) if the dominant religion in a destination is not the same as the religion of visiting tourists (Fourie et al., 2015: 52). Vietze (2012) investigated the effects of cultural factors, especially religious factors, on tourist flows to the United States, which is an important tourism destination. As a result of his research, it has been determined that cultural and religious similarity has positive effects on tourism flows between these countries (Vietze, 2012: 121).

Reflections of Cultural Distance on Tourism

Cross-cultural studies in the tourism and hospitality industries show that cultural norms and values play an important role in the assessment of service quality expectations and the nature and strength of customer relationships. Research also shows that culture shapes people's values, preferences, attitudes, and behaviors. For instance, the interaction between service providers and service recipients in service encounters can be significantly affected by the cultural characteristics of the service provider (Koç, 2013: 3682).

Hofstede's cultural dimensions are exceedingly major for the research of travel motivations, since they influence all aspects of human life, with the inclusion cultural values, personal factors like lifestyle, and psychological factors like motivation (Manrai & Manrai, 2011: 31). For instance, Kim and Lee (2000) stated that individual tourists are more probably to look for innovation, and that the collectivist tourists' motivation is firstly to be with family (Kim & Lee, 2000: 167). Countries with individualistic social structures are generally better prepared to adopt a diversity of strategies to increase travel and tourism competitiveness and to work towards such goals at an individual level. Whereas, collectivist societies may seek approval from all members before implementing any strategy. This can take a time and boring process. In addition, individualist countries welcome discussions to increase their travel and tourism competitiveness and are willing to utilize newer technologies that can increase their competitiveness (Kumar & Dhir, 2020: 4).

As a result of their research, Ahn and McKrecher (2015: 96) stated that cultural distance has an effect on destination choice, satisfaction, demand, expenditure/consumption, travel profile and behaviors. In addition, other studies reveal that tourists' cultural value judgments affect their destination choices (Mercan & Kazancı, 2019; Köroğlu & Güzel, 2013; Vela, 2009). The culture of the destination also affects the tourists' destination choice. Cultural differences and similarities also affect tourists' intentions to visit destinations (Pappas, 2014: 390).

In the field of tourism, tourists with a high uncertainty avoidance culture tend to spend more time planning their travels, especially to reduce the uncertainties of traveling to a new destination. Therefore, they spend more time searching for information about the destination they travel to and use different sources of information such as travel agencies. In addition, they stay at destinations for a shorter time on average and visit fewer destinations because they avoid traveling alone. However, tourists with a higher degree of uncertainty avoidance do not spend much time making travel decisions or airline reservations (Money & Crotts, 2003).

Tourist behavior studies show that the power distance culture dimension is positively related to the tourists' tipping behavior. Tourists with a high power distance culture tip the service provider when traveling more than those with a low power distance culture. In general, tourists with high power distance culture prefer to take a short trip compared to tourists with low power distance culture (Farahani & Mohamed, 2011: 575).

According to the research of Kumar and Dhir (2020), members of long-term oriented societies are disposed to learn strategies from other countries to increase competitiveness and invest in building a tourism infrastructure to provide long-term benefits. In addition, societies with a long-term focus draw inspiration from other societies for travel and tourism policies that can help increase travel and tourism competitiveness (Kumar & Dhir, 2020: 5).

It can be said that the research of tourist cultures has important managerial and marketing effects on the tourism industry. Consumer behavior studies in the field of tourism examine many aspects by trying to explain what tourists do and why. Differences in tourist behavior among various cultures and/or nationalities are seen in characteristics of tourist motivation such as length of travel, importance of food, service requirement, complaint behavior and many other aspects. In addition, these differences are also seen in the decision-making process, image and perception, satisfaction and travel characteristics, including destination selection, information seeking, and planning process (Meng, 2010: 341–3421).

Figure 17.1 was created in the context of research on social distance and cultural distance. Therefore, according to the situations where ethnic and racial discrimination based on social distance is seen or not seen in the destinations and Hofstede's cultural dimensions, the reflections of social and cultural distance on tourism can develop negatively or positively. While tourist'

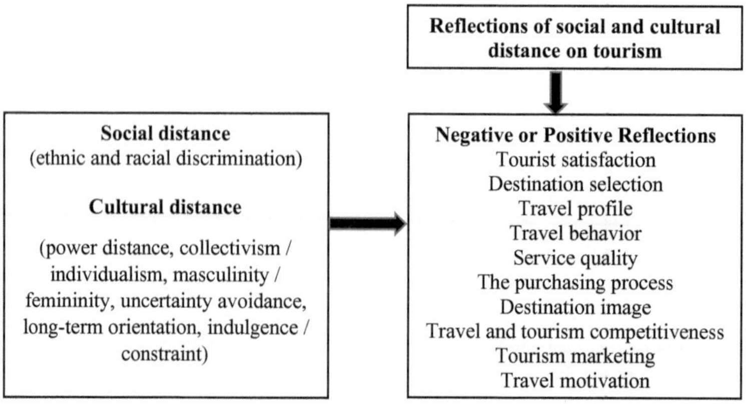

Figure 17.1 Reflections of Social and Cultural Distance on Tourism.
Source: Created By Author.

satisfaction, destination selection, travel profile, travel behaviors, service quality, purchasing process, destination image, travel and tourism competitiveness, tourism marketing and travel motivation for destinations with high cultural distance and social distance may develop negatively, they can develop positively in destinations where cultural and social distance is low.

Conclusion and Recommendations

With the disappearance of borders with globalization, the rapid development of the tourism industry causes the number of people participating in tourism activities to increase day by day and the displacement of millions of people every day. As a result of this situation, tourism can provide societies with the opportunity to get to know other cultures better, on the other hand, it may cause either the cohesion or polarization of societies that encounter people with different social and cultural characteristics who are not informed about their own beliefs, cultures, and lifestyles. In this case, positive interaction may occur between tourists and locals who show indulgence and understanding towards differences, while negative interactions may occur between tourists and locals who are not sensitive and respect for differences.

Social and cultural differences such as traditions and customs, color, language, religion, race, nation, ethnicity, individualist/collectivist societies, and masculine/feminine societies have significant effects on social and cultural distance. When evaluated in terms of the tourism industry, a tourist's desire to wonder, know, and experience these social and cultural differences can be perceived as a travel motivation. For this reason, destinations should introduce their social and cultural characteristics to the world public, and create a competitive advantage by maintaining their marketing and promotional activities effectively as a destination attraction by creating a positive image. Thus, the positive image created in the world public opinion will affect a tourist's choice of destination and purchasing process, thereby improving his/her travel behavior positively. In addition, the local people's positive perception towards tourism is important. The local people's behavior based on hospitality and indulgence towards tourists is effective on tourist satisfaction. A tourist who leaves satisfied from the destinations visits likely to revisit the destinations and recommend them to his/her surroundings.

On the other hand, social and cultural differences can create a feeling of hatred and hostility based on racial and ethnic discrimination in a tourist or local population. While this situation may create a prejudice regarding the preference of destinations in terms of tourists, it may cause a racist behavior towards tourists by the local people. Unfortunately, ethnic and religious minorities and many similar groups can be victims of racial discrimination, despite the fundamental obligations in international human rights conventions to preclude racism. Therefore, people who experience such grievances do not include destinations where they have suffered or may experience victimization in their destination choices and they prefer destinations that are shorter, less numerous, and similar to their own cultures. In this case, it may be a travel profile based on ethnic and racial discrimination. In addition, the local people's negative behaviors towards tourists can negatively affect the tourists' destination choices, their intention to revisit and recommend, their satisfaction levels, the perception of service quality and their purchasing processes. This situation will damage the competitiveness of the destinations and the sustainability of the tourism industry, which may cause negative images on a global scale. However, in order for destinations to gain competitive advantage and ensure the sustainability of tourism activities, it is a prerequisite that they can easily enter markets with different ethnic origins and cultures and get a larger share from the market. Therefore, destination managers need to be able to manage these differences and turn this situation into a competitive advantage.

In order to ensure the sustainability of the tourism industry, which has an important economic contribution to destinations, destination managers should be more careful about discrimination that may cause social and cultural distance. It is important for the destination managers to organize awareness activities and widespread social responsibility campaigns on the subject, which can enable the local people to act consciously in maintaining their relations with tourists, taking into account ideological, religious, and ethnic sensitivities. In addition, the awareness that social and cultural differences that will harm the relations between local people and tourists should be removed from taboo and that the different should be respected.

By creating the awareness of the local people by the destination managers that each culture has different characteristics and values and should be respected, local people should be enabled to take an active role in equality and fairness issues, and they should be encouraged to engage in indulgent, natural, and sensitive communication. Thus, it may be possible to minimize or eliminate language and discriminatory discourses, prejudices, and stereotypes that may harm communication between tourists and local people.

Destination managers should develop and implement policies that are able to unite tourists and local people with different social and cultural characteristics around common and shared values, adopting a unifying approach to different values and their own values by avoiding the ethnocentric view, transforming cultural differences into the destination's most important competitive tool, and having the flexibility to effectively implement different strategies related to travel motivation, contain human values.

References

Abdullah, N. H., Hassan, H., Ali, M. H., & Karim, M. S. A. (2014). Cultural values (power distance) impact on the stakeholders' engagement in organizing the monsoon cup international sailing event. *Procedia Social and Behavioral Sciences, 144*, 118–126. doi:10.1016/j.sbspro.2014.07.280

Ahn, M. J., & McKercher, B. (2015). The effect of cultural distance on tourism: a study of international visitors to hong kong, *Asia Pacific Journal of Tourism Research, 20*(1), 94–113. doi: 10.1080/10941665.2013.866586

Altınay Özdemir, M. & Kızılırmak, İ. (2019). Tüketicilerin destinasyon seçim tutumları ile etnosentrik tutumları arasındaki ilişkinin değerlendirilmesi. *Güncel Turizm Araştırmaları Dergisi, 3*(2), 175–201.

Bogardus, E. S. (1925). Measuring social distances. *Journal of Applied Sociology, 9*, 299–308.

Bogardus, E. S. (1926). Social distance in the city. *Proceedings and Publications of the American Sociological, 20*, 40–46.

Bogardus, E. S. (1940). Scales in social research. *Sociology and Social Research, 24*, 69–75.

Bogardus, E. S. (1947). Measurement of personal-group relations. *Sociometry, 10*(4), 306–311.

Buafai, T., & Khunon, S. (2016). Relationship between Hofstede's cultural dimensions and tourism product satisfaction. *International Scholarly and Scientific Research & Innovation, 10*(8), 2994–2998.

Carter, P. L. (2008) Coloured places and pigmented holidays: Racialized leisure travel. *Tourism Geographies, 10*(3), 265–284, DOI: 10.1080/14616680802236287

Cameron, K. S., & Quinn, R. E. (1992). *Diagnosing and changing organizational culture.* Massachusetts: Adison Wesley.

Çelik, S. (2019) Does tourism reduce social distance? A study on domestic tourists in Turkey. *Anatolia, 30*(1), 115–126, DOI: 10.1080/13032917.2018.1517267

Çöllü, E. F. & Öztürk, Y. E. (2006). Örgütlerde inançlar-tutumlar, tutumların ölçüm yöntemleri ve uygulama örnekleri, bu yöntemlerin değerlendirilmesi. *Selçuk Üniversitesi Sosyal Bilimler MYO Dergisi*, 9(1–2), 373–404.

Dahl, R. A. (1957).The concept of power. *Behavioral Science, 2*(3), 201–215.

De Vries, M. F. K., & Miller, D. (1986). Personality, culture, and organization. *Academy of Management Review, 11*(2), 266–279.

Farahani, B. M., & Mohamed, B. (2011). The influence of national culture on tourists' behavior towards environment. *WIT Transactions on Ecology and the Environment, 148*, 573–582.

Fourie, J., Rossello, J., & Santana-Gallego, M. (2015). Religion, religious diversity and tourism. *KYKLOS, 68*, 51–64.

Geisinger, K. F. (2010). Bogardus social distance scale. *The Corsini Encyclopedia of Psychology.* doi:10.1002/9780470479216.corpsy0135

Guo, Q., Liu, Z., Li, X., & Qiao, X. (2018), Indulgence and long term orientation influence prosocial behavior at national level, *Frontiers of Psychology, 9*, 1798. doi: 10.3389/fpsyg. 2018.01798.

Harrison, R. (1972). Understanding your organization's character. *Harvard Business Review*, 3, 119–128.

Hodgetts, D., & Stolte, O. (2014). Social distance. *Encyclopedia of Critical Psychology*, 1776–1778. doi:10.1007/978-1-4614-5583-7_559

Hofstede, G. (1980). Culture and organizations. *International Studies of Management & Organization, 10*(4), 15–41. doi:10.1080/00208825.1980.11656300.

Hofstede, G., & Bond, M. H. (1988). The Confucius connection: From cultural roots to economic growth. *Organizational Dynamics, 16*(4), 5–21. doi:10.1016/0090-2616(88)90009-5

Hofstede, G. (1994). *Cultures and organizations: Software of the mind.* London: Harper Collins Business.

Hofstede, G. (2001). *Culture's consequences: Comparing values, behaviors, institutions, and organizations across nations* (2nd ed). Thousand Oaks, CA: Sage.

Hofstede, G. (2011). Dimensionalizing cultures: The Hofstede model in context. *Online Readings in Psychology and Culture, 2*(1). https://doi.org/10.9707/2307-0919.1014

Hoşgörür, V. (1997). Bogardus, Gutman ve Likert ölçekleri [Scales of Bogardus, Gutman and Likert]. *Ondokuz Mayıs Üniversitesi Eğitim Fakültesi Dergisi, 10*, 346–358.

Hudson, S., Fung So, K. K., Meng, F., Cárdenas, D., & Li, J. (2018): Racial discrimination in tourism: The case of African-American travellers in South Carolina. *Current Issues in Tourism*, DOI: 10.1080/13683500.2018.1516743

Jung, T., Tom Dieck, M. C., Lee, H., & Chung, N. (2020). Moderating role of long-term orientation on augmented reality adoption. *International Journal of Human–Computer Interaction,* 36(3), 239–250, DOI: 10.1080/10447318.2019.1630933

Karakayali, N. (2009). Social distance and affective orientations. *Sociological Forum, 24*(3), 538–562.

Kemikkıran, N. (2015). Güç mesafesi yüksekse eşitsizlik mi istenir? *Ankara Üniversitesi SBF Dergisi, 70*(2), 317–344.

Kim, C., & Lee, S. (2000). Understanding the cultural differences in tourist motivation between Anglo-American and Japanese tourists. *Journal of Travel and Tourism Marketing,* 9(1/2), 153–170.

Kluckhohn, C. (1951a). The study of culture. In D. Lerner and H. D. Lasswell (Eds.), *The policy sciences* (pp. 86–101). Stanford, CA: Stanford University Press.

Koç, E. (2013). Power distance and its implications for upward communication and empowerment: Crisis management and recovery in hospitality services. *International Journal of Human Resource Management, 24*(-19), 3681–3696.

Köroğlu, Ö., & Güzel, F. Ö. (2013). Kültürel değerlerin destinasyon imajına etkisi: Eskişehir 2013 türk dünyası kültür başkentine yönelik bir araştırma, *İşletme Araştırmaları Dergisi, 5*(4), 191–209.

Krippendorf, J. (1987). *The holiday makers: understanding the impact of leisure and travel.* London: Heinemann.

Kumar, S., & Dhir, A. (2020). Associations between travel and tourism competitiveness and culture. *Journal of Destination Marketing & Management, 18*, 1–11.

Lee, K. J., & Scott, D. (2017). Racial discrimination and African Americans' travel behavior. The utility of habitus and vignette technique. *Journal of Travel Research, 56*(3), 381–392.

Leino, M., & Himmelroos, S. (2019). How context shapes acceptance of immigrants: The link between affective social distance and locational distance. *Ethnic and Racial Studies,* DOI: 10.1080/01419870.2019.1665696

Magee, J. C., & Smith, P. K. (2013). The social distance theory of power. *Personality and Social Psychology Review, 17*(2), 158–186.

Manrai, L. A., & Manrai, A. K. (2011). Cross-cultural and cross-national consumer research in the global economy of the twenty-first century. *Journal of International Consumer Marketing, 23*(3–4), 167–180, DOI: 10.1080/08961530.2011.578056

Martin (1971). The concept of power. *The British Journal of Sociology,* Sep., *22*(3), 240–256.

Mather, D. M., Jones, S. W., & Moats, S. (2017). Improving upon Bogardus: Creating a more sensitive and dynamic social distance scale. *Survey Practice, 10*(4), 1–10.

Meng, F. (2010). Individualism/collectivism and group travel behavior: A cross-cultural perspective. *International Journal of Culture, Tourism and Hospitality Research, 4*(4), 340–351.

Mercan, Ş. O., & Kazancı, M. (2019). Kültürel değerlere yönelik destinasyon seçimi: Çanakkale'ye gelen yerli ziyaretçiler üzerine bir araştırma. *Turizm Akademik Dergisi, 6*(2), 115–125.

Minkov, M. (2007). *What makes us different and similar: a new interpretation of the world values survey and other cross -cultural data.* Sofia: Klasika i Stil.

Mobley, M., Slaney, R. B., & Rice, K. G. (2005). Cultural validity of the Almost Perfect Scale--Revised for African American college students. *Journal of Counseling Psychology, 52*(4), 629–639. doi:10.1037/0022-0167.52.4.629

Money, R. B., & Crotts, J. C. (2003). The effect of uncertainty avoidance on information search, planning, and purchases of international travel vacations. *Tourism Management, 24*, 191–202.

Öncül, M. S., Deniz, M., & İnce, A. R. (2016), Hofstede'nin örgüt kültürü modelinin potansiyel girişimcilerin yetiştiği çevresel özellikler kapsamında değerlendirilmesi. *Akademik Yaklaşımlar Dergisi, 7*(1), 255–269.
Öztürk, Y. (2020). Tüketici etnosentrizminin turistik satın alma tercihine etkisi. *İşletme Araştırmaları Dergisi, 12*(1), 304–313.
Pappas, N. (2014). The effect of distance, expenditure and culture on the expression of social status through tourism. *Tourism Planning & Development, 11*(4), 387–404, DOI: 10.1080/21568316.2014.883425
Parrillo, V. N., & Donoghue, C. (2005). Updating the Bogardus social distance studies: A new national survey. *The Social Science Journal, 42*(2), 257–271. https://doi.org/10.1016/j.soscij.2005.03.011
Podrug, N., Pavičić, J., & Bratić, V. (2006). Cross-cultural comparison of Hofstede's dimensions and decision-making style within CEE context. In The Program Committee ICES2006 (Ed.), *From transition to sustainable development: The path to European integration* (pp. 339–343). Sarajevo: School of Economics and Business, University of Sarajevo.
Sharlamanov, K., & Jovanoski, A. (2013). The ethnic relations in the Macedonian society measured through the concept of affective social distance. *American International Journal of Social Science, 2*(3), 33–39.
Sin, L., Cheung, G., & Lee, R. (1999). Methodology in cross-cultural consumer research: A review and critical assessment. *Journal of International Consumer Marketing, 11*(4), 75–96.
Sorensen, K., Okan, O., Kondilis, B., & Levin-Zamir, D. (2021). Rebranding social distancing to physical distancing: Calling for a change in the health promotion vocabulary to enhance clear communication during a pandemic. *Global Health Promotion, 28*(1), 5–14.
Stephenson, M. L., & Hughes, H. L. (2005). Racialised boundaries in tourism and travel: A case study of the UK black Caribbean community. *Leisure Studies, 24*(2), 137–160, DOI: 10.1080/0261436052000308811
TDK (2021). Türk Dil Kurumu Sözlükleri. Retrieved from https://sozluk.gov.tr/ (accessed on 09.06.2021)
Thyne, M., Woosnam, K. M., Watkins, L., & Ribeiro, M. A. (2020). Social distance between residents and tourists explained by residents' attitudes concerning tourism. *Journal of Travel Research*, 61(1) https://doi.org/10.1177/0047287520971052
Türker, N., & Karadağ, D. (2019). Kültürel farklılıklar: Hofstede'nin kültürel boyutları üzerine Trabzon ve Şanlıurfa illerinde bir uygulama. *Journal of Economy Culture and Society, 61*, 271–295. https://doi.org/10.26650/JECS2019-0033
Usal, A., & Kuşluvan, Z. (1998). *Davranış bilimleri: Sosyal psikoloji*. 2. Baskı, İzmir: Barış Yayın.
Uzuntarla, Y., Fırat, İ., & Ceyhan, S. (2016). Kolektivizm ve belirsizlikten kaçınma davranışı arasındaki ilişkinin incelenmesi. *The Journal of International Educational Sciences, 3*(6), 206.
Vela, M. R. (2009). Rural-cultural excursion conceptualization: A local tourism marketing management model based on tourist destination image measurement. *Tourism Management, 30*(3), 419–428.
Vietze, C. (2012). Cultural effects on inbound tourism into the USA: A gravity approach. *Tourism Economics, 18*, 121–138.
Wallace, C., Vandevijvere, S., Lee, A., Jaacks, L. M., Schachner, M., & Swinburn, B. (2019). Dimensions of national culture associated with different trajectories of male and female mean body mass index in countries over 25 years. *Obesity Reviews*. doi:10.1111/obr.12884
Yıldırım, Y., & Gültaş, P. (2016). Farklı etnik kimliklere sahip tüketicilerin etnosentrizm düzeylerinin ve ürün tercihlerindeki tutumlarının incelenmesi: Malatya ili örneği, *International Journal of Academic Value Studies, 2*(6), 18–34. (ISSN:2149–8598).

18

INBOUND TOURISM AND ALTERATION IN SOCIAL CULTURE, NORMS, AND COMMUNITY ATTITUDES IN THE TOURISM INDUSTRY

The South Asian Experiences

Sakib Bin Amin, Farhan Khan, Shah Zahidur Rahman and Birsen Bulut Solak

Introduction

The tourism industry is considered as an amalgamation of activities, services, and initiatives that provide transportation, accommodation, food, shopping, entertainment, and various other hospitality services to travelers near and far across the globe. According to the statistics of the World Travel & Tourism Council (WTTC, 2020), the tourism industry created 330 million jobs worldwide in 2019, which is one in every ten jobs. The industry also contributed 10.4% of the global GDP, which is equivalent to USD 9170 billion.

However, not long ago, tourism was seen as small-scale industry and immensely underrated. The relationship with the tourism industry and other relevant stakeholders such as governments, private sector, and monetary organizations was not visible and prominent. There were many myths regarding the different attributes of tourism. For instance, people believed that the tourism industry could not significantly contribute to poverty alleviation and employment generation.

During 1970 and 1980, the criticizers declared that countries should avoid tourism unless tourism is a small industry, indigenously owned, and does not harm the traditional norms and cultures as well as the surrounding environment (Amin et al., 2020; Zhang & Zhang, 2020). As time elapsed, the underlying myths regarding the tourism industry started changing, and the industry emerged as the fastest growing global economic sector of the world economy. Tourism can play a vital role in economic development for many countries around the world by offering jobs, generating income, reducing poverty, generating foreign exchange earnings, and improving the standard of livelihood. In 2005, the United Nations (UN) declared the tourism industry one of the most influential determinants of sustainable growth, especially for poor, developing, and emerging countries. Figure 18.1 shows the significance of international tourism as the industry is ranked 3rd largest globally.

The UN has also acknowledged the expansion of tourism as one of the drivers to meet the Sustainable Development Goals (SDGs). The tourism industry is associated with about 109 sectors

DOI: 10.4324/9781003161868-18

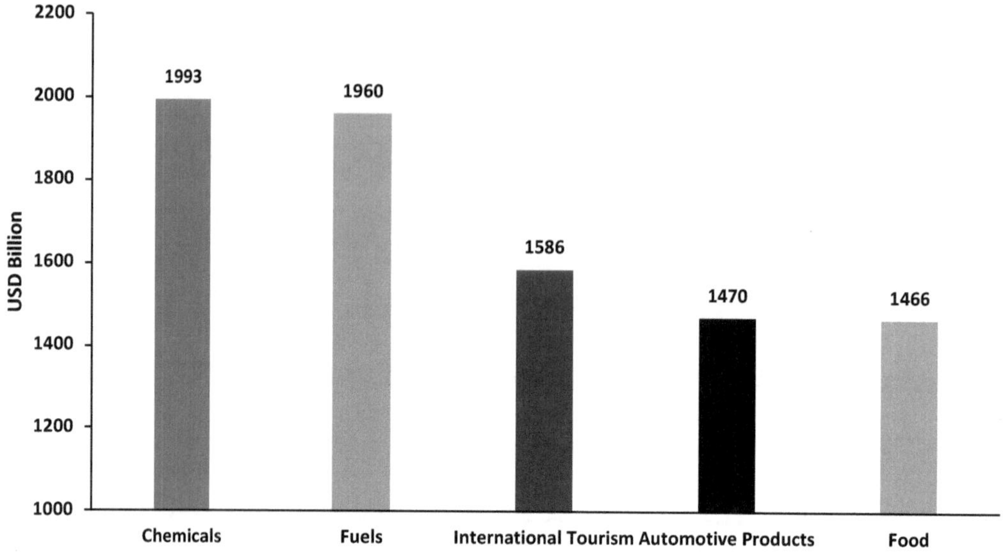

Figure 18.1 Export Earnings by Product Category, 2017 (USD billion)[1] (UNWTO, 2019).

and can spread its socio-economic benefits to all social levels (Liang & Bao, 2015; Amin, 2021). The tourism industry can also be viewed as a major industry in the fight against poverty (Mitchell & Ashley, 2010; Brida et al., 2016; Njoya & Seetaram, 2017; Zhang & Yang, 2018). Therefore, this industry's sustainable development can ensure 17 goals that the United Nations has set for achieving the SDGs by 2030. With its ability to produce flexible labor markets and offer varied working prospects, the tourism industry is directly linked to SDG-8, SDG-10, and SDG-12. Because tourism implicitly promotes the protection of historical, archaeological, and religious shrines; and stimulates local folklore, traditions, arts and crafts, and cuisine, it enables the transfer of people, culture, and ideas; the other 14 goals have cross-cutting involvement with tourism. Therefore, it is popularly believed that tourism development can play a significant role in implementing SDGs.

Apart from the discussion of tourism's role in economic development, investigation of social psychological effects of tourism is an area of literature. Even though several studies[2] have been done to understand the underlying correlation between tourism development and social parameters through the lenses of sociological and psychological theories; however, the area has received relatively less attention in the developing countries like South Asian region. Therefore, in-depth conceptual and theoretical analysis in this area can provide constituent information for future tourism sustainability.

Tourism activities start with interaction between host and tourists, which is a natural phenomenon that causes an exchange of values, norms, and cultural practices, leading to a change in the social atmosphere of a tourist destination (Nawijn & Mitas, 2012; Woo et al., 2015; Liang & Hui, 2016; Chen & Rahman, 2018). With the continuous development of the tourism industry, it is evident that traditional social cultures have changed vividly in many tourist destinations worldwide. The changes in the local community start to occur with the tourist arrival and interaction with the residents of the host community. These changes include the impact of tourism development on inhabitants' traditional values, lifestyles, and interpersonal relationships both in rural and urban localities. However, sometimes the mixing of wealthy foreign tourists and relatively poor residents can increase the likelihood of a community backlash against tourists and tourism development.

Therefore, it is imperative to scrutinize the host residents' perception of tourists and their attitude towards them as the residents of many tourist destinations are the fundamental elements of the tourism 'product' and have a sizable impact on its sustainability (Gursoy et al., 2002; Gursoy et al., 2009; Filimonau et al., 2018).

South Asian countries witnessed a growth in the tourism industry as the region attracted 40% of the total international tourists in the Asia Pacific region in 2018.[3] Furthermore, 18% of the tourism industry's contribution to the Asia Pacific region's GDP in 2018 came from South Asia, which is worth USD 157,420 million (Calderwood & Soshkin, 2019). In a broader picture, according to the WTTC (2019) statistics, tourism's total contribution in 2018 was USD 224.10 billion, about 6.6% of South Asian GDP. The tourism industry supported 46,318 jobs in 2018, equal to 7.3% of South Asian employment. This industry also generated USD 39 billion in the export sector in 2018, which is 6.2% of the South Asian exports. Given the momentum in the South Asian tourism industry, it is crucial to analyze how social cultures, values, norms, and attitudes occur as host communities interact with tourists and tourism development. We construct a novel conceptual framework, considering no other earlier attempts have been made focusing on the South Asian region. Based on the discussion, we propose policy advices for the policymakers to prepare guidelines for integrated sustainable tourism management and development in the South Asian region.

The rest of the chapter has the following structure. The "Overview of South Asian Tourism Development" section provides an overview of the South Asian tourism industry and its impact on the regional economy. The "Social Factors of Tourism Development" section focuses on discussing the social aspects of the tourism industry, followed by the conceptual framework in the "The Conceptual Framework" section linking with previous studies. Finally, the "Conclusion" section brings a concussion of the chapter.

Overview of South Asian Tourism Development

Tourism is rapidly becoming a popular leisure activity across the globe. In recent years it has gained significant importance as a driver of economic growth and development. According to UNWTO (2019), tourism is one of the largest industries in the world economy. A thriving tourism sector is essential for both developed and developing economies. The tourism industry directly contributes to economic growth and development by creating employment opportunities, generating tax revenues, and producing foreign exchange earnings. 2019 marked a strong year of growth for the tourism industry as the total international tourist arrivals reached 1.46 billion, a 4% increase over the last year, and total global tourism receipts amounted to USD 1.481 trillion, a 3% increase from last year. For many destinations worldwide, tourism is a vital source of foreign revenues, representing up to 90% of their exports and even leading to a tourism trade surplus (UNWTO, 2020).

South Asia is a unique region with vast adjoining land, varied landscapes from marshlands to deserts, grasslands to forests, seaside areas to mountains, a huge and attractive range of natural resources, scenic beauty, and the coastal regions to mountains (Rasul & Manandhar, 2009). Two of the highest mountain peaks (Everest and K2)[4] belong to this region. The coral reefs of the Maldives,[5] and the longest sandy sea beach, Cox's Bazar in Bangladesh are some of the most eye-catching maritime destinations in South Asia. Biologically diverse areas such as Sinharja in Sri Lanka[6] and the Sundarbans in Bangladesh are home to this region, not to mention wondrous structures like the Taj Mahal[7] is also located here. Ancestral customs and traditions are deep-rooted in the people of this region. Countless Sultans, Kings, Emperors, and Viceroys have ruled over this region, leaving their mark on it in the form of extravagant palaces, monuments, and mausoleums. The cultural and religious diversity, the exquisite cuisines, and the warm and hospitable people set South Asia apart from the rest of the world.

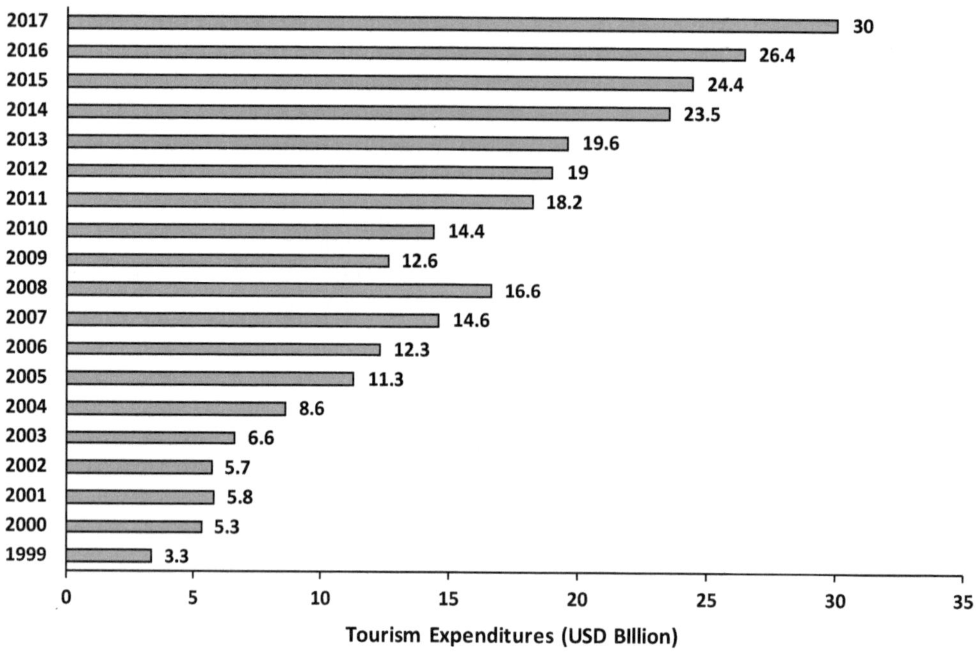

Figure 18.2 South Asian Tourism Expenditures (UNWTO, 2020).

Although tourism growth has risen in South Asia, it is still lagging far behind other regions. The share of international tourist arrivals and tourism receipts is relatively minimal compared to the rest of the world. In 2019, it only had a 2.4% share of total international tourist arrivals and only 9.7% in the Asia and Pacific region. Thailand alone had 40 million arrivals in the same year. It is also evident that, in 2019, total international tourist arrivals reached 35.1 million (2.4% of total arrivals), and international tourism receipts reached USD 46.2 billion (3.1% of total receipts). What is heartening to note is that South Asia achieved a commendable 7% growth in tourist arrivals and a 5% growth in tourism receipts in 2019. This shows South Asia's latent potential to become a tourism powerhouse even though it hasn't established a foothold in the tourism industry yet due to social, economic, and political reasons. So, the question arises as to why South Asia, which is so rich in flora and fauna, biodiversity, cultures, cuisines, and monuments, has such a small proportion of the tourism industry.

Like tourist arrivals, tourism receipts and tourism expenditure follow a similar trend in South Asia (Figure 18.2). In the early 2000s, it showed low growth and stagnancy and a sharp dip in 2009 before recovering. Since the mid-2010s, the growth has started to increase, almost reaching double digits. The effects of increased globalization and the ease of travel restrictions by the South Asian governments have enabled this growth spurt to occur.

Many vital factors constrain tourism promotion in South Asia. Travel restrictions in most countries in South Asia act as the first barrier. Travel procedures such as border formalities, visas, permits, airline access, and currency are quite complicated and restrictive. It ultimately becomes a big hassle for both international and regional visitors. Simplifying the visa requirements and making the documentation procedures more streamlined can open up South Asian tourist destinations to a much broader population. Along with travel restrictions, the connectivity between South Asian countries (air, rail, and road) also remains in lackluster conditions. Air connections have improved over the years, but land connections remain disjointed, making many tourist destinations difficult to access. Cross-border connectivity in countries like India, Pakistan, Nepal, and

Table 18.1 Travel and Tourism Competitiveness Index[8] (2019)

Country	*Overall Index*	*Human Resources and Labor Market*	*Safety and Security*	*Ground and Port Infrastructure*	*Tourist Service Infrastructure*	*Price Competitiveness*	*Natural and Cultural Resources*
BGD	120	120	105	60	133	85	105
IND	34	76	122	28	109	13	9
NPL	102	83	91	131	126	15	56
PAK	121	135	134	73	112	37	87
SLK	77	84	78	52	92	74	52

Source: World Economic Forum (2019)
Note: BGD, IND, NPL, PAK, and SLK indicate Bangladesh, India, Nepal, Pakistan, and Sri Lanka, respectively.

Bangladesh is also quite restricted as there are limited road and rail links between these countries, and they are used quite rarely for passenger travel. Driving a vehicle across borders also requires special permits, which further discourages travelers.

Tourism facilities and services are also somewhat inadequate for tourists. There is a distinct lack of hotels, resorts, restaurants, and other facilities in many tourist destinations of South Asian countries. Most South Asian countries cannot compete with South-East Asian countries regarding price, amenities, and facilities. All South Asian countries are ranked very low in the Tourist service infrastructure sub-index (Table 18.1). The South Asian tourism sector has been neglected for many years with limited investment from the public sector and a lack of human capital. Moreover, the region has also been unable to attract foreign investment due to the region's economic turmoil, frequent policy changes, and a generally negative image associated with the region.

Safety and security are another grave area of concern for South Asia. Tourists are often discouraged from traveling to South Asia because of the dangers and uncertainty surrounding this region. Political unrest, frequent strikes, border disputes, terrorist attacks over the years have tarnished South Asia's image as a tourist destination. Some developed countries have even labeled certain parts of South Asia as high-risk zones. As a result, South Asian countries are ranked poorly in the safety and security sub-index, with Sri Lanka being ranked the highest and Pakistan the lowest. Due to these reasons, South Asian countries have been ranked poorly in the overall competitive index (World Economic Forum, 2019). Only India has been able to command a respectable rank of 34 out of 140 countries.

Although South Asian countries are lagging in certain aspects, but the region has performed well in terms of natural and cultural resources and price competitiveness. In terms of price competitiveness, high-ranking countries Spain and France are ranked lower than all South Asian countries. This shows that South Asia can be a good and affordable alternative tourist destination if its tourist infrastructure and security conditions are improved.

Globalization and economic liberalization have been paramount in overcoming national boundaries and integrating national economies into the global economy. This has stimulated regional cooperation among neighboring countries worldwide to use their pool of resources better and formulate standard tourism policies. South Asian countries are in a unique position to form a Regional Tourism Integration (RTI) and generate employment, increase foreign exchange earnings and reduce poverty through tourism. These neighboring nations have similar terrain, common cultural and religious heritage, and similar physical and economic infrastructure. Regional cooperation can help neighboring countries benefit from economies of scale in the supply of tourist goods and services. They can, for example, allow tourists to visit the entire South Asian region on a single visa instead of acquiring multiple visas. Through the formation of the RTI, South Asian countries can share resources, formulate common tourism policies, overcome geographic barriers and benefit from a collective competitive advantage. Therefore, regional cooperation between the South Asian nations can help establish a new tourism hub that can redirect some tourist traffic from the European, Middle Eastern, and American countries.

Social Factors of Tourism Development

The nature and interaction of both the visitors and the host communities determine the magnitude of tourism's sociocultural impacts. In South Asian nations, where there is a large cultural and religious contrast between the hosts and the visitors, the effects will likely be greatest. In recent years, there have been some studies focused on the host communities' perceived impacts and attitude towards tourism development (Yoon et al., 2001; Gursoy et al., 2002; Sirakaya et al., 2002; Gursoy & Rutherford, 2004; Chuang, 2013; Mirzaei, 2013). The support and goodwill of residents are paramount to the development, successful operation, and sustainability of tourist destinations Jurowski (1994). Once an area becomes a tourist destination, the residents' quality of life is affected by this development, whether good or bad (McGehee & Andereck, 2004; Acharya & Halpenny 2013; Altinay & Taheri, 2019; Olya et al., 2020).

The primary benefits of tourism development as perceived by residents are its economic opportunities in employment and revenue generation (Amin et al., 2020; Amin, 2021). In developing countries, it can also bring about changes in employment from traditional agriculture to service-based industries, which often leads to higher wages and better job prospects. This can revitalize poor or non-industrialized rural regions. Other benefits include the renewed prevalence of traditional arts and trades and local cultural festivities. The government will take notice of architectural landmarks that require rebuilding and take steps to preserve conservation areas (Amin, 2021).

However, tourism can also lead to harmful effects on the host communities. Tourism during the peak season can lead to overcrowding and traffic congestion which is undesirable for both hosts and visitors. The overcrowding and overuse of resources can lead to environmental degradation and cause wear and tear of local infrastructure. Another significant negative sociocultural impact of tourism is the "demonstration" effect when there are visible differences between tourists and hosts. It is argued that the locals will take note of the luxurious possessions of the tourists and aspire to acquire them, Williams (2002). Local people, particularly the younger generation, become resentful when they are unable to replicate the lifestyle of the tourists. This can lead to conflict between the two groups for which tourists are often blamed for creating a societal division between the youth and the elderly. In communities with strong religious and cultural codes, the different values and norms exhibited by tourists are viewed as largely undesirable. This effect is observed in South Asian nations, where most tourist destinations are located in rural areas and are visited by wealthy western tourists.

An existing difficulty in gauging sociocultural impacts is that it is challenging to measure them. However, certain factors can give an insight into the magnitude and extent of these impacts. The religious and cultural factor depicts the social concerns of host communities in response to tourism development. Zamani-Farahani and Musa (2010) studied the effects of Islamic spiritual matters on the perception of host communities in Iran. The study revealed that locals showed greater concern for moral standards and being conservative among religious people. Residents with solid Islamic beliefs perceived tourism negatively because of specific religious differences such as alcohol consumption and western attires. Other studies show a general intolerance of religious people to those of different religions (Guiso et al., 2003; Daniels & von der Ruhr, 2005).

The host community's knowledge about tourism is another crucial factor that affects their perception of tourism's costs and benefits. Davis et al. (1988) shows that hosts' knowledge about tourism and the local economy can influence their attitudes toward tourism development. Knowledgeable locals tend to embrace tourism as they positively perceive tourism's impact on the local economy (Andereck et al., 2005). The intrinsic motivation factor of residents also supports tourism development. Kayat (2002) gave some examples of intrinsic motivation such as welcoming and hosting tourists, the possibility to learn about different cultures, to work and live as a community, to feel needed, and to gain self-respect.

Another important factor that affects hosts' support for tourism development, particularly in rural areas, is their attachment to the community. Several studies have examined locals' community attachment and their support for tourism, but the results have been contradictory. Huttasin (2008) and Um and Crompton (1987) have concluded that residents attached to their community and have lived there for a long time are more likely to be against tourism development. Other studies have suggested that greater attachment to the community creates a positive perception of tourism (Gursoy et al., 2002; Látková & Vogt, 2012; Stylidis, 2016). McCool and Martin (1994) could not establish a link between hosts' community attachment their support for tourism development. Although the findings have been inconsistent, it is evident that attachment to the community greatly affects locals' perception of tourism's impact and their support for tourism development.

The Conceptual Framework

This section aims to conceptualize a simple framework to explain the vicissitudes in social cultures, values, norms, and attitudes in the emerging South Asian countries due to the region's recent momentum in tourism development (Figure 18.3). The core idea for the conceptualization is based on the arguments of social exchange theory (Ap, 1992), likelihood model (Petty et al.,1981), presumed influenced model (Gunther & Storey, 2003; Truong & King, 2014; Yoo et al., 2016). Constructing such a framework in the context of tourism literature is immensely important because, as Bargeman and Richards (2020) highlight, this type of frameworks are optimal to expose the underlying interaction rather than focusing on only consumption and individualism. Following the analysis of Ajzen and Fishbein (1977), Arai and Pedlar (2003), Ren et al. (2018), Soica (2016), Bargeman and Richards (2020) further argue that too much focus on individualism and consumption pattern of tourism goods and services may lead to a wrong direction and provide distorted inferences of the structure and role of the individual actors play in the tourism industry and further sustainability.

The proposed conceptual framework of this section starts with the notion mentioned by Jafari (1987). We place tourists at the beginning of the framework since tourists are considered the homogeneous denominator of tourism activities worldwide. Then the framework is expanded following the arguments of the mentioned theories/models earlier. It is worth pointing out that the proposed conceptual framework is simple; however, it can deliver a dynamic causal relationship among the main stakeholders of interest. It is worth mentioning that particular propositions characterize each stakeholder's decision-making process. We will elaborate on the recommendations as we logically structure the framework.

Before planning any trip, tourists first choose a destination. This choice is highly dominated by the preconceived attitude of the tourists regarding the potential destinations in their lists (Farmaki et al., 2019). Information regarding the possible destinations is obtained from the perception of the host country's national attributes, an exogenous factor in our framework. The information can be obtained by different modes such as social media and e-tourism tourism agents. Given the information, dynamics of the tourists' attitude can change through different channels (Petty & Cacioppo, 1986; Becken et al., 2019; Nunkoo et al., 2020).

For example, their perceptions of destination and cultural values and norms, religious values, and environmental concerns. These perceptions influence tourists' attitudes in different magnitudes that ultimately confirm the decision of making the trip to a particular destination, and constructs the behavioral pattern while staying in the destination for consuming tourism-related goods and services. It is worth mentioning that the key aspect of the recent tourism momentum in the South Asian region is the targeted policies for facilitating digital tourism marketing or, in other words, E-tourism. According to Ma et al. (2003), E-tourism is an ideal tool for maximizing competence and efficacy through various practices and value chains in the tourism industry. Therefore, the first proposition is as follows.

***Proposition 1**: Perception of the host country's national attributes tends to exogenously influence the preconceived attitude regarding destination choice through numerous channels via different modes.*

A key aspect of the tourism destination is the local residents or the host. Perception of the host has been a major discussion point in the existing literature of tourism industry development. Cook et al. (2006) define a host community in the most stress-free way. They define a host community as a place where tourists are welcomed and given desired goods and services that comfort their stay. On the other hand, Smith (2001), Chen and Rahman (2018), and Amin et al. (2020) advocate that those host communities are the group of people who live beside different tourist destinations and their daily lives are affected (positively or negatively) by the tourism activities both explicitly and implicitly.

Considering both the positive and negative effects, behavior of the host can be divided into two sub-groups, namely, pro-tourism and anti-tourism (Gunther & Storey, 2003). The behavioral dynamics are found to be correlated with the level of tourism development and sustainability. Among others, Gursoy et al. (2002), Gursoy et al. (2009), and Gursoy et al. (2019) point out that the host's behavioral attitude is paramount in shaping the tourism industry since it evolves in a socially compatible way. Given the discussion, the second proposition can be sketched by the following statement.

***Proposition 2**: The behavioral dynamics of the host depend on the existing structure of the society. It also correlates with tourism development and sustainability.*

Moreover, the intensity of the behavioral dynamics comes from the perception regarding the tourism industry development advocated in the social exchange theory, further elaborated in Amin (2021), apart from heterogeneous cultural and religious values, environmental protection, and so on. It is argued that the expansion of the tourism industry can contribute to economic development through globalization by increasing the trade of various commodities and attracting foreign investment. However, digging deeper, such an outcome only occurs when the host participates in an exchange process from the tourism industry development. Attitude towards tourism development certainly depends on the potential benefits, which are coming from the consequence of tourism activities and interaction with the tourists (Ngo & Pham, 2021). Once the benefits exceed the costs, the host trades-off between benefits and future tourism development in direct and indirect contributions.

As a result, the behavioral pattern changes as attitude changes due to the trade-offs (Amin et al., 2019). From the recent trends of inbound tourism in the South Asian context, it can be argued that the effect trade-off is bringing net benefits for the host communities and the countries too. Some of the benefits that tourism has brought in the South Asian region are infrastructural development in the touristic areas, generates employment generation through backward-linkages, helps to reduce societal inequality, introduces cultural blend, and improves the living standard and lifestyle, and many more (Nepal et al., 2019; Amin, 2021; Ngo & Pham, 2021).

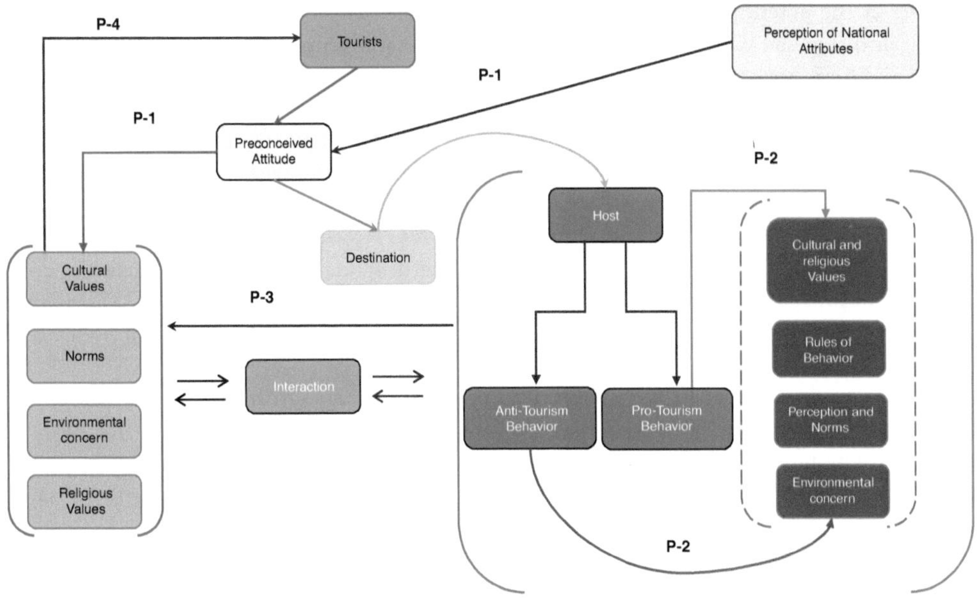

Figure 18.3 Conceptual Framework, *Authors' Own Elaboration.*

Nevertheless, the behavior pattern sometimes is not as expected for many reasons in South Asian countries. Among others, religious values sometimes restrict the optimal behavior pattern given the presence of benefits. Besides, enhanced tourism activities also lead to the environmental problem given the concept of carrying capacity.[9] The influx of tourists more than the carrying capacity can lead to different problems, such as exploiting resources, leading to community's future goals at risk. (Burke et al., 2001; Garrod & Wilson, 2003). It is worth mentioning that probability of anti-tourism behavior increases as future rick escalates. So, the third proposition can be written as follows.

Proposition 3*: Interaction behavior intensity and possible changes in the host community depend on the degree trade-offs between benefits and future tourism development.*

As the host understands the beneficial impact of the tourism development, it can be predicted that host communities will welcome the tourists in the upcoming future as well. Also, from the other side of the picture, the interaction pattern changes the tourists' perception that ultimately facilitates tourism development through a snowball effect. Thus, the fourth proposition is stated by the following statement.

Proposition 4*: The beneficial impact translates into a feedback causal chain that creates a snowball effect and attracts more tourists.*

Conclusion

Since the tourism industry is one of the potential industries in the South Asian region, the main aim of this chapter has been to analyze the alteration in the host communities in terms of social and psychological perspectives. In this regard, following the previous literature on tourism-related social and psychological theories, we have constructed a dynamic and straightforward feedback conceptual framework that links various factors that influence dynamic interactions of tourists and hosts of the potential destinations within South Asia. The conceptual framework provides a logical intuition and clarifies the direct and indirect relationships that explain the outcomes of tourists' and hosts' interactions.

From the conceptual framework, we argue that tourists' prior perceptions define their interaction behavior with the hosts. On the other hand, the behavioral pattern of the hosts depends on the rooted social structures. Besides, hosts' interaction with the tourists (pro and anti) certainly depends on the possible degree trade-offs between benefits and future tourism development in the South Asian context. We have also highlighted that pro-tourism behavior that emerges from the visible positive trade-off indicates an alteration in different social factors; however, it may not be the optimal change due to distinct religious values and carrying capacity. Also, the prevailing behavioral pattern translates into a causal chain and attracts more tourists in South Asian countries.

Community development is very important to enhance the tourism benefits by minimizing the restrictions mentioned earlier. Andrades and Dimanche (2017) argue that creating awareness among the community people of the host communities regarding the expectations of the foreign tourists can help local firms to be more competitive. Therefore, the tourism policy should emphasize the development of community-level education and awareness programs. It should be noted that a low level of community education and awareness are deterrent to tourism development. Local community representatives, representatives from the government officials and the non-governmental organizations (NGOs), local educators, and the politicians can create awareness among the community people to make them understand how to preserve the cultures, religious values, and traditions while at the same time to become more tourism-friendly for reaping the benefits of tourism development. On the issue of carrying capacity, governments should focus on managing tourist influx as well as diversify the tourist areas from the host communities. This is not only reducing pressure from the host communities for supplying good and services but also improves environmental aspects (Amin, 2021).

The conceptual framework can be further elaborated by augmenting more factors to yield different dynamics on the interaction between tourists and hosts. Also, country specific conceptual frameworks can become very handy for policy makers because of incorporating country level heterogeneity. Since the proposed conceptual framework is based on intuitive logics and recent tourism trends of the South Asian region, we welcome empirical analysis to check the framework's effectiveness. Besides, appropriate structural changes can be introduced based on the empirical outcomes if necessary for stimulating ideas for future research.

Notes

1 Published in International Tourism Highlights (2019) Edition

2 The "The Conceptual Framework" section discusses more about these studies.

3 It is worth noting that eight emerging countries make up the South Asian region. These countries include India, Maldives, Nepal, Pakistan, Sri Lanka, Bhutan, Afghanistan, and Bangladesh.

4 Mount Everest is the highest mountain globally, located in Nepal and China, with 29,029 feet. On the other hand, K2 is the second heightened mountain within the Himalayas range, having 28,251 feet.

5 For more details, see: https://www.themaldivesexpert.com/2406/coral-reefs-of-the-maldives/

6 For more details, see: https://whc.unesco.org/en/list/405/

7 For more details, see: https://whc.unesco.org/en/list/252/

8 Travel and Tourism Competitiveness Index is a measurement of the factors that make it attractive to develop business in individual countries' travel and tourism industry based on 14 sub-indices. Out of 14, 6 sub-indices are present here. The report gives each country a score of 1–7 on each specific sub-index. The 2019 report covers 140 countries in total.

9 For more details, see: Amin (2021); Ministry of Environment, Forests and Climate Change Wildlife Division (2018); Dar et al. (2016)

References

Acharya, B. P., & Halpenny, E. A. (2013). Homestays as an alternative tourism product for sustainable community development: A case study of women-managed tourism product in rural Nepal. *Tourism Planning & Development, 10*(4), 367–387.

Ajzen, I., & Fishbein, M. (1977). Attitude-behavior relations: A theoretical analysis and review of empirical research. *Psychological Bulletin, 84*(5), 888.

Altinay, L., & Taheri, B. (2019). Emerging themes and theories in the sharing economy: A critical note for hospitality and tourism. *International Journal of Contemporary Hospitality Management, 31(1)*, 180–193

Ap, J. (1992). Residents' perceptions on tourism impacts. *Annals of Tourism Research, 19*(4), 665–690.

Amin, S. (2021). *The economy of tourism in Bangladesh: Prospects, constraints, and policies.* Palgrave Macmillan.

Amin, S. B., Kabir, F. A., & Khan, F. (2020). Tourism and energy nexus in selected South Asian countries: A panel study. *Current Issues in Tourism, 23*(16), 1963–1967.

Amin, S. B., Kabir, F. A., Khan, F., & Rahman, S. Z. (2019). Impact of seaside tourism on host community in Bangladesh: The case of Cox's Bazar. *North South Business Review, 10*(1), 69–89.

Andereck, K. L., Valentine, K. M., Knopf, R. C., & Vogt, C. A. (2005). Residents' perceptions of community tourism impacts. *Annals of Tourism Research, 32*(4), 1056–1076.

Andrades, L., & Dimanche, F. (2017). Destination competitiveness and tourism development in Russia: Issues and challenges. *Tourism Management, 62*, 360–376.

Arai, S., & Pedlar, A. (2003). Moving beyond individualism in leisure theory: A critical analysis of concepts of community and social engagement. *Leisure Studies, 22*(3), 185–202.

Becken, S., Alaei, A. R., & Wang, Y. (2019). Benefits and pitfalls of using tweets to assess destination sentiment. *Journal of Hospitality and Tourism Technology, 11*(1), 19–34.

Bargeman, B., & Richards, G. (2020). A new approach to understanding tourism practices. *Annals of Tourism Research, 84*, 102988.

Brida, J. G., Cortés-Jiménez, I., & Pulina, M. (2016). Has the tourism-led growth hypothesis been validated? A literature reviews. *Current Issues in Tourism, 19*(5), 394–430.

Burke, L. Kura, Y., Kassem, K., Revenga, C., Spalding, M., & McAllister, D. (2001). *Pilot analysis of global ecosystems: Coastal ecosystems.* Washington, DC: World Resources Institute.

Calderwood, L. U., & Soshkin, M. (2019). *The travel & tourism competitiveness report 2019.* Geneva: World Economic Forum.

Cook, R.A., Yale, L. J., & Marqua, J. J. (2006). *Tourism: The Business of Travel* (3rd ed.). New Jersey: Prentice Hall.

Chen, H., & Rahman, I. (2018). Cultural tourism: An analysis of engagement, cultural contact, memorable tourism experience and destination loyalty. *Tourism Management Perspectives, 26*, 153–163.

Chuang, S. T. (2013). Residents' attitudes toward rural tourism in Taiwan: A comparative viewpoint. *International Journal of Tourism Research, 15*(2), 152–170.

Daniels, J. P., & Von Der Ruhr, M. (2005). God and the global economy: Religion and attitudes towards trade and immigration in the United States. *Socio-Economic Review, 3*(3), 467–489.

Dar, S. N., Wani, M. A., & Shah, S. A. (2016). Tourism carrying capacity assessment for Leh town of Ladakh region in Jammu and Kashmir. *International Journal of Current Research, 8*(02), 26403–26410.

Davis, D., Allen, J., & Cosenza, R. M. (1988). Segmenting local residents by their attitudes, interests, and opinions toward tourism. *Journal of Travel Research, 27*(2), 2–8.

Farmaki, A., Khalilzadeh, J., & Altinay, L. (2019). Travel motivation and demotivation within politically unstable nations. *Tourism Management Perspectives, 29*, 118–130.

Filimonau, V., Matute, J., Mika, M., & Faracik, R. (2018). National culture as a driver of pro-environmental attitudes and behavioural intentions in tourism. *Journal of Sustainable Tourism, 26*(10), 1804–1825.

Garrod, B., & Wilson, J. C. (Eds.). (2003). *Marine ecotourism: issues and experiences.* Channel View Publications.

Guiso, L., Sapienza, P., & Zingales, L. (2003). People's opium? Religion and economic attitudes. *Journal of Monetary Economics, 50*(1), 225–282.

Gunther, A. C., & Storey, J. D. (2003). The influence of presumed influence. *Journal of Communication, 53*(2), 199–215.

Gursoy, D., Chi, C. G., & Dyer, P. (2009). Locals' attitude toward mass and alternative tourism: The case of Sunshine Coast, Australia. *Journal of Travel Research, 49*(3), 381–394.

Gursoy, D., Jurowski, C., & Uysal, M. (2002). Resident attitudes: A structural modeling approach. *Annals of Tourism Research, 29*(1), 79–105.

Gursoy, D., Ouyang, Z., Nunkoo, R., & Wei, W. (2019). Residents' impact perceptions of and attitudes towards tourism development: A meta-analysis. *Journal of Hospitality Marketing & Management, 28*(3), 306–333.

Gursoy, D., & Rutherford, D. G. (2004). Host attitudes toward tourism: An improved structural model. *Annals of Tourism Research, 31*(3), 495–516.

Huttasin, N. (2008). Perceived social impacts of tourism by residents in the OTOP tourism village, Thailand. *Asia Pacific Journal of Tourism Research, 13*(2), 175–191.

Jafari, J. (1987). Tourism models: The sociocultural aspects. *Tourism Management, 8*(2), 151–159.

Jurowski, C. A. (1994). *The interplay of elements affecting host community resident attitudes toward tourism: A path analytic approach*. Virginia Tech.

Kayat, K. (2002). Exploring factors influencing individual participation in community-based tourism: The case of Kampung Relau homestay program, Malaysia. *Asia Pacific Journal of Tourism Research,* 7(2), 19–27.

Látková, P., & Vogt, C. A. (2012). Residents' attitudes toward existing and future tourism development in rural communities. *Journal of Travel Research, 51*(1), 50–67.

Liang, Z. X., & Bao, J. G. (2015). Tourism gentrification in Shenzhen, China: Causes and socio-spatial consequences. *Tourism Geographies, 17*(3), 461–481.

Liang, Z. X., & Hui, T. K. (2016). Residents' quality of life and attitudes toward tourism development in China. *Tourism Management, 57*, 56–67.

Ma, J. X., Buhalis, D., & Song, H. (2003). ICTs and Internet adoption in China's tourism industry. *International Journal of Information Management, 23*(6), 451–467.

McCool, S. F., & Martin, S. R. (1994). Community attachment and attitudes toward tourism development. *Journal of Travel Research, 32*(3), 29–34.

McGehee, N. G., & Andereck, K. L. (2004). Factors predicting rural residents' support of tourism. *Journal of Travel Research, 43*(2), 131–140.

Ministry of Environment, Forests and Climate Change Wildlife Division. (2018). *Policy for eco-tourism in forest and wildlife areas*. New Delhi, Government of India.

Mirzaei, R. (2013). Modeling the socioeconomic and environmental impacts of nature-based tourism to the host communities and their support for tourism. *Perceptions of Local Population: Mazandaran, North of Iran*. Licentiate thesis. University of Giessen.

Mitchell, J., & Ashley, C. (2010). *Tourism and poverty reduction: Pathways to prosperity*. Earthscan.

Nawijn, J., & Mitas, O. (2012). Resident attitudes to tourism and their effect on subjective well-being: The case of Palma de Mallorca. *Journal of Travel Research, 51*(5), 531–541.

Nepal, R., Irsyad, M. I. A., & Nepal, S. K. (2019). Tourist arrivals, energy consumption and pollutant emissions in a developing economy-implications for sustainable tourism. *Tourism Management, 72*, 145–154.

Ngo, T., & Pham, T. (2021). Indigenous residents, tourism knowledge exchange and situated perceptions of tourism. *Journal of Sustainable Tourism*, 1–18. DOI: https://doi.org/10.1080/09669582.2021.1920967

Njoya, E. T., & Seetaram, N. (2017). Tourism contribution to poverty alleviation in Kenya: A dynamic computable general equilibrium analysis. *Journal of Travel Research, 57*(4), 513–524.

Nunkoo, R., Gursoy, D., & Dwivedi, Y. K. (2020). Effects of social media on residents' attitudes to tourism: Conceptual framework and research propositions. *Journal of Sustainable Tourism*, 1–17. DOI: https://doi.org/10.1080/09669582.2020.1845710

Olya, H., Altinay, L., Farmaki, A., Kenebayeva, A., & Gursoy, D. (2020). Hotels' sustainability practices and guests' familiarity, attitudes and behaviours. *Journal of Sustainable Tourism, 29*(7),1063–1081.

Petty, R. E., & Cacioppo, J. T. (1986). The elaboration likelihood model of persuasion. In *Communication and persuasion* (pp. 1–24). New York: Springer.

Petty, R. E., Cacioppo, J. T., & Goldman, R. (1981). Personal involvement as a determinant of argument-based persuasión. *Journal of Personality and Social Psychology, 41*(5), 847.

Rasul, G., & Manandhar, P. (2009). Prospects and problems in promoting tourism in South Asia: A regional perspective. *South Asia Economic Journal, 10*(1), 187–207.

Ren, C., James, L., & Halkier, H. (Eds). (2018). Practices in and of tourism. In *Theories of practice in tourism* (pp. 1–9). Routledge.

Sirakaya, E., Teye, V., & Sönmez, S. (2002). Understanding residents' support for tourism development in the central region of Ghana. *Journal of Travel Research, 41*(1), 57–67.

Smith, S. L. (2001). Measuring the economic impact of visitors to sport tournament and special events. *Annals of Tourism Research, 28*(3), 829–831.

Soica, S. (2016). Tourism as practice of making meaning. *Annals of Tourism Research, 61*, 96–110.

Stylidis, D. (2016). The role of place image dimensions in residents' support for tourism development. *International Journal of Tourism Research, 18*(2), 129–139.

Truong, T. H., & King, B. (2006). Comparing cross-cultural dimensions of the experiences of international tourists in Vietnam. *Journal of Law and Governance, 1*(1).

Um, S., & Crompton, J. L. (1987). Measuring resident's attachment levels in a host community. *Journal of Travel Research, 26*(1), 27–29.
Williams, S. W. (2002). *Tourism geography*. Routledge.
Woo, E., Kim, H., & Uysal, M. (2015). Life satisfaction and support for tourism development. *Annals of Tourism Research, 50*, 84–97.
World Economic Forum (2019). *The travel & tourism competitiveness report 2019: Travel and tourism at a tipping point*. Geneva, Switzerland.
World Tourism Organization (UNWTO). (2019). *UNWTO tourism highlights*. 2019 Edition.
World Tourism Organization (UNWTO). (2020). *UNWTO tourism highlights*. 2020 Edition.
WTTC (2019). *World travel and tourism council. Travel and tourism economic impact 2019: World*. London: World Travel and Tourism Council.
WTTC (2020). *World travel and tourism Council. Travel and tourism economic impact 2020: World*. London: World Travel and Tourism Council.
Yoo, W., Yang, J., & Cho, E. (2016). How social media influence college students' smoking attitudes and intentions. *Computers in Human Behavior, 64*, 173–182.
Yoon, Y., Gursoy, D., & Chen, J. S. (2001). Validating a tourism development theory with structural equation modeling. *Tourism Management, 22*(4), 363–372.
Zamani-Farahani, H., & Musa, G. (2012). The relationship between Islamic religiosity and residents' perceptions of socio-cultural impacts of tourism in Iran: Case studies of Sare'in and Masooleh. *Tourism Management, 33*(4), 802–814.
Zhang, H., & Yang, Y. (2018). Prescribing for the tourism-induced Dutch disease: A DSGE analysis of subsidy policies. *Tourism Economics, 25*(6), 942–962.
Zhang, J., & Zhang, Y. (2020). Tourism, economic growth, energy consumption, and CO2 emissions in China. *Tourism Economics,* 1354816620918458.

19

CULTURE SHOCK EXPERIENCES OF TOURISTS

A Transformative Perspective

Nagihan Cakmakoglu Arici

Introduction

The developments in communication and transportation with the reduction in costs have facilitated travel to be easier, cheaper, and more often for the people, but especially for the ones, who always have desire to explore far away destinations and to seek different cultures. The stimuli behind this sojourning can also include converting, conquering, trading, teaching, learning, vacationing, and settling (Furnham, 2019). Due to the nature of these movements, they are investigated within the scope of several social sciences, such as anthropology, economics, education, psychiatry, psychology, and sociology. However, all these mobilities share a joint result, which is the challenge of adapting to the new culture, more specifically culture shock as stated in the cross-cultural adaptation theory (Lyon, 2002). Experiencing novel types of behavior, feelings, and ways of thinking, individuals from an unfamiliar culture confront a kind of confusion and stress. Culture shock has been regarded as a negative notion in the literature (e.g., Hottola, 2004); however, as Adler (1975) claimed culture shock stimulates cultural learning and personal development. This occasion can be referred to as a positive transformation, by which an individual can reach another novel and positive identity by virtue of disintegrating (Moufakkir, 2013). Relatedly, Coghlan and Gooch (2011) regard this as the notion of disorienting dilemmas that stimulate tourists to find the missing piece of their life puzzle. Mezirow's (1991) transformative learning theory sheds light on both the dysphoria in culture in the host destination and the new awareness with novel connections with the host culture. From several perspectives the concepts of culture shock, transformative learning, and tourism have been investigated regarding the relationship between culture shock and volunteers (Coghlan & Gooch, 2011; Lee, 2020; Liu & Leung, 2019), migrants (Oberg, 1960), students (Brown, 2009; Furnham & Bochner, 1982), and sojourners (Furnham, 2019; Hottola, 2004).

Taking this background into account, this chapter aims to unveil the relation between the concept of tourists' culture shock and the theory of transformative learning in building permanent behavior, attitude, and belief change. While it reveals extend literature on subjects of culture shock experiences of tourists, its stages, symptoms, and impacts of culture shock, it concentrates on the transformative learning process of tourists in a new environment in constructing perpetual attitudes and behaviors. More specifically, in this study the following questions are addressed.

1 Does culture shock lead to a transformational behavior change in tourist behavior?
2 Does the theory of transformative learning clarify the change through culture shock?

 DOI: 10.4324/9781003161868-19

3 What are the positive outcomes of the culture shock on the transformational change in tourist behavior?
4 What are the negative outcomes of the culture shock on the transformational change in tourist behavior?

To answer these research questions, the hermeneutic approach has been utilized and the studies regarding culture shock and transformative tourism studies have been reviewed to examine the potential permanent behavior change after experiencing culture shock with a transformative impact on tourists. After reviewing the stages of both culture shock and transformative learning, they have been comparatively evaluated and the matching steps have been discussed with their potential positive and negative outcomes regard to their impacts on permanent behavior change and this process has been illustrated within a conceptual framework. As a result of the conceptual model, this study supports the positive outcome of the culture shock, mostly accepted as a negative phenomenon, in developing perpetual tourist behavior change.

Literature Review

Culture Shock

Introduced by Oberg in the 1950s, culture shock refers to a kind of state faced by persons, who enter a new and unfamiliar cultural environment resulting in the feelings of being confused, anxious, stressful, and disoriented (Furnham & Bochner, 1986; Hottola, 2004). According to Oberg (1960, p. 177), "culture shock is precipitated by the anxiety that results from losing all our familiar signs and symbols of social intercourse". The lack of familiar symbols in the new environment and commonly perceived concepts in the previous culture may cause several reflections, such as irritation, pain, panic, and even psychological illnesses (Cupsa, 2018). The new concepts at the outset do have no meaning for the newcomer and cause misunderstandings like the emotion of being deceived or ignored by the hosts (Mingli, 2015). Culture shock experience can cause both physical and psychological impacts on a persons' body and mind at different levels. Oberg (1960) refers to these symptoms as washing hands, drinking water, eating and sleeping more often than normal, avoiding contact with others, absent-mindedness, fatigue, insomnia, drug or alcohol abuse, losing temper more easily, and at last a strong yearn for home. Furthermore, some detrimental psychological impacts of culture shock can be alienation, isolation, boredom, sorrow, depression, loss of identity, and overrating home (Smolina, 2012; Xia, 2009).

Stages of Culture Shock

The adaptation to the new culture necessitates the tourist to learn the novel lifestyle, which comes off in some stages as defined in is the "U-curve theory" by Lysgaard (1955). After studying on Norwegian scholars in the United States, he concluded with a four-staged U curve of culture shock. In the first stage, called *honeymoon*, the newcomers feel euphoria, a kind of fascination and optimism to the new environment. They immediately start to learn the new language, search for similar concepts between the origin and host environment and get the assumption that everything is not so difficult in this new life. However, this assumption does not last long and the second stage starts, which is referred to as *crisis*, *shock*, or *downstage*. The newcomers in this stage commence to recognize the different aspects of this new life, which are challenging and making them feel ambiguous and hostile to the new environment. Nevertheless, in the stage of recovery, positive feelings, comfort, and familiarity to the new culture show up and the newcomer can overcome some challenges. At the last stage called *adjustment*, people feel more self-confident and personal development starts in a bicultural way (Winkelman, 1994).

Like Lysgaard, Oberg (1960) also investigated the notion culture shock and presented the course of adaptation with a U-curve model with the phases of *honeymoon, culture shock, adjustment,* and *mastery*. Furthermore, Gullahorn and Gullahorn (1963) investigated about the return experiences of several scholars and defined reverse culture shock with a W-curve model, which is illustrated as a W covering the re-versions of all these stages honeymoon, culture shock, recovery, and adjustment in the original country. What is troublesome in this re-entry is that newcomers generally expect to turn back to the same home as the same persons, which is unfortunately impossible (Wang, 1997). Later, Lewis and Jungman (1986) further investigated about the U-Curve theory and they extended it with five stages of culture shock: the preliminary phase, the spectator phase, the increasing participation phase, the adaptation phase, and the reentry into original culture.

Impact of Culture Shock

The impact of a culture shock experience is of great importance for the tourists or the newcomers of a new cross-cultural environment in that the adaptation level shapes the personal development of the individual. In the literature, these impacts have been classified into two: *disease model and growth model* (Dongfeng, 2012; Muecke et al., 2011; Pedersen, 1994; Sulaiman & Saputri, 2019). The disease model, also referred to as the psychological impact of culture shock, is mainly about the unfavorable impact of culture shock like mental diseases, emotional disorders, and at least psychological confusion, which start at the very beginning of the interaction with the new culture. Due to stress and confusion newcomers behave different from the hosts and make more mistakes, which drive them to feel culture fatigue (Jones, 1973; Solway, 2011). This, in turn, results in ambiguity, confusion, stress, frustration, anxiety, and alienation. Coping with emotional reactions consists of a different side of culture shock. The way to manage this challenge is to understand the underlying consideration of this mental and physical state. Anxiety, loss of control, and uncertainty play the main role and are to be overcome with the help of some factors, such as equality, cooperation, and interaction with locals (Lombard, 2014). The lack of family ties, limited or no communication with the host and the identity dilemma challenge the newcomers emotionally. According to this model, emotional disorders caused by culture shock emerge with the above-mentioned symptoms and effects, and could be overcome (Dongfeng, 2012). On the other hand, the growth model also called intellectual impact of culture shock can be evaluated as a positive aspect of this phenomenon. Adaptation to a new culture can be stressful, but it can have a stimulative, motivating, and creative effect on the personal development of the newcomers through competence on cross-cultural communication and cultural intelligence. The aspect of cultural intelligence, the capability to adjust competently to a strange environment is composed of physical, cognitive, emotional, and motivational elements while interpreting unfamiliar behaviors of new individuals belonging to that new environment (Earley & Mosakowski, 2004). The key point before reacting is to think and interpret, which stimulates motivation and later the physical behavior, namely, head, body, and heart. Relatedly, cultural intelligence is of great importance to more easily adapt to the new culture through acknowledging the host lifestyle and its necessities like language and social life, which, in turn, brings about adjustment, personal development, and vision in this new world. Similarly, Adler (1975) emphasizes the positive effect of culture shock on intercultural learning, inducing awareness and self-confidence. Furthermore, Furnham and Bochner (1986)'s advocations are consistent with the constructive side of culture shock claiming that social skills, norms, and roles learnt by the newcomers regarding the new culture accelerate the competence, learning, and development process. Moreover, the theory of transformative learning gives us more detailed information about this particular change and the answers of the research questions asserted in the introduction part.

Transformative Learning Theory

The transformative learning theory, first theorized as a teaching method by Jack Mezirow (1978), addresses a kind of perceptional change in the understanding and belief system of an individual, by mainly focusing on the promotion of change via a kind of critical questioning and evaluates one's own present suppositions in order to apply them into the environment (Mezirow & Taylor, 2009). To this theory, persons live through a great deal of experience like relationships, ideas, perceptions, values, and senses, which defines their environment (Mezirow, 1991). Mezirow's research was influenced by some concepts like consciousness, frames of reference, mindsets, presumptions, suppositions, and expectations, which all shape individual's acknowledging, sense, emotions, and assumptions, and later design behavior. This theory claims that individuals have a tendency to disapprove ideas, beliefs, and assumptions that do not match with their previous perception (Mezirow, 1991). Transformational learning emerges once individuals start to question their previous actions and afterwards they unexpectedly recognize that their previous dispositions and actions hamper their understanding of new environment (Elkins, 2003). At that time, they commence to adjust their ideas and actions properly to the novel necessities of the new context. All in all, the theory of transformative learning requires persons to look from a broad perspective bringing about new competences (Brookfield, 2012; Mezirow, 1991).

The research of Mezirow (1991, 1994, 2003, 2004) and Brookfield (1987) contribute to the field with the guidelines of transformative learning theory in developing new aspects for the requirements of new environments like tourism and hospitality focused in this study. Moreover, Taylor (1998) and O'Sullivan (1999) further developed Mezirow's research, with a ten-staged process. These stages are (Mezirow, 1994, p. 224),

1. Disorienting dilemma,
2. Self-examination with feelings of guilt or shame, sometimes turning to religion for support,
3. A critical assessment of assumptions,
4. Recognition that one's discontent and the process of transformation are shared and others have negotiated a similar change,
5. Exploration of options for new roles, relationships and actions,
6. Planning a course of action,
7. Acquiring knowledge and skills for implementing one's plans,
8. Provisionally trying out new roles,
9. Renegotiating new relationships and
10. A reintegration into one's life on the basis of conditions dictated by one's new perspective.

(Mezirow, 1994, p. 224)

These phases are identified in the following part of this chapter due to their importance in building the final conceptual framework in a comparative and integrative manner with the stages of culture shock in order to emphasize the effect of culture shock on tourists' permanent behavior change adoption.

Transformative Learning and Culture Shock

Culture shock is acknowledged as a kind of crisis with a negative effect on an individual, but it should be also accepted as a stimulus driving individuals to survive and adapt to the new culture with high levels of cultural intelligence and speed up their self-advancement. An overview of related literature on culture shock, transformative learning theory, and their applications on tourism includes several studies. For example, Walter (2016) studied on three different disorienting dilemmas in community-based ecotourism in Southeast Asia as encouraging elements in tourists' transformative learning process. Müller et al. (2020)'s study investigated volunteer tourism and related critical events leading to transformative learning in social-linguistic, psychological, and moral-ethical perspectives. They also suggested the contribution of volunteer tourism on career sustainability. Similarly, Coghlan and Gooch (2011) studied volunteer tourism and ten phases of transformative learning theory and concluded that volunteer tourism could create opportunities for tourists promoting the transformative process. Stone and Duffy (2015) also reviewed related literature about the transformational learning and travel and tourism. More recently, Pung et al. (2020) evaluated tourist transformation and discussed key dimensions, suggesting that liminality, cultural shock, and difficulties experienced in the new environment inaugurate transforming through contextual stimuli, which make tourists to evaluate their experience and get competence.

The conceptual model in this chapter mainly focuses on both the stages of culture shock and the phases of transformative learning process and their intersections. Even though transformative learning is a broad concept and culture shock is a temporary or not a life-long process, experienced in a new environment; transformative learning can be evaluated as a learning process in one's life, the process during which a tourist overcomes the crisis and ends up with a transformation and a new point of view. Thus, the heart of the study is how culture shock cherishes transformational learning process with its challenges and unexpected gifts.

Drawing on this background, this study proposes a conceptual framework of tourists' transformative culture shock experiences, including six stages: fancy, disorienting dilemma, self-evaluation, liminality, coping strategies, and transformation (see Figure 19.1). Before illustrating the conceptual model, it is of importance to compare and discuss both the stages and clarify the similarities, differences, and connections. As known from Lysgaard (1955)'s U-curve model of culture shock, the first stage is called honeymoon, when tourists are fascinated with the new setting. With high levels of optimism, they right away get ready to learn the language heavily spoken there and see the similarities between home and host. They fancy that everything seems not so challenging as said in this new environment with a high imagination of success and wealth. Hence, the first step is labeled as fancy in the proposed model. As an example, you look at the fabulous buildings of an old city with a long history and every passer-by greets you with a sincere smile and this makes you very engaged in that new city. However, this first fancy stage does not take too long and it ends up with a disorienting dilemma. Imagine that a police officer approaches and gives you a fine for wrong parking of your rented bike or car because there are different and strange rules in this country. The fancy ends and here starts the real story for the tourist.

As Mezirow (1990) stated, transformative learning is precipitated with a disorienting dilemma exposed externally. Disorienting dilemmas, also referred to as trigger events in the literature, are disturbing occurrences or events that an individual goes through and right after the necessity to cease and think dawns on that individual (Taylor, 1994). There is an abrupt response to this trigger event, which is a milestone for a person exposed to culture shock in his/her adjustment to a new environment (Furnham, 2019). For example, as a previous method which works at hometown, you try to say that you were there for just a couple of seconds, but the police officer does not even listen to you. This dilemma between home and host makes tourist stop to think. Thus, the second stage of the model is named as disorienting dilemma.

With that shock, the tourist starts a kind of self-examination together with the emotions of guilt or shame as Mezirow (1994) stated. In this third stage, tourists commence questioning on their own whether they have carried out a wrong behavior and have a guilty conscience about the situation. At the same time or immediately after, they assess the situation critically and make a judgment for reaching a conclusion, and thus this stage is called as self-evaluation in the proposed model.

At the fourth stage, the tourist recognizes that s/he is discontented with the situation and discusses with others who have gone through similar processes regarding this feeling change.

The fifth stage of Mezirow (1994)'s theory and the third stage of Lysgaard (1955)'s U-curve approach, named crisis/down stage coincide with this fourth stage of the conceptual framework, which is then called as liminality stage. It is referred to in this way because the tourist bottoms out in this new environment and hereafter s/he starts to think that s/he is to transform due to the fact that previous knowledge does not work anymore.

Following the liminality, the stage of recovery according to U-Curve approach begins. As to transformative learning theory, the processes of exploring novel role alternatives, within relationships; planning a course of action; and obtaining knowledge and competence to carry out the plans are the coping strategies to overcome the crisis. This phase includes discovering new and different ways of figuring the crisis out, which causes stress for the tourist. Therefore, it is referred to as the stage of finding coping strategies in the model. Lastly, the tourist starts to adjust to the new culture through temporary trials of new roles, discussing new relationships, and integration into a new way of life from a new perspective. This new way in time becomes a part of tourist's behaviors and actions. Thus, this last stage is called this framework as transformation.

Conclusion

This chapter aims to investigate the relationship between tourists' culture shock experiences and the role of transformative learning in the process of adjustment to a new cultural setting and permanent behavioral change. By doing a critical review of culture shock and transformative learning process of tourists with a hermeneutic approach, a conceptual framework proposing an integrative approach, has been developed to explain the transformative process of tourists' culture shock experiences. This model has discussed and illustrated the key stages and the possible outcomes of tourist transformation, with an important contribution to the knowledge on both tourist adjustment and transformative learning. Lysgaand (1955)'s U-curve approach regarding to the adjustment of a new society and Mezirow (1991)'s transformative learning theory has added to the knowledge regarding tourist transformation, as they scrutinize the authentic experience of tourists in a novel setting. In the light of these two approaches, the inquiries of whether culture shock leads to a transformational behavior change in tourist behavior, the theory of transformative learning clarifies the change through culture shock, as well as the positive and negative outcomes of the culture shock on the transformational change in tourist behavior have been investigated. Thus, this chapter has contributed to tourist transformation through framing the main concepts of transformative tourist adjustment in the proposed model with the dimensions of fancy: the first fascinating stage of tourist encounter with the new cultural environment, followed by a disorienting dilemma, confusing the tourist's mind with a trigger event; liminality through recognizing the depth of the process with a crisis and exploring new roles; finding coping strategies for recovery; and lastly the transformation embodying adjustment of the tourist with perpetual behavioral change. This study supports the positive outcome of the culture shock, mostly acknowledged as a negative situation in constructing perpetual tourist behavioral change.

This study recommends practitioners to take the constructive transformative experiences tourists can encounter during their stay into consideration in their strategical decision-making processes like branding of a destination. The more support is given by the destination practitioners to the tourist to recover from their crisis stage with high liminality, the easier a tourist adjusts to that new environment. It can be achieved with some encounters, such as learning the lifestyles of local people, the regional language, and consultation with the newcomer about the disorienting dilemmas and challenging situations faced.

The study limitations include first the methodology of hermeneutic approach in that it could not contain all the reasons and outcomes of transformation. It has only investigated the notion of culture shock and its effects on tourist behavior. Second, the final proposed conceptual framework

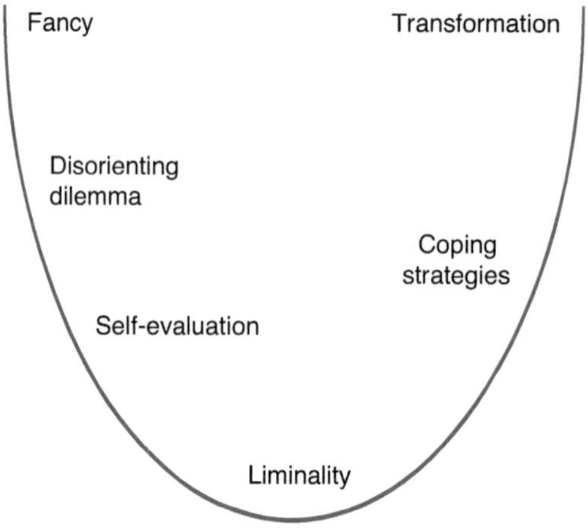

Figure 19.1 Proposed Conceptual Framework of Transformative Process of Tourist Culture Shock.

with the evaluation of the related literature has focused on the subjective interpretations. Lastly, this chapter has investigated the transformation of a general tourist; thus, future studies can concentrate on specific tourist types like scholars temporarily abroad for research.

References

Adler, P. S. (1975). The transitional experience: An alternative view of culture shock. *Journal of Humanistic Psychology, 15*(1), 13–23.

Brookfield, S. D. (1987). *Developing critical thinkers: Challenging adults to explore alternative ways of thinking and acting*. Jossey-Bass.

Brookfield, S. D. (2012). Critical theory and transformative learning. *The handbook of transformative learning: Theory, research, and practice*, 131–146.

Brown, L. (2009). The transformative power of the international sojourn: An ethnographic study of the international student experience. *Annals of Tourism Research, 36*(3), 502–521. https://doi.org/10.1016/j.annals.2009.03.002

Coghlan, A., & Gooch, M. (2011). Applying a transformative learning framework to volunteer tourism. *Journal of Sustainable Tourism, 19*(6), 713–728. https://doi.org/10.1080/09669582.2010.542246

Cupsa, I. (2018). Culture shock and identity. *Transactional Analysis Journal, 48*(2), 181–191. https://doi.org/10.1080/03621537.2018.1431467

Dongfeng, L. I. (2012). Culture shock and its implications for cross-cultural training and culture teaching. *Cross-Cultural Communication, 8*(4), 70–74. http://dx.doi.org/10.3968/j.ccc.1923670020120804.1433

Earley, P. C., & Mosakowski, E. (2004). Cultural intelligence. *Harvard Business Review, 82*(10), 139–146.

Elkins, J. (2003). *Visual studies: A skeptical introduction*. Psychology Press.

Furnham, A. (2019). Culture shock: A review of the literature for practitioners. *Psychology, 10*(13), 1832–1855. https://doi.org/10.4236/psych.2019.1013119

Furnham, A., & Bochner, S. (1982). Social difficulty in a foreign culture: An empirical analysis of culture shock. *Cultures in Contact: Studies in Cross-cultural Interaction, 1*, 161–198.

Furnham, A., & Bochner, S. (1986). *Culture shock. Psychological reactions to unfamiliar environments*. Methuen & Co. Ltd.

Gullahorn, J. T., & Gullahorn, J. E. (1963). An extension of the u-curve hypothesis 1. *Journal of Social Issues, 19*(3), 33–47. https://doi.org/10.1111/j.1540-4560.1963.tb00447.x

Hottola, P. (2004). Culture confusion: Intercultural adaptation in tourism. *Annals of Tourism Research, 31*(2), 447–466. https://doi.org/10.1016/j.annals.2004.01.003

Jones, D. J. (1973). Culture fatigue: The results of role-playing in anthropological research. *Anthropological Quarterly, 46*(1), 30–37. https://www.jstor.org/stable/3694292

Lee, H. Y. (2020). Understanding community attitudes towards volunteer tourism. *Tourism Recreation Research, 45*(4), 445–458. https://doi.org/10.1080/02508281.2020.1740959

Lewis, T. J., & Jungman, R. E. (1986). *On being foreign: Culture shock in short fiction: An international anthology.* Intercultural Press.
Liu, T.-M., & Leung, K.-K. (2019). Volunteer tourism, endangered species conservation, and aboriginal culture shock. *Biodiversity and Conservation, 28*(1), 115–129. https://doi.org/10.1007/s10531-018-1639-2
Lombard, C. A. (2014). Coping with anxiety and rebuilding identity: A psychosynthesis approach to culture shock. *Counselling Psychology Quarterly, 27*(2), 174–199. https://doi.org/10.1080/09515070.2013.875887
Lyon, C. R. (2002, May 24–26). Trigger event meets culture shock: Linking the literature of transformative learning theory and cross-cultural adaptation. *The 43rd annual meeting of the adult education research conference,* Raleigh, NC. https://files.eric.ed.gov/fulltext/ED472072.pdf
Lysgaard, S. (1955). Adjustment in a foreign society: Norwegian Fulbright grantees visiting the United States. *International Social Science Bulletin, 7,* 45–51.
Mezirow, J. (1978). Perspective transformation. *Adult Education, 28*(2), 100–110. https://doi.org/10.1177/074171367802800202
Mezirow, J. (1991). *Transformative dimensions of adult learning.* Jossey Bass.
Mezirow, J. (1994). Understanding transformation theory. *Adult Education Quarterly, 44*(4), 222–232.
Mezirow, J. (2003). Transformative learning as discourse. *Journal of Transformative Education 1*(1), 58–63. https://doi.org/10.1177/1541344603252172
Mezirow, J. (2004). Forum comment on Sharan Merriam's "The role of cognitive development in Mezirow's transformational learning theory". *Adult education quarterly, 55*(1), 69–70.
Mezirow, J., & Taylor, E. W. (Eds.). (2009). *Transformative learning in practice: Insights from community, workplace, and higher education.* John Wiley & Sons.
Mingli, W. A. N. G. (2015). Culture shock-one of common problems in intercultural communication. *Cross-Cultural Communication, 11*(8), 72–75. http://dx.doi.org/10.3968/7399
Moufakkir, O. (2013). Culture shock, what culture shock? Conceptualizing culture unrest in intercultural tourism and assessing its effect on tourists' perceptions and travel propensity. *Tourist Studies, 13*(3), 322–340. https://doi.org/10.1177/1468797613498166
Muecke, A., Lenthall, S., & Lindeman, M. (2011). Culture shock and healthcare workers in remote Indigenous communities of Australia: What do we know and how can we measure it?. *Rural and Remote Health, 11*(2), 1–13. https://doi.org/10.22605/RRH1607
Müller, C. V., Scheffer, A. B. B., & Closs, L. Q. (2020). Volunteer tourism, transformative learning and its impacts on careers: The case of Brazilian volunteers. *International Journal of Tourism Research, 22*(6), 726–738. https://doi.org/10.1002/jtr.2368
Oberg, K. (1960). Cultural shock: Adjustment to new cultural environments. *Practical Anthropology,* 7(4), 177–182.
O'Sullivan, E. (1999). *Transformative learning. Educational Vision for the 21st Century.* Zed.
Pedersen, P. (1994). *The five stages of culture shock: Critical incidents around the world.* Greenwood Press.
Pung, J. M., Gnoth, J., & Del Chiappa, G. (2020). Tourist transformation: Towards a conceptual model. *Annals of Tourism Research, 81,* 102885.
Smolina, T. L. (2012). Symptoms of culture shock: Overview and classification. *Psychological-Educational Studies,* 4(3), 1–107.
Solway, J. (2011). "Culture Fatigue": The state and minority rights in Botswana. Indiana *Journal of Global Legal Studies, 18*(1), 211–240. https://doi.org/10.2979/indjglolegstu.18.1.21
Stone, G. A., & Duffy, L. N. (2015). Transformative learning theory: A systematic review of travel and tourism scholarship. *Journal of Teaching in Travel & Tourism, 15*(3), 204–224. https://doi.org/10.1080/15313220.2015.1059305
Sulaiman, M., & Saputri, K. (2019). Culture shock among foreign students: A case-study of Thai Students studying at Universitas Muhammadiyah Palembang. *English Community Journal, 3*(1), 295–306. https://doi.org/10.32502/ecj.v3i1.1695
Taylor, E. W. (1994). Intercultural competency: A transformative learning process. *Adult Education Quarterly, 44*(3), 154–174. https://doi.org/10.1177/074171369404400303
Taylor, E. W. (1998). *The theory and practice of transformative learning: a critical review.* Information Series No. 374.
Walter, P. G. (2016). Catalysts for transformative learning in community-based ecotourism. *Current Issues in Tourism, 19*(13), 1356–1371. https://doi.org/10.1080/13683500.2013.850063
Wang, M. M. (1997). Reentry and reverse culture shock. *Multicultural Aspects of Counseling Series, 1,* 109–128.
Winkelman, M. (1994). Cultural shock and adaptation. *Journal of Counseling & Development, 73*(2), 121–126.
Xia, J. (2009). Analysis of impact of culture shock on individual psychology. *International Journal of Psychological Studies, 1*(2), 97.

20

TOURIST-TO-TOURIST INTERACTION (TTI)

A Social Distance Perspective

Seda Sökmen and Medet Yolal

Introduction

The centre of tourism constitutes encounters as tourism makes it possible for the tourist to encounter other people, places, objects, or meanings. Encounters and interactions of tourists with other people, in particular, are very important in terms of giving the meaning of the similarities and differences between the parties of tourism (Gillen & Mostafanezhad, 2019). As a matter of fact, the determinants of the perceptions, attitudes, behaviours, and experiences of the parties are these similarities and differences as such that these have the potential to influence even the development of tourism in a destination. The members of the local people are the first of the people that the tourist encounters during the realisation of the tourism activity and the most frequently mentioned in the literature. Early studies addressing the host-guest encounters have focussed on the structure of this encounter and emphasised that the interaction between tourists and local people is transitory and that perceptions of both parties regarding a particular encounter may be different from each other (Sharpley, 2014). Another type of encounter takes part between tourists and service providers in the tourism sector. Customer-to-customer interaction (CCI), which started to be mentioned in the tourism field in the early 1980s, is a very important matter particularly for the field of service marketing in terms of affecting satisfaction and loyalty (Wu, 2007). The last one of the types of encounters but the main subject of this section is the encounters that tourists have with other people in the role of tourists like themselves.

Tourist-to-tourist interaction (TTI) is the encounter that takes part at the level of individual-individual, individual-group, or group-group interacting directly or indirectly throughout the tourism activity (Chang, 2017). TTI is affected by individual characteristics and environmental conditions. For instance, the social identities of tourists can be attributed to individual characteristics, which is an important factor as it can determine the course of interaction in TTI. Environmental conditions can be exemplified through a music festival. Hence, TTI can be affected by the preferences of the participants of the festival regarding the musical style, as well as by the social or physical environment of the festival. The situation is similar for backpackers, cruise ship customers, or group tour participants. From another standpoint, TTI can take place at different levels in restricted spaces such as hotels. In their study, Bosio and Lewis (2008) claimed that TTI occurs in hotels in three different ways: the exchange of greetings and pleasantries, engaging in mutual moans, and offering physical and informational assistance. The general social structure of the group or the proximity to the group, and the quantity and physical proximity of other tourists in terms of environmental conditions, are also among the effective factors (Sun et al., 2019).

 DOI: 10.4324/9781003161868-20

TTI studies have started to become a current state with the activities in the field of recreation. The first studies carried out in the context of TTI are the studies that draw attention to the effect of the number of people in recreational activities on the satisfaction levels of the participants (Pearce, 2005; Yin & Poon, 2016). The presence of other tourists in cruise ship travel, group tours, or festivals where tourists have to share limited space and time with others is not only inevitable but also indispensable. Therefore, studies on the causes, consequences, and effects of TTI provide important contributions to the tourism literature. The gathering of tourists with others during tourism activities can be addressed from the perspectives of different disciplines. Nevertheless, both complex and deep, and rich meanings are reached when trying to interpret TTI from the perspective of social psychology. Tourism studies, guided by different social psychology theories and conceptual frameworks, have only just started to evolve (Tang, 2014). In this context, social distance is also an important phenomenon that can make it possible to give meaning to TTI.

Studies using social distance concepts have reached a certain saturation in tourism studies focussing on host-guest relations. Nevertheless, it is seen that very few studies have referred to the concept of social distance to make sense of the tourist-tourist interaction (Adam et al., 2020; Bai & Chang, 2021). This chapter, written to draw attention to this gap, will discuss the similarities and differences that underlie many theories in social psychology. The relationship between ethnicity, culture, and others; parameters that determine similarities or differences; and the concept of social distance will be explained, and it will be discussed whether tourists can see the similarities between them due to social distance (Ahmed, 2007; Jenkins, 2014; Johnston, 2001). Then, the concept of social distance will be associated with Allport's contact hypothesis, thus establishing the theoretical and conceptual foundations for TTI. This theory is essential in understanding the concept of social distancing as it formalises the belief that through contact, people will get to know and understand each other better in the mid-twentieth century (Katz, 1991; Tomljenovic, 2010). In this sense, prejudices that are seen as the main reason for social distance, the conflict that may occur after the increase in the social distance between tourists, hostility, and concepts such as peace and empathy may arise with the decrease of social distance also be included (Çelik, 2019a; Farmaki, 2017).

Progress of Social Distance in TTI Research

In the 1920s, when the concept of social distancing was first introduced, it meant the degree of understanding and intimacy that characterised person-to-person or intergroup relationships or between a person and a group (Wark & Galliher, 2007; Yilmaz & Tasci, 2013). Robert Park, a student of Georg Simmel and who used this concept for the first time, asked Emory Bogardus to design a quantitative indicator of social distancing (Bogardus, 1933; Harvey, 1987; Park, 1924). In 1925, the first form of the Social Distance Scale was developed by Bogardus, thus pioneering the statistical measurement in the field of race and ethnic relations. Bogardus (1925) used the Social Distance Scale every ten years until 1966, following the evolution of America for 40 years (Bogardus, 1967), thus introducing one of the most famous social psychological measurement tools to the literature. Not only in sociology or social psychology but also in studies conducted by researchers from many different fields and countries in different languages, this scale was used in contexts such as race, nationality, and culture (Joo et al., 2018; Thyne & Zins, 2003; Wark & Galliher, 2007; Weinfurt & Moghaddam, 2001). One of these fields in question is tourism, where there are many encounters at the individual-individual, individual-group, and group-group levels.

The first studies to explore the explanatory power of the concept of social distance in tourism belong to Thyne and Lawson (2001) and Thyne and Zins (2003). In most of the studies that followed these pioneering studies, the concept of social distance was used to understand the attitude and behavioural tendencies in host-guest encounters. For example, Thyne et al. (2006) conducted an experimental study that questioned the level of New Zealand residents accepting tourists from Australia, Germany, and the USA. The main finding of this study is that the more culturally different and socially distant a tourist is to the residents, the less accepted by them. In another study conducted by Sinkovics and Penz (2009), it was suggested that social distancing might reveal the behavioural intentions of the hosts, and it was found that Austrian residents are more willing to interact with German tourists than Japanese tourists. In Tasci's (2009) study, which he carried out on tourists instead of the local community, he investigated the social distance perceptions of his students in the USA as potential tourists from a different perspective. This study, which was conducted by controlling the visual stimuli and the country name, was focussed on virtual contact established through visual information, and it was found that this contact could reduce the perceived social distance of tourists towards a destination. Unlike the others, the study conducted by Carson et al. (2013) has shown that the concept of social distance can be used not only in international tourism activities but also in host-guest encounters in domestic tourism activities. This study is about the group of tourists called long grassers, who are seen as problem tourists in Australia, and their anti-social behaviour in public places is also one of the first studies to address tourist-tourist interaction in the context of social distance. The study results are also significant in drawing attention to strategies that may be useful to reduce social distance.

As the number of studies on social distance in tourism literature increases, it is possible to say that these studies gain more and more depth. For example, the study conducted by Yilmaz and Tasci (2013) in Muğla revealed that social distance was associated with both quantitative and qualitative determinants of host-guest encounters. It was determined by this study that the perceived social distance levels of tourists who have close friendships with local service providers and who have visited Turkey several times before decrease. In the study conducted by Nyaupane et al. (2015) as another example, it was emphasised that apart from culture and nationality, beliefs can also be a determining factor in social distance. It was determined that the social distance between people of the same religion is less, and the social distance between people of different religions is more significant, with this study conducted on Buddhists, Hindus, and Christians. Shtudiner et al. (2018), the Jerusalem version of this study in Lumbini, which also examines beliefs in the context of social distance, first asked the tourists which of the categories they considered themselves to belong to within the scope of the research. 200 out of 1776 Jews surveyed described themselves as ultra-Orthodox, 994 Orthodox, and 496 seculars. According to the findings of the study, the highest level of social distance towards tourists belongs to members of the ultra-Orthodox category. The researchers note that this finding is not surprising because ultra-Orthodox residents live in separate neighbourhoods, attend different schools, and often lead a closed life within themselves.

The prevailing view in these studies, which interpret the host-guest interaction from a social distance perspective, is as follows: Local communities are friendlier towards tourists who are more culturally/racially similar. On the other hand, in the study conducted by Thyne et al. (2018), contrary findings were reached. In the study conducted on Japanese residents, it was revealed that the local community attributes less social distance to tourists from the USA and Australia than to tourists from China and Taiwan. The fact that the Chinese have the highest perception of social distance between Asian and Western nations is a significant finding in the literature, and it is estimated by the researchers that there may be many underlying reasons for this finding. Likewise, in the study conducted by Yankholmes and Timothy (2017), findings contrary to the prevailing view were obtained. The findings obtained in this study, which investigates the social distance between the residents of Ghana and Afro-Americans who have settled in Ghana since the 1960s, are pretty remarkable for the tourism literature. When Afro-American root seekers visit Ghana, they are not welcomed as diaspora brothers or even considered foreigners, even if their intention is not just to visit but to live in Ghana. The reason for this, in other sayings, the social distance between expatriates and locals, is shown as the excessive attachment of expatriates to their slave heritage areas and overemphasis on slavery. In his work, which addresses the concept of emotional solidarity and contact theory, together with the concept of social distance, Joo et al. (2018) has investigated what kind of interaction enhances emotional solidarity and reduces social distance and how, in turn, emotional solidarity affects social distance. With this study conducted on USA residents and evaluated in the context of domestic tourism, it was understood that the different types of activities that residents do with tourists explain both emotional solidarity and social distance. Çelik's (2019a) study, which is another current tourism study, investigated the host-guest interaction in the context of domestic tourism, and in consequence of the study, it was determined that the social distance perceptions of the local tourists visiting the provinces of Şanlıurfa and Gaziantep decreased due to tourism. Aleshinloye et al. (2020), who analysed the relationality of social distance phenomenon and another phenomenon (place attachment), as suggested by Joo et al. (2018), investigated the extent to which place attachment affects tourists' perception of social distance. The main finding of this study is that emotional solidarity mediates the relationship between place attachment and social distance. Another critical study in the literature, investigating the relations of residents with each other with the help of social distance, was carried out on four different local communities in China. Chen et al. (2020), who developed a 16-item scale consisting of three dimensions in consequence of this study, claimed that familiarity, degree of interaction, and degree of support are predictors of a healthy social distance. In all of the studies mentioned so far, social distance has been considered a reason for host-guest interaction, but what may result from a social distance has never been investigated. In the study conducted by Thyne et al. (2020), it was noted that the cause-effect relationship in question could also operate in the opposite direction. This study builds on the study of Thyne et al. (2018), showing that Japanese residents' attitudes towards tourism development can be a determinant of their perceived social distance from international tourists. The existing literature in this field expanded it by focussing on how tourism attitudes can predict social distancing.

Placing the TTI between Similarity and Difference

Since the first studies on social distance, it was claimed that the determinants of social distance are race, social class, religion, nationality, and so on. These factors are mainly based on similarities and differences. As it is known, when a person tries to identify him/herself or someone else, to define him/herself and the other, he/she sets out from similarities and differences (Hsu & Chen, 2019; Jenkins, 2014). Identity contains both the aspects that we are different from others and the same aspects of us as others (Lawler, 2015). Hence, in groups that we belong to with our cultural, ethnic, national, and similar identities, we leave by constructing other identities or reinforcing our existing identities. It is assumed that these similarities and differences inevitably affect the social distances of individuals or groups from each other (Ahmed, 2007; Johnston, 2001). As the Chinese search for similar cultural characteristics and lifestyles in choosing a destination, for example, the more cultural similarities between a tourist's homeland and the destination he/she will visit, the higher the probability of the tourist to choose that destination (Wei et al., 2021).

On the contrary, cultural differences rather than similarities can also be the determinant of destination selection (Ng et al., 2007). To exemplify with another study, Özdemir and Yolal (2017), who describe the behaviours of American, French, German, English, Italian and Spanish tourists during the tour through the eyes of tourist guides, were stated that cultural similarities and differences are the determining factors in the interaction of these tourist groups with each other. In this sense, it is clear that tourists evaluate other tourists according to whether they are similar or different from themselves.

The origins of tourist-tourist interaction studies go back to the recreational research carried out in North America in the 1980s (Pearce, 2005). Nevertheless, in these early studies, the concept of social distance was not included. As Triandis and Triandis (1962) pointed out, although many studies are about social distancing, researchers have not chosen to use this concept. For example, Yagi (2001) investigated TTI preferences based on nationality in the study conducted on the concept of a familiar stranger. Familiar strangers are tourists that a tourist encounters repeatedly in many tourism environments, especially on long trips and act closer than the tourists they know, even though they have never met. Thyne et al.'s (2018) findings on social distancing correspond to Yagi's (2001) finding that Japanese people may prefer familiar strangers to tourists of their nationality. In the first studies in the field of recreation, it is stated that tourist reactions in tourist-tourist interaction can vary depending on both the environment and the number of other visitors present. As can be seen from here, tourists can positively or negatively affect the destination experience and recreational experience of other tourists (Guthrie & Anderson, 2007; Yagi & Pearce, 2007).

Some of the TTI studies are also related to the service experience (Lin et al., 2019). For example, Wu's (2007) suggestion that travel companies can group customers with similar characteristics and actively manage customer-customer interaction by presenting an appropriate orientation and code of conduct before the trip begins in terms of service experience is an essential suggestion about social distance because the homogenisation mentioned by the researcher means reducing social distance. Considering TTI from another perspective, Reichenberger (2017), as for that, focussed on the collaborative creation of experiences among visitors. In consequence of the increasing number of visitors, according to the researcher, the increase in the use of infrastructure and common areas leads to more social contact, and co-creation is seen as a solution at this point. In the study, which accepts that not all visitors will be in the same affinity to this interaction, three interaction levels were determined among tourists: *communitas* refers to the feeling of togetherness among tourists sharing an experience; social bubble refers to the social practice among tourists who have met before; detached tourist refers to not contacting unfamiliar social actors. Indeed, these three levels of interaction are due to the differentiation of social distance perception among tourists.

During group tours, festivals, backpacking, or cruising, where TTI is intense, also make significant contributions to the literature (Bui & Wilkins, 2016; Huang & Hsu, 2009; Murphy, 2001; Papathanassis, 2012; Sun et al. 2019; Torres, 2015; Yin & Poon, 2016). For example, in their study of the interaction between cruise passengers, Huang and Hsu (2009) reveal that three situations can occur regarding the interaction between passengers: a situation where there is no or minimal interaction with other passengers, a situation where you interact with other passengers in a spontaneous and fun way, and a friendship situation where close interaction with other passengers goes beyond the cruise. A similar classification is made in the festival context, and it is stated that three different types of interaction can be experienced among tourists, namely entertainment, mutual aid, or conflict (Sun et al., 2019). Regardless of festival participants or cruise tourists, the basis of all these classifications is the social distance that tourists perceive towards each other. The study conducted by Bai and Chang (2021), which shows that philosophical beliefs can also be considered a determinant of social distance in the context of TTI, states that Confucian values may be practical on the social distance perceptions of Asian tourists. Researchers who stated that these values make Asian tourists more compatible in intergroup interaction assumed that tourists with a Confucianism-oriented background would not be interested in seeing or observing other tourists but would acknowledge their presence. Although Taiwan and Chinese tourists with this orientation were addressed in the study, it was found that Chinese tourists act following the assumptions in question, while Taiwanese tourists are more inclined to conflict. The findings of this study correspond to the findings of the study by Yagi and Pearce (2007), which revealed that the Japanese are not bothered by being outnumbered by other tourists and can tolerate other tourists at a higher level than Western tourists.

Interaction between tourists is seen as positive when it rewards and enhances experiences and negatively when it harms experiences. In the study conducted by Adam et al. (2020), in which the dimensions of the negative TTI were determined, these dimensions were divided into four: (1) negative interpersonal directed TTI, (2) negative interpersonal non-directed TTI, (3) negative site directed TTI, and (4) negative interpersonal interaction. An important conclusion is reached from these dimensions: Negative TTI is not always initiated deliberately to do away with other tourists' values. For example, while in the first dimension, an intentionally directed interaction with other tourists in the same consumption area is expressed (e.g. abusing or pushing and jumping of queues), this situation changes in the second dimension. Here, because tourists consider their behaviour as a vacation right for themselves (smoking, interrupting tour guides constantly and unnecessarily, taking photos without permission, etc.), interaction is not directed at other tourists or attractions. In the study, findings on European and North American discomfort with African chatter have been interpreted in culture and associated with differences between individual-collective societies. At this point, the importance of the concept of social distance in TTI interaction comes to the fore.

In the study conducted by Carson et al. (2013), it is argued that some of the problematic tourists arise as a result of the inability to reduce the social distance between different tourist groups. However, to reduce social distance, it must first be understood correctly. At this point, the intratourist gaze concept proposed by Hollaway et al. (2011) provides an essential framework for researchers. Intratourist gaze developed based on Urry's (1990) tourist gaze is a functional concept that expresses the determination of acceptable and unacceptable behaviour in TTIs and the discipline/regulation of the behaviour of other tourists within this framework. For example, backpackers often see themselves as representatives of a better form of tourism, and they separate a backpacker (us) from a tourist (other) (Sorensen, 2003). Similarly, the study conducted by Schwarz (2018) on volunteer tourists found that tourists can make judgments about the behaviour of other tourists by taking into account the putative results (stereotypes) without first-hand observation. Understanding this concept, which states that other tourists are critically evaluated and separated from the group, is critical for reducing social distancing in TTI. As can be seen from the studies in the literature, for tourist groups to realise the similarities between each other, it is crucial to reduce the social distance they perceive towards each other first. Nevertheless, the critical precondition for this is primarily social contact between groups.

Rethinking TTI through Social Contact and Social Distance

The subject of intergroup social contact first came to the fore with the Contact Hypothesis, which Allport (1954) put forward, and this is a critical theory for social psychology. This attempt of Allport is considered the starting point in the formation of social contact theories. Subsequently, Pettigrew (1998) developed the intergroup contact theory based on the claims of the contact hypothesis and, in a way, laid the groundwork for new social contact theories. Thus, imagined contact theory (Turner et al., 2007) was suggested with the extended contact theory (Wright et al., 1997). Social contact theories have begun to be used in the social psychology literature to regulate intergroup relations positively. While intergroup attitudes become more positive with social contact with these theories, it is assumed that negative attitudes will also decrease.

The central assumption in Allport's (1954) contact hypothesis is that the lack of information about people from the other group leads to prejudice, and it is estimated that this bias will decrease if people from different groups get to know each other and gain information. However, certain conditions must be met for this: Equal status, sharing the same purpose, cooperative relationship, and authority support. If these conditions are met, positive attitudes towards the outgroup can be increased. Pettigrew (1998), as for that, proposed the intergroup contact theory by questioning the explanatory power of Allport's (1954) hypothesis about the psychological processes underlying the contact-bias relationship. Besides, Pettigrew and Troop (2006) report that Allport's (1954) conditions were not required in positive contact situations, but might be a requirement in negative contact situations. Extended contact theory claims that recognising an ingroup member who has contact with an outgroup member, even without actual social contact, has a bias-reducing effect (Wright et al., 1997). Imagined contact theory is proposed for situations where there is no contact between outgroup members and includes an imaginary interaction to improve intergroup relations (Turner et al., 2007). The mentally constructed form of social interaction involving outgroup members is called imaginary contact. In contact studies, participants are asked to imagine having had positive contact with anyone from the outgroup. It is ideal for groups that do not live in the same country or groups that live in the same country, but are subject to discrimination due to specific characteristics.

Although the mentioned new contact types are not encountered in the tourism literature, some studies use the contact hypothesis and intergroup contact theory and mainly deal with host-guest relationships (Fan, 2020; Pearce, 2005). These theories play an important role in understanding the relations between residents and tourists, attitudes, and behaviours towards each other (Fan et al., 2017a; Fan et al., 2017b; Nyaupane et al., 2008; Pizam et al., 2000). For this reason, it is seen that social contact theories are applied in studies investigating the attitude changes of residents or tourists due to the tourism phenomenon (Çelik, 2019b; Ming, 2018; Paris et al., 2014). From the studies conducted in both tourism and social psychology, it can be deduced that if there is positive social contact between the residents and the tourists, the residents have a positive attitude towards tourism development and the levels of understanding and empathy of the groups towards each other increase. Otherwise, it can be deduced that the social conflict, anxiety, insecurity in the destination intensifies, and social tolerance and acceptance decrease (Carneiro et al., 2018; Page-Gould et al., 2008; Pizam et al., 2000; Vezzali et al., 2012).

In the study by Fan (2020) emphasising the importance of social contact theory for tourism research, attention was drawn to the antecedents of social contact in tourism based on the fundamental assumption in Allport's (1954) contact hypothesis. Fan (2020) divided the antecedents of social contact, which he determined based on leisure and tourism behaviour restrictions, into three categories: internal, interpersonal, and structural barriers. Internal barriers describe individual psychological states and qualities related to an individual's preferences (personality, language, purpose of travel, personal role in travel). Interpersonal barriers are the result of interpersonal interaction or the relationship between characteristics of individuals (ingroup contact, stereotype, discrimination, common goals, intergroup cooperation, personal interaction). Structural barriers refer to environmental factors between behavioural choice and active participation (equal status, support of authorities, travel mode, length of stay, cultural differences, cultural/political sensitivity, destination maturity, types of attraction, destination security). To exemplify the importance of social contact precursors, the fact that North Korean residents feel threatened when social contact with tourists due to restrictions (social isolation between residents and tourists in areas such as hotels, shops, restaurants, and train compartments) in the country has caused them to avoid contact with Chinese tourists (Li & Wang, 2020). This finding will be more accurately interpreted when considering the antecedents of social contact rather than from a similarity/difference perspective.

A conceptual model was designed by the researchers based on the work of Fan (2020) and Li and Wang (2020), which emphasises the antecedents of social contact and the positive or prejudiced changes such as empathy and understanding, and adverse changes such as hostility (see Figure 20.1).

The general opinion in most TTI studies mentioned in the previous title is that the decreased perception of social distance supports positive attitude change among tourist groups, and the increased perception of social distance causes negative attitude changes (Bai & Chang, 2021; Huang & Hsu, 2009; Özdemir & Yolal, 2017; Reichenberger, 2017; Sun et al., 2019). In this sense, intergroup social contact is a prerequisite for the perception of social distance and an influential

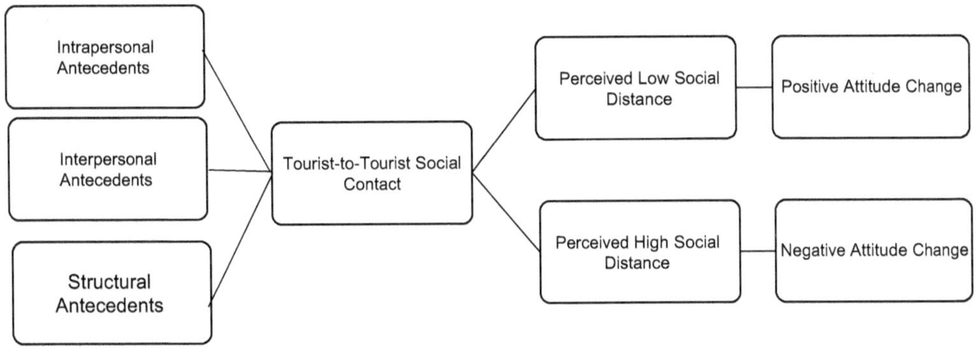

Figure 20.1 Tourist-to-Tourist Social Distance.
Source: Adapted from the Fan (2020) and Li and Wang (2020).

factor in reducing or increasing the level of perceived social distance (Triandis & Triandis, 1960; Segall et al., 1999). Yilmaz and Tasci (2015) stated that increased social contact reduces perceived social distance in host-guest relationships and Joo et al. (2018) found that the frequency of social contact and the types of activities performed during contact are associated with approach and avoidance in the context of social distancing; these results show that the relationships in the model are established in the right direction. It is also true that there are still many aspects of social distancing in TTI that remain to be explored.

Conclusion

Social contact theories and the concept of social distance, which constitute the theoretical and conceptual structure of tourist-to-tourist interaction, are significant in understanding the complex TTI. Nevertheless, the explanatory power of new contact theories has not yet been discovered in the field of tourism. For example, imagined contact theory could be applied when international tourist groups who do not live in the same country cannot make face-to-face contact due to various restrictions (Li & Wang, 2020) or understand domestic tourists who live in the same country but are discriminated against due to various restrictions to specific characteristics (Carson et al., 2013). For the concept of social distance with a 20-year history in tourism, empirical studies that investigate causality rather than relationality and focus on the antecedents of social distance are needed. Therefore, to ensure this causality, more experimental designs can be included in social distance studies (Thyne et al., 2006), or research findings can be expanded and enriched with different methods such as social priming, which is used methodologically in social psychology (Molden, 2014).

The explanatory power of the concept of social distance can be increased by using other theories or concepts other than social contact theories. For example, emotional solidarity (Joo et al., 2018), place attachment (Aleshinloye et al., 2020) is clearly and directly associated with social distancing; on the other hand, the relationship between Hollaway and associates' (2011) intratourist gaze and social distance concept has not been investigated yet. Even the social identity theory, which states that increasing social distance can affect the social identities of groups through social categorisation, has not received enough attention in the literature yet (Sinkovics & Penz, 2009; Ye et al., 2014). Social identity theory based on intergroup relations can offer researchers many alternative explanations as it mainly focusses on the role of social identity in intergroup conflict and cohesion (Tajfel & Turner, 1979). For example, the difference between Japanese interaction with the Chinese as a local position and their interaction with the Chinese as a tourist is a complex issue that requires alternative explanations. Herein, it becomes essential to apply different methodological approaches, perhaps more than one theory/concept.

The scope or validity of the conceptual model proposed by the researchers within the scope of the section can be empirically addressed in future studies. Particular attention can be given to negative attitude changes resulting from a perceived high social distance. For example, in the study conducted by Adam et al. (2020), in which it is stated that negative TTIs are not always initiated deliberately to destroy the values of other tourists, it is pointed out that tourists may see their behaviour as a vacation right for themselves. This description is essential to prevent possible negative attitude changes among tourist groups. Even the study of Sinkovics and Penz (2009) is also essential in this sense, which states that minimising TTI to prevent negative attitude change may cause more superficial relationships and exacerbate mutual misunderstandings and reduce cultural learning opportunities. In particular, the study conducted by Carson et al. (2013) generated significant findings for social distance research in TTIs, which states that seeing othering groups as problematic tourists and as who would harm the experiences of other tourists may be a crucial strategic mistake, and suggests that different tactics and strategies should be applied to understand these groups. These findings need to be expanded with further research.

References

Adam, I., Taale, F., & Adongo, C. A. (2020). Measuring negative tourist-to-tourist interaction: Scale development and validation. *Journal of Travel & Tourism Marketing, 37*(3), 287–301.

Ahmed, A. M. (2007). Group identity, social distance and intergroup bias. *Journal of Economic Psychology, 28*(3), 324–337.

Aleshinloye, K. D., Fu, X., Ribeiro, M. A., Woosnam, K. M., & Tasci, A. D. (2020). The influence of place attachment on social distance: Examining mediating effects of emotional solidarity and the moderating role of interaction. *Journal of Travel Research, 59*(5), 828–849.

Allport, G. (1954). *The nature of prejudice.* Addison-Wesley.

Bai, S., & Chang, H. H. (2021). Effect of tourist-to-tourist encounters: increased conflict or reduced social distance? *Journal of Hospitality & Tourism Research,* https://doi.org/10.1177/10963480211014938

Bogardus, E. S. (1925). Measuring social distance. *Journal of Applied Sociology, 9,* 299–308.

Bogardus, E. S. (1933). A social distance scale. *Sociology and Social Research, 17,* 265–271.

Bogardus, E. S. (1967). *A forty year racial distance study.* University of Southern California Press.

Bosio, E., & Lewis, B. (2008). Customer-to-customer interactions: Examining consumer behaviour in hotels in Cyprus. *International Journal of Management Cases, 10*(3), 345–354.

Bui, H. T., & Wilkins, H. C. (2016). Social interactions among Asian backpackers: Scale development and validation. *Current Issues in Tourism, 21*(10), 1097–1114.

Carneiro, M. J., Eusebio, C., & Caldeira, A. (2018). The influence of social contact in residents' perceptions of the tourism impact on their quality of life: A structural equation model. *Journal of Quality Assurance in Hospitality & Tourism, 19*(1), 1–30.

Carson, D., Carson, D., & Taylor, A. (2013). Indigenous long grassers: Itinerants or problem tourists? *Annals of Tourism Research, 42,* 1–21.

Çelik, S. (2019a). Does tourism reduce social distance? A study on domestic tourists in Turkey. *Anatolia, 30*(1), 115–126.

Çelik, S. (2019b). Does tourism change tourist attitudes (prejudice and stereotype) towards residents? *Journal of Tourism and Services, 10*(18), 35–46.

Chang, H. H. (2017, June 20–22). *Coping and co-creating strategy to overcome tourist-to-tourist encounter by using Critical Incident Technique* [Conference presentation]. Annual International Conference of the Travel and Tourism Research Association, Quebec City, Quebec, Canada. https://scholarworks.umass.edu/cgi/viewcontent.cgi?article=2054&context=ttra

Chen, M., Zhang, J., Sun, J., Wang, C., & Yang, J. (2020). Developing a scale to measure the social distance between tourism community residents. *Tourism Geographies,* https://doi.org/10.1080/14616688.2020.1765012

Fan, D. X. (2020). Understanding the tourist-resident relationship through social contact: progressing the development of social contact in tourism. *Journal of Sustainable Tourism,* https://doi.org/10.1080/09669582.2020.1852409

Fan, D. X. F., Zhang, H. Q., Jenkins, C. L., & Lin, P. M. C. (2017a). Does tourist–host Social contact reduce perceived cultural distance? *Journal of Travel Research, 56*(8), 998–1010.
Fan, D. X. F., Zhang, H. Q., Jenkins, C. L., & Tavitiyaman, P. (2017b). Tourist typology in social contact: An addition to existing theories. *Tourism Management, 60*, 357–366.
Farmaki, A. (2017). The tourism and peace nexus. *Tourism Management, 59*, 528–540.
Gillen, J., & Mostafanezhad, M. (2019). Geopolitical encounters of tourism: A conceptual approach. *Annals of Tourism Research, 75*, 70–78.
Guthrie, C., & Anderson, A. (2007). Tourists on tourists: The impact of other people on destination experience. In C. Guthrie & A. Anderson (Eds.), *Developments in tourism research* (pp. 159–170). Routledge.
Harvey, L. (1987). *Myths of the Chicago school of sociology*. Gower.
Hollaway, D., Green, L., & Holloway, D. (2011). The intratourist gaze: Grey nomads and 'other tourists'. *Tourist Studies, 11*(3), 235–252.
Hsu, C. H., & Chen, N. (2019). Resident attribution and tourist stereotypes. *Journal of Hospitality & Tourism Research, 43*(4), 489–516.
Huang, J., & Hsu, C. H. (2009). Interaction among fellow cruise passengers: Diverse experiences and impacts. *Journal of Travel & Tourism Marketing, 26*(5–6), 547–567.
Jenkins, R. (2014). *Social identity*. Routledge.
Johnston, L. (2001). (Other) bodies and tourism studies. *Annals of Tourism Research, 28*(1), 180–201.
Joo, D., Tasci, A. D., Woosnam, K. M., Maruyama, N. U., Hollas, C. R., & Aleshinloye, K. D. (2018). Residents' attitude towards domestic tourists explained by contact, emotional solidarity and social distance. *Tourism Management, 64*, 245–257.
Katz, I. (1991). Gordon Allport's "The nature of prejudice". *Political Psychology, 12*(1), 125–157.
Lawler, S. (2015). *Identity: sociological perspectives*. John Wiley & Sons.
Li, F. S., & Wang, B. (2020). Social contact theory and attitude change through tourism: Researching Chinese visitors to North Korea. *Tourism Management Perspectives, 36*, https://doi.org/10.1016/j.tmp.2020.100743
Lin, H., Zhang, M., Gursoy, D., & Fu, X. (2019). Impact of tourist-to-tourist interaction on tourism experience: The mediating role of cohesion and intimacy. *Annals of Tourism Research, 76*, 153–167.
Ming, H. (2018). Cross-cultural difference and cultural stereotypes in tourism—Chinese tourists in Thailand. *Journal of Hotel & Business Management, S1*, 1–5.
Molden, D. C. (2014). Understanding priming effects in social psychology: What is "social priming" and how does it occur?. *Social Cognition, 32*(Supplement), 1–11.
Murphy, L. (2001). Exploring social interactions of backpackers. *Annals of Tourism Research, 28*(1), 50–67.
Ng, S. I., Lee, J. A., & Soutar, G. N. (2007). Tourists' intention to visit a country: The impact of cultural distance. *Tourism Management, 28*(6), 1497–1506.
Nyaupane, G. P., Teye, V., & Paris, C. (2008). Innocents abroad. *Annals of Tourism Research, 35*(3), 650–667.
Nyaupane, G. P., Timothy, D. J., & Poudel, S. (2015). Understanding tourists in religious destinations: A social distance perspective. *Tourism Management, 48*, 343–353.
Özdemir, C., & Yolal, M. (2017). Cross-cultural tourist behavior: An examination of tourists' behavior in guided tours. *Tourism and Hospitality Research, 17*(3), 314–324.
Page-Gould, E., Mendoza-Denton, R., & Tropp, L. R. (2008). With a little help from my cross-group friend: Reducing anxiety in intergroup contexts through cross-group friendship. *Journal of Personality and Social Psychology, 95*(5), 1080–1094.
Papathanassis, A. (2012). Guest-to-guest interaction on board cruise ships: Exploring social dynamics and the role of situational factors. *Tourism Management, 33*(5), 1148–1158.
Paris, C. M., Nyaupane, G. P., & Teye, V. (2014). Expectations, outcomes and attitude change of study abroad students. *Annals of Tourism Research, 48*(Supplement C), 275–277.
Park, R. E. (1924). The concept of social distance as applied to the study of racial attitudes and racial relations. *Journal of Applied Sociology, 8*, 339–344.
Pearce, P. L. (2005). *Tourist behavior: Themes and conceptual schemes*. Channel View Publications.
Pettigrew, T. F. (1998). Intergroup contact theory. *Annual Reviews of Psychology, 49*, 65–85.
Pettigrew, T. F., & Troop, L. R. (2006). A meta-analytic test of intergroup contact theory. *Journal of Personality and Social Psychology, 90*, 751–786.
Pizam, A., Uriely, N., & Reichel, A. (2000). The intensity of tourist–host social relationship and its effects on satisfaction and change of attitudes: The case of working tourists in Israel. *Tourism Management, 21*(4), 395–406.
Reichenberger, I. (2017). C2C value co-creation through social interactions in tourism. *International Journal of Tourism Research, 19*(6), 629–638.
Schwarz, K. C. (2018). Volunteer tourism and the intratourist gaze. *Tourism Recreation Research, 43*(2), 186–196.

Segall, M.H., Dansen, P.R., Berry, J.W. and Poortinga, Y.H. (1999). *Human behavior in global perspective: An introduction to cross-cultural psychology*. Allyn & Bacon.
Sharpley, R. (2014). Host perceptions of tourism: A review of the research. *Tourism Management, 42*, 37–49.
Shtudiner, Z. E., Klein, G., & Kantor, J. (2018). How religiosity affects the attitudes of communities towards tourism in a sacred city: The case of Jerusalem. *Tourism Management, 69*, 167–179.
Sinkovics, R. R., & Penz, E. (2009). Social distance between residents and international tourists—Implications for international business. *International Business Review, 18*(5), 457–469.
Sorensen, A. (2003). Backpacker Ethnography. *Annals of Tourism Research, 30*(4), 847–867.
Sun, H., Wu, S., Li, Y., & Dai, G. (2019). Tourist-to-tourist interaction at festivals: A grounded theory approach. *Sustainability, 11*(15), 4030.
Tajfel, H. and Turner, J. C. (1979). An integrative theory of intergroup conflict. In W. G. Austin & S. Worchel (Eds.), *The social psychology of intergroup relations* (pp. 33–47). Brooks/Cole.
Tang, L. R. (2014). The application of social psychology theories and concepts in hospitality and tourism studies: A review and research agenda. *International Journal of Hospitality Management, 36*, 188–196.
Tasci, A. (2009). Social distance. The missing link in the loop of movies, destination image, and tourist behaviour?. *Journal of Travel Research, 47*(4), 494–507.
Thyne, M. A., & Lawson, R. (2001). The design of a social distance scale to be used in the context of tourism. In P. M. Tidwell & T. E. Muller (Eds.), *AP-Asia Pacific advances in consumer research* (pp. 102–107). Association for Consumer Research.
Thyne, M., Lawson, R., & Todd, S. (2006). The use of conjoint analysis to assess the impact of the cross-cultural exchange between hosts and guests. *Tourism Management, 27*(2), 201–213.
Thyne, M., Watkins, L., & Yoshida, M. (2018). Resident perceptions of tourism: The role of social distance. *International Journal of Tourism Research, 20*(2), 256–266.
Thyne, M., Woosnam, K. M., Watkins, L., & Ribeiro, M. A. (2020). Social distance between residents and tourists explained by residents' attitudes concerning tourism. *Journal of Travel Research*, https://doi.org/10.1177/0047287520971052
Thyne, M., & Zins, A. H. (2003). Designing and testing a Guttman-type social distance scale for a tourism context. *Tourism Analysis, 8*(2), 129–135.
Tomljenovic, R. (2010). Tourism and intercultural understanding or contact hypothesis revisited. In O. Moufakkir & I. Kelly (Eds.), *Tourism, progress and peace* (pp. 17–34). CABI.
Torres, E. N. (2015). The influence of others on the vacation experience: An ethnographic study of psychographics, decision making, and group dynamics among young travelers. *Journal of Hospitality Marketing & Management, 24*(8), 826–856.
Triandis, H. C., & Triandis, L. M. (1962). A cross-cultural study of social distance. *Psychological Monographs: General and Applied, 76*(21), 1.
Triandis, L.M., & Triandis, H.C. (1960). Race, social class, religion, and nationality as determinants of social Distance. *Journal of Abnormal and Social Psychology, 61*(1), 110–118.
Turner, R. N., Crisp, R. J., & Lambert, E. (2007). Imagining intergroup contact can improve intergroup attitudes. *Group Processes and Intergroup Relations, 10*, 427–441.
Urry, J. (1990). *The tourist gaze*. Sage.
Vezzali, L., Capozza, D., Stathi, S., & Giovannini, D. (2012). Increasing outgroup trust, reducing infrahumanization, and enhancing future contact intentions via imagined intergroup contact. *Journal of Experimental Social Psychology, 48*(1), 437–440.
Wark, C., & Galliher, J. F. (2007). Emory Bogardus and the origins of the social distance scale. *The American Sociologist, 38*(4), 383–395.
Wei, S., Ng, S. I., Lee, J. A., & Soutar, G. N. (2021). Similarity-attraction cluster of outbound Chinese tourists: Who belongs there? *Journal of Hospitality & Tourism Research*, https://doi.org/10.1177/1096348021996441
Weinfurt, K. P., & Moghaddam, F. M. (2001). Culture and social distance: A case study of methodological cautions. *The Journal of Social Psychology, 141*(1), 101–110.
Wright, S. C., Aron, A., McLaughlin-Volpe, T., & Ropp, S. A. (1997). The extended contact effect: Knowledge of cross-group friendships and prejudice. *Journal of Personality and Social Psychology, 73*, 73–90.
Wu, C. H. J. (2007). The impact of customer-to-customer interaction and customer homogeneity on customer satisfaction in tourism service—the service encounter prospective. *Tourism Management, 28*(6), 1518–1528.
Yagi, C. (2001). How tourists see other tourists: Analysis of online travelogues. *Journal of Tourism Studies, 12*(2), 22–31.
Yagi, C., & Pearce, P. L. (2007). The influence of appearance and the number of people viewed on tourists' preferences for seeing other tourists. *Journal of Sustainable Tourism, 15*(1), 28–43.

Yankholmes, A., & Timothy, D. J. (2017). Social distance between local residents and African American expatriates in the context of Ghana's slavery-based heritage tourism. *International Journal of Tourism Research, 19*(5), 486–495.
Ye, B. H., Zhang, H. Q., Shen, J. H., & Goh, C. (2014). Does social identity affect residents' attitude toward tourism development? An evidence from the relaxation of the individual visit scheme. *International Journal of Contemporary Hospitality Management, 26*(6), 907–929.
Yilmaz, S. S., & Tasci, A. D. (2013). Internet as an information source and social distance: Any relationship? *Journal of Hospitality and Tourism Technology, 4*(2), 188–196.
Yilmaz, S. S., & Tasci, A. D. (2015). Circumstantial impact of contact on social distance. *Journal of Tourism and Cultural Change, 13*(2), 115–131.
Yin, C. Y., & Poon, P. (2016). The impact of other group members on tourists' travel experiences. *International Journal of Contemporary Hospitality Management, 28*(3), 640–658.

21

VALUE TYPOLOGY IN THE CONTEXT OF THE TOURISM SECTOR

Üzeyir Kement

Introduction

With the dominance of relationship-based marketing activities in the market, tourism industry has begun to give importance to macro/micro-environment interactions. The realization of business objectives such as increasing the profitability rate and market share, prolonging the business life cycle and branding are based on relationship-based marketing planning. In particular, the fact that customer-centered communication networks such as consumer experience sharing activities and usage of the power of electronic word of mouth through online communities (Cheung & Lee, 2012; Dwyer, 2007; Yüksel & Kılıç, 2016) have become an understanding that will make a difference in terms of competition in the market has led businesses to define the concept of value. Customer-oriented marketing practices are based on valuing the customer. In this respect, it becomes the key to success for businesses to offer value to their existing and potential customers.

It is important for the customer that tourism businesses make a difference in terms of products, services, and image in the market. In this way, the preferences of tourists can be guided more easily. In order to make a difference, the needs, and expectations of the customers should be investigated. In line with the results of the research, a customer value proposition should be created, and this proposition should be delivered to the customer. Such a marketing process should continue continuously in the business-customer relationship. However, this can make a difference in the market. It includes interaction based on the concept of customer value. Therefore, it has started to be discussed since the 1950s and has been the subject of marketing research and industry applications with the development of relational marketing understanding. The tourism sector is also one of the sectors where interaction is most intense. Therefore, it is important that the perceived customer value is high in order to develop the attitudes and behaviors of customers towards purchasing. In this section, customer value is discussed in terms of conceptual, historical development and value typology.

The Concept and Development of Customer Value

According to Kotler, Saliba and Wrenn (1991: 221) "marketing is a management process that facilitates the transaction between the buyer and the seller". From this point of view, it is possible to say that both the buyer and the seller give up a value in exchange for a greater gain. Therefore, it should be said that the concept of customer value is at the center of marketing activities and is an important concept for every customer research.

DOI: 10.4324/9781003161868-21

When the transaction between the customer and the seller in the tourism sector is handled by ignoring external variables (the product creates social problems, etc.), it can be mentioned that a value is created for both parties. Value, in this case, is the mutual creation of value through a transaction, with both parties giving up something for it. In the tourism sector, accommodation businesses, food and beverage businesses, travel businesses or recreation businesses must improve their relations with customers in order to continue their activities and to be in a better position in the market. By using marketing mix elements, tourism businesses aim to gain advantage in the service they offer. This advantage is that the business reaches the customer by placing itself above the competitors. All this process can happen with the value of the relevant business in the minds of the consumer. Therefore, the concept of value is a concept that plays a role in the center of marketing activities in the tourism sector.

According to the developments in marketing science, with the importance of relationship marketing activities since the 1990s, macro/micro-environment relations have become important for businesses. In addition, with the customer in the center of relationship marketing, direct customer-oriented managerial processes have started. These processes, on the other hand, required the realization of production and sales processes in accordance with customer demands and expectations. In the tourism sector, it is important for businesses to be able to distinguish themselves from their competitors with their services or the image they have. It has become a very important issue for them to create a value in order to ensure this and be preferable for customers. When the differences between the marketing understandings of the past and today are examined, it is understood how great a necessity is to create value.

Mahajan (2007: 41) expresses the concept of value as the value (appropriate/worthy) of a product to the customer and then states the following: "Value is the balance between price and quality (goods, services, brand/relationship status, etc.) perceived by a customer. Value is a perception, but a synthesis of economic, functional and psychological factors that are important to the customer". Many tourism businesses do not expect their customers to be satisfied only with the service, but they want customers to get the best experience from that service. This is the most important point of relationship marketing. Experiences that exceed customer expectations are needed to create a superior customer value. From this point of view, the concept of value is a strategic power that businesses in the tourism sector use to differentiate themselves from their competitors in the minds of their customers (Weinstein, 2018: 4).

Value is an important term for business and marketing activities such as quality, service, and excellence. It is possible for many people to have different standards. This can change according to many situations (product, price, quality, purchasing process, or after, etc.). For example, the expectations of two customers in an accommodation business may not be the same. While one gives importance to the details of the room, the other may give more importance to the interaction of the employees.

Customer value can also be associated with barter commercial transactions made in ancient times. It can be explained by value that buyers listen to sellers and decide whether the cost is fair or not. In this direction, the concept of value can be explained as "the satisfaction of purchase, ownership and use at the lowest cost in meeting customer needs" (Hollensen, 2010: 34). Value means giving relative importance. It can also be expressed as being able to address customer desires or excellence based on utility. It should not be forgotten that value is an abstract concept considering that it is relative (Duchessi, 2002; Holbrook, 2002). According to Weinstein (2018) value can be defined as the difference between the service offered in the best way from the customer's point of view, the payment or effort (money, time, stress, etc.) and the benefit obtained by the customer. Value creation is possible as a result of the customer receiving the service. Therefore, the customer has a perception of value as a result of every purchase they make.

Gale and Wood (1994: 318) define the concept of customer value as "perceived quality for determining the relative price of the product in the market". According to the author, quality is a valid feature in all areas of the customer. From this point of view, it can be said that customer opinions, quality and price-oriented perceptions are the focus of marketing plans.

Tourism businesses should attach importance to being a leader in terms of cost in today's market where the competitive environment has increased. In order to be a leader in cost, it is important to analyze customer needs and expectations well and to develop management strategies in this direction. Customers may not care about the size of a business, how it is financed, or where it is in the market. Customers care that the value propositions offered for them are compatible with their expectations. Therefore, tourism businesses should aim to serve their customers with value propositions that can differentiate them from their competitors, in addition to being a cost leader.

It is expected that tourism enterprises aim to offer the highest quality product at the lowest price to their customers in their strategic management studies. In addition, if they attach importance to differentiation while doing this, they can create a successful customer value process (Duchessi, 2002). The diversity of the products offered in the hotel or the relationship that the employees establish with the customers are in a different position from other businesses are the determining factors in the formation of value. In doing so, it is necessary to create an atmosphere in which customers will feel special.

Customer value has three basic components: product quality, service quality, and price. Duchessi (2002: 83) explains these components with the customer value cube. Any or all of the product quality, service quality, and price components that are meant to be explained by the customer value cube, ensure that customer value increases by meeting or exceeding customer expectations. When businesses achieve what is desired in all three value components, they can deliver innovative customer value. Today, just having a strong brand or having quality products is not enough to create customer value. In addition, customers expect low prices and support services from businesses.

Customer value varies depending on the product's feature. Value can be expressed as a force that motivates people. Therefore, it is necessary to put people and their wishes on two different sides here. People and their desires are two important factors in the formation of value. Environmental factors that push people to realize their wishes are mediated in the formation of value. A person may have a need or expectation, but environmental conditions are important in determining the value he will derive from this need or expectation. These conditions are the price or the level of alternatives to meet the needs or expectations. Shillito and De Marle (1992) define the customer value formation process as "value power diagram".

Figure 21.1 shows the diagram where value is treated as power. In the diagram, the customer is placed in the center and around the different options from A to H. These letters represent the needs of the individual. The size of the circle of each letter expresses the level of meeting the needs. Therefore, large apartments are more effective in meeting the needs than small apartments. In addition, apartments that are close to the person represent less cost, while those that are far away represent higher costs. Not only is the cost mentioned here of a monetary nature, but it can also be a temporal, physical, or psychological sacrifice. There are two different (F and C) alternatives on the dotted circle in the diagram. However, the level of meeting the needs of both is not the same. Because the circle representing C is larger than the circle representing F. Therefore, apartment C is more successful in meeting the need than apartment F. What is meant by this diagram is that there is a power that can measure the value. Thus, value is a power. The magnitude of the power depends on the interplay between needs, benefits, and costs. When the value is strong, it can be said that the customer is motivated to buy the product that he believes will meet his needs with the help of cost-benefit analysis.

Holbrook Customer Value Content and Examples from the Tourism Industry

Value philosophy, which was put forward in parallel with the developments in marketing science, has been discussed by many researchers since the 1950s and the relationship between businesses

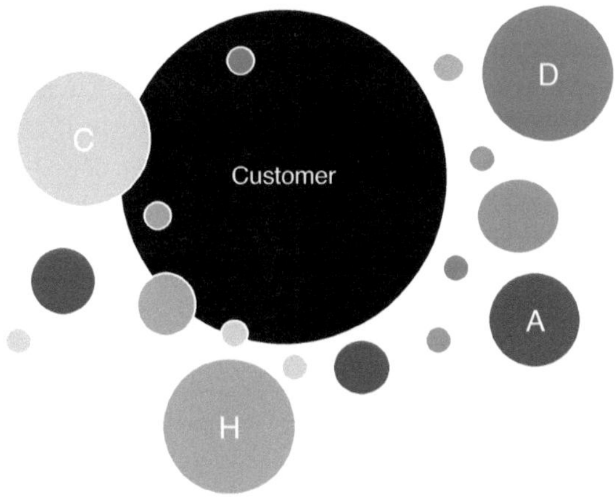

Figure 21.1 Customer Value as Power.
Source: Adapted from Shillito and De Marle, 1992: 5.

Table 21.1 Holbrook's Customer Value Concept Content

"*Customer value is an interactive, relative and preferential experience.*"	
Interaction	Customer value requires interaction between the customer and the product.
Relativity	Customer value is comparable, individual and may vary from situation to situation.
Preference	Customer value includes a choice decision.
Experiential	Customer value is formed by consumption experiences from products, brands and properties.

Source: Adapted from Schröder (2003: 17)

and customers has been tried to be explained (Holbrook, 2002). It can be said that Holbrook (1994, 1996) made great contributions to the in-depth analysis of the theory of value and its introduction to the marketing literature. Issues such as the importance of the value theme, its antecedents, and its use in tourism businesses have been covered in many research works by the author (Holbrook, 1985). Holbrook (2002: 5), inspired by the work of Hilliard (1950: 42), defines customer value as "a relative choice experience". In a broader expression, it states that customer value is formed by the exposure of some objects to some subjective evaluations. The customer, who makes subjective evaluations in the formation of value, is defined as any service (a good or service, a political candidate, a holiday destination, a musical piece, painting, photograph, etc.) (Taylor, 1961; Holbrook, 2002).

Holbrook (2002) uses four different concepts (interaction, relativism, preference, and experiential) to explain customer value. It also explains that these concepts are interrelated and not independent from each other. He states that it is possible to talk about customer experience as a result of combining these four concepts with each other. The general definitions of the four concepts are shown in Table 21.1.

Interaction argues that customer value should be examined from an objective and subjective point of view. Here, objectivity is about the product, while subjectivity is about the customer. Over the years, many studies have been conducted and discussed in the literature on the importance of subjectivity or objectivity or the dominance of any party. While some researchers (Perry, 1954; Frondizi, 1971; Donner, Gohier & de Vries, 2020; Zerjav, 2021) argued that value formation depends on subjectivity, some researchers (Osborne, 1933; Bahm, 1993; Moore & Baldwin, 1993) argued that value formation depends on objectivity (Holbrook, 2002). For example, those who defend subjectivity have tried to prove this with statements such as "man is the measure of all things" (Riley, 2005: 34), "there is no arguing in matters of taste" (Nerlich, 1989) or "beauty is in the eye of the beholder" (Nozick, 1981: 400). In addition, Levitt (1960) states in his study in the field of marketing that a product can only be valuable if it satisfies a customer. On the other hand, Tuchman (1980) argued that "quality in a work is intrinsic (of the product)", Osborne (1933) argued that beauty is "a formal property of [the] beautiful", and Adler (1981) considered "admirable beauty" as "objective, not subjective" (Holbrook, 2002). In the field of marketing, characterizing value as a feature of the product increases the offer value of the product. Karl Marx explains that "the value of a product depends on the amount of work spent producing it" with his "labor theory of value" (Holbrook, 2002). Contrary to all these approaches, some researchers (Weinstein, 2018; Wiggins, 1998) state that value is based on the interaction between an object and subjective evaluations. Therefore, it is argued that in the formation of value, besides the physical and cognitive properties of the product, it can take place depending on a subjective evaluation (Pepper, 1958). For example, the feeling, thought or pleasure that a tourist feels, who goes to a hotel for summer vacation, is as important as the room he stays in, the food he eats in the restaurant or the cleanliness of the beach.

Relativity draws attention to three different points in the formation of value. The first of these is the case of "value being comparative". Value can be created when the customer attributes value to a product by comparing it with another product (Holbrook, 2002). For example, when choosing a hotel for vacation, tourists decide by comparing many aspects such as the service, size, and location of the hotels. Therefore, in the customer's value judgment, there is a relative evaluation made as a comparative (between two similar products) rather than a utility analysis. What is meant by the comparative evaluation here is the personal preference obtained by comparing two different products in the same category by the same customer (Pap, 1946). For example, if an individual who will go on holiday in the summer says that he prefers the Aegean Sea more than the Mediterranean for swimming, he determines his value judgment by comparing two different touristic experiences. However, the statement that I like swimming in the Aegean Sea more than my mother does not mean a comparative value judgment. With another example, instead of summer tourism versus winter tourism preference; the comparison of winter tourism in different destinations against winter tourism offers a comparison opportunity in the same category. As a result of all these evaluations, it can be said that in the formation of value judgment, comparisons may arise from evaluations made over the same categories.

The second view is that "value is personal because it can vary from one individual to another" (Bahm, 1993). Although it is claimed by marketers that the value of products is equal for everyone, these value judgments may not be the same for every individual. Therefore, it can be said that there are absolute subjective evaluations in the relative value (Osborne, 1933; Wiggins, 1998). However, it should not be forgotten that marketers can create different objective values for individuals (Bond, 1983). For example, although the bungee jumping activity for adventurers is seen as recreationally valuable, this activity may not have the same value for another type of tourist. As a result, it can be said that this situation is a big factor in the fact that marketing is a need in the world and this situation is the cornerstone of marketing. The first of the marketing principles, the customer difference, explains this situation. For this reason, marketing strategy issues such as market segmentation and positioning have been developed. Thirdly, the judgment that "customer value occurs depending on the current situation when making the evaluation decision" explains the situation of relativity (Bahm, 1993; Holbrook, 2002). In the formation of the value judgment, conditions, time frame or any other issue may affect the customer's decision. An example is the situation where hotels on the seaside in summer and a hotel on the top of the mountains in winter can be more valuable for people.

Preference refers to the effect of customers' preferences on the formation of customer value (Fine & Peters, 2016; Leroi-Werelds, 2021; Pap, 1946). In the field of axiology, which explains the philosophy of value, the reasons for preference of individuals are examined by many researchers within the scope of the "Interest Theory of Value" (McFee, 2004). Axiology, also referred to as the philosophy of value, focuses on the reasons why people choose. As stated by many axiologists, the concept of preference is associated with different value terms (tolerance-discontentment, good-bad, positive-negative, bias-reaction, positive-negative, etc.) that stand out in different disciplines. For example, it can be explained as a good-bad evaluation, while positive-negative tendency is explained as like-dislike attitude, tolerance-discontent as emotion, and positive-negative as value. Each of these preference types expresses different preference features (Pap, 1946). Evaluation of preferences can be made in two different ways as singular and plural. However, multi-state (multi-choice hosting) has revealed evaluation judgments such as standards (Kahle, 1983), rules (Arrow, 1967), criteria (Baylis, 1958; Fine & Peters, 2016; Pepper, 1958), norms (Pepper, 1958), goals (Veroff, 1983), or ideals (Cowan, 1964; Pepper, 1958). Before determining the value, it is necessary to distinguish well between single and multiple assessments. Because while single value formation depends on individual preferences, multiple value formation occurs within the scope of many evaluation issues (norm, attitude, standard, etc.). In addition, individual differences (personality, education, culture, etc.) of customers can also be important in the formation of customer value (singular). Therefore, individual differences should also be taken into account when evaluating preferential situations (Holbrook, 2002).

Experientialism means that customer value is derived from consumption experiences, not from the product purchased, the object chosen or owned (Holbrook, 1982; Weinstein & Johnson, 1999). At the core of experientialism is the idea that all products serve to create needs or satisfying experiences. Therefore, marketing can be expressed as "service marketing" in its entirety. Experience plays a central role in creating customer value (Holbrook, 2002). In addition, Chambers, Kouvelis, and Semple (2006) support this view by claiming that what people desire is not the product, but rather satisfying experiences. In addition, according to the author, experiences can occur through activities. Physical facilities and human services are needed to carry out the activities. This makes service marketing the cornerstone of marketing. Because when people want products, they want services that provide the experience they expect through the products. The tourism sector is a sector that provides services in its entirety. Therefore, providing experience is an important issue for all tourism sub-sectors. From the ambiance of the environment to the bliss of the food, almost every touristic service is essentially aimed at providing an experience.

Customer Value Typology in the Context of the Tourism Sector

Different types of value can emerge as a result of customers' experiences. These value types are expressed by Holbrook (2002) as customer value typology. The customer value typology is described in three different dimensions and these dimensions are shown in Table 21.2.

Table 21.2 The Typology of Customer Value

Categories		*Extrinsic*	*Intrinsic*
Self-Oriented	Active	Efficiency (O/I, Convenience)	Play (Fun)
	Reactive	Excellence (Quality)	Aesthetics (Beauty)
Other-Oriented	Active	Status (Success)	Ethics (Justice, Virtue, Morality)
	Reactive	Esteem (Reputation, Possessions)	Spirituality (Faith, Magic)

Source: Holbrook (2002: 12).

Customer value dimensions are considered in two ways. But each has various transitions from one extreme to the other. The first of the customer value dimensions is *extrinsic-intrinsic value.* Extrinsic value means that consumption is functional or useful to achieve utility, purpose or objects (Diesing, 1962). For example, planes that provide transportation services for a holiday abroad create an external value by ensuring that the destination is reached for the holiday. Intrinsic value, on the other hand, is that the individual directly serves his main purpose and in the center of consumption, the individual seeks to consume for himself (Baylis, 1958; Pepper, 1958). For example, spending the day at the beach is a situation that only benefits that individual and has the opportunity to provide little benefit other than himself. However, while the bus that enables the individual to go to the beach is an external value, spending the day on the beach is an internal value for the individual.

Self-oriented and other-oriented is the second dimension of customer value. What is meant in this dimension is that the value created by the individual within his own wishes and desires is self-oriented, while the value he creates within the others is other-oriented (Fromm, 1941). When an individual acts selfishly or prudently by considering the value of his own self, he creates a self-directed value. For example, it is a self-directed value that the individual participates in the hunting activity without considering the social harm and focuses on the pleasure he will get from it. However, it is a value other-oriented, since the individual pays attention to the green star while choosing the hotel where he will stay in the destination, and because it has the purpose of contributing to the sustainability of the world. The concept of others, at the micro level; family, friends, and co-workers, at intermediate level; society, country, and the world, at the macro level includes the universe, nature, and divine thought (spirituality). Also, to the other-oriented value; it is also viewed as the invisible side of the iceberg (super consciousness) explained in Freud's psychoanalytic theory (Holbrook, 2002).

Active-reactive value means that the individual does the action actively and the action is made by the individual through an effect outside the individual. Active value includes actions taken by an individual with a product, either directly or as part of a consumption experience, when a tangible or intangible object needs to be physically or mentally manipulated.

Active value occurs when a tangible object is physically manipulated (driving a car), an intangible object is mentally manipulated (crossword puzzle solving), an intangible object is physically manipulated (taking mind-altering drugs), and a tangible object is mentally manipulated (telekinesis) (Holbrook, 2002). Diesing (1962) states that active value involves moving the object by the individual. In the reactive value, on the other hand, the passive state is the opposite of the active value. That is, value is a reaction state that includes situations such as appreciation, admiration, comprehension (education) when an object is part of some consumption experience or includes things made by another individual (painting, music, etc.). In the active, the individual (subject) is doing something (object) to him, while in the reactive, something (object) is made by another person/object/situation (subject). In other words, value is active if the individual acts on himself, but reactive if it activates the individual. In the literature, it is seen that there are studies on the subjects of opposition (opposite direction) indicating that active and reactive values can occur. For example, there are studies on the contrast between activity and passivity (Pepper, 1958), between control and dependence (Cobley, 2009), between being dominated and being dominated (Mehrabian & Russell, 1974), and between acting and being moved (Harré & Secord, 1973). However, in the literature, the distinction between active and reactive value is not very common in value-oriented studies.

There are eight categories in the customer value typology. These are efficiency (convenience), excellence (quality), status (success, impression management), esteem (reputation, materialism, possessions), aesthetics (beauty), ethics (virtue, justice, morality), and spirituality (faith, ecstasy, sacredness, magic).

Efficiency refers to the extrinsic value of active use of a product or consumption experience as a means to achieve a self-focused goal (Bond, 1983). Keys that people use to open doors or tokens to get drinks from vending machines represent the efficiency of exogenous value. The method of dividing inputs by outputs (I/O) is used to determine productivity (Pepper, 1958; Diesing, 1962). Here, the perception of productivity towards the relevant product is important for individuals to achieve an external goal. Convenience or convenience is an important input in making an input-output comparison in the mind of the customer. For example, in a food and beverage business that provides fast food service, the fact that customers think fast food is unhealthy and is fast and easy changes their view of efficiency. Therefore, it provides the opinion that the product is efficient. An example of efficiency can be the act of taking credit when individuals take out loans from banks and achieve the desired external purpose. In another example, it can be said that the refrigerator can store food and enable us to use it for a long time. These examples do not mean that the customer enjoys the product, but that he can benefit from it because of its efficiency. Finally, the fact that restaurants provide a playground for their children and people in charge of the park today creates an opportunity for people to spend their time comfortably and is productive for the individual. In other words, it provides convenience to the customer.

Excellence is a reactive evaluation of an object or experience's ability to serve as an external tool. In other words, excellence is when the product is at its best in fulfilling a goal or a specific function. In other words, excellence is when the product fulfills a purpose or a specific function. In the customer's view of the product as the best; the effect of the concepts of satisfaction and quality, which is based on the comparison of product performance and expectations, can be mentioned (Bond, 1983; Çavuşoğlu, Demirağ & Durmaz, 2020; Zeithaml, 1988). For example, the equipment of an accommodation company in the room, the service provided in the restaurant, the food prepared in the kitchen at an ideal location for the customer explains the perfection.

Status refers to the active orientation of one's own consumption behavior as an external tool to obtain a positive response from another. The situational formed value can basically be evaluated as a political attitude (Nozick, 1981). In other words, contingency can be described as a political behavior, as it is an attitude to influence others as a result of consumption (Holbrook, 2002). The clothes, jewelry, and so on that individuals wear while going to a job interview. It is an example of contingency because it is carried out to influence the manager to be interviewed on the basis of product use (Fiore & Kim, 1997). In summary, customers can choose the products they consume and the consumption experiences they follow, partly as a set of symbols to build a person who has achieved success in the eyes of others as "status". Contingency is seen as an ongoing process. Because if the individual is trying to create a social positioning for himself, he first researches and determines the position and performs consumption activities that will support gaining the position. Therefore, the efforts to prove oneself to the environment is a significant motivation (Yüksel & Bekar, 2017) which shapes individuals' behaviors. Thus, contingency includes a positioning strategy for people. The contingency-value process can consist of several stages. These are (i) a social character is determined or an attempt to adapt to a social character, (status description) (ii) necessary products or experiences can be purchased to achieve the goal (case research), (iii) evaluation of the efficiency of consumption activity towards the chosen character (state verification) form.

Esteem is very similar to the previous typology subject, contingency. Axiologists and economists liken it to the concept of reactive, conspicuous consumption (Veblen, 2017), which is tried to create external appreciation for respectability (Holbrook, 2002). In a broader expression, when the tourist's conspicuous consumption experience is aimed at gaining appreciation and admiration from others, it achieves prestige. While the individual realizes it only as an indicator in order to receive dignity from others (he is in a passive state), he can also perform it to gain his own self-esteem. This can be seen as the individual's effort to reach the inner self, which is expressed as "other-oriented". The tendency of people to have an elite reputation in society or prestigious property in a way that is consistent with boosting self-esteem indicates that it is "others" they want to influence. The difference between reactive esteem and active contingency is paralleled by the difference between two different self-presentation styles, which Clark, Slama, and Wolfe (1999) refer to as "passing" and "advancing" (Celuch & Slama, 1995). While reactive prestige and similar evasion involve a tendency to conform (adapting to others, adapting to society, etc.), active contingency and similar progress include the tendency to direct the event for a gain (consumption activities for the job application, the manager to get it). While it is sufficient for the individual to adapt in reactive dignity, in active contingency it is necessary for the individual to have an achievement or direction (Slama & Wolfe, 1997). An example of gaining prestige is when a tourist goes to a seafood business and orders a lobster to act like famous names he takes as an example, although he has never eaten before or does not have such a culture and photographing this experience and sharing it on his social account.

After examining the external value dimensions (efficiency, excellence, status and esteem) in the customer value typology, the internal value dimensions are examined in the next section. The subject of internal evaluation and at the same time self-directed first value typology is play. Play, as a self-directed experience, is when an individual engages, enjoys, and has fun in his spare time, in a motivated way, actively for his own good (Bond, 1983). For many years, the nature of play has been expressed by many axiologists as an experience maintained as an end in itself (Santayana, 1896). Therefore, it can be stated that play is a self-directed experience (Miller, 1973). However, the difference between gaming and productivity should be well understood here. There is a difference between an individual who constantly trains on the tennis court to achieve success in a tennis match and an individual who goes to play tennis with his family on the weekend. While the individual playing tennis for training purposes works on the basis of productivity, the individual playing tennis with his family on the weekend as a leisure activity has the purpose of game (entertainment). Activities that people do for recreational purposes in their spare time can be evaluated in this group.

Aesthetics refers to the evaluation of the consumption experience by the individual in terms of aesthetics (beauty, etc.). An individual's expressing his admiration with his personal evaluations or pointing out beauty is an aesthetic value. Aesthetic value serves no purpose other than subjective meaning to the individual. Therefore, it is an intrinsic value (Moore & Baldwin, 1993). Attribution aesthetic value to an individual's work of art is simply an appreciation of the work. On the other hand, attaching a work of art to the door of a house as an ornament is to use it for utility. In this case, the work of art turns into efficiency rather than aesthetic value.

Therefore, the aesthetic value of the individual passively admiration, beauty, etc. to any consumption experience or product, as an example of its intrinsic value. In other words, the value loaded on the product; instead of ease of use (efficiency) and quality (excellence), there should be beauty. In this context, tourists can examine historical buildings, antique objects, or visit museums on cultural trips.

Ethics refers to doing something for the good of others. Therefore, it is active and other-oriented (Nozick, 1981). By attributing an ethical meaning, individuals create an internal value with their consumption experiences (Bahm, 1993; Bond, 1983). It shows that an individual is virtuous with dispositions on the line between righteousness and justice. The tendencies of the individual, which have consequences that increase the welfare of the society, express ethical value (Holbrook, 2002). An example of ethical value is an individual's behavior, such as a blood donation and granting scholarships to students. Here, the individual should not expect an external gain. If a politician kisses a small child to win votes in an election, if an individual donates blood to gain social approval, such experiences show that the individual has an external expectation. Therefore, it cannot be expressed as an ethical value. In another example, in a restaurant where waste is sent to animal shelters, if an individual puts napkin, toothpicks, etc. into the food waste to look good to their friends with whom they come to dinner. If they take care not to throw their garbage away, this situation is not an ethical value. On the other hand, if it cares about sustainability and exhibits nature-friendly behaviors in areas such as national parks, this can be an example of ethics.

Spirituality is reactive rather than active. It is also another oriented value to receive acceptance of some divine or mystical being. The fact that the individual aims to receive divine appreciation (doing charity, earning merit, etc.) while adopting shows that the spiritual value is a reactive, other-oriented inner value. There is a form of devotion or worship in spiritual value, and these orientations are reactive and other oriented. However, if the individual is inclined to direct the behavior by evaluating it as a result of his own thoughts (blood donation, etc.), he represents ethical value by being active and other-oriented (Holbrook, 2002). Axiologists state that the spiritual orientation of individuals is an end in itself (Pepper, 1958). Praying as a religious requirement of individuals can be shown as an example of spirituality value. However, if the individual meditates or prays in order to achieve some goals, this can be cited as an example of contingency or efficiency (Holbrook, 2002). Therefore, the individual should perform the action in line with the understanding of holiness and should not use it to achieve another purpose. Participation in the pilgrimage as religious tourism is considered under the title of spirituality because it is an action that people do in line with their spiritual values.

Conclusion

In this research, Holbrook's value typology was evaluated through tourism sector. In addition, the definition of customer value and its characteristics are discussed in detail, considering the literature. When the value typology is examined in terms of the service industry, it is especially important for the interaction of the enterprises operating in the tourism sector with their customers. In this way, businesses can learn about the expectations of their customers or the value attributions of their products. Thus, they can create the right plans and programs for their marketing activities.

Value typology sees service marketing as a fundamental component of marketing. Because the service offered to customers creates an experience. This experience improves the formation of value over time, and value improves the flow of marketing activities. In this respect, the explanation and elaboration of the value typology is important both for the tourism sector and for the development of marketing science in the literature.

The research is limited to value typologies and the evaluation of these typologies for the tourism sector. In this context, value typology can be evaluated by considering different sectors in future studies. In addition, comparisons can be made between service industries within the scope of value attributions.

References

Adler, M. J. (1981). *Six great ideas.* New York: Macmillan.

Arrow, K. J. (1967). Public and private values. In S. Hook (Ed.), *Human values and economic policy* (pp. 3–21). New York: New York University Press.

Bahm, A. J. (1993). *Axiology: The science of values.* Atlanta-Amsterdam: Rodopi.

Baylis, C. A. (1958). Grading, values, and choice. *Mind, 67*(268), 485–501.

Bond, E. J. (1983). *Reason and value.* UK: Cambridge University Press.

Çavuşoğlu, S., Demirağ, B., & Durmaz, Y. (2020). Investigation of the effect of hedonic shopping value on discounted product purchasing, *Review of International Business and Strategy, (ahead-of-print)*, ahead-of-print.

Celuch, K., & Slama, M. (1995). "Getting along" and "getting ahead" as motives for self-presentation: Their impact on advertising effectiveness. *Journal of Applied Social Psychology, 25*(19), 1700–1713.

Chambers, C., Kouvelis, P., & Semple, J. (2006). Quality-based competition, profitability, and variable costs. *Management Science, 52*(12), 1884–1895.
Cheung, C. M., & Lee, M. K. (2012). What drives consumers to spread electronic word of mouth in online consumer-opinion platforms. *Decision Support Systems, 53*(1), 218–225.
Clark, T., Slama, M., & Wolfe, R. (1999). Consumption as self-presentation: A socioanalytic interpretation of Mrs. Cage. *Journal of Marketing, 63*(4), 135–138.
Cobley, P. (Ed.). (2009). *The Routledge companion to semiotics*. UK: Routledge.
Cowan, A. (1964). *Quality control for the manager*. Oxford: Pergamon Press.
Diesing, P. (1962). *Reason in Society: Five types of decisions and their social conditions*. Urbana: University of Illinois Press.
Donner, M., Gohier, R., & de Vries, H. (2020). A new circular business model typology for creating value from agro-waste. *Science of the Total Environment, 716*, 137065.
Duchessi, P. (2002). *Crafting customer value: The art and science*. Indiana: Purdue University Press.
Dwyer, P. (2007). Measuring the value of electronic word of mouth and its impact in consumer communities. *Journal of Interactive Marketing, 21*(2), 63–79.
Fine, M., & Peters, J. (2016). *The nature of health: How America lost, and can regain, a basic human value*. New York: CRC Press.
Fiore, A. M., & Kim, S. (1997). Olfactory cues of appearance affecting impressions of professional image of women. *Journal of Career Development, 23*(4), 247–263.
Fromm, E. (1941). *Escape from Freedom*. Oxford: Farrar.
Frondizi, R. (1971). *What is value? An introduction to axiology*. Second Edition, La Salle, IL: Open Court Publishing Company.
Gale, B. T., & Wood, R. C. (1994). *Managing customer value: Creating quality and service that customers can see*. New York: The Free Press.
Harré, R., & Secord, P. F. (1973). *The explanation of social behavior*. Littlefield, NJ: Adams & Co.
Hilliard, A. L. (1950). *The forms of value: The extension of hedonistic axiology*. New York: Columbia University Press.
Holbrook, M. B. (1982). *The experiential aspects of consumption: Consumer fantasies, Feelings, and Fun*. 9(September).
Holbrook, M. B. (1985). Quality and value in the consumption experience: Phaedrus Rides Again. In J. Jacoby and J. C. Olson (Ed.), *Perceived quality: How consumers view stores and merchandise, in* (pp. 31–57). Lexington, MA: D.C. Heath and Company.
Holbrook, M. B. (1994). The nature of customer value: An axiology of services in the consumption experience. In R. T. (Ed.), *Service quality: New directions in theory and practice* (pp. 21–71). Thousand Oaks, CA: Sage Publications.
Holbrook, M. B. (1996). Customer value-a framework for analysis and research. In J. K. P. Corfman and J. G. Lynch (Eds.), *Advances in consumer research, provo* (pp. 138–142). Vol. 23, Provo, UT: Association for Consumer Research.
Holbrook, M. B. (2002). *Introduction to consumer value in Consumer value*. London: Routledge.
Hollensen, S. (2010). *Marketing management: A relationship approach (2nd ed.)*. England: Pearson Education.
Kahle, L. R. (1983). A theory and a method for studying values. In, L. R. (ed.), *Social values and social change* (pp. 43–69). New York: Praeger.
Kotler, P., Saliba, S., & Wrenn, B. (1991). *Marketing management: Analysis, planning, and control: Instructor's manual*. London: Prentice-Hall.
Leroi-Werelds, S. (2021). Conceptualising customer value in physical retail: A marketing perspective. In, K., Petermans, A., Melewar, T.C. and Dennis, C. (Ed.), *The Value of Design in Retail and Branding*, (pp. 9-24), UK, Bingley: Emerald Publishing Limited.
Levitt, T. (1960). Marketing myopia. *Harvard Business Review, 38*(4), 45-56.
Mahajan, G. (2007). *Customer value investment: Formula for sustained business success*. New Delhi: SAGE Publications.
McFee, G. (2004). *Sport, rules and values: Philosophical investigations into the nature of sport*. London: Routledge.
Mehrabian, A., & Russell, J. A. (1974). *An approach to environmental psychology*. Cambridge, MA: The MIT Press.
Miller, S. (1973). Ends, means, and galumphing: Some leitmotifs of play 1. *American Anthropologist, 75*(1), 87–98.
Moore, G. E., & Baldwin, T. (1993). *Principia ethica*. Cambridge: Cambridge University Press.
Nerlich, G. (1989). *Values and valuing: Speculations on the ethical life of persons*. New York: Oxford.
Nozick, R. (1981). *Philosophical explanations*. Cambridge, MA: Harvard University Press.
Osborne, H. (1933). *Foundations of the philosophy of value*. Cambridge: Cambridge University Press.

Pap, A. (1946). The verifiability of value judgments. *Ethics, 56*(3), 178–185.
Pepper, S. C. (1958). *The sources of value*. Berkeley: University of California Press.
Perry, R. B. (1954). *Realms of value*. Cambridge, MA: Harvard University Press.
Porter, M. E. (1985). *Competitive advantage: Creating and sustaining superior performance*. New York: Free Press.
Riley, M. W. (2005). *Plato's cratylus: Argument, form, and structure*. Holland: Rodopi.
Santayana, G. (1896). *The sense of beauty*. New York: Dover Publications.
Schröder, M. J. (2003). *Food quality and consumer value: Delivering food that satisfies*. New York: Springer Science & Business Media.
Shillito, M. L., & De Marle, D. J. (1992). *Value: Its measurement, design, and management*. New York: John Wiley & Sons.
Slama, M., & Wolfe, R. (1997). Consumption as Self-Presentation: A Socioanalytic Interpretation of Mrs. Cage, *Working Paper*, College of Business, Illinois State University, Normal, IL.
Taylor, P. W. (1961). *Normative discourse*. Englewood Cliffs, NJ: Prentice-Hall.
Tuchman, B. W. (1980). The decline of quality. *New York Times Magazine, 104*(2), 38–41.
Veblen, T. (2017). *The theory of the leisure class*. UK: Routledge.
Veroff, J. (1983). *Introduction*. In L. R. Kahle (Ed.), *Social values and social change* (pp. xiii–xviii). New York: Praeger.
Weinstein, A. (2018). *Superior customer value: Finding and keeping customers in the now economy*. Routledge.
Weinstein, A., & Johnson, W. (1999). *Designing and delivering superior customer value: Concepts, cases, and applications*. USA: CRC press.
Wiggins, D. (1998). *Needs, values, truth: Essays in the philosophy of value*. USA: Oxford University Press.
Yüksel, F., & Bekar, A. (2017). Küçük ve orta ölçekli yiyecek-içecek işletmesi sahiplerinin girişimci kişilik özellikleri ve girişimcilik motivasyonları (Entrepreneurial personality traits and entrepreneurship motivations of small and medium-sized food and beverage business owners). *Journal of Tourism and Gastronomy Studies, 5*(4), 33–46.
Yüksel, F., & Kılıç, B. (2016). Elektronik ağızdan kulağa iletişimin (e-wom) turistik destinasyon seçimi üzerine etkisi. *Journal of International Social Research, 9*(46).
Zeithaml, V. A. (1988). Consumer perceptions of price, quality, and value: A means-end model and synthesis of evidence. *Journal of Marketing, 52*(3), 2–22.
Zerjav, V. (2021). Why do business organizations participate in projects? Toward a typology of project value domains. *Project Management Journal, 52*(3), 287–297.

22

IMPACT OF OVERTOURISM ON RESIDENTS

Sebastian Amrhein, Gert-Jan Hospers, and Dirk Reiser

Introduction

Global tourism, as one of the world's biggest industries has tremendous effects on the environment, the economy as well as on the people involved. For tourists, travelling provides joy and relaxation, for others, it offers a possibility to earn a living. However, as recent research on overtourism (OT) has demonstrated, for residents of frequently visited destinations, masses of visitors also represent a burden and negatively influence their everyday lives. Consequently, a growing number of residents has joined forces and raised their voices against the increasing influx of tourists. In various cities, such as Barcelona, Venice, Palma de Mallorca, Berlin or Amsterdam rather spontaneous neighbourhood actions developed within only a few years into activist groups and coordinated social movements. They provide a platform for affected parties to interact, express their concerns and demands in form of meetings, demonstrations and, in some cases, led to the formation of international networks (e.g. SET – Ciudades del Sur de Europa ante la Turistización/Southern European Cities against touristication). Political measures to counter the problem of OT were rather hastily, not embedded in a coordinated vision, and consequently not yet thriving. At the same time, affected residents were becoming politically active, getting involved in social movements and potentially even questioning the growth-driven capitalist system by calling for degrowth.

In general, residents' dissatisfaction and a rather negative notion towards tourism are not novel. Tourism scholars have been examining the development of tourism and its occasionally negative effects on residents' attitudes for decades (e.g. Doxey's Irritation Index, 1975). Previous research has focussed almost exclusively on the opinions of residents regarding tourism itself. Recent events initiated by social movements provide evidence that OT not only affects residents' attitudes regarding visitors and tourism but are more profound. The demands towards politics for degrowth demonstrate the politicising effect of OT and give rise to the assumption that OT is even impacting socio-political attitudes and world views which the hitherto existing investigations are not able to display. A broader socio-political approach, which considers tourism as one of the largest industries within a neoliberal system – having a massive impact on people's life rather than simply changing attitudes towards tourism – is therefore necessary. The consideration of substantial effects of OT will open up a new perspective on tourisms' socio-psychological impacts. It will thus contribute to Dodds and Butler's (2019) statement that "overtourism has the power to influence decision-makers and change the state of affairs" (p. 273).

A useful framework to systematically analyse and understand significant changes in adults' attitudes might be provided by the transformative learning theory (TLT), developed in 1978 by

DOI: 10.4324/9781003161868-22

the American sociologist Jack Mezirow. Mezirow's theory assumes that the experience of a disorienting dilemma can trigger a transformation process, which can result in changes in one's own attitudes and values, worldview and social behaviour. In this chapter, it is argued that experiencing the negative effects of mass tourism can be considered such a disorienting dilemma leading to the aforementioned transformative process.

This chapter will outline the interrelations between OT and its impact on residents. It will draw the attention to progressive demands of resident's resistance organisations and propose a possible theory for analysing the underlying socio-psychological processes triggered by OT. It therefore provides an interdisciplinary conceptual framework of how a socio-political transformation of residents can be explored and extends the discussion about the transformative power of tourism. Research's state of the art of each specific field will be displayed before establishing the interrelations between them. The particular circumstances caused by the Covid-19 pandemic will be considered by means of Box No. 1, while Box No. 2 briefly discusses two examples of social movements from Spain that developed in the field of OT.

Overtourism and Residents' Reaction

The term OT, which was presumably first used in the context of mass tourism in 2012 (Goodwin, 2017), is defined by Goodwin (2017) as "destinations where hosts or guests, locals or visitors, feel that there are too many visitors and that the quality of life in the area or the quality of the experience has deteriorated unacceptably" (p. 1). The World Tourism Organization (2018) largely concurs with Goodwin's definition by describing OT as "the impact of tourism on a destination, or parts thereof, that excessively influences perceived quality of life of citizens and/or quality of visitors' experiences in a negative way" (p. 4). Interestingly, both definitions see the negative impacts of tourism, but avoid weighting the negative effects to either the side of the visited or that of the visiting. By doing so, the impression arises that OT effects the visited as well as the visitors similarly. In fact, visitors mainly suffer from decreasing travel experiences due to overcrowding, limited service quality or over-advertisement (Pechlaner et al., 2020; Żemła, 2020). Negative consequences for residents and destinations seem to be more severe, for example, increasing real estate and rental prices, accelerated gentrification processes, pollution (e.g. noise, environmental), loss of identity and sense of community and loss of shops for daily supply. Many tourism scholars have paid attention to these issues and the demands of residents to tackle it (e.g. Antunes et al., 2020; Butler, 2019; Cocola-Gant et al., 2020; Diaz-Parra & Jover, 2020; Koens et al., 2018; Mansilla, 2018; Mihalic, 2020; Milano et al., 2019; Novy & Colomb, 2019; Romero-Padilla et al., 2019; Séraphin et al., 2020; Vollmer, 2018). In their quantitative analysis, Antunes et al. (2020) found proof for physical changes of urban environments as well as increasing costs for real estate in certain areas in Barcelona. Vollmer (2018) confirms the effects on real estate prices and even points out interrelations of OT and gentrification processes in Berlin. By applying a qualitative approach, Mansilla (2018) reveals the daily challenges of residents due to changes of their surroundings, for example, supermarkets, which hardly stock any goods for locals but only for tourist needs, the increasing number of bars and expensive restaurants as well as crowded streets, noise and pollution. Similarly, Cocola-Gant et al. (2020) disclose negative effects after conducting research in Barcelona, Lisbon and Seville. Furthermore, Diaz-Parra and Jover (2020) portray the residents' risk of alienation from their domiciles due to the high influx of tourists in Seville, and Butler (2019) reminds of the negative effects OT can create even in rural areas. A definition recognising the mentioned circumstances is given by Peeters et al. (2018), who describe OT as "[...] the situation in which the impact of tourism, at certain times and in certain locations, exceeds physical, ecological, social, economic, psychological, and/or political capacity thresholds" (p. 22).

Box No. 1 Covid-19 Impacts on Global (Over-)Tourism

The pandemic and its impacts on global mobility make OT and the resident-resistance seem long forgotten. However, the problem is not yet solved. Tourism scholars such as Hall et al. (2020) or Haywood (2020) already point towards the risk that the current crisis might be used by the tourism industry to even expand pre-Covid-19 growth. Media reports on the tourism industry and politicians who already advocate for an ease of travel restrictions to get international tourism up and running again, underline these concerns (Deutsche Welle, 2020; Wilson, 2020). Other scholars however, see the unexpected Covid-19 outbreak and its subsequent lockdown as a chance for a redefinition and reorientation of global tourism (Higgins-Desbiolles, 2020). To take this opportunity, as mentioned by Higgins-Desbiolles, social movements must reassert their influence on local politics. But, is this still the aim of most residents after experiencing the two extreme situations? Media reports allow the assumption that this is not clearly said. While some people are excited about the current situation (Benz, 2020; Riverine Herald, 2020), others are not enthusiastic at all and rather worry about the negative social and economic consequences (Henley & Smith, 2020; Sharma & Nicolau, 2020; Qiu et al., 2020).

At the point of writing, a reliable prediction for the future development of Covid-19 and its effects for global tourism cannot be made. Until the effects can be determined, the chapter will focus on overtourism and its effects on residents.

Alternatively, Koens et al. (2018) define OT as "an excessive negative impact of tourism on the host communities and/or natural environment" (p. 2).

Although the negative effects seem to weigh heavily on the shoulders of those affected, there are scholars who criticise the attention the negative effects receive. Claiming that, "Overtourism is revealed as contested and also as a plastic phenomenon that can be molded to fit the assumptions of the user" (Butcher, 2020, p. 85). He doubts the existence and scope of OT. Instead he explicitly points out the positive outcomes, including mainly economic benefits and job opportunities, which, in his view, outweighing the costs (ibid.). Butcher and supporters of the "tourism growth leads to prosperity" narrative often justify their point of view with economic figures.

For residents in frequently visited, mainly urban destinations in the contrary, those economic figures are seen as a statistical underpinning of their daily struggle with the negative effects of tourism against which some were forming resistance. In Berlin, for example, the leftist organisation "Interim" released a publication in which they identified "tourists to be legitimate targets in the fight against gentrification and encouraged readers to steal phones and wallets from visitors and engage in all sorts of other hostile and intimidating activities so as to scare them away" (Novy, 2017, p. 61). These events spread rapidly around the globe with some accusing Berlin to be a place of "tourist hate" (ibid.). A further example given by Novy (2017, p. 63) is a community meeting organised by the local green party in Berlin in 2011 with the motto "Help, the tourists are coming!", which tempted journalists to portray this particular district as the source of "tourist-haters" (ibid.). Another group that attracted much international media attention was "Arran", a self-called Marxist youth organisation in the Catalan countries fighting for national, social and gender equality of their (Catalan) people (Arran, 2020). With slogans such as "tourists go home" or "tourism kills neighbourhoods" and even physically violent attacks on bike-rents (videos uploaded on YouTube[1]) and on tourist busses, they covered newspapers all around the globe (Burgen, 2017; Hunt, 2017; Spiegel, 2017). Yet, as Colomb and Novy (2017) make clear, events like this are rather individual and conducted by unorganised groups. They are gladly taken up by media, but do not reflect the whole spectrum of the protests. Here too, other opinions prevail. In his aforementioned publication, Butcher (2020) draws on such media reports and paints a one-sided picture of residents, describing them as violent, nationalistic and anti-tourism orientated. Blanco-Romero et al. (2019) allude that those accusations are created as well from the tourism industry. In a qualitative research of social movements in Barcelona, Blanco-Romero et al. (2019) cite members of the ABTS[2] (Assemblea de Barris per un Turisme Sostenible – ABTS, Assembly of Neighbourhoods for a Sustainable Tourism):

> The origin of the term tourism-phobia as 'a propaganda campaign, to exchange the roles between victim and executioner, to put it very dramatically, between aggressor and assaulted, let's say, between oppressor and oppressed.' They suggest that the term is a 'corporate creation,' seeking to link with xenophobia, thereby framing the phenomenon as a personal opposition against the tourists, not the industry [and to be aware, that] we are all tourists at some point in our lives.
>
> *(p. 11)*

The ABTS, as one of the largest and most prominent movements against OT and part of the International Network SET, thus clearly distance themselves from being nationalist or anti-tourist. Instead, they demand their right to the city and "working on collective action against tourism and, in particular, against the Barcelona model of urban entrepreneurialism" (Cocola-Gant & Pardo, 2017). Milano et al. (2019a) confirm these findings by stating that the social movements blame the present growth-driven economic model responsible for the negative effects of tourism and demand profound socio-economic changes. Radical groups such as Arran leave no doubt about their opinion by labelling themselves as capitalism-phobic (Novy & Colomb, 2019) with which they underline their position and refer to the accusation of being tourism-phobic. As Novy and Colomb (2019) point out, those demands are not only formulated by radical-leftists but people who are often "critical of current, neoliberal forms and practices of urban development" (p. 8). The resulting demands towards politics are both ambitious and unequivocal and often include the call for degrowth (Fletcher et al., 2019; Milano et al., 2019a; Valdivielso & Moranta, 2019).

The increasing numbers of dissatisfied residents and the public expression of their discontent put pressure on the industry and politics to take action. The initiated measures embraced efforts to distribute tourists from the centre to peripheral attractions, smart solutions such as apps warning visitors from crowding or the implementation of tourist taxes (or the increase of current taxes) (Coffey, 2017; McKinsey & Company and World Travel & Tourism Council, 2017; Peeters et al., 2018; Redazione ANSA, 2020; Sendlhofer, 2018). However, the measures taken to better manage tourist flows did not have the short-term effects the decision makers had hoped for, as the largely unhindered increasing numbers of visitors portrayed. Critical tourism scholars were sceptical about their success from the start (García-Hernández et al., 2019; Milano, 2018). For example, the recommendations from McKinsey & Company and World Travel & Tourism Council (2017) or the World Tourism Organization (2018) to develop points of interests in the periphery of highly frequented tourism hotspots or to focus on tourists of higher quality are questionable from a critical perspective, as Blanco-Romero et al. (2019) state clearly. Mansilla and Milano (2019) even argue that the proposed measures primarily serve to keep the capitalist wheel turning instead of tackling the problem by its origin. Similarly, to many social movements and resisting residents, critical scholars see the roots of the problem in the prevailing socio-economic system, characterising OT as a symptom of a much bigger, systemic error which is rooted in the dominant growth-driven, neoliberal model (Büscher & Fletcher, 2017; Higgins-Desbiolles et al., 2019). Both sides therefore bring degrowth into play as an alternative to the current neoliberal system.

Degrowth and Tourism

Degrowth is the English translation of the original French word *décroissance*. It can be explained as a critical evaluation of the current development predominance (Petridis et al., 2015). Demaria et al. (2013) see it as "an attempt to re-politicise debates about desired socio-environmental futures" (p. 191). Kallis and March (2015) describe it as a "project of radical socioecological transformation calling for decolonizing the social imaginary from capitalism's pursuit of endless growth" (p. 360). Degrowth recognises the natural and social resource bases that make infinite growth on a finite planet impossible and offers "a frame constituted by a large array of concerns, goals, strategies and actions" (Demaria et al., 2013, p. 192) to transform the actual growth-driven capitalist world order.

The different explanations of degrowth outline the complexity of the concept. However, supporters of the degrowth idea agree on the urgent need of a holistic alternative to the existing growth driven system, which is currently dominant in most parts of the planet. Simultaneously they point out that the successful application of the degrowth approach requires "a whole reorientation of paradigm" (March, 2018, p. 1695, as cited in Higgins-Desbiolles et al., 2019, p. 13), or, as Demaria et al. (2013) put it,

> Degrowth only makes sense when its sources are taken into account, meaning not just ecology and bioeconomics, but also meaning of life and well-being, anti-utilitarianism, justice and democracy. Taken independently they can lead to incomplete and reductionist projects fundamentally incompatible with the ideas of the degrowth movement.
>
> *(p. 206)*

In such a holistic approach, tourism, as one of the world's biggest industries, needs to be considered. Yet this has not happened widely. Before social movements merged degrowth and tourism (e.g. in Barcelona and Berlin, as mentioned above), the latter has hardly been taken into account in the early degrowth discussions, and vice versa. Possibly the earliest combination of tourism and degrowth was made by Hall (2009), who stated that "sustainable tourism development is tourism development without growth" (p. 53). A few years from then, the topic was discussed among critical tourism scholars (Boluk et al., 2019; Fletcher et al., 2019; Milano, 2018) as an alternative to the present and unjust tourism practice.

By projecting Latouch's eight r's[3] for a degrowth transition on the tourism model, Higgins-Desbiolles et al. (2019, p. 13) even elaborated how tourism degrowth could be applied in practice.

As stated above, degrowth is a holistic concept and the application on tourism alone would not be promising. It would rather require substantial changes of the accustomed western lifestyles and resistance against neoliberal and patriarchal structures many are benefitting from (Berberoglu, 2019). However, "are we ready for those fundamental changes?" as Higgins-Desbiolles et al. (2019, p. 16) ask. Dodds and Butler (2019) might be optimistic when claiming that "overtourism has the power to influence decision makers and change the state of affairs" (p. 273). The characterised social movements and their demands give also reason to believe that profound changes are desired, but it remains unclear how deep rooted those demands are. Are people marching for degrowth in tourism aware of the fundamental changes necessary? Do they intend a change of the state of the art and are they aware what it would mean for themselves? According to Appiah and Bischoff (2011), such far-reaching systematic changes need to be initiated from below and require a moral or cultural transformation of people. But can tourism be a driver of, or even initiate those far-reaching transformative effects?

At this point, it is crucial to distinguish between 'attitude change' and 'transformation', in this case human transformation. Both are terms of social psychology and have been the subject of research and discussion in their fields for many years (and are still ongoing) (Bohner & Dickel, 2011). A conclusive, unambiguous definition is therefore not possible. However, it can be stated that an attitude change is described as the processing of information, resulting in an assessment of an "object of thought" (Bohner & Dickel, 2011, p. 397). Additionally, the stableness of an attitude can vary due to factors, for example, implicitness or explicitness of the attitude (Crano & Prislin, 2008). Human transformation can be described as changes of previous assumptions and beliefs about people and the world, accompanied by behavioural changes, prompted by new experiences (Mezirow, 1978; Reisinger, 2013).

From Irritation to Transformation

This distinction can be found in tourism research, even if it is often not explicitly mentioned. Doxey's famous irritation index model (1975) was one of the earliest approaches to identify residents attitudes on tourism development. Doxey describes those attitudes as four stages of euphoria, apathy, and if numbers of visitors exceed a certain threshold, irritation and finally antagonism. Ever since, people's attitudes and reactions on tourism have received much attention (Andereck et al., 2005; Ap & Crompton, 1993; Bertocchi & Visentin, 2019; Brougham & Butler, 1981; Garau-Vadell et al., 2014; Szromek et al., 2020). Szromek et al. (2020) investigated residents' attitudes towards tourism development and increasing number of visitors in Kraków. Garau-Vadell et al. (2014) examined residents' perceptions of tourism's impacts on the economy, culture, society and the environment in Tenerife and Mallorca and Andereck et al. (2005) for example determined the conditions for community's positive or negative attitudes towards tourism.

However, those investigations mainly focus on residents' attitudes towards the object 'tourism'. The aspect of human or personal transformation has not yet been sufficiently considered in tourism studies (Reisinger, 2015; Ross, 2010), even though the pictured examples of social movements and their demands towards politics (see also Box No. 2) give rise to the assumption that OT not only affects their attitudes towards tourism but are more fundamental. Furthermore, existing literature debating tourisms' transformative effects largely focusses on tourists (Kottler, 1997; Lean, 2012; Reisinger, 2013; Stone & Duffy, 2015). Stone and Duffy (2015) for example examined 53 publications on transformative tourism, of which none was targeting the effects on residents even though the few existing investigations confirm transformative effects of tourism for the host community (Burrai & Cuevas, 2015; Deville, 2015; Reisinger, 2015; Schweinsberg et al., 2015).

Box No. 2 Examples of Social Movements and Their Demands Towards Politics and Economy

In Madrid, for example, the movement Lavapiés Dónde Vas (Lavapiés – a neighbourhood in Madrid – where do you go, translated by the author) recognises tourism and its negative effects for residents as a symptom of global capitalism (Assemblea de Barris pel Decreixement Turístic 2018a, translated by the author). A further example is the Mallorcan movement Ciutat per a qui l'habita (City for the inhabitants, translated by the author) that sees the ever increasing tourist industry as a main driver of social inequality as well as the environmental destruction of the island. This movement points at structural problems, benefitting only a few privileged people at the expense of many others who suffer from the negative effects. Ciutat per a qui l'habita calls for fundamental changes towards "a self-sufficient economy and horizontal and solidary relations" (Ciutat per a qui l'habita, 2017, no page, translated by the author).

Burrai and Cuevas (2015) for example attest the transformative potential of volunteer tourism based on investigations in host communities in Peru and Thailand. Deville (2015), examining the experiences of WWOOF (Willing Workers on Organic Farms) hosts in Australia, comes to a similar conclusion.

In addition, studies targeted on the transformation of visitors or affected residents have so far examined the transformative effects of rather niche products, often referred to as transformative travel. According to Stone and Duffy (2015) transformative travel embraces educational travel, cross-cultural travel, volunteer tourism/voluntourism or tour guides, operators and hospitality students. In tourism practice, transformative travel even seems to develop into a trend, suggested by the travel blog/website "Transform Me Travel" (2020), explaining transformative travel as "travel that changes you. Transformative travel has positive, long term impacts on your life. It can include group travel or going solo; think guided retreats or tours versus self-taught and self-directed travel" (n.p.). Ross (2010) also mentions different forms of positive transformations for travellers, for example, for the spirit or the heart or through a physical challenge, etc. It can be noted that both, in tourism research as well as in tourism practice, transformative travel is considered to create rather positive outcomes, mainly for the traveller. Transformative effects of mass tourism, which Bruner (1991) already estimated as being more extensive for hosts and host communities than for tourists, considering the masses of visitors as well as the temporal and spatial intensity, are instead scarce.

As outlined above, the behaviour as well as the political demands of the social movements against OT indicate that years of mass tourism have led to more profound consequences than just a changed attitude towards tourism. To the researchers' knowledge a corresponding scientific investigation is not yet available and therefore necessary. This assumption is confirmed by Gössling et al. (2020), stating that "As a phenomenon associated with residents' negative views of tourism development outcomes, socio-psychological foundations of overtourism have so far been insufficiently considered" (p. 1).

Conclusion

This chapter has demonstrated the interrelations between the negative effects of mass tourism and resident resistance. Building on statements of scholars, research findings as well as publications of social movements and their demands, it is claimed that the effects of tourism on residents are more profound than hitherto debated in tourism practice and academia. Furthermore, it is criticised that existing research is mainly focussed on the positive transformative effects of tourism for travellers, ignoring the effects of mass tourism even though the spatial and temporal influences it has on residents are obvious and have been pointed out by scholars such as Bruner (1991), Dodds and Butler (2019) or Higgins-Desbiolles et al. (2020). Those assessments are supported by examples of two social movements from southern European cities against OT. They additionally demonstrate their understanding of tourism in a broader socio-economic context that needs to be changed – supporting the claim that hitherto research is not considering the socio-psychological effects of mass tourism in adequate profoundness.

Despite the outlined interrelations and indications for the transformative effects of OT, research analysing these effects have not been carried out to date (to the best of the authors' knowledge). Consequently, many questions still remain unanswered. What are the effects of the frequent exposure to high amounts of visitors on adult thinking and acting? Are transformative processes noticeable and are these processes supported – or even triggered by OT? It remains unclear whether people who call for far-reaching changes are aware of the scope of their demands. Do they see themselves as part of the system and are they themselves willing to change their habits and behaviour, or perhaps have already done so? Furthermore, it is important to clarify the role of people's personal and socio-political position and its scope of influence. Empirical studies are needed to answer these questions, to gain clarity about the profundity of the socio-psychological impact of mass tourism and to contribute to the above mentioned discussion raised by Dodds and Butler (2019) about tourisms' power to initiate fundamental changes. Further research should therefore be done. A theoretical framework to assess the transformative effects of OT might be provided by Mezirow's TLT.

Notes

1 https://www.youtube.com/watch?v=fcVNwg8rVK8&feature=emb_logo (retrieved 03.11.2020)

2 ABTS is one of the most famous and active social movements in the investigated context. The movement has been formed by Barcelona residents in 2015 (Assemblea de Barris pel Decreixement Turístic (ABDT).

3 The eight r's (Latouche, 2012)
   1 re-evaluate and shift values;
   2 re-conceptualize entrenched capitalist concepts;
   3 restructure production;
   4 redistributions at the global, regional and local scale;
   5 re-localize the economy;
   6 reduce;
   7 re-use; and
   8 recycle resources.

References

Andereck, K. L., Valentine, K. M., Knopf, R. C., & Vogt, C. A. (2005). Residents' perceptions of community tourism impacts. *Annals of Tourism Research, 32*(4), 1056–1076. https://doi.org/10.1016/j.annals.2005.03.001

Antunes, B., March, H., & Connolly, J. J.T. (2020). Spatializing gentrification in situ: A critical cartography of resident perceptions of neighbourhood change in Vallcarca, Barcelona. *Cities, 97*, 102521. https://doi.org/10.1016/j.cities.2019.102521

Ap, J., & Crompton, J. L. (1993). Residents' strategies for responding to tourism impacts. *Journal of Travel Research, 32*(1), 47–50. https://doi.org/10.1177/004728759303200108

Appiah, K. A., & Bischoff, M. (2011). *Eine Frage der Ehre: Oder Wie es zu moralischen Revolutionen kommt*. C.H. Beck.

Arran. (2020). *Organització juvenil de l'Esquerra Independentista*. Arran. https://arran.cat/

Assemblea de Barris pel Decreixement Turístic. (2018). *Assemblea de Barris pel Decreixement Turístic (ABDT)*. Assembleabarris. https://assembleabarris.wordpress.com/

Assemblea de Barris pel Decreixement Turístic. (2018a). *2nd Fòrum Veïnal sobre Turisme*. Assembleabarris. https://assembleabarris.wordpress.com/resumen-del-2n-forum-veinal-sobre-turisme/

Benz, M. (2020, May 19). *Tourismus in der Coronakrise: Hallstatts Zukunft ohne Chinesen*. Neue Zürcher Zeitung. https://www.nzz.ch/wirtschaft/tourismus-in-der-coronakrise-hallstatts-zukunft-ohne-chinesen-ld.1556255

Berberoglu, B. (Ed.). (2019). *The Palgrave handbook of social movements, revolution, and social transformation*. Springer International Publishing. https://doi.org/10.1007/978-3-319-92354-3

Bertocchi, D., & Visentin, F. (2019). "The Overwhelmed City": Physical and social over-capacities of global tourism in Venice. *Sustainability, 11*(24), 6937. https://doi.org/10.3390/su11246937

Blanco-Romero, A., Blàzquez-Salom, M., Morell, M., & Fletcher, R. (2019). Not tourism-phobia but urban-philia: understanding stakeholders' perceptions of urban touristification. *Boletín De La Asociación De Geógrafos Españoles*(83), 1–30. https://doi.org/10.21138/bage.2834

Bohner, G., & Dickel, N. (2011). Attitudes and attitude change. *Annual Review of Psychology, 62*, 391–417. https://doi.org/10.1146/annurev.psych.121208.131609

Boluk, K. A., Cavaliere, C. T., & Higgins-Desbiolles, F. (2019). A critical framework for interrogating the United Nations Sustainable Development Goals 2030 Agenda in tourism. *Journal of Sustainable Tourism, 27*(7), 847–864. https://doi.org/10.1080/09669582.2019.1619748

Brougham, J. E., & Butler, R. W. (1981). A segmentation analysis of resident attitudes to the social impact of tourism. *Annals of Tourism Research, 8*(4), 569–590. https://doi.org/10.1016/0160-7383(81)90042-6
Bruner, E. M. (1991). Transformation of self in tourism. *Annals of Tourism Research, 18*(2), 238–250. https://doi.org/10.1016/0160-7383(91)90007-X
Burgen, S. (2017, August 1). Barcelona anti-tourism activists vandalise bikes and bus. *The Guardian*. https://www.theguardian.com/world/2017/aug/01/barcelona-anti-tourism-activists-vandalise-bikes-and-bus
Burrai, E., & de las Cuevas, J. I. (2015). Transformation of local lives through volunteer tourism: Peruvian and Thai case studies. In Y. Reisinger (Ed.), *Transformational tourism: Host perspectives* (pp. 117–128). CABI.
Büscher, B., & Fletcher, R. (2017). Destructive creation: Capital accumulation and the structural violence of tourism. *Journal of Sustainable Tourism, 25*(5), 651–667. https://doi.org/10.1080/09669582.2016.1159214
Butcher, J. (2020). The Construction of 'Overtourism': The case of UK media coverage of Barcelona's 2017 tourism protests and their aftermath. In H. Séraphin, T. Gladkikh, & T. Vo Thanh (Eds.), *Overtourism: Causes, implications and solutions* (pp. 69–88). Palgrave Macmillan.
Butler, R. W. (2019). Overtourism in rural settings: The Scottish highlands and islands. In R. Dodds & R. Butler (Eds.), *Overtourism: Issues, realities and solutions* (pp. 199–213). De Gruyter.
Ciutat per a qui l'habita. (2017). *Manifest*. Ciutat per a qui l'habita. https://ciutatperaquilhabitablist.blog/prova/
Cocola-Gant, A., Gago, A., & Jover, J. (2020). Tourism, gentrification and neighbourhood change: An analytical framework– reflections from Southern European Cities. In J. Oskam (Ed.), *The overtourism debate: NIMBY, nuisance, commodification* (pp. 121–135). Emerald Publishing Limited.
Cocola-Gant, A., & Pardo, D. (2017). Resisting tourism gentrification: The experience of grass-roots movements in Barcelona. *Journal of Urban Design and Planning, 13*(5), 39–48.
Coffey, H. (2017, October 23). Amsterdam has a new solution for overtourism. *The Independent*. https://www.independent.co.uk/travel/news-and-advice/amsterdam-overtourism-solution-tourists-technology-van-gogh-museum-canal-boat-rides-a8015811.html
Colomb, C., & Novy, J. (Eds.). (2017). *Contemporary geographies of leisure, tourism and mobility. Protest and resistance in the tourist city*. Routledge Taylor & Francis Group.
Crano, W. D., & Prislin, R. (Eds.). (2008). *Frontiers of social psychology. Attitudes and attitude change*. Psychology Press.
Demaria, F., Schneider, F., Sekulova, F., & Martinez-Alier, J. (2013). What is degrowth? From an activist slogan to a social movement. *Environmental Values, 22*(2), 191–215. https://doi.org/10.3197/096327113X13581561725194
Deutsche Welle. (2020, May 26). Germany calls for opening of internal EU borders by June. *Deutsche Welle*. https://www.dw.com/en/germany-calls-for-opening-of-internal-eu-borders-by-june/a-53567535
Deville, A. (2015). Transformation and the WWOOF exchange: The host experience. In Y. Reisinger (Ed.), *Transformational tourism: Host perspectives* (pp. 141–164). CABI.
Diaz-Parra, I., & Jover, J. (2020). Overtourism, place alienation and the right to the city: insights from the historic centre of Seville, Spain. *Journal of Sustainable Tourism, 18*(3), 1–18. https://doi.org/10.1080/09669582.2020.1717504
Dodds, R., & Butler, R. (Eds.). (2019). *Overtourism: Issues, realities and solutions*. De Gruyter.
Doxey, G. V. (1975). A Causation theory of visitor–resident irritants, methodology and research inferences: The impact of tourism. *Travel Research Association, 6th Annual Conference Proceedings*, 195–198.
Fletcher, R., Murray Mas, I., Blanco-Romero, A., & Blázquez-Salom, M. (2019). Tourism and degrowth: An emerging agenda for research and praxis. *Journal of Sustainable Tourism, 27*(12), 1745–1763. https://doi.org/10.1080/09669582.2019.1679822
Garau-Vadell, J. B., Díaz-Armas, R., & Gutierrez-Taño, D. (2014). Residents' perceptions of tourism impacts on Island Destinations: A comparative analysis. *International Journal of Tourism Research, 16*(6), 578–585. https://doi.org/10.1002/jtr.1951
García Hernández, M., Baidal, J. I., & Mendoza de Miguel, S. (2019). Overtourism in urban destinations: The myth of smart solutions. *Boletín De La Asociación De Geógrafos Españoles* (83), 1–38. https://doi.org/10.21138/bage.2830
Goodwin, H. (2017). The challenge of overtourism. *Responsible Tourism Partnership Working Paper*. http://www.millennium-destinations.com/uploads/4/1/9/7/41979675/rtpwp4overtourism012017.pdf
Gössling, S., McCabe, S., & Chen, N. C. (2020). A socio-psychological conceptualisation of overtourism. *Annals of Tourism Research, 84*, 102976. https://doi.org/10.1016/j.annals.2020.102976
Hall, C. M. (2009). Degrowing tourism: Décroissance, sustainable consumption and steady-state tourism. *Anatolia, 20*(1), 46–61. https://doi.org/10.1080/13032917.2009.10518894
Hall, C. M., Scott, D., & Gössling, S. (2020). Pandemics, transformations and tourism: Be careful what you wish for. *Tourism Geographies, 22*(3), 577–598. https://doi.org/10.1080/14616688.2020.1759131

Haywood, K. M. (2020). A post COVID-19 future – tourism re-imagined and re-enabled. *Tourism Geographies, 22*(3), 599–609. https://doi.org/10.1080/14616688.2020.1762120
Henley, J., & Smith, H. (2020). Covid-19 throws Europe's tourism industry into chaos. *The Guardian*. https://www.theguardian.com/world/2020/may/02/covid-19-throws-europes-tourism-industry-into-chaos
Higgins-Desbiolles, F. (2020). Socialising tourism for social and ecological justice after COVID-19. *Tourism Geographies, 22*(3), 610–623. https://doi.org/10.1080/14616688.2020.1757748
Higgins-Desbiolles, F., Carnicelli, S., Krolikowski, C., Wijesinghe, G., & Boluk, K. (2019). Degrowing tourism: rethinking tourism. *Journal of Sustainable Tourism, 27*(12), 1926–1944. https://doi.org/10.1080/09669582.2019.1601732
Hunt, T. (2017, August 5). Spain's tourism RIOTS: Militant group promise new ATTACKS using 'dangerous' tactics. *Express*. https://www.express.co.uk/news/world/837251/holiday-spain-arran-terror-attack-militant-Laura-Flores-alvaro-Nadal-Barcelona-riots
Kallis, G., & March, H. (2015). Imaginaries of hope: The Utopianism of degrowth. *Annals of the Association of American Geographers, 105*(2), 360–368. https://doi.org/10.1080/00045608.2014.973803
Koens, K., Postma, A., & Papp, B. (2018). Is overtourism overused? Understanding the impact of tourism in a city context. *Sustainability, 10*(12), 4384. https://doi.org/10.3390/su10124384
Kottler, J. A. (1997). *Travel that can change your life: How to create a transformative experience*. Jossey-Bass.
Latouche, S. (2012). *La sociedad de la abundancia frugal*. Icaria.
Lean, G. L. (2012). Transformative travel: A mobilities perspective. *Tourist Studies, 12*(2), 151–172. https://doi.org/10.1177/1468797612454624
Mansilla, J. A., & Milano, C. (2019). Becoming centre: Tourism placemaking and space production in two neighborhoods in Barcelona. *Tourism Geographies, 18*(2), 1–22. https://doi.org/10.1080/14616688.2019.1571097
Mansilla, J. A. L. (2018). Vecinos en peligro de extinción. Turismo urbano, movimientos sociales y exclusión socioespacial en Barcelona. *Pasos. Revista De Turismo Y Patrimonio Cultural, 16*(2), 279–296. https://doi.org/10.25145/j.pasos.2018.16.020
McKinsey & Company and World Travel & Tourism Council. (2017). *Coping with success: Managing overcrowding in tourism destinations*. McKinsey & Company and World Travel & Tourism Council. https://www.mckinsey.com/industries/travel-logistics-and-infrastructure/our-insights/coping-with-success-managing-overcrowding-in-tourism-destinations
Mezirow, J. (1978). Perspective transformation. *Adult Education, 28*(2), 100–110. https://doi.org/10.1177/074171367802800202
Mihalic, T. (2020). Conceptualising overtourism: A sustainability approach. *Annals of Tourism Research, 84*, 103025. https://doi.org/10.1016/j.annals.2020.103025
Milano, C. (2018). Overtourism, malestar social y turismofobia. Un debate controvertido. *Pasos. Revista De Turismo Y Patrimonio Cultural, 18*(3), 551–564. https://doi.org/10.25145/j.pasos.2018.16.041
Milano, C., Novelli, M., & Cheer, J. M. (2019a). Overtourism and degrowth: A social movements perspective. *Journal of Sustainable Tourism, 27*(12), 1857–1875. https://doi.org/10.1080/09669582.2019.1650054
Milano, C., Novelli, M., & Cheer, J. M. (2019). Overtourism and tourismphobia: A journey through four decades of tourism development, planning and local concerns. *Tourism Planning & Development, 16*, 353–357. https://doi.org/10.1080/21568316.2019.1599604
Novy, J. (2017). The selling (out) of Berlin and the de- and re-politicization of urban tourism in Europe's 'Capital of Cool'. In C. Colomb & J. Novy (Eds.), *Contemporary geographies of leisure, tourism and mobility. Protest and resistance in the tourist city* (pp. 52–72). Routledge Taylor & Francis Group.
Novy, J., & Colomb, C. (2019). Urban tourism as a source of contention and social mobilisations: A critical review. *Tourism Planning & Development, 16*(4), 358–375. https://doi.org/10.1080/21568316.2019.1577293
Pechlaner, H., Innerhofer, E., & Erschbamer, G. (2020). *Overtourism: Tourism management and solutions. Contemporary geographies of leisure, tourism and mobility*. Routledge Taylor & Francis Group.
Peeters, P., Gössling, S., Klijs, J., Milano, C., Novelli, M., Dijkmans, C., Eijgelaar, E., Hartman, S., Heslinga, J., Isaac, R., Mitas, O., Moretti, S., Nawijn, J., Papp, B., & Postma, A. (2018, October). *Research for TRAN committee – Overtourism: Impact and possible policy responses*. European Parliament. https://www.google.com/url?sa=t&rct=j&q=&esrc=s&source=web&cd=&cad=rja&uact=8&ved=2ahUKEwiQ6pbSg67yAhXUBGMBHaalB9sQFnoECAUQAQ&url=https%3A%2F%2Fwww.europarl.europa.eu%2FRegData%2Fetudes%2FSTUD%2F2018%2F629184%2FIPOL_STU(2018)629184_EN.pdf&usg=AOvVaw3zwYG35ZMAnfcMgx9JY_Gc
Petridis, P., Muraca, B., & Kallis, G. (2015). Degrowth: between a scientific concept and a slogan for a social movement. In J. Martinez-Alier & R. Muradian (Eds.), *Handbook of ecological economics* (pp. 176–200). Edward Elgar Publishing.
Qiu, R. T. R., Park, J., Li, S., & Song, H. (2020). Social costs of tourism during the COVID-19 pandemic. *Annals of Tourism Research, 84*, 102994. https://doi.org/10.1016/j.annals.2020.102994

Redazione ANSA (2020, February 7). Venice readies visitor counting system – Lifestyle. *ANSA*. https://www.ansa.it/english/news/lifestyle/travel/2020/02/07/venice-readies-visitor-counting-system_50c51fd2-f952-47af-9334-4c374b219350.html

Reisinger, Y. (Ed.). (2013). *Transformational tourism: Tourist perspectives*. CABI.

Reisinger, Y. (Ed.). (2015). *Transformational Tourism: Host Perspectives*. CABI.

Riverine Herald. (2020, May 5). *Indigenous leaders praise virus travel ban*. Riverine Herald. https://www.riverineherald.com.au/national/2020/05/05/1158617/indigenous-leaders-praise-virus-travel-ban

Romero-Padilla, Y., Cerezo-Medina, A., Navarro-Jurado, E., Romero-Martínez, J. M., & Guevara-Plaza, A. (2019). Conflicts in the tourist city from the perspective of local social movements. *Boletín De La Asociación De Geógrafos Españoles*. Advance online publication. https://doi.org/10.21138/bage.2837

Ross, S. L. (2010). Transformative travel: An enjoyable way to foster radical change. *ReVision*, *32*(1), 54–61. https://doi.org/10.4298/REVN.32.1.54-62

Schweinsberg, S., Wearing, S., & Wearing, M. (2015). Transforming nature's value – cultural change comes from below: Rural communities, the 'othered' and host capacity building. In Y. Reisinger (Ed.), *Transformational Tourism: Host Perspectives* (pp. 102–113). CABI.

Sendlhofer, T. (2018, May 13). Zu viele Touristen: Hallstatt zieht Notbremse. Kurier.At. https://kurier.at/chronik/oesterreich/zu-viele-touristen-hallstatt-zieht-notbremse/400034545

Séraphin, H., Gladkikh, T., & Vo Thanh, T. (Eds.) (2020). *Overtourism: Causes, implications and solutions* (1st ed.). https://doi.org/10.1007/978-3-030-42458-9

Sharma, A., & Nicolau, J. L. (2020). An open market valuation of the effects of COVID-19 on the travel and tourism industry. *Annals of Tourism Research*, *83*, 102990. https://doi.org/10.1016/j.annals.2020.102990

Spiegel (2017, August 2). *Proteste in Spanien: "Tourist go home!"*. Spiegel. https://www.spiegel.de/reise/europa/barcelona-und-mallorca-proteste-gegen-massentourismus-werden-vehementer-a-1161072.html

Stone, G. A., & Duffy, L. N. (2015). Transformative learning theory: A systematic review of travel and tourism scholarship. *Journal of Teaching in Travel & Tourism*, *15*(3), 204–224. https://doi.org/10.1080/15313220.2015.1059305

Szromek, A. R., Kruczek, Z., & Walas, B. (2020). The attitude of tourist destination residents towards the effects of overtourism—Kraków case study. *Sustainability*, *12*(1), 228. https://doi.org/10.3390/su12010228

Transform Me Travel. (2020). *Transformative travel*. Transform Me Travel. https://transformmetravel.com/transformative-travel/

Valdivielso, J., & Moranta, J. (2019). The social construction of the tourism degrowth discourse in the Balearic Islands. *Journal of Sustainable Tourism*, *27*(12), 1876–1892. https://doi.org/10.1080/09669582.2019.1660670

Vollmer, L. (2018). *Strategien gegen Gentrifizierung*. Schmetterling Verlag.

Wilson, A. (2020, June 19). Which European countries are easing travel restrictions? *The Guardian*. https://www.theguardian.com/travel/2020/may/18/europe-holidays-which-european-countries-are-easing-coronavirus-travel-restrictions-lockdown-measures

World Tourism Organization (UNWTO). (2018). *'Overtourism'? – Understanding and managing urban tourism growth beyond perceptions, executive summary*. World Tourism Organization. https://doi.org/10.18111/9789284420070

Żemła, M. (2020). Reasons and consequences of overtourism in contemporary cities—knowledge gaps and future research. *Sustainability*, *12*(5), 1729. https://doi.org/10.3390/su12051729

23

THE DYADIC INFLUENCE OF PERSONAL AND CULTURAL FACTORS ON TOURISM AND HOSPITALITY

Erdogan Koc, Elif Yolbulan Okan, and Fulya Acikgoz

Introduction

Travel forms the basis of tourism activity and it has evolved from the old French word *travailen* meaning "to toil, labor". Later it was used "to make a journey" via the notion of "go on a difficult journey" (Online Etymology Dictionary, 2021). Before the modernization of transportation methods, travel was associated with hardship, difficulty, and adventure. While the technological advancements in transportation have made travel easier, speedier, safer, and more comfortable, technological advancements in media and communication have reduced fear of foreign lands and people and increased the desire to experience different places in the world (Koc, 2020a).

With the increasing number of people participating in tourism from different countries and cultures (Mihalič & Fennell, 2015; Koc, 2020a), both as customers and service providers, tourism has become an intercultural activity necessitating the need to understand the intercultural characteristics of customers and service providers, as well as their personal or individual characteristics.

As social psychology is about the formation of people's thoughts, feelings, beliefs, intentions, and goals within a social context (Kenrick et al., 2010), analysis and discussion on the influence of personal factors such as gender and personality characteristics, representing the psychological perspective and sociological perspective, would be extremely relevant in a book titled *Social Psychology of Tourism*.

Tourism and hospitality services occur in a social servicescape (Tombs & McColl-Kennedy, 2003; Koc, 2019) within which there is intense and continuous social contact and interaction between customers and employees, and among employees themselves (Dolnicar et al., 2011; Koc, 2013; Koc, 2020a). Because of this intense and continuous social contact and interaction, tourism and hospitality businesses are often referred to as people businesses.

Social psychology, which emphasizes the individuals within the social interaction paradigm, has been widely used as a theoretical foundation to explain human behavior in tourism and hospitality (Huan & Beaman, 2004; Pearce, 2013; Tang, 2014). Researchers (Gibson & Zhong, 2005; Lieberman & Gamst, 2015; Koc, 2020a) argue that tourism and hospitality are likely the two most relevant areas to investigate an individual's behavior in intercultural social contexts.

Tourism has shown consistent growth after the 1960s and has become one of the largest industries in the world, in terms of both revenues and employment, comprising more than 10% of world GDP and employment in 2018 (UNWTO, 2019). Although the Covid-19 pandemic the world has been through between 2020 and 2021 has brought the tourism and hospitality activities to a halt, a quick recovery is expected after the pandemic (Koc, 2020a). However, some of the

DOI: 10.4324/9781003161868-23

precautions learned during the pandemic mainly in the form of shunning social/physical interaction may continue to influence the social psychology of tourism and hospitality customers and service providers in years to come.

Against this backdrop, this chapter explores the influence of personal and cultural factors with a dyadic perspective. The intensity and frequency of social interactions taking place in tourism and hospitality service context require studies elaborating the social servicescape (Tombs & McColl-Kennedy, 2003; Koc, 2021) including factors influencing both customers' and employees' attitudes and behaviors that exist in the tourism and hospitality consumption settings (Kim & Baker, 2017; Rosenbaum & Montoya, 2007).

Influence of Customers' Cultural, Psychological, and Personal Background on Tourism and Hospitality Activities

Understanding consumer behavior in tourism and hospitality can be more complex and important than any other context of consumption due to the multitude of factors that may influence an individual's choice and satisfaction. Consumer decisions regarding tourism and hospitality activities are generally high-involvement and extensive decision-making purchases, due to the relatively high costs and risks involved. Thus, the need for an interdisciplinary approach involving disciplines such as psychology, sociology, and anthropology in studying customer behavior from the perspective of tourism and hospitality appears to be important. Psychology, as the study of the human mind; anthropology, as the study of culture; and sociology, as the study of human society, could provide essential knowledge underpinning the theoretical background investigating customer behavior in tourism and hospitality.

Influence of Customers' Culture on Tourism and Hospitality Activities

"Culture can be defined as a set of control mechanisms for governing behavior" (Geertz, 1973, p. 144) or "the collective mental programming of the people in an environment" by Hofstede (1991, p. 5). The famous quote by Peter Drucker, "Culture eats strategy for breakfast" (as cited in Koc, 2020a, p. ix) shows the dominant influence of culture on most business and management activities.

Thus, managers operating in tourism and hospitality need to measure the intercultural competence of their staff and continuously develop their intercultural skills (Koc, 2021). Pinto et al. (2014), who studied the frequency of the use of the intercultural paradigms in top international business and management journals, found that more than 75% of the studies were based on Hofstede et al. (2010) and Hall's (1989) paradigms.

Hall (1989) developed a paradigm based on context, space, time, and information flow to classify, compare, and understand cultures. According to Hall (1989), in high-context cultures communication is composed of little explicit content as it relies significantly on the physical context, whereas in low-context cultures communication tends to be more explicit and straightforward. Reisinger and Turner (2002) who studied cultural differences between Asian tourists and Australian service providers found that as members of high-context cultures, Asian people communicated in an indirect, implicit way, and preferred using non-verbal cues such as body language and facial expressions, as opposed to the Australians, a low-context culture, communicated in a more direct, explicit way by emphasizing words and verbal expressions.

Hofstede (1984) who views culture as the collective programming of the mind has developed six cultural dimensions to explain, understand, and compare cultures (Table 23.1).

Hofstede et al.'s (2010) cultural dimensions are the most widely used in various disciplines, including business, management, marketing, tourism, and hospitality. Litvin et al. (2004) and

Table 23.1 Hofstede's Five Dimensions of Culture

Cultural Variable	*Explanation*
Power distance	The extent to which society accepts inequality in power and the way in which interpersonal relationships develop in a hierarchical society
Uncertainty avoidance	The extent to which culture encourages risk-taking and tolerates uncertainty and the extent to which people feel threatened by ambiguous situations
Individualism-collectivism	The extent to which culture encourages individuals to be concerned about their own goals and needs as opposed to collective goals and needs
Masculinity-femininity	The extent to which "masculine" values such as assertiveness, materialism, and lack of concern for others prevail over the "feminine" values such as quality of life, concern for others, and harmonious human relations
Long-term/short term orientation	Long-term orientation stands for the fostering of virtues oriented towards future rewards, in particular perseverance and thrift. Short-term orientation is to do with the fostering of virtues related to the past, and present; respect for tradition; the preservation of "face", and fulfilling social obligations.
Indulgence-Restraint	Indulgence refers to a tendency to allow free gratification of basic and natural desires and enjoyment in life, while restraint, that is, a lack of indulgence, is the belief that basic and natural desires and enjoyment in life need to be controlled and regulated by social norms.

Source: Hofstede, G. Hofstede, G. J., & Minkov, M. (2010). *Cultures and Organizations: Software of the Mind*, Third Edition. McGraw-Hill, New York.

Litvin and Kar (2003) found that tourists participated in tourism activities that are congruent with their self-image, which, in turn, largely shaped by their cultural orientations. Crotts and Erdmann (2000) argued that, in general, tourists thought and behaved in parallel with their mental program, shaped by their culture. In fact, the above findings show the interconnected nature of the psychology and sociology components of social psychology (Koc, 2020). Pizam and Fleischer (2005)'s study in 11 countries showed that Hofstede's cultural dimensions influence the tourism preferences of the customers. For instance, tourists from masculine cultures (e.g., Ireland, Germany, Italy, the USA, and South Africa) preferred more dynamic and active tourist activities than tourists from feminine cultures (e.g., South Korea, Gabon, and Spain).

With respect to the individualism and collectivism dimension, Kim and Lee's (2000) study discovered that North American tourists showed more preference for separation from groups (i.e., avoiding package holidays) and emotional detachment than the collectivistic Japanese tourists did. There are also various other studies that investigated the influence of other cultural dimensions on tourism and hospitality. For instance, uncertainty avoidance as a cultural variable may influence various aspects of tourism and hospitality from the destination choice to responses of customers towards the various elements of the marketing mix (Money & Crotts, 2003). Koc (2020a) summarizes how culture may influence customers in tourism and hospitality during the pre-purchase, purchase and consumption, and post-purchase and consumption phases (Table 23.2).

Understanding complaint behavior during the post-purchase phase in tourism and hospitality consumption settings may be critically important due to the high level of interaction between service providers and customers. By understanding the cultural orientations of customers, businesses may understand how customers respond and do not respond to service failures and develop their service design and delivery (Wong, 2004). Understanding the complaint behaviors of customers according to their cultural characteristics may also increase the success rate in service recoveries (Mueller et al., 2003a).

Table 23.2 Influence of Culture on Tourism and Hospitality Customers

Examples of Potential Influences in the Pre-purchase and Consumption Phase	*Examples of Potential Influences in the Purchase and Consumption Phase*	*Examples of Potential Influences in Post-purchase and Consumption Phase*
Customers' cultural characteristics may influence:	*Customers' cultural characteristics may influence:*	*Customers' cultural characteristics may influence:*
• how they collect information and where they collect information from. • their overall approach towards and response to advertisements, and the cues used in advertisements. • their evaluations of corporate social responsibility activities of the tourism and hospitality business. • the types of tourism and hospitality products, destinations, their various features, how they are designed and delivered. • the distribution channels they use. • their response to tools and types of sales promotions • their evaluations of location and physical evidence. • their evaluations of tourism and hospitality staff. • their perceptions of reservation, booking, and payment and purchase processes.	• how they perceive payment systems. • their perceptions of reservation, booking and payment, purchase, and service delivery processes. • how they evaluate tourism and hospitality products. • how they evaluate their interactions with products, physical evidence, and service staff. • how they evaluate other customers. • how they evaluate service encounters, their satisfaction, and dissatisfaction. • how they evaluate service quality and the service quality dimensions they view as important. • their evaluations of service failures (e.g., attribution) and service recovery attempts. • whether they make complaints or not, and how they make complaints.	• what they remember from their tourism and hospitality experiences. • their satisfaction and dissatisfaction regarding the individual elements of tourism and hospitality products and processes. • how they evaluate the service and the feedback they provide. • their repurchase intentions. • their loyalty. • whether they engage in WOM and the type of WOM they engage in.

Source: Koc, E. (2020a). *Cross-Cultural Aspects of Tourism and Hospitality: A Services Marketing and Management Perspective.* London: Routledge.

For instance, research shows that dissatisfied tourism and hospitality customers from individualistic countries are more likely to complain and engage in negative electronic word-of-mouth communication (eWOM) than customers from collectivistic countries (Patterson et al., 2006; Yuksel et al., 2006; Koc, 2019). Moreover, there are studies suggesting that customers from individualistic cultures prefer personal gains, while customers from collectivistic cultures are more interested in avoiding losses (Briley & Wyer, 2002; Markus & Kitayama, 1991).

People from the same culture tend to share similar norms, values, traditions, and customs which define and guide appropriate and inappropriate attitudes and behaviors. For a tourism service provider, it is vital to understand acceptable patterns of behavior to create satisfactory experiences. Koc (2020a) provides several examples which may influence service encounter effectiveness significantly when dealing with customers from these countries:

"In general, people in Brazil tend to be relatively more patient. They do not tend to be angry when they are kept waiting for a certain amount of time".

"Indonesians hate loud voice, even when it is used to attract attention or call someone at a distance".
"While in many countries children eat plenty of burgers, hotdogs, chicken fingers, and fries, in France it is more common to see children eating 'adult' food".
"In the UK raising a hand to call a waiter in a restaurant is considered inappropriate. The customer waits for the waiter to come".
"In Finland, water and milk are the most popular drinks with meals. Finnish people may drink milk with their lunch and dinner".

Influence of Customers' Personal and Psychological Characteristics

Personal factors such as age and gender are believed to be the most important personal characteristics influencing consumer behavior in marketing and tourism. Since it may not be possible to satisfy customers from different age and gender groups with the same standard marketing mix elements, marketing managers need to provide variations in their offerings according to the personal characteristics of their customers. A study by Baykal and Ayyildiz (2020) showed that personal factors influenced hotel customers risk perceptions significantly. Studies (e.g., Song, 2017; Jeffrey, 2018; Liu-Lastres et al., 2021) show that personal factors such as gender plays a significant role in determining tourism and hospitality customers' decisions. The outdated view of "woman's place is in the home" was a barrier for single women for decades to travel freely for leisure (Rybczynski, 1991). However, in recent years, female tourists have increased in number, mirroring their changing societal and economic roles (UNTWO, 2019). The Global Report on women in tourism (2019) refers to UN Sustainable Development's fifth goal "achieving gender equality and empowering all women and girls" and underline the means to eliminate inequality and use tourism's potential to improve gender equality and women's empowerment throughout the world.

There are several studies (e.g., Koc, 2002; Jiménez-Esquinas, 2017; Pritchard, 2018; Alarcón & Cole, 2019; Russen et al., 2021b) exploring gender differences regarding their perceptual differences reflected in different preferences and consumption patterns and decisions in tourism and hospitality. For instance, Koc (2002) showed that while women used both emotional and rational cues, also in larger amounts when making holiday decisions, while men depended on limited information and made their decisions based on shortcuts, or heuristics. In another study that compared attitudes of male and female hotel guests towards service robots, Ayyildiz and Baykal (2021) found that female guests were interested in services delivered by the service robots, although, in general, tend to be more likely to adopt new technology.

As in other purchasing decisions, price, fashion, and convenience have been pointed as key decision drivers for female customers in tourism and hospitality contexts (Liu-Lastres et al., 2021). Male and female tourists' preferences may show variations in terms of motivations, reasons for choosing a destination, and the various aspects of the services involved in their experiences (Song, 2017; Liu-Lastres et al., 2021). Lin and Lehto (2006) argue that more than 70% of all travel decisions are made by women. Koc (2002) argues that as women tend to collect and analyze more information regarding the family holidays due to their communal orientations, they are more likely to filter the information and for the purchase repertoire.

Women's overwhelming influence in tourism and hospitality decisions has encouraged tourism businesses to create female-specific products, tours, and marketing programs. Although female tourists are not homogenous with respect to socio-economic characteristics, life cycle stage, lifestyles, etc., businesses engage in the development of holidays for women (Junek et al., 2006). According to Junek et al. (2006) women-only holidays, in general, tend to be more relaxed, comfortable (Junek et al., 2006) due to the most popular common motivation for relaxation among women (Lin & Lehto, 2006).

In addition to their own tourism and hospitality preferences, as stated above (e.g., Koc, 2002; Lin & Lehto, 2006) women also tend to influence tourism and hospitality decisions as they are more aware of the needs of the family members, and they tend to collect and analyze more information about the options available due to their communal orientations. Also, in general, women may be perceived as more sophisticated customers as they tend to collect more information and pay more attention to details (Koc, 2002; Sengupta, 2010). Furthermore, women tend to have a higher level of concern for the environment and sustainable development (OECD, 2008) and be more ethical customers than men (Han et al., 2019b). However, due to female consumers' communal orientations, higher needs for affiliation, and harmonious relationships with others (Carlson, 1972; Yuksel & Yuksel, 2001; Omar et al., 2016; Xu & Zhang, 2021), they generally tend to provide higher ratings when evaluating their service experiences and tend to behave more sensitively than male customers. Male customers tend to be more goal and outcome-oriented in their feedbacks (Omar et al., 2016).

Like gender groups, age, as another personal/individual factor, influences tourism and hospitality decisions (Batra, 2009; Lehto et al., 2009; Hung & Lu, 2016; Ketter, 2021). Demographic changes show that the share of people in the global population over 60 years of age is expected to nearly double by 2040 (Dwyer et al., 2008).

With the increase in average life expectancy, senior citizens are expected to constitute a more significant chunk of the tourism and hospitality market. As the senior consumer segment is recently developing, there is an urgent need to study senior travelers' tourism preferences in terms of destinations, period of travel, length of stay, type of accommodation, and expenditure habits. For a start, an orthodox categorization of seniors needs to be established as yet there is still no clear definition of senior customers (Le Serre, 2008; Chen, 2009). Among the various categorizations, Tarlow's (2019) categorizations of senior travelers could be used as a reference (Table 23.3).

Studies investigating the travel motivations of senior groups found that rest, relaxation, socialization, nostalgia as the common motivations for travel (Huang & Tsai, 2003). Moreover, senior tourists tended to be more influenced by the attitudes, behaviors, and interactions of the tour guides (Hwang & Lee, 2019).

In addition to the seniors, millennials are also a fast-growing segment in the tourism and hospitality market (Cavagnaro et al., 2018; OECD, 2018). According to GLOBETRENDER (2017), although millennials travel more frequently, their trips tend to be shorter in duration compared

Table 23.3 Age Groups of Senior Travelers

Category	*Age Range*	*Salient Characteristics*
Early old	50–65	• May or may not be retired • Tend to be finishing raising family • Tend to worry about large expenses such as university education • Just retired/the young old
Just retired/the young old	65–75	• Children are now grown • Tend to have expendable income • Tend to travel both for enjoyment and to see family and friends • Realize that they may soon no longer be well enough to travel easily
The old elderly	75+	• Are more fragile • Tend to worry about future • Find it easier to pay for the loved ones to come to them than themselves traveling to visit them

Source: Tarlow, D. P. (2019). Safe senior citizen travel and the aging travel population. *International Journal of Safety and Security in Tourism and Hospitality*, (20), 1–17.

with other age groups. Young travelers consider travel more important than buying cars or other possessions, and they tend to value authentic experiences relatively more. The popularity of sharing economy among millennials, and their preferences to save accommodation costs over unique experiences caused several significant changes in tourism such as the emergence of alternative accommodation systems like sharing systems (e.g., Airbnb) and glamping, nature associated luxury camping (Cavagnaro et al., 2018). Personalization and digitalization are two major trends influencing tourism and hospitality service providers targeting the millennials. It appears that creating customized and unique experiences incorporating technology and providing connectivity will be the key to success factors in the tourism and hospitality industry (Moreno-Izquierdo et al., 2019; Pencarelli et al., 2020).

As another factor, the concept and the types of personality have been extensively researched in tourism and hospitality (Matzler et al., 2005; Schneider & Vogt, 2012; Ryu et al., 2020). Personality can influence people's consumption preferences, decision-making processes, interaction with others, and emotions (Carver & Scheier, 2008; Esfahani et al., 2021). For example, Schneider and Vogt's (2012) study found that the need for arousal was found to be a common personality trait among hard adventure travelers while competitiveness was a common personality trait among soft adventure travelers.

Many personality theories (e.g., Five Factors Model, 3M Model of Personality, and the A/B Personality types model) have been extensively studied over the past decades to provide a structure to explain and understand human behavior (Lin et al., 2001).The "Big Five" taxonomy is the most studied model on personality traits, and comprises five broad dimensions of extraversion, agreeableness, conscientiousness, neuroticism, and openness (Kvasova, 2015). Kvasova (2015) examined personality traits and their relationship with tourists' attitudes towards the environment and found that agreeableness, conscientiousness, extraversion, and neuroticism were positively associated with environmental tourist behavior. In another study of personality in relation to tourism, Tobin (2000) found that people with higher levels of neuroticism, due to their tendency to have a more negative mood, are more likely to engage in conflicts in service encounters. Also, people who score higher in agreeableness tend to be easy-going people and prefer calm and relaxing destinations and peaceful environments (Çelik & Dedeoglu, 2019). Similarly, conscientious people prefer relaxing destinations to get away from the stress in their normal lives, as conscientious people tend to be ambitious and more success-oriented (McCrae & Costa, 1989; Çelik & Dedeoglu, 2019).

The Influence of Service Providers' Culture on Tourism and Hospitality Activities

As explained above, multicultural service encounters in tourism and hospitality involve continuous and frequent social contact and interactions between customers and service providers, and among service providers themselves (Koc, 2020a). These multicultural service interactions can be so important that the overall service quality and satisfaction are usually determined based on these interactions (Babakus et al., 2008; Lo et al., 2015). Hence, the cultural characteristics of service providers are important not only for establishing high-quality service encounters but also in terms of managing and motivating them.

Based on the above background, there have been many studies investigating the influence of culture in tourism from the supply, that is, service providers' perspective as well (e.g., Mueller et al., 2003b; Wong, 2004; Ringberg et al., 2007; Tsang & Ap, 2007; Swanson & Hsu, 2011; Tam et al., 2014; Tsang, 2011).

Koc (2013) found that a lack of empowerment among employees and a lack of direct communication between subordinates and managers were more prevalent among service employees from high power-distance cultures. Koc (2013) concluded that a high level of power distance among service providers caused delays in the recovery of service failures in tourism and hospitality, while a low level of power distance among service providers resulted in immediate response to service failures as the subordinates were able to communicate directly with the superiors.

Similarly, Vukonjanski et al.'s (2012) research in the Serbian hospitality industry has found that a high level of uncertainty avoidance and power distance resulted in a tendency to agree with company rules, valued employment stability, though the employees experienced higher levels of job stress and dissatisfaction.

In the same vein, Magnini et al. (2013) showed that employees in collectivistic societies tended to be uncomfortable with empowerment. Koc (2020a) also argues that low employee turnover rates in high uncertainty avoidance cultures should not be confused with job satisfaction. Employees may stay in their jobs due to the high level of risk aversion, although they may be highly dissatisfied with their jobs.

Zheng et al. (2020) studied the motivation of tourism and hospitality employees from the perspective of their cultural orientations. Their study showed that while the employees from low-power distance cultures valued the intrinsic factors such as achievement, recognition, the work itself, responsibility, advancement, and growth more, the employees from high-power distance cultures valued the extrinsic factors such as achievement, recognition, company policy, relationships with supervisors, working conditions, relationships with peers and subordinates, salary and benefits, and job security more.

The Influence of Service Providers' Personal and Psychological Characteristics on Tourism and Hospitality Activities

Personality characteristics significantly influence service providers' attitudes and behaviors due to the role they play in information collection, decision-making, and learning processes (Papathanassis, 2020). Since there is a strong relationship between personality traits and tourism and hospitality employees' work outcomes such as performance, satisfaction (Papathanassis, 2020), talent management which is a combination of human resource management practices such as recruitment, selection, retention, performance management and career development, has gained strategic importance for the companies in the tourism and hospitality industry.

Employees' personality traits are considered an important individual predictor efficient working conditions which leads to customer satisfaction, trust, and loyalty (Sohn & Lee, 2012). Thus, there is a need for further studies concentrating on the relationship between employees' personality and job performance in tourism field. Deep understanding of personality can provide critical insights for managers in tourism sector to provide better service (Leung & Law, 2010). Moreover, Basoda (2012) points the critical importance of person-job fit, which is a precondition for creating delightful experiences in tourism and hospitality services.

Managers in tourism companies can create teams that are diversely composed in terms of personality traits to utilize the strengths of each personal trait. For example, introverts' listening skills could be critical for service recovery since they can pick up details more easily than others.

Research (e.g., Eysenck & Eysenck, 1967; Pazda & Thorstenson, 2018) shows that introvert employees may be more likely to pay external stimuli and pay more attention to the tangibles element (Hatipoglu, 2021). Hence, introverts can play an important role in the design and implementation of physical evidence or tangibles as one of the marketing mix elements or the service quality dimensions respectively.

Additionally, there are several studies in the literature pointing out to the positive relationships between agreeableness, extroversion, openness to experience, conscientiousness, and service orientation (Williams & Sanchez, 1998; Costen & Barrash, 2006; Lanjananda & Patterson, 2009). Among those variables, extroversion, openness to experience, and agreeableness are the most key personality traits that conduce to service orientation individually (Chait et al., 2000).

Since extroverted people are known as friendly and social individuals, they provide better quality service compared to others and they are more appropriate employees for tourism sector (Liao & Chuang, 2004; Teng, 2008).

Along with personality characteristics, age and gender have also an impact on service providers' attitudes and behavior in the tourism and hospitality area. Malina and Schmidt (1997) asserted that the service provider's gender is pivotal to figure out the quality of servicescape experience. In Koc's (2020b) article an interdisciplinary literature review is provided to support the view that women can be regarded as better equipped service providers than men. Through the biological, anthropological, social and psychological differences, women tend to have better characteristics to ensure service quality (Koc, 2020b).

However, against this above background, women's employment in tourism and hospitality is both horizontally and vertically segregated in tourism (Koc, 2020b). Although women may have relatively more employment opportunities in tourism than some other industries, there are many problems and challenges to do with women's employment in tourism (Ntanjana et al., 2018). Although the number of women working in tourism sector may be high, in general, women in tourism are likely to work as receptionists, waitresses, room attendants, cleaners, travel agency salespersons, and in lower-level positions with fewer opportunities with upward mobility (Cave & Kilic, 2010; Carvalho et al., 2018; Remington & Kitterlin-Lynch, 2018; Mooney, 2020; Russen et al., 2021a). On the other hand, men tend to be recruited for higher-status jobs, more often with higher pay (Nickson, 2007).

The scenarios about post-Covid-19 recovery for tourism and hospitality sector imply a need for gender-just human rights approach indicating the need for a perspective taking women at the center. Besides all other qualifications, women have the power to care, collaborate, and heal, which will be extremely important to restructure the sector. Women's employment in higher-status jobs in tourism is necessary not only for sustainability reasons but also for efficiency and effectiveness reasons. At micro level, tourism businesses may provide more jobs for women, while at macro level government organizations may offer more support and incentives for female entrepreneurs.

Tourism is known as a labor-intensive industry that is generally reliant on younger and flexible employees due the seasonality factor of the sector. Many young people find their first jobs in tourism (Golubovskaya et al., 2019), thus the industry depends on, and benefits from, young labor especially in countries that include Australia, Turkey, Germany, and France (WTTC, 2013). Young employees differ from adults with respect to biological, cognitive, psychological, and social changes they experience. The dynamism and energy young people have can leverage the quality in service settings if managed effectively. According to Kusluvan et al. (2010) older employees in tourism sector may be more tolerant, emotionally mature, and sympathetic, whereas younger employees may be more energetic. To transform young employees' energy into workplace efficiency, talent management to equip young service providers with necessary skills and attributes is very critical. Moreover, considering the high rates of turnover among young employees in the tourism and hospitality industry, Tews et al. (2020) suggests that there is a need to design an entertaining workplace atmosphere and training climate supporting co-worker socialization.

Since organizations with diverse age groups can attract diverse customer groups, there is a need to hire older individuals with knowledge and experience. Therefore, in building a balanced composition in work groups is one of the key success factors for tourism companies.

Conclusion

Tourism and hospitality processes and outcomes are influenced by customers' and service providers' socio-psychological characteristics. This chapter explained how factors within the domain of social psychology such as culture, personality, age, and gender characteristics may influence both customers and service providers in tourism and hospitality supported by empirical research findings. For instance, from the demand perspective, the chapter shows customers' socio-psychological characteristics may influence the design and implementation of marketing mix elements. Also from the supply side, the chapter shows how socio-psychological characteristics of service providers may influence the operation of certain processes, performance, efficiency, and effectiveness in tourism and hospitality.

Consequently, this chapter provides useful recommendations for both the practitioners and researchers. The practitioners may integrate the socio-psychological characteristics into the design of their offerings and operations (e.g., in the development of marketing mix elements) to meet the needs of their target market in a more appropriate manner. Based on customers' cultural characteristics, offerings and operations can be established accordingly, enabling the tourism and hospitality business to establish competitive advantage. Similarly, as stated above, the personal characteristics of the customers may influence their perceptions, attitudes, and behaviors towards tourism and hospitality offerings and operations. Also, the practitioners may take customers' personal characteristics into account both in the design and implementation of their offering and operations. For instance, as shown by Koc (2020b) female employees may be better at ensuring a high-quality service across all service quality (SERVQUAL) dimensions.

From a theoretical perspective, the chapter also provides cues for the researchers in designing future research. The researchers may better understand the socio-psychological characteristics of the target group they base their study on, and hence develop more appropriate methods and implement more suitable research design.

The chapter also underlines the importance of developing an understanding of socio-psychological factors among service employees and managers. For instance, as tourism and hospitality activities are increasingly becoming international and intercultural, from both demand and supply perspectives, managers in these businesses need to build intercultural competence as one of the required competences into their human resource management processes.

References

Alarcón, D. M., & Cole, S. (2019). No sustainability for tourism without gender equality. *Journal of Sustainable Tourism*. https://doi.org/10.1080/09669582.2019.1588283.

Ayyildiz, Yazici, A., & Baykal, M. (2021), Attitudes of hotel customers towards the use of service robots in hospitality service encounters. *Journal of Hospitality and Tourism Technology*. Article in Press.

Babakus, E., Yavas, U., & Karatepe, O. M. (2008). The effects of job demands, job resources and intrinsic motivation on emotional exhaustion and turnover intentions: A study in the Turkish hotel industry. *International Journal of Hospitality & Tourism Administration, 9*(4), 384–404.

Basoda, A. (2012). *The impact of service orientation as a personality trait on job satisfaction and intention to leave: A research in the lodging industry.* (publication No. 280574099) [Master's Thesis, Nevsehir University].

Batra, A. (2009). Senior pleasure tourists: Examination of their demography, travel experience, and travel behavior upon visiting the Bangkok metropolis. *International Journal of Hospitality & Tourism Administration, 10*(3), 197–212.

Baykal, M., & Ayyildiz, Y. A. (2020). Kişilik özelliklerinin algılanan risk üzerine etkisi: Kuşadası'ndaki 5 Yıldızlı Otel Müşterileri Örneği. *Elektronik Sosyal Bilimler Dergisi, 19*(75), 1371–1392.

Briley, D. A., & Wyer Jr, R. S. (2002). The effect of group membership salience on the avoidance of negative outcomes: Implications for social and consumer decisions. *Journal of Consumer Research, 29*(3), 400–415.

Carlson, R. (1972). Understanding women: Implications for personality theory and research. *Journal of Social Issues, 28*(2), 17–32.

Carvalho, I., Costa, C., Lykke, N., Torres, A., & Wahl, A. (2018). Women at the top of tourism organizations: Views from the glass roof. *Journal of Human Resources in Hospitality & Tourism, 17*(4), 1–26.

Carver, C. S., & Scheier, M. F. (2008). Feedback processes in the simultaneous regulation of action and affect. In J. Y. Shah & W. L. Gardner (Eds.), Handbook of motivation science (pp. 308–324). New York: Guilford Press.

Cavagnaro, E., Staffieri, S., & Postma, A. (2018). Understanding millennials' tourism experience: Values and meaning to travel as a key for identifying target clusters for youth (sustainable) tourism. *Journal of Tourism Futures, 4*(1), 31–42.
Cave, P. & Kilic, S. (2010). The role of women in tourism employment with special reference to Antalya, Turkey. *Journal of Hospitality Marketing & Management, 19*(3), 280–292.
Çelik, S., & Dedeoğlu, B. B. (2019). Psychological factors affecting the behavioral intention of the tourist visiting Southeastern Anatolia. *Journal of Hospitality and Tourism Insights, 2*(4), 425–450.
Chait, H. N., Carraher, S. M., & Buckley, M. R. (2000). Measuring service orientation with biodata. *Journal of Managerial Issues, 12*(1), 109–120.
Chen, D. (2009). Innovation of tourism supply chain management. In *2009 International conference on management of e-commerce and e-government* (pp. 310–313). IEEE.
Costen, W. M., & Barrash, D. I. (2006). ACEing the hiring process: A customer service orientation model. *Journal of Human Resources in Hospitality & Tourism, 5*(1), 35–49.
Crotts, J. C., & Erdmann, R. (2000). Does national culture influence consumers' evaluation of travel services? A test of Hofstede's model of cross-cultural differences. *Managing Service Quality, 10*(6), 410–419.
Dennis, N. (2007). *Human resource management for the hospitality and tourism industries.* United Kingdom: Elsevier.
Dolnicar, S., Grabler, K., Grün, B., & Kulnig, A. (2011). Key drivers of airline loyalty. *Tourism Management, 32*(5), 1020–1026.
Dwyer, L., Edwards, D. C., Mistilis, N., Roman, C., Scott, N., & Cooper, C. (2008). *Megatrends underpinning tourism to 2020: Analysis of key drivers for change.* Sustainable Tourism Pty Ltd., Brooke Pickering.
Esfahani, M., Khoo, S., Musa, G., Heydari, R., & Keshtidar, M. (2021). The influences of personality and knowledge on safety-related behaviour among climbers. *Current Issues in Tourism, 23*(24), 1–13.
Eysenck, S. B., & Eysenck, H. J. (1967). Salivary response to lemon juice as a measure of introversion. *Perceptual and Motor Skills, 24*(3_suppl), 1047–1053.
Geertz, C. (1973). Thick description: Toward an interpretive theory of culture. *Turning Points in Qualitative Research: Tying Knots in a Handkerchief, 3,* 143–168.
Gibson, D., & Zhong, M. (2005). Intercultural communication competence in the healthcare context. *International Journal of Intercultural Relations, 29*(5), 621–634.
Globetrender (2017). *From Boomers to Gen Z: Travel trends across the generations.* https://www.globetrender.com.
Golubovskaya, M., Solnet, D., & Robinson, R. N. (2019). Recalibrating talent management for hospitality: A youth development perspective. *International Journal of Contemporary Hospitality Management, 31*(10), 4105–4125.
Hall, S. (1989). Cultural identity and cinematic representation. *Framework: The Journal of Cinema and Media, 36,* 68–81.
Han, H., Yu, J., & Kim, W. (2019b). Investigating airline customers' decision-making process for emerging environmentally responsible electric airplanes: Influence of gender and age. *Tourism Management Perspectives, 31,* 85–94.
Hatipoglu, S. (2021). *Konaklama İşletmelerinde Algılanan Hizmet Kalitesinin Müşteri Memnuniyetine Etkisinde Kişilik Özelliklerinin Düzenleyici Rolü.* [Doctorate Thesis, Bandırma Onyedi Eylul University].
Hofstede, G. (1984). *Culture's consequences: International differences in work-related values.* Newbury Park, CA: Sage Publications.
Hofstede, G. (1991). *Cultures and organizations — software of the mind.* New York: McGraw Hill.
Hofstede, G. (2001). *Culture's consequences.* 2nd ed. Thousand Oaks, CA: Sage Publications.
Hofstede, G., Hofstede, G. J., & Minkov, M. (2010). *Cultures and Organizations: Software of the Mind* (3rd ed.). New York: McGraw-Hill.
Huan, T., & Beaman, J. (2004). Context and dynamic of social interaction and information search in decision making for discretionary travel. In: A. Woodside & J. Mazanec (Eds.), *Consumer psychology of tourism, hospitality and leisure* (pp. 155–163). Cambridge, MA: CAB International.
Huang, L., & Tsai, H. T. (2003). The study of senior traveler behavior in Taiwan. *Tourism Management, 24*(-5), 561–574.
Hung, K., & Lu, J. (2016). Active living in later life: An overview of aging studies in hospitality and tourism journals. *International Journal of Hospitality Management, 53,* 133–144.
Hwang, J., & Lee, J. (2019). Relationships among senior tourists' perceptions of tour guides' professional competencies, rapport, satisfaction with the guide service, tour satisfaction, and word of mouth. *Journal of Travel Research, 58*(8), 1331–1346.
Jeffrey, H. L. (2018). Tourism and gendered hosts and guests. *Tourism Review, 74*(5), 1038–1046.
Jiménez-Esquinas., & Guadalupe. (2017). This is not only about culture": On tourism, gender stereotypes another affective fluxes. *Journal of Sustainable Tourism, 25*(3), 311–326.

Junek, O., Binney, W., & Winn, S. (2006). All-female travel: What do women really want?. *Tourism: An International Interdisciplinary Journal, 54*(1), 53–62.

Kenrick, D. T., Neuberg, S. L., Cialdini, R. B., & Cialdini, Robert B. (2010). *Social psychology: Goals in interaction*. Boston, MA: Pearson.

Ketter, Eran (2021). Millennial travel: Tourism micro-trends of European Generation Y. *Journal of Tourism Futures*. https://doi.org/10.1108/JTF-10-2019-0106.

Kim, K. A. W. O. N., & Baker, M. A. (2017). *The influence of other customers in service failure and recovery.* Service Failures and Recovery in Tourism and Hospitality: A Practical Manual, CABI Publishing, 122–134.

Kim, C. & S. Lee (2000). Understanding the cultural differences in tourist motivation between Anglo-American and Japanese tourists. *Journal of Travel & Tourism Marketing, 9*(1-2), 153-170.

Koc, E. (2002). The impact of gender in marketing communications: The role of cognitive and affective cues. *Journal of Marketing Communications, 8*(4), 257–275.

Koc, E. (2013). Power distance and its implications for upward communication and empowerment: Crisis management and recovery in hospitality services. *The International Journal of Human Resource Management, 24*(19), 3681–3696.

Koc, E. (2017). *Service failures and recovery in tourism and hospitality a practical manual.* Wallingford, Oxford: CABI.

Koc, E. (2019). Service failures and recovery in hospitality and tourism: A review of literature and recommendations for future research. *Journal of Hospitality Marketing & Management, 28*(5), 513–537.

Koc, E. (2020a). *Cross-cultural aspects of tourism and hospitality: A services marketing and management perspective.* London: Routledge.

Koc, E. (2020b). Do women make better in tourism and hospitality? A conceptual review from a customer satisfaction and service quality perspective. *Journal of Quality Assurance in Hospitality & Tourism, 21*(4), 402–429.

Koc, E. (2021). Intercultural competence in tourism and hospitality: Self-efficacy beliefs and the Dunning Kruger effect. *International Journal of Intercultural Relations, 82.* 175–184.

Kusluvan, S., Kusluvan, Z., Ilhan, I., & Buyruk, L. (2010). The human dimension: A review of human resources management issues in the tourism and hospitality industry. *Cornell Hospitality Quarterly, 51*(2), 171–214.

Kvasova, O. (2015). The Big Five personality traits as antecedents of eco-friendly tourist behavior. *Personality and Individual Differences, 83*, 111–116.

Lanjananda, P., & Patterson, P.G. (2009).Determinants of customer-oriented behavior in a health care context. *Journal of Service Management, 20*(1), 5–32.

Le Serre, D. (2008). Who is the senior consumer for the tourism industry. *Amfiteatru Economic, 10*(special 2), 195–206.

Lehto, X. Y., Choi, S., Lin, Y. C., & MacDermid, S. M. (2009). Vacation and family functioning. *Annals of Tourism Research, 36*(3), 459–479.

Leung, R., & Law, R. (2010). A review of personality research in the tourism and hospitality context. *Journal of Travel & Tourism Marketing, 27*(5), 439–459.

Liao, H., & Chuang, A. (2004). A multilevel investigation of factors influencing employee service performance and customer outcomes. *Academy of Management Journal, 47*(1), 41–58.

Lieberman, D. A., & Gamst, G. (2015). Intercultural communication competence revisited: Linking the intercultural and multicultural fields. *International Journal of Intercultural Relations, 48*, 17–19.

Lin, N. P., Chiu, H. C., & Hsieh, Y. C. (2001). Investigating the relationship between service providers' personality and customers' perceptions of service quality across gender. *Total Quality Management, 12*(1), 57–67.

Lin, Y. C., & Lehto, X. Y. (2006). A study of female travelers' needs trajectory and family life cycle. *Journal of Hospitality & Leisure Marketing, 15*(1), 65–88.

Litvin, S. W., Crotts, J. C., & Hefner, F. L. (2004). Cross-cultural tourist behaviour: A replication and extension involving Hofstede's uncertainty avoidance dimension. *International Journal of Tourism Research, 6*(1), 29–37.

Litvin, S. W., & Kar, G. H. (2003). Individualism/collectivism as a moderating factor to the self-image congruity concept. *Journal of Vacation Marketing, 10*(1), 23–42.

Liu-Lastres, B., Mirehie, M., & Cecil, A. (2021). Are female business travelers willing to travel during COVID-19? An exploratory study. *Journal of Vacation Marketing*, 27(3), 252–266. Doi: https://doi.org/JVM.10.1177/1356766720987873.

Lo, A., Wu, C., & Tsai, H. (2015). The impact of service quality on positive consumption emotions in resort and hotel spa experiences. *Journal of Hospitality Marketing & Management, 24*(2), 155–179.

Magnini, V.P., Hyun, S.(S)., Kim, B.(P)., & Uysal, M. (2013). The influences of collectivism in hospitality work settings. *International Journal of Contemporary Hospitality Management, 25*(6), 844–864.
Malina, D., & Schmidt, R.A. (1997). It's business doing pleasure with you: Sh! A women's sex shop case. *Marketing Intelligence & Planning, 15*(7), 352–360.
Markus, H.R., & Kitayama, S. (1991). Culture and the self: Implications for cognition, emotion, and motivation. *Psychological Review, 98*, 224–253.
Matzler, K., Faullant, R., Renzl, B., & Leiter, V. (2005). The relationship between personality traits (extraversion and neuroticism), emotions and customer self-satisfaction. *Innovative Marketing, 1*(2), 32–39.
McCrae, R. R., & Costa Jr, P. T. (1989). Reinterpreting the Myers-Briggs type indicator from the perspective of the five-factor model of personality. *Journal of Personality, 57*(1), 17–40.
Mihalič, T., & Fennell, D. (2015). In pursuit of a more just international tourism: The concept of trading tourism rights. *Journal of Sustainable Tourism, 23*(2), 188–206.
Money, R. B., & Crotts, J. C. (2003). The effect of uncertainty avoidance on information search, planning, and purchases of international travel vacations. *Tourism Management, 24*(2), 191–202.
Mooney, S. (2020). Gender research in hospitality and tourism management: Time to change the guard. *International Journal of Contemporary Hospitality Management, 32*(5), 1861–1879.
Moreno-Izquierdo, L., Ramón-Rodríguez, A. B., Such-Devesa, M. J., & Perles-Ribes, J. F. (2019). Tourist environment and online reputation as a generator of added value in the sharing economy: The case of Airbnb in urban and sun-and-beach holiday destinations. *Journal of Destination Marketing & Management, 11*, 53–66.
Mueller, R. D., Palmer, A., Mack, R., & McMullan, R. (2003a). Service in the restaurant industry: An American and Irish comparison of service failures and recovery strategies. *International Journal of Hospitality Management, 22*(4), 395–418.
Mueller, S., Testa, M. R., L., & Thomas, A. S. (2003b). Cultural fit and job satisfaction in a global service environment. *MIR: Management International Review, 43*(2), 129–148.
Nickson, D. (2007). *Human resources management for the hospitality and tourism industries.* Oxford: Butterworth-Heinemann.
Ntanjana, A., Maleka, M., Tshipala, N., & Du Plessis, L. (2018). Employment condition differences based on gender: A case of adventure tourism employees in Gauteng, South Africa. *African Journal of Hospitality, Tourism and Leisure*, 7(4).
OECD (2008).*Women's economic empowerment.*, 47561694.pdf (oecd.org)
OECD (2018). Megatrends shaping the future of tourism. https://www.oecdilibrary.org/docserver/tour-2018.en.pdf?expires=1619973412&id=id&accname=guest&checksum=31E9F6C77438972DA43D88A07F88C96D
Omar, M. S., Ariffin, H. F., & Ahmad, R. (2016). Service quality, customers' satisfaction and the moderating effects of gender: A study of Arabic restaurants. *Procedia-Social and Behavioral Sciences, 224*, 384–392.
Online Etymology Dictionary (2021). available at: https:// www.etymonline.com/ (accessed May 9th, 2021).
Papathanassis, A. (2020). Cruise tourism 'brain drain': Exploring the role of personality traits, educational experience and career choice attributes. *Current Issues in Tourism,* 24(14), 2028–2043. doi: 10.1080/13683500.2020.1816930
Patterson, P. G., Cowley, E., & Prasongsukarn, K. (2006). Service failure recovery: The moderating impact of individual-level cultural value orientation on perceptions of justice. *International Journal of Research in Marketing, 23*(3), 263–277.
Pazda, A. D., & Thorstenson, C. A. (2018). Extraversion predicts a preference for high-chroma colors. *Personality and Individual Differences, 127*, 133–138.
Pearce, P. L. (2013). The social psychology of tourist behaviour: *International series in experimental social psychology* (Vol. 3). Elsevier.
Pencarelli, T., Gabbianelli, L., & Savelli, E. (2020). The tourist experience in the digital era: The case of Italian millennials. *Sinergie Italian Journal of Management, 38*(3), 165–190.
Pinto, C. F., Serra, F. R., & Ferreira, M. P. (2014). A bibliometric study on culture research in International Business. *BAR-Brazilian Administration Review, 11*(3), 340–363.
Pizam, A., & Fleischer, A. (2005). The relationship between cultural characteristics and preference for active vs. passive tourist activities. *Journal of Hospitality & Leisure Marketing, 12*(4), 5–25.
Pritchard, A. (2018). Predicting the next decade of tourism gender research. *Tourism Management Perspectives, 25*, 144–146.
Reisinger, Y., & Turner, L. W. (2002). Cultural differences between Asian tourist markets and Australian hosts, Part 1. *Journal of Travel Research, 40*(3), 295–315.
Remington, J., & Kitterlin-Lynch, M. (2018). Still pounding on the glass ceiling: A study of female leaders in hospitality, travel, and tourism management. *Journal of Human Resources in Hospitality & Tourism, 17*(1), 22–37.

Ringberg, T., Odekerken-Schröder, G., & Christensen, G. L. (2007). A cultural models approach to service recovery. *Journal of Marketing, 71*(3), 194–214.
Rosenbaum, M. S., & Montoya, D. Y. (2007). Am I welcome here? Exploring how ethnic consumers assess their place identity. *Journal of Business Research, 60*(3), 206–214.
Russen, M., Dawson, M., & Madera, J. M. (2021a). Gender discrimination and perceived fairness in the promotion process of hotel employees. *International Journal of Contemporary Hospitality Management*. Article in Press.
Russen, M., Dawson, M., & Madera, J. M. (2021b). Gender diversity in hospitality and tourism top management teams: A systematic review of the last 10 years. *International Journal of Hospitality Management, 95,* 102942.
Rybczynski, W. (1991). *Waiting for the weekend*. New York: Viking.
Ryu, K. H., Youn, S. H., & Moon, S. J. (2020). A study on the recommendation of tourism using topic map according to personality types based on MBTI and big five models. *The Society of Convergence Knowledge Transactions, 8*(2), 21–32.
Schneider, P. P., & Vogt, C. A. (2012). Applying the 3M model of personality and motivation to adventure travelers. *Journal of Travel Research, 51*(6), 704–716.
Sengupta, I. (2010). The economy of desire: Globalisation, gender and sex tourism in the Caribbean. *Gender and Sex Tourism in the Caribbean*, July 28, 2010.
Sohn, H. K., & Lee, T. J. (2012). Relationship between HEXACO personality factors and emotional labour of service providers in the tourism industry. *Tourism Management, 33*(1), 116–125.
Song, H. (2017). Females and tourism activities: An insight for all-female tours in Hong Kong. *Journal of China Tourism Research, 13*(1), 83–102.
Swanson, S. R., & Hsu, M. K. (2011). The effect of recovery locus attributions and service failure severity on word-of-mouth and repurchase behaviors in the hospitality industry. *Journal of Hospitality & Tourism Research, 35*(4), 511–529.
Tam, J., Sharma, P., & Kim, N. (2014). Examining the role of attribution and intercultural competence in intercultural service encounters. *Journal of Services Marketing, 28*(2), 159–170.
Tang, L. R. (2014). The application of social psychology theories and concepts in hospitality and tourism studies: A review and research agenda. *International Journal of Hospitality Management, 36,* 188–196.
Tarlow, D. P. (2019). Safe senior citizen travel and the aging travel population. *International Journal of Safety and Security in Tourism and Hospitality,* (20), 1–17. https:// www.palermo.edu/Archivos_content/ 2017/Economicas/ journal-tourism/edicion20/00_SafeSeniorCitizenTravel. pdf
Teng, C. (2008). The effects of personality traits and attitudes on student uptake in hospitality employment. *International Journal of Hospitality Management, 27*(1), 76–86.
Tews, M. J., Hoefnagels, A., Jolly, P. M., & Stafford, K. (2020). Turnover among young adults in the hospitality industry: Examining the impact of fun in the workplace and training climate. *Employee Relations: The International Journal, 43*(1), 245–261.
Tobin, J. J. (2000). *Good guys don't wear hats": Children's talk about the media*. New York: Teachers College Press.
Tombs, A., & McColl-Kennedy, J. R. (2003) Social-servicescape conceptual model. *Marketing Theory, 3*(4), 447–475.
Tsang, N. K. (2011). Dimensions of Chinese culture values in relation to service provision in hospitality and tourism industry. *International Journal of Hospitality Management, 30*(3), 670–679.
Tsang, N. K. F., & Ap, J. (2007). Tourists' perceptions of relational quality service attributes: A cross-cultural study. *Journal of Travel Research, 45*(3), 355–363.
UNWTO (2019). Global report on women in tourism. *Global Report on Women in Tourism | UNWTO*.
Vukonjanski, J., Nikolić, M., Hadžić, O., Terek, E., & Nedeljković, M. (2012). Relationship between GLOBE organizational culture dimensions, job satisfaction and leader-member exchange in Serbian organizations. *Journal for East European Management Studies, 17*(3), 333–368.
Williams, M., & Sanchez, J. I. (1998). Customer service-oriented behavior: Person and situational antecedents. *Journal of Quality Management, 3*(1), 101–116.
Wong, N. Y. (2004). The role of culture in the perception of service recovery. *Journal of Business Research, 57*(9), 957–963.
WTTC [World Tourism & Travel Council] (2013). Gender equality and youth employment: Travel & Tourism as a key employer of women and young people. Retrieved from https://www.wttc.org/-/media/files/reports/policy%20research/gender_equality_and_youth_employment_final.pdf.
Xu, W., & Zhang, X. (2021). Online expression as Well-be (com) ing: A study of travel blogs on Nepal by Chinese female tourists. *Tourism Management, 83,* 104224.

Yuksel, A., & Yuksel, F. (2001). The expectancy-disconfirmation paradigm: A critique. *Journal of Hospitality & Tourism Research, 25*(2), 107–131.
Yuksel, A., Kilinc, U., & Yuksel, F. (2006). Cross-national analysis of hotel customers' attitudes toward complaining and their complaining behaviours. *Tourism Management, 27*(1), 11–24.
Zheng, T. M., Zhu, D., Kim, P. B., & Williamson, D. (2020). An examination of the interaction effects of hospitality employees' motivational and cultural factors in the workplace. *International Journal of Hospitality & Tourism Administration*, 1–29.https://doi.org/10.1080/15256480.2020.1805088

24

SOCIAL-PSYCHOLOGICAL ISSUES IN TOURISM BUSINESS

Emrah Özkul and Gozde Turktarhan

An Overview of Social Psychology

Psychology is a social science that studies human behavior, experience, attitude, and relationships. On the other hand, social psychology examines people affected by different groups in society (Simkova, 2014). It is possible to say that social psychology, which is based on psychology, has a more specific field of study.

Tang (2014: 188) defines social psychology "as a social science that aims to examine and comprehend the impact of real or imagined thoughts, experiences and behaviors of other individuals or groups on others". Social psychology examines the effect of the environment on individuals and directs sociological studies and information-sharing behaviors (Yiu & Law, 2012). Social psychology, with its interdisciplinary feature, investigates the effects on people socially and individually. It is evaluating social psychology from different perspectives. Wood (2017) examined internal and external factors such as social acceptance in society and self-esteem and reached different results. Dhont et al. (2019) discussed social psychology in terms of animal-human interaction. In this context, they focused on the factors that made people think of animals as a social group. They examined the complexities of valuing animals (living in the same house, feeding, raising, caring) and devaluing (seeing them as food, thinking they were created for oneself). Finally, they focused on the interactions between human groups as a result of human-animal relations. As a result, it has been concluded that animal-human interaction causes many behaviors between humans to be understood. Based on the explanations, it is possible to say that social psychology is a science that examines the levels and results of individuals or groups in society being affected by others.

Dimensions of Social Psychology

Researchers have discussed two different concepts in social psychology that affect different disciplines. The first is individual psychology (Baldwin, 1897; McDougall, 1909), and the other is community psychology (Tarde, 1903). According to the first understanding, social psychology is seen as a branch of psychology, while in the second understanding, it is defended as a branch of sociology (Ross, 1909). Similarly, Poggi (2021) states that there are two separate research traditions in social psychology. These sociological and psychological schools are examined in two different categories, although they are sometimes confused. Social psychology can be classified into four groups (Tang, 2014).

DOI: 10.4324/9781003161868-24

- *Social Cognition:* It is acquiring a particular part of the knowledge gained in society by observing others through social interaction.
- *Social Comparison:*. Social comparison theory is basically a behavioral strategy in which people compare themselves to other people and try to better understand their situation regarding abilities, opinions, emotional response, and more. People constantly make social comparisons in different areas throughout their lives. As a result of this comparison, they identify where others are and at what point they associate themselves.
- *Social Empowerment:* It refers to the feedback people receive from their environment, such as being appreciated, respected, approved, complimented, and congratulated.
- *Essence:* The fourth and last category is the essence of the person, namely, the self. Here, it is stated that the person's experience takes place in his mind, and the self makes him realize that the background belongs to him.
- In addition to these concepts, social influence, social cohesion, and attitude can be considered essential components of social psychology.
- *Social Influence:* Some rules and behaviors have developed as people live in groups to survive. As mentioned, these behaviors have some social effects. Social impact can be explained as the effect of action, project, and movement. It is aimed to see possible problems, reduce problems and increase benefits for society. The process is based on managing (Wilson, 2017). The main reason for evaluating social impact is understanding whether the aim carries out the activities and actions. Results may vary depending on the evaluation. For example, if a model is developed, its preliminary study should be done in detail. Social impact assessment is done with quantitative and qualitative methods (observation, structured interviews, case studies). Problems and alternatives are identified, predictions are made to solve problems, and impact assessment and an interdisciplinary approach are developed. (Antonie, 2010).
- *Social Cohesion:* People started to live as a community to maintain their existence in life. Although written rules have been established to ensure this, it is impossible to say sufficient. For this reason, societies have developed a set ofbehaviors called norms that change according to the regions and ensure harmony. Social cohesion, in general, is defined as "the ability of individuals to adapt their behavior according to society and their environment" (Mualifah et al., 2019). It is complicated to say the entirely positive and completely negative results of achieving social adaptation. It can be noted that these results may differ according to demographic variables.
- *Attitude:* Attitude is one of the most fundamental subjects of behavioral sciences that examine behavior. Since it is used in many subjects and different fields, it is evaluated from different perspectives. According to Arul (1977), attitude is a term used to see essential reactions; according to Altmann (2008), it is a tendency that organizes thoughts, feelings, and behaviors that people develop through cumulative experience, consciously or unconsciously. In other words, it is a state of mental readiness. In support of this idea, Albarracín, Chan, and Jiang (2018) stated that attitudes lead to the intention to act somehow, resulting in behavior. They stated that the process proceeds in belief, attitude, goal, and, finally, behavior. While defining the attitude, it is vital to be inclusive in the literature and generalize and conceptualize all the studies done and developed. Attitude can encompass an object, person, or even an idea.

Social-Psychology in Tourism Business

Tourism is a labor-intensive industry where human interaction is much more intense. People's behaviors, family relations, group movements, and attitudes are sociocultural effects in the tourism system. These effects constitute indirect or direct communication between tourists and locals

(Pizam & Milman, 1986; Ramkissoon, 2020). The relationship between psychology and tourism management has been discussed in the literature. Especially in terms of internal customers (employees), human resources and psychology in consumer behavior are seen as criteria for businesses operating in the tourism industry (Tang, 2014). The concept of social psychology in terms of tourism is a multidimensional issue (Villamira, 2001).

The literature review is limited to typologies and local-people interaction. As often stated in this chapter, different theories in many different disciplines can also be the subject of research. A study by Tang (2014) examined social psychology theories and concepts in publications in tourism journals. The main theories are as follows: "cognitive dissonance", "attribution theory", "balance theory", "elaboration likelihood model", "perception theory", "justice theory", "equity theory", "selectivity theory", "interdependence theory", "self-categorization theory", "identity theory", "social exchange theory", "symbolic interaction theory", "role theory", "impression management", "conflict management", "congruity theory", "disconfirmation theory", and "social learning theory". These theories and concepts show which fields the subject can be associated with. It can be said that these approaches, which are very difficult to explain and generalize, will vary from culture to culture. The highlights of some of these theories are as follows: conflicting situations, behaviors, and beliefs; the process of explaining the causes of events and behaviors; providing balance in motivation psychology; options for influencing human attitudes; observation of behaviors, a fair distribution of resources; cost-benefit analysis and human action.

In this chapter, social psychology in the tourism industry is examined only in terms of employees. The employees' communication among themselves and with the managers, customers, and other stakeholders is the subject of this chapter. To make a general evaluation, in the next part of the chapter, social psychology issues will be evaluated as positive/negative. These topics are organizational justice, counterproductive work behavior, positive psychological capital, nepotism, discrimination, and cultural differences.

Organizational Justice

Organizational justice is a term used to show the justice mechanism within the organization. It is a concept based on the employees' perceptions, showing how fair the owners or managers of the workplace treat their employees (Colquitt et al., 2013). Greenberg (1987), who was the first to use the theory of organizational justice in history, stated that organizational justice is the creation of people's perceptions of the just attitudes of the organization by applying the rules by the theory of justice, the events in the environment in which the people were working in the organization work. Again, according to Greenberg, organizational justice is shaped by performance appraisal, the right choice of working people, rewards, excess salary, or other benefits (Greenberg, 1987). When managers need to distribute tasks fairly to the people working in the organization, evaluating their thoughts helps the employees create the perception that there is organizational justice (Ehrhart, 2004).

Organizational justice is the result of the moral and fair practices of managers within the organization. In other words, in a justice organization, employees evaluate employers' attitudes reasonably, morally, and rationally (Farh et al., 1997). While fair behaviors within the organization have a positive effect and benefit organizational outputs, unfair behaviors can cause aggression, restlessness, silence, theft, etc., in the organization (Khaola & Coldwell, 2019). Organizational justice also plays a vital role in positively affecting the job satisfaction of institutions and employees working in the institution and shaping it positively in terms of behavior (Lavelle et al., 2007). In this case, the positive behavior and processes in the organization allow the working people to develop harmonious relations in terms of trust and loyalty among themselves and with their managers (Lehmann-Willenbrock et al., 2013).

On the other hand, perceived organizational justice is based on the perception of distribution, functioning, and influence according to the working knowledge and competence of the employees in the environment in which they work. Since this is a perceptual situation, it may occur differently in each of the employees (Niehoff & Moorman, 1993). If working individuals are faced with unfair problems within the organization, it is inevitable to experience a decrease in both motivation and performance levels. It has been observed that perceived organizational injustice leads to the emergence of situations such as weakening of the workforce, introversion, experiencing stress, being tense, lack of job satisfaction, and perhaps the intention to leave the job.

Organizational justice means that all persons in an institution are morally and materially fair to each other and that the deserved is given their rights to their persons. Institutions should know that they can achieve the desired performance with effort and devotion since they adopt and implement this perception in motivating the people working at the workplace (Thibaut & Walker, 1975). On the contrary, in cases where there is no right and law, internal reluctance, relaxation, and low morale are avoided in the employees. Organizational justice covers the perceptions of employees about the correctness of organizational practices and decisions and the effects of these perceptions on employees. The fact that employees do not evaluate their organization or managers fairly may cause employees to develop reactive behaviors. From this point of view, employees' perceptions of justice can emerge in different ways. In this case, it directly reflects negatively on the institution (Tyler, 1989). It causes the organization's targeted goals and objectives to not be on time. In a world that is developing and growing every day, it causes employees to lag in terms of competition with organizations in other sectors due to their situation both technologically and economically (Walumbwa et al., 2006). The concept of organizational justice is related to job satisfaction, internal commitment, work performances of working people, etc. These situations need to be valued for the organization to continue its life in a meaningful sense.

Organizational justice includes an organization concept where open and intense relations are experienced in tourism businesses. This structure, in which there are formal and informal relations, is based on the teamwork of leaders, managers, and employees with a system understanding. It is recommended that employees feel organizational justice to provide quality service and achieve goals (Erkılıc et al., 2018).

In the tourism business, distribution justice is mentioned, primarily seen in seasonal businesses. Comparing what is presented to them with other employees, and as a result of this comparison, various actions are taken in case of possible injustice. For example, in an accommodation business, employees in one department receive low wages or tips from other department employees. The results of learning this situation can change the perception of the business. All procedures within the tourism business and their implementation also affect organizational justice. While applying a method (such as working hours, ethical principles), fairness must be observed. In addition to all these, in the tourism industry, where human relations in the working environment are extremely high, it is expected that the same concerns will be between the employee and the manager. Here, it can be said that each tourism business has a different organizational climate, but some features are expected to be in every business.

Counterproductive Work Behavior

The concept of "counterproductive work behavior", or in other words, "subjective well-being", which was first used by Bradburn (1969) in the literature, basically refers to well-being at the psychological level, being happy and satisfied, and psychological, mental, and physical health. Although counterproductive work behavior stands out in modern psychology, it has an increasingly rich field of study in the literature. According to Nawaz et al. (2018: 209), counterproductive work behavior can be defined as psychological functionality and experience, emphasizing concepts such as autonomy, development, originality, and existential engagement with a meaningful life. Counterproductive work behavior means that individuals experience positive emotions more than negative emotions in their emotional lives, and positive emotions dominate negative emotions (Spector et al., 2010). Deng and Zhang (2016) defines counterproductive work behavior as the personal experiences of individuals and states that the counterproductive work behavior levels of individuals who believe that they can achieve the goals they set will increase.

On the other hand, Fox and Spector (2006) argue that people's self-belief that they can succeed in the face of challenging situations in their natural lives will affect their counterproductive work behavior. The abilities of individuals, autonomy, and establishing social relations are psychological needs for people, and meeting these needs explains well-being (Nawaz et al., 2018). Spector et al. (2010) define counterproductive work behavior as establishing quality relationships with other individuals, attributing meaning to life activities, and managing situations such as personal development while defining the individual's way of being in life.

The counterproductive work behavior of the employees in the organizations is about themselves, the managers, and the organizations they work for. Because happy employees in good psychological conditions can be more productive in the organization, the stress in organizations can lead to anxiety, depression, and psychological fatigue, reducing productivity (Siu et al., 2007). Therefore, these negative situations should be identified and eliminated by the managers. Research on counterproductive work behavior has provided evidence for the physiological effects of the individual's mood and emotional state. If the happiness hormones such as serotonin, dopamine, endorphin, and noradrenaline are secreted at a sufficient level, the individual defines himself as happy. The individual experiences a state of being energetic, joyful, calm, and peaceful in happiness states, while hormones are secreted less, and depression, weakness, and pessimism increase in negative mood states. In addition, it is known that happiness hormones are secreted in the presence of sufficient sunlight, in clean environments with oxygen, and when habits that make the individual feel good are maintained (Siu et al., 2007).

The study results of Kara et al. (2013), which questioned the effect of leadership styles on the psychological well-being of the accommodation establishments, indicate that transformational leadership has a positive effect on the psychological well-being of the personnel. The result that transactional leadership also has a positive effect on employee performance is among the unexpected consequences of the study. Therefore, the hypothesis that "transactional leadership has a negative effect on employee performance" was rejected as a result of the analysis. In addition, the necessity of training the managers, who are seen as leaders, to use the transformational leadership style effectively, is among the other results of the study.

Nyberg et al. (2011) stated that the destructive leadership approach has a negative impact on the psychological well-being of the accommodation tourism industry employees and suggested that this negative effect can only be positive by optimizing the leadership understanding in the organization. In another tourism field study on the subject, it was emphasized that psychological well-being has a negative effect on the intention to leave (Amin & Akbar, 2013). While it has been revealed that the job stress in the tourism industry is higher in the managers working in the leadership position, it is among the study results that job stress negatively affects psychological well-being (O'Neill & Davis, 2011).

Roothman et al. (2003) suggested that counterproductive work behavior can be conceptualized for physical, emotional, cognitive, personal, spiritual, and social processes. Since happiness is an abstract concept, some components are taken as a basis while defining it. Spector et al. (2010) expressed the cognitive and affective dimensions of happiness. While he associates the affective dimension of happiness with emotions, he associates the cognitive dimension with reflective judgments such as satisfaction. Waterman (2007) defined dimensions for counterproductive work behavior as "positive emotions, negative emotions, and cognitive dimensions". Positive emotions refer to the positive state of counterproductive work behavior, negative emotions refer to the negative form, and the cognitive dimension refers to the holistic perception of life. Ryff (1989) proposes six dimensions in the counterproductive work behavior model.

- Self-acceptance
- Positive relationships with others
- Personal development
- Purpose of life
- Environmental control
- Autonomy

Positive Psychological Capital

Psychological capital was first used scientifically by Luthans et al. (2007). However, when the annual developmental stages are investigated, it is seen that the term positive psychological capital was first used in the literature as positive organizational behavior capacity. In the process, the concept of positive organizational behavior has developed and started to be used as "positive psychological capital" in the literature (Avey et al., 2010).

The concept of psychological capital differs from human capital that focuses on what one knows, economic capital that indicates what one has, and social capital that defines who one knows by expressing "who he is, what he is and what he aims to be" (Avey et al., 2011). Thus, positive psychological capital behavior is examined as an approach that aims to improve performance within the organization by discovering the strengths and positive sides of the employees, instead of the weak and deficient aspects, and by developing, measuring, and managing the psychological factors behind their strengths (Cheung et al., 2011). Luthans interprets psychological capital as "the individual's positive psychological development status,". Besides this, he depicts that psychological capital refers the question of "Who am I ?" (Luthans & Youssef, 2004).

Employee positive psychological capital includes a disposition that motivates individuals to work success; it is not a solid trait like a personality trait. It is an essential element from which performance can be enhanced when measured and developed. Employees' psychological abilities play a significant role in motivation, where they can be developed as a measure to improve performance (Luthans et al., 2007). Positive psychological capital is an important variable that can contribute to the positive behavior of the individual in the organization. Therefore, organizations can ensure that employees continue to improve their organizational performance by maintaining positive psychological conditions.

Although the human factor is an essential element in the organization, the most important power within the human is the psychological state. In other words, the strong positive psychological capital of the individual enables him to use his potential effectively and reflect it positively to his organization (Rego et al., 2012). Focusing on the positive and strong sides of all employees in organizations has become an element that gains value in developing new organizational strategies and gaining competitive advantage (Tosten & Toprak, 2017). Human resource, when used correctly, is expressed as the great power of organizations with the spread of positive psychology capital day by day.

The characteristics that need to be measured while trying to answer how the organization's employees define and perceive themselves constitute the main dimensions of "positive psychological capital". According to Luthans, one of the reasons why elements other than positive psychological capital components (hope, optimism, self-efficacy, resilience) are not included in the scope of positive psychological capital is the ability of these four elements to measure performance; another is since these elements can be developed (Luthans et al., 2007).

Historically, businesses have tried many ways to attract talented employees to their workforce; however, many, if not most, organizations still fail to realize the full potential of their human resources. Institutions can gain a significant advantage over their competition by investing, developing, managing, and using positive psychological capital, a renewable, cumulative, context-specific, and hard-to-imitate resource (Avey et al., 2011).

Positive psychological capital is a high-level construct of self-efficacy, hope, resilience, and optimism. The conceptual dependence of these four constructs is supported both theoretically and empirically in the literature. However, while they are different constructs, there are enough similarities between self-efficacy, hope, resilience, and optimism to show that a common, fundamental bond operates between them and binds them together—for example, Toor and Ofori (2010) found that resilience in the face of negativity was more likely for individuals with high self-efficacy than those with low self-efficacy, and Rego et al. (2012) found that self-efficacy tended to be higher in individuals with high hope than those with low hope (Tosten & Toprak, 2017). Due to the intense working conditions of the tourism industry, there is a need for hope, self-efficacy, resilience, and optimism. It is stated that these dimensions of psychological capital have significant effects on job satisfaction and organizational continuity in tourism enterprises. It is a complicated process to see the positive aspects of the employees living in challenging conditions and bring this to an organizational understanding. For example, it is seen that demographic characteristics such as gender, age, education level of employees, and determinants such as culture-subculture are evaluated in terms of positive psychology.

Nepotism

Nepotism is derived from the Latin word *nepos* meaning 'nephew.' During the Middle Ages, the concept of nepotism was noticed when the Popes appointed their nephews to important positions in the church (Arasli & Arici, 2020). The opening of business doors to nephews by the papal authority caused the concept of nepotism to be perceived negatively and damaged the effectiveness of the church (Iqbal, & Ahmad, 2020). Nepotism is the real support and perceived preference a family member gives to other family members (Block & Fisch, 2020). By taking advantage of the power provided by their office and position, managers can engage in discriminatory practices in favor of their relatives, employ and empower their close relatives regardless of their merit (Ignatowski et al., 2020).

Iqbal and Ahmad (2020) defined nepotism as the state of focusing on blood ties with politicians and influential people in society, rather than focusing on knowledge, education and experience. It is generally characterized as a biased management system that favors some employees over others. Nepotist behaviors are usually perceived as preferring the bad over the good and putting incompetent people in important positions (Kawo & Torun, 2020). In organizations where nepotism is seen, the influence of the family can be felt in personnel policy, and the interests of family members may be protected rather than the interests of the organization (Iqbal & Ahmad, 2020). Privileges of individuals with kinship relations with rulers, regardless of their usefulness, may cause family priorities and interests to precede the interests and rules of the organization (Ford & McLaughlin, 1986). Nepotism is effective in employment, promotion or managerial appointments, belonging, and loyalty; it can lead the organization to prevent economic losses or profits (Kawo & Torun, 2020). Considering the kinship ties with senior managers by ignoring their suitability for the job, experience, sector, or department knowledge shows that nepotism is not a professional practice and has an amateurish management approach. This management approach is based on kinship relations, ignores the universal principle of merit, and gives family members; provides more privileges than other employees in employment, promotion, wage, or working conditions (Iqbal & Ahmad, 2020). Individuals in employment or relocation; regardless of whether they have criteria such as knowledge, skills, education, talent, and experience, acting based on kinship relations with politicians, bureaucrats, or public officials (Block & Fisch, 2020) and opening new positions or new positions for individuals; can prevent the institutionalization of organizations.

Nepotism is closely related to the social and cultural structure of society. Family, kinship, sociability perception, and many societal factors affect nepotism (Ford & McLaughlin, 1986). Although nepotism is a phenomenon seen in every culture, traditions, customs, symbols, and rules can shape nepotist behaviors. In underdeveloped countries, where traditional ties and relations are more important, society can respect and appreciate individuals who engage in favoritism. This situation may cause nepotism to be seen more frequently (Kawo & Torun, 2020). Nepotism can be viewed more moderately by society than other types of favoritism (Iqbal & Ahmad, 2020). This is because nepotism is based on kinship relations rather than financial interests. Communities can tolerate and adapt favoritism based on kinship relations (Arasli & Arici, 2020).

Nepotism is frequently felt in service businesses, and this situation does not differ for accommodation businesses. Due to its labor-intensive characteristics, the human element is crucial in accommodation businesses. The importance of the human factor also increases the importance of human resources practices. However, considering that nepotism is a practice that benefits only close friends and family members rather than the entire enterprise, it prevents the regular performance of human resources activities in enterprises. It reduces the perception of justice among employees (Arasli et al., 2006).

Due to the negative impact of nepotism practices on organizational justice, tourism businesses should focus on this issue sensitively. For this reason, it is thought that businesses should avoid nepotism practices because these practices alienate employees from the business and create a bad business culture. At this point, especially business owners and senior managers have important roles. Giving opportunities to deserving employees rather than family members in recruitment and promotion can be crucial in terms of showing other employees that they have this right, increasing the perception of justice and loyalty, and keeping talented employees at work. According to Notzke (2004), the tourism sector has been a sector where nepotism is encountered quite frequently, and it is stated that this situation can be overcome by avoiding including the politics into the working process. It is stated that nepotism practices are carried out in the private sector and the planning and regulations related to tourism in public institutions (Larsen et al., 2011: 482). From this point of view, it can be said that nepotism and partisanship practices such as nepotism should be avoided for the effective development of tourism, especially in human resources planning. Nepotism practices in an enterprise make employees think that there is an unfair management approach; injustice can also cause many negative outcomes by affecting employee motivation. Aryee et al. (2002) also reached results supporting this view in their study. According to the authors, there is a positive relationship between organizational justice and job satisfaction, commitment to the organization, and trust in the manager. In addition, the intention to leave the job increases in businesses where unfair practices are present.

Discrimination

One of the goals of the emergence of modernism by feeding on processes such as the nation-state, urbanization, and bringing rationality to the fore is to create a society where equal and free individuals live (Bayl-Smith & Griffin, 2014). However, with the emergence of some dominant groups, minorities, immigrants as a result of immigration due to work or life concerns, religious groups due to various terrorist activities, and so on, discrimination in many areas shows that modern society has not yet reached its goal (same). Looking from the past, this issue has been emphasized with various studies. Since the first publication of *The Journal of Social Issues* in 1945 on the subject of "Racial and religious prejudice in everyday living" edited by Gene Weltfish, Gordon Allport, Kenneth Clark, Margaret Mead, and some prominent social scientists of the time, such as Kurt Lewin, were involved. Articles were published on anti-Semitism, discrimination against women in business life, and discrimination against blacks, which are still talked about today (Colella et al., 2012).

When we look at the root of the word "discrimination", it is seen that it comes from the Latin word *Discriminato*, which means "Separation". When this concept is carried to the field of society, it means "segregation of some disadvantaged social segments for reasons such as skin color, name difference, gender, religion". Discrimination in the discipline of social psychology, which means that an individual is exposed to negative treatment and behavior only because of belonging to a particular group; it can be expressed as "transformation of prejudices into actions". Prejudices can exist without discrimination as a fact, and discrimination can live in the absence of prejudice. Considering this, the term discrimination has a huge negative impact on individuals or communities and its scope is extremely wide. Boehm et al. (2014), this includes all unfair acts whether intentionally or not, based on race, physical appearance, religious beliefs, gender, physical or mental disability, marital status, sources of income, or sexual orientation.

It is seen that the phenomenon of discrimination has been made by the majority against the characteristics of the minority group, such as ethnic structure, nationality, color, or physical disability, from past to present. Rather than interpreting people as unique beings, they tend to label people according to their common characteristics such as race, age, gender, religious belief, and so on. This type of labeling causes individuals to be excluded from their individual qualities and treated unequally compared to other groups due to the attributes attributed to the class they belong to. In such a case, it creates discrimination. Several individuals or individuals enter the field of "categorical discrimination", and because of this, they are socially placed in a certain classification "either because of race, religion, gender or any naming used to distinguish members of a society" (Wegge et al., 2008). In general, discrimination is often seen together with the concepts of prejudice and stereotype. Although social scientists vary in their explanations of "prejudice", they are united in the opinion that prejudice is generally an inadequate preliminary assessment against a group or its members (Colella et al., 2012).

The concept of discrimination is divided into two as direct and indirect discrimination (Bayl-Smith & Griffin, 2014). Direct discrimination constitutes the best-known form of discrimination, based on not showing the same behavior under the same conditions, especially in fundamental rights and freedoms protected by law (Colella et al., 2012). Indirect discrimination is based on the fact that differences between individuals can create inequality in fundamental rights and freedoms, and it causes a concrete inequality rather than a formal one.

When the situation is evaluated in terms of tourism businesses, it is possible to say that discrimination based on gender is the most common type of discrimination. Less than 40% of all managerial and supervisory positions in the international hospitality industry are women. 20% of general management roles belong to women. Women are less than 20% in the ownership of hospitality businesses, and this rate drops to 10% in hotel ownership worldwide (Obadic, 2016). According to Arlı (2013), the qualifications required in the tourism sector also limit women's capabilities working in this sector. Because most jobs in the sector do not require special skills, they are paid for a low wage, and there is an intense workforce circulation resulting from part-time work. The units where women work most intensively in the hospitality sector are the front office and housekeeping units. In Ng and Pine's (2003) research on the gender distribution of managers in accommodation businesses in Hong Kong, it was determined that general managers were primarily men. It has been determined that women mainly undertake managerial duties in housekeeping and front office. Thrane (2008) examined the effects of education, experience, and demographic characteristics on the wages of tourism workers. Changes in annual salaries were observed in the Norwegian tourism sector in the 1994–2002 period, and four of the results were found to be significant. Men working in the tourism sector earn an average of 20% higher wages per year than women. Being married and having children affects wages, but this effect differs between male and female employees. There was a positive development in the salaries of all tourism employees in this period.

Cultural Differences

Cultural differences are an essential organizational management element for international businesses. However, culture is an issue that needs to be well managed for almost every company in the age of globalization. The human element, one of the most important capitals of organizations, spends most of its life in the institution it works (Aneas & Sandin, 2009). As a result of specialization and interaction in working forms with developing technology, there is now a dynamic organizational model. With the innovations brought by globalization and technology, the management of differences has gained importance for the harmony of both employees from different countries in multinational enterprises and those who are citizens of the same country and have different social identities. Religion, language, race, gender, region, age, and similar factors have become essential for individuals (Humes, 1993). Differently, being a vegetarian is now even more critical for a person than in the past. It has become an upper identity for many people and may have based its life on being a vegetarian. This naturally affects his way of doing business. Companies also develop some policies to ensure the coordination of people from different cultures in their corporate structures. In this context, diversity management aims to enable individuals with various internal and external qualities to operate effectively and efficiently. The different characteristics of people are now seen as an issue that needs to be managed and paid attention to within the organization. Because it is natural and expected that people who have come together for common purposes, separate in working life, where financial elements are at the forefront due to their differences. Here, institutions have turned to the policy of managing differences to prevent both harmonization and possible problems. People who come together to realize the institution's goals, such as efficiency, performance, effectiveness, and profitability, expect to adapt to the organization on the one hand and respect their colleagues for the differences they have on the other hand (Hofstede, 2000). Employees from different cultures in the working environment necessarily reflect their cultural values and attitudes when communicating with others. In intercultural interaction, especially in face-to-face communication, the use of voice, expression, way of speaking, and the chosen words are sometimes affected by their cultural roots. It is necessary to add to this the past experiences of the individual.

Individuals also attach importance to the communication style, managers, and the general organization in the working environment. In other words, they can be affected by exclusion, an unequal attitude, or rules arising from cultural factors in the interaction process (Luijters et al., 2008). For this reason, in the institutions' human resources and corporate culture management policies, importance is given to the management of the interaction of people from different segments. The issue of differences in administrative processes emerged in America and spread to Europe and Japan. It has been discussed frequently, especially after the discriminatory policies towards Black and Mexican people in the USA. This phenomenon, which was examined under the titles of "diversity" and "diversity management" in the United States, is now one of the main agenda items in the business literature. In this sense, diversity management reflects an understanding that ensures that all people have equal opportunities while respecting everyone's differences (Tharpp, 2009).

The primary purpose of diversity management is based on a business strategy that prioritizes equality of opportunity. Especially with the increase of employees belonging to different groups in multinational enterprises, these enterprises have gained a heterogeneous dimension. Diversity management is an essential strategy for organizations' performance and corporate goals by embracing pluralism in institutions and adding it to management policies (Humes, 1993). In this context, the existence of different and separate characteristics of the employees has come to a vital point in terms of commitment to the institution and performance. It is crucial to create a harmonious whole with the colors of today's world, rather than standardizing the employee profile that has become complex in the global world (Aneas & Sandin, 2009).

The multicultural nature of tourism establishments brings with it some problems arising from cultural differences. It is known that there are many cultural differences, especially in chain tourism businesses consisting of multicultural owners. According to Reisinger and Turner (1997), cultural differences between countries are significant for international markets, as culture varies from region to region, from country to country. Therefore, each country's traditions, customs, attitudes, habits, and behaviors, the development, distribution, religion, language, race, social classes, family systems, social values, and norms of the population should be considered, especially in international markets. One of the most important duties of business managers who appeal to consumers from different cultures; is to analyze the pure culture of its own country first, and then the culture of the countries in which it operates, and organize business activities accordingly (Park & Reisinger, 2012). Countries with similar economic structures may have very different cultural forms from each other (Crotts & Erdmann, 2000); it is observed that people living in different cultures can react differently to similar issues or conditions (Jansen-Verbeke, 1996). Therefore, businesses should examine countries according to their cultural characteristics in international investment and avoid generalizing only their economic characteristics.

Conclusion

In this chapter, the employee relations of social psychology in tourism enterprises are discussed. After the general evaluation of social psychology, its dimensions were discussed. The concepts of organizational justice, counterproductive work behavior, positive psychological capital, nepotism, discrimination, and cultural differences, which will be associated with the field of organizational behavior, are explained with examples in the subject of social psychology in tourism business, which forms the basis of the chapter.

Tourism makes it necessary to examine the social-psychological structure of employee relations in many aspects. The first is that tourism occurs in the services sector, where social relations are at the highest level. Unwritten rules and behaviors can affect motivation and attendance. On the other hand, many positive/negative results affect the relations among the employees, between the employees and the managers, and between the employees and the customers.

Understanding the social-psychological dynamics of tourism and scientific research on this subject are necessary to achieve sustainable tourism goals with ethical principles. Measuring employees' attitudes, behaviors, and expectations towards different environments (examining internal and external factors) will also help explain their behavior. Achieving success in this regard is possible with an interdisciplinary approach. Obtaining results with field research by examining the theories and discussing these results in the literature.

The conclusion from this book chapter can be summarized as that social-psychological employee relations and results in the tourism business are not based on a single cause, and different theories/approaches should be used to explain them. Different views can be provided in this field of study by examining all the factors affecting group relations with my employee's attitudes, beliefs, and demographic characteristics.

References

Albarracín, D., Chan, M. S., & Jiang, D. (2018). Attitudes and attitude change: Social and personality considerations about specific and general patterns of behavior. In K. Deux & M. Snyder (Eds.), *The Oxford Handbook of Personality and Social Psychology* (pp. 439–463). New York: Oxford University Press.

Altmann, T. (2008). Attitude: A concept analysis. *Nursing Forum*, 43(3), 144–150.
Amin, Z. & Akbar, K. P. (2013). Analysis of psychological well-being and turnover intentions of hotel employees: An empirical study. *International Journal of Innovation and Applied Studies*, 3(3), 662–671
Aneas, M. A. & Sandin, M. P. (2009). Intercultural and cross-cultural communication research: Some reflections about culture and qualitative methods. *Qualitative Social Research*, 10(1), 43–61.
Antonie, R. (2010). Social impact assessment models. *Transylvanian Review of Administrative Sciences*, 29E, 22–29.
Arasli, H., Bavik, A., & Ekiz, E. H. (2006). The effects of nepotism on human resource management: The case of three-, four- and five-star hotels in Northern Cyprus. *International Journal of Sociology and Social Policy*, 26(7–8), 295–308.
Arasli, H. & Arici, N. C. (2020). The effect of Nepotism on tolerance to workplace incivility: Mediating role of psychological contract violation and moderating role of authentic leadership. *Leadership & Organization Development Journal*, 41(4), 597–613.
Arlı, E. (2013). Deniz Turizm Sektöründe Algılanan Cinsiyet Ayrımcılığı ve Cinsiyet Önyargısı: Karamürsel Meslek Yüksekokulu Öğrencileri Üzerine Bir Araştırma [Perceived Gender Discrimination and Gender Bias in the Marine Tourism Sector: A Study on Karamürsel Vocational School Students]. *Çalışma ve Toplum : Ekonomi ve Hukuk Dergisi*, 38(3), 283–301.
Arul, M. J. (1977). Measurement of attitudes. https://www.arulmj.net/atti2-a.htm.
Aryee, S., Budhwar, P. S. & Chen, Z. X. (2002). Trust as a mediator of the relationship between organizational justice and work outcomes: Test of a social exchange model. *Journal of Organizational Behavior*, 23(3), 267–285.
Avey, J. B., Luthans, F., Smith, R. M. & Palmer, N. F. (2010). Impact of positive psychological capital on employee well-being over time. *Journal of Occupational Health Psychology*, 15 (1), 17–28.
Avey, J. B., Reichard, R. J., Luthans, F. & Mhatre, K. H. (2011). Meta-analysis of the impact of positive psychological capital on employee attitudes behaviors, and performance. *Human Resource Development Quarterly*, 22, 127–152.
Baldwin, J. M. (1897). *Social and Ethical Interpretations in Mental Development: A Study in Social Psychology*. New York: MacMillan Co. https://doi.org/10.1037/12907-000
Bayl-Smith, P. H. & Griffin, B. (2014). Age discrimination in the workplace: Identifying as a late-career worker and its relationship with engagement and intended retirement age. *Journal of Applied Social Psychology*, 44, 588–599.
Block, J. H. & Fisch, C. (2020). Eight tips and questions for your bibliographic study in business and management research. *Management Review Quarterly*, 70(3), 307–312.
Boehm, S. A., Kunze, F. & Bruch, H. (2014). Spotlight on age-diversity climate: The impact of age-inclusive HR practices on firm-level outcomes. *Personnel Psychology*, 67, 667–704.
Bradburn, N. M. (1969). *The Structure of Psychological Well-being*. Chicago: Aldine Publishing Company.
Cheung, F., Tang, C. S. K. & Tang, S. (2011). Psychological capital as a moderator between emotional labor, burnout, and job satisfaction among school teachers in China. *International Journal of Stress Management*, 18(4), 348–371.
Colella, A., McKay, P., Daniels, S. & Signal, S. (2012). Employment discrimination. In S. Kozlowski (Ed.), *The Oxford Handbook of Organizational Psychology* (pp. 1034–1102). Oxford and New York: Oxford University Press.
Colquitt, J. A., Scott, B. A., Roddell, J. B., Long, D. M., Zapata, C. P., Conlon, D. E. & Wesson, M. J. (2013). Justice at the millennium, a decade later: A meta-analytic test of social exchange and affect based perspectives. *Journal of Applied Psychology*, 98(2), 199–236.
Crotts, J. C. & Erdmann, R. (2000). Does national culture influence consumers' evaluation of travel services? A test of Hofstede's model of cross-cultural differences. *Managing Service Quality: An International Journal*.
Deng, Y. & Zhang, L. (2016). Guanxi with supervisor and counterproductive work behavior: The mediating role of job satisfaction. *Journal of Business Ethics*, 134(3), 413–427.
Dhont, K., Hodson, G., Loughnan, S. & Amiot, C. E. (2019). Rethinking human-animal relations: The critical role of social psychology. *Group Processes and Intergroup Relations*, 22(6), 769–784.
Ehrhart, M. G. (2004). Leadership and procedural justice climate as antecedents of unit-level organizational citizenship behaviour. *Personnel Psychology*, 57(1), 61–94.
Erkılıc, E., Gazeloglu, C. & Aytekin, E. (2018). Organizational justice perceptions of hospitality business employees in the scope of demographic characteristics: A study in Rize. *Tourism Human Rights & Sustainable Environment*. UK: IJOPEC Publication Limited.
Farh, J. L., Earley, P. C. & Lin, S. C. (1997). Impetus for action: A cultural analysis of justice and organizational citizenship behavior in Chinese society. *Administrative Science Quarterly*, 42(3), 421–444.

Ford, R. & McLaughlin, F. (1986). Nepotism: Boon or bane. *Personnel Administrator,* 31(11), 78–89.
Fox, S., & Spector, P. E. (2006). The many roles of control in a stressor-emotion theory of counterproductive work behavior. In P. L. Perrewé & D. C. Ganster (Eds.), Employee health, coping and methodologies (pp. 171–201). Elsevier Science/JAI Press. https://doi.org/10.1016/S1479-3555(05)05005-5Greenberg, J. (1987). A taxonomy of organizational justice theories. *Academy of Management Review,* 12(1), 9–22.
Hofstede, G. (2000). *Cultures and Organizations: Software of the Mind.* London: McGraw-Hill Book Company.
Humes, S. (1993). *Managing the Multinational: Conflicting the Global-Local Dilemma.* Hemel Hempstead: Prentice Hall.
Ignatowski, G., Sułkowski, Ł. & Stopczyński, B. (2020). The perception of organisational Nepotism depending on the membership in selected Christian churches. *Religions,* 11(1), 47.
Iqbal, Q. & Ahmad, N. H. (2020). Workplace spirituality and nepotism-favouritism in selected ASEAN countries: The role of gender as moderator. *Journal of Asia Business Studies,* 14(1), 31–49.
Jansen-Verbeke, M. (1996). Cross-cultural differences in the practices of hotel managers: A study of Dutch and Belgian hotel managers. *Tourism Management,* 17(7), 544–548.
Kara, D., Uysal, M., Sirgy, M. J. & Lee, G. (2013). The effects of leadership style on employee well-being in hospitality. *International Journal of Hospitality Management,* 34, 9–18.
Kawo, J. W. & Torun, A. (2020). The relationship between nepotism and disengagement: The case of institutions in Ethiopia. *Journal of Management Marketing and Logistics,* 7(1), 53–65.
Khaola, P. & Coldwell, D. (2019). Explaining how leadership and justice influence employee innovative behaviours. *European Journal of Innovation Management,* 22(1), 193–212.
Larsen, R. K., Calgaro, E. & Thomalla, F. (2011). Governing resilience building in Thailand's tourism-dependent coastal communities: Conceptualising stakeholder agency in social–ecological systems. *Global Environmental Change,* 21, 481–491.
Lavelle, J. J., Rupp, D. E. & Brockner, J. (2007). Taking a multifoci approach to the study of justice, social exchange, and citizenship behavior: The target similarity model. *Journal of Management,* 33(6), 841–866.
Lehmann-Willenbrock, N., Grohmann, A. & Kauffeld, S. (2013). Promoting multifoci citizenship behaviour: Time-lagged effects of procedural justice, trust, and commitment. *Applied Psychology,* 62(3), 454–485.
Luijters, K., Zee, K. & Otten, S. (2008). Cultural diversity in organizations: Enhancing identification by valuing differences. *International Journal of Intercultural Relations,* 32(2), 154–163.
Luthans, F. & Youssef, C. M. (2004). Human, social, and now positive psychological capital management: Investing in people for competitive advantage. *Organizational Dynamics,* 3, 143–160.
Luthans, F., Avolio, B. J., Avey, J. B. & Norman, S. M. (2007). Positive psychological capital: Measurement and relationship with performance and satisfaction. *Personnel Psychology,* 60(3), 541–572.
Luthans, F., Youssef, C. M. & Avolio, B. J. (2007). Psychological capital: Investing and developing positive organizational behavior. *Positive Organizational Behavior,* 1(2), 9–24.
McDougall, W. (1909). *Introduction to Social Psychology.* London: Methuen and Co.
Mualifah, A., Barida, M. & Farhana, L. (2019). The effect of self-acceptance and social adjustment on senior high school students' self-concept. *International Journal of Educational Research Review,* Special Issue, 719–724
Nawaz, R., Zia-UD-Din, M., Nadeem, M. T. & Din, M. (2018). The impact of psychopathy on counterproductive work behavior. *International Journal of Academic Research in Business and Social Sciences,* 8(7), 208–220.
Ng, C. W. & Pine, R. (2003). Women and men in hotel management in Hong Kong, perceptions of gender and career development issues. *Hospitality Management,* 22, 85–102.
Niehoff, B. P. & Moorman, R. H. (1993). Justice as a mediator of the relationship between methods of monitoring and organizational citizenship behaviour. *Academy of Management Journal,* 36, 527–556.
Notzke, C. (2004). Indigenous tourism development in Southern Alberta, Canada: Tentative engagement. *Journal of Sustainable Tourism,* 12(1), 29–54.
Nyberg, A., Holmberg, I., Bernin, P. & Alderling, M. (2011). Destructive managerial leadership and psychological well-being among employees in Swedish, Polish, and Italian hotels. *Work,* 39(3), 267–281.
Obadic, A. (2016). Gender discrimination and pay gap on tourism labor market. *International Scholarly and Scientific Research & Innovation,* 10(3), 808–813.
O'Neill, J. W. & Davis, K. (2011). Work stress and well-being in the hotel industry. *International Journal of Hospitality Management,* 30(2), 385–390.
Park, S. & Reisinger, Y. (2012). Cultural differences in tourism web communication: A preliminary study. *Tourism Analysis,* 17(6), 761–774.
Pizam, A. & Milman, A. (1986). The social impacts of tourism. *Tourism Recreation Research,* 1(11), 29–33
Poggi, G. (2021). Social psychology of tourism. https://slideplayer.com/slide/16347285/
Ramkissoon, H. (2020). Perceived social impacts of tourism and quality-of-life: A new conceptual model. *Journal of Sustainable Tourism,* DOI: 10.1080/09669582.2020.1858091

Rego, A., Sousa, F., Marques, C. & e Cunha, M. P. (2012). Authentic leadership promoting employees' psychological capital and creativity. *Journal of Business Research*, 65(3), 429–437.
Reisinger, Y. & Turner, L. (1997). Cross-cultural differences in tourism: Indonesian tourists in Australia. *Tourism Management*, 18(3), 139–147.
Roothman, B., Kirsten, D. K. & Wissing, M. P. (2003). Gender differences in aspects of psychological well-being. *South African Journal of Psychology*, 33(4), 212–218.
Ross, E. A. (1909). What is social psychology? *Psychological Bulletin,* 6, 409–411.
Ryff, C. D. (1989). Happiness is everything, or is it? Explorations on the meaning of psychological well-being. *Journal of Personality and Social Psychology*, 57(6), 1069.
Simkova, E. (2014). Psychology and its application in tourism. *Procedia – Social and Behavioral Sciences,* 114, 317–321.
Siu, O. L., Lu, C. Q. & Spector, P. E. (2007). Employees' well-being in greater China: The direct and moderating effects of general self-efficacy. *Applied Psychology: An International Review*, 56(2), 288–301.
Spector, P. E., Bauer, J. A. & Fox, S. (2010). Measurement artifacts in the assessment of counterproductive work behavior and organizational citizenship behavior: Do we know what we think we know? *Journal of Applied Psychology*, 95(4), 781–790.
Tang, L. (2014). The application of social psychology theories and concepts in hospitality and tourism studies: A review and research agenda. *International Journal of Hospitality Management*, 36, 188–196.
Tarde, G. (1903). *The Laws of Imitation* (E., Parsons, Trans.; French ed., 1880). New York: Henry Holt and Company.
Tharpp, B. M. (2009). Defining culture and organizational culture: from anthropology to the office. *Haworth*, http://www.thercfgroup.com/files/resources/Defining-Culture-and-Organizationa-Culture_5.pdf.
Thibaut, J. & Walker, L. (1975). *Procedural Justice: A Psychological Analysis*. New York: Springer.
Thrane, C. (2008). Earnings differentiation in the tourism industry: Gender, human capital and socio-demographic effects. *Tourism Management*, 29, 514–524.
Toor, S.- & Ofori, G. (2010). Positive psychological capital as a source of sustainable competitive advantage for organizations. *Journal of Construction Engineering and Management*, 136(3), 341–352.
Tosten, R. & Toprak, M. (2017). Positive psychological capital and emotional labor: A study in educational organizations. *Cogent Education*, 4(1), 1–11.
Tyler, T. R. (1989). The psychology of procedural justice: A test of the group-value model. *Journal of Personality and Social Psychology*, 57(5), 830–838.
Villamira, M. A. (2001). *Psicologia del viaggio e del turismo*. Torino: UTET.
Walumbwa, F. O., Wu, C. & Orwa, B. (2006). Leadership, procedural justice climate, work attitudes, and organizational citizenship behavior. *Academy of Management Proceedings*, 2006(1), i-C6.
Waterman, A. S. (2007). Doing well: The relationship of identity status to three conceptions of well-being. *Identity: An International Journal of Theory and Research*, 7(4), 289–307.
Wegge, J., Roth, C., Neubach, B., Schmidt, K.-H. & Kanfer, R. (2008). Age and gender diversity as determinants of performance and health in a public organization: The role of task complexity and group size. *Journal of Applied Psychology*, 93, 1301–1313.
Wilson, E. (2017). What is social impact assessment? *Indigenous Peoples and Resource Extraction in the Arctic: Evaluating Ethical Guidelines*. Arran (January), 1–19.
Wood, W. (2017). Habit in personality and social psychology. *Personality and Social Psychology Review,* 21(4), 389–403
Yiu, M. & Law, R. (2012). Factors influencing knowledge sharing behavior: A social-psychological view in tourism. *Service Science,* 3(2), 11–31.

25

ATTITUDES AND BEHAVIORS OF TOURISM EMPLOYEES AT WORK AND AMONG CO-WORKERS

Irene Huertas-Valdivia

Introduction

The tourism industry differs from all other industries due to the characteristics of the product delivered; this product is based mostly on *experiences*. The tourist experience is intangible and highly dependent on the people who deliver the service (on their attitude, behavior, personality, mood, customer orientation, etc.) and the people who receive it. In fact, exactly the same service provided by the same agent can be perceived and appraised very differently by different customers due to customers' cultural differences, previous experiences, dissimilar expectations, and varied sociodemographic backgrounds. In all cases, however, it is the tourism employee's performance (promptness, resolution, diligence, proactivity, kindness, etc.) that shapes customers' perceptions of service quality, which, in turn, impact overall organizational performance in the long run.

Given their close, frequent interaction with customers, frontline employees play a crucial role in the success of tourism service delivery and in building customer loyalty. Motivated and engaged customer service providers are truly instrumental in creating and maintaining a tourism firm's value. Although customer-contact employees represent the face of their organization to clients, the service displayed in every tourism business is the result of intertwined teamwork *backstage*.

Each "moment of truth" (situation in which staff are actually providing the service and customers receiving it) is conditioned by the degree of cooperation and coordination among the business's co-workers and thus by the climate in the work unit. Given the intense interdependence among different workers' actions, no tourism employee can work successfully in isolation from colleagues. In a hotel, for example, the reservations agent must wait for the chef to confirm a special guest's dietary request; the front desk agent assigning a room during the check-in depends on the housekeeper's promptness in cleaning; the maintenance technician is informed of specific repairs needed in guests' rooms through reports provided by the Room Division, and the waiter's delivery of good food in the restaurant depends on the kitchen's pace and professionalism. Service delivery performance – and customer satisfaction with it – are, in fact, highly dependent on the effectiveness of employees' teamwork, collaboration, and cohesion.

Interactions between workers are determined not only by their psychological functioning but also by the social context in which workers are immersed. Individuals belonging to one group (e.g., a work unit or department) interact, collectively or individually, with another group or its members.

The social psychology paradigm can help to explain tourism workers' behavior by understanding how both their social context and their emotional and mental state influence their social interactions (Tang, 2014).

DOI: 10.4324/9781003161868-25

Group Dynamics in Tourism Businesses

Work teams in tourism businesses resemble group dynamics as theorized in social psychology.

Lewin (1943) believed that individuals are better understood in a group context, since groups have the power to influence individuals. He therefore studied "group dynamics", how small groups and individuals act and react in different circumstances, and how both internal and external forces affect the group's behavior.

Within the same department or work unit, individuals can develop a sense of in-group belonging because they share the same hobbies or interests, or have common enemies (i.e., an abusive boss). Employees in different departments within the same organization may be perceived as members of an out-group, creating conflicts with members of other departments, who are thus perceived as a threat to the in-group and evoke different emotional reactions.

Furthermore, the organization's members frequently have different thoughts, values, interests, and goals, and different ways to achieve them (Jung & Yoon, 2018). Tourism businesses teams are frequently composed of a heterogeneous multicultural workforce of different ages with different professional and educational profiles. Work values vary based on culture, society, and personality. Baby boomers, millennials, and members of Generation Z and Y also show significant intergenerational differences in their understanding of and approach to job tasks (Gursoy et al., 2013). Given the intense human interaction within a service organization, conflict is inevitable.

As Stein (1976, p. 165) notes, "external conflict does increase internal cohesion under certain conditions". Although the institution may not create or endorse antagonism between groups, individuals in each group tend to prefer members of their own, out of motivation to protect a "positive group distinctiveness" important to their group's social identity.

Importance of Identity Construction

Bitner (1992) explored how physical spaces and environments influence employees' and customers' behavior. This line of research suggests that the particular physical and sociocultural working conditions in some tourism businesses, such as cruise ships, impact employees' experiences, well-being and identity construction. Identity emerges as a reflection of and reaction to external stimuli, as the person's immediate physical and social living space interacts with their self-identity (Dennett, 2018).

Dennett (2018, p. 231) sheds some light on "the development of harmonious communities in a transient workplace setting", arguing that individuals derive their identities from the social categories (nationality, occupational community, position) of the society on the ship that employs them. Dennett identifies several conditions that promote a shared experience of belonging and contribute to developing a ship-based identity.

1. Cruises involve a high degree of social and hierarchical control. Clear rules and formal and informal systems shape conventions of language, behavior, and social interaction. The company's management of its heterogeneous workforce – characterized by diverse backgrounds and cultural beliefs – is very formal and rests on a strong hierarchical system and chain of command.
2. Working on a transient vessel that is both home and workplace involves prolonged physical and social separation from mainland society, fostering a unique cultural atmosphere shared by all members. It is therefore possible that idiosyncratic working conditions on board (Radic et al., 2020) and some labor practices may clash with practices considered as common or acceptable in other workplaces: work time is intense (all day long) and shift-based; staff have no days off (Radic et al., 2020), and it is difficult socially to divide work and life. Moreover, "labor becomes ethnically and racially stratified" (Terry, 2014, p. 75).

Authors such as Terry (2009) criticize the flexible work regimes common in the cruise industry, which make workers (30% of them Filipino seafarers) more vulnerable to the fluctuations of the marketplace. Further, Terry's research (2014) identifies stereotypes as the primary reason for the prevalence of Filipinos among cruise ship employees. Framed as "docile and compliant yet industrious and inexpensive", they are viewed as suited to low-level positions in the cruise industry (Terry, 2014, p. 75).

3 The cruise industry presents a lucrative opportunity for the estimated 250,000 crew members and officers the sector employs worldwide (Radic et al., 2020). Some positions are highly reliant on tips for income. This situation encourages employees to provide their best service but also generates competition. For example, in Dennett's (2018) study, waiters report conflicts over who gets to serve the tables with the best location and cases of "sabotaging" co-workers, for example, by stealing cutlery and glassware from rival tables.

4 On-board relationships are very important to workers' happiness and to the ship's ability to retain personnel. Although transitory and sometimes portrayed as superficial, relationships made on board are strong, and workers consider themselves as part of a "family". While this response gives them a sense of psychological safety and of belonging to a community, other perceptions of relationships – for example, with management – are clearly influenced by occupational position. Dennett (2018, p. 239) writes that "pursers spoke very highly of the management, suggesting they were fair and supportive", but waiters "describe their relationship with managers as being difficult".

5 Social identity boundaries are reaffirmed in two shared physical spaces on the ship: the staff canteen (where officers, staff and crew have different mess areas) and the cabins (coordinated by hierarchy, department, occupation, and gender). Workers' on-board jobs are therefore very important to their sense of individual and social identity, and this identity is dependent on the ship's established social structures.

In Dennett's (2018) study, all participants express a ship-based identity that differs from how they perceive themselves on land (resulting from their adjustment to the physical and social aspects of the ship's unique environment and system). Furthermore, cruise workers believe that "outsiders" do not really understand their work. Social Identity Theory (Tajfel, 1982a; Tajfel, 1982b; Tajfel & Turner, 1986) can help to explain the attachment cruise workers feel to their occupational role and the ways individuals perceive different groups depending on whether they belong to the group in question.

According to Social Identity Theory, individuals create in- and out-groups to protect themselves and avoid ambiguity in social interaction. People tend to view their in-group as good or superior and out-groups as bad or inferior, while also protecting their group's identity against negative forces, such as stigma (Mejia et al., 2021).

Occupational Stigma

As Mejia et al. (2021) state, specific low-wage hourly jobs in services industries (i.e., hotel housekeeping, taxi/transportation services, retail grocery, foodservice employees, delivery drivers) have traditionally carried the stigma of low-quality jobs or even "dirty work", profoundly affecting these workers' identity and well-being.

In the tourism industry, some of the jobs that hold low status and prestige may lead to stereotyping. For example, people who work in restaurants are often assumed to have few educational qualifications and may become targets of marginalization, disrespect, and stigmatization.

Goffman's (1963) theory of stigma distinguishes three types of stigma: bodily (e.g., deformity), of personal "character" (not conforming to moral or social norms) and "tribal" (belonging to a collective (ethnic, religious, etc.) to which negative qualities are attributed). Link and Phelan

(2001) subsequently build on this classification to enrich understanding of the origin of stigma. They attribute stigma to the interaction of multiple factors that come into play when individuals or groups are marginalized, classified as "other", treated unequally or stereotyped. Whether the source of the stigma is visible or invisible, being or feeling stigmatized harms people's well-being.

Shigihara (2018) examines how restaurant workers perceive, anticipate and respond to occupational stigma while holding their jobs and identifies stigma management strategies. Her analysis draws on an identity work framework to argue that the felt connection between job and self-concept can cause people working in stigmatized jobs to disidentify (e.g., profess a desire to change jobs), disengage, and feel disrespected.

Ashforth and Kreiner (1999) proposed three ways to change the negative identities associated with stigmatized occupations: "reframing", "recalibrating", and "refocusing". Shigihara (2018) documents the refocusing strategies that some of the restaurant workers interviewed use to highlight the positive aspects of their work over the negative ones.

Such strategies, termed "forever talk", involved various ways of reinforcing and realigning self-concept with more positive, socially accepted characteristics and identities to protect the individual from the damage of stigma. Whether fantasized or real, these strategies were integral to the workers' perceptions of their identities and careers.

Shigihara (2018) also reveals the ambivalence that social identity creates in workers in stigmatized occupations. Because they both embrace and resist popular (out-group) negative stereotypes of their work, they may feel both shame and pleasure at work or simultaneously value and disparage their jobs.

Mejia et al. (2021) remark how the Covid-19 pandemic has led society to value some workers more highly, elevating previously stigmatized jobs to "essential" status and altering social attitudes toward them. This sudden shift from historically undervalued to "essential" and even heroic status has raised the social standing of supermarket employees, foodservice workers, and food delivery drivers. The unprecedented change in the cultural narrative has increased the value and recognition that members of the out-group show service workers, resulting in more respect and gratitude toward them.

Negative Social Behaviors and Conduct at Work

Mistreatment in the workplace – whether by managers, co-workers, and/or customers – is the number one cause of burnout (Wigert & Agrawal, 2018). Tourism employees must frequently deal with low-quality interpersonal treatment from the various people with whom they are in frequent contact.

Shum et al. (2020a) show that racial discrimination is still prevalent in hospitality. Based on a sample of hospitality students working in the industry, the authors conclude that people of color suffer more discrimination than Whites. Furthermore, a high level of racial discrimination results in a lower level of hospitality career satisfaction.

Appraising others as more capable or virtuous can spark envy among workmates; competitive goal interdependence also fosters co-worker envy. Workplace envy can lead some co-workers or supervisors to hinder employees' career advancement, for example, to block a team member's promotion. Envy engenders low-quality relationships among workmates and can produce a sense of isolation and discomfort in the envied individual. Co-worker envy and other such stressful situations can cause anger to flare and create negative feeling, leading tourism employees to display less organizational citizenship behavior (OCB) (Ye et al., 2021).

Deviant workplace behaviors are behaviors that contravene organizational norms, threatening service quality and potentially decreasing the company's revenue. Scholars have reported that 75–85% of all employees engage in some misconduct at work, and that organizational factors (such as a harmful organizational contexts or organizational injustice) and individual characteristics (employees' inclination to take risks and need for social approval by colleagues) determine such misconduct (Chen et al., 2018; Harris & Ogbonna, 2002). These intentional dysfunctional behaviors include embezzlement, theft, lateness attitude, absences from work, and service sabotage. Karatepe et al. (2021) identified the propensity to be late for work and 'the failure to report for scheduled work' as deviant behaviors in the hospitality industry.

Deviant behaviors threaten the well-being of the organization and its members, generating frustration, nervousness, grumpiness, and interpersonal tension. For example, employees come to suspect each other, eroding trust, cooperation, and healthy coexistence. Such misconduct must be implicitly or explicitly identified and sanctioned. Moreover, different forms of control – direct surveillance and subtle control exerted through organizational culture – may reduce the incidence of service sabotage (Harris & Ogbonna, 2002). Organizational justice (fair treatment, fair compensation, and fair rewards) is necessary to avoid deviant behavior among employees. Tourism companies must also focus on developing a favorable training climate through ethics training and communication, monitoring pursuit of a code of ethics, and applying specific policies and procedures (such as instigating a rewards system) (Chen et al., 2018).

Other negative workplace behaviors include ostracism and bullying. Workplace ostracism is "a dark phenomenon within organisations" (Zhu et al., 2017, p. 63) that isolates workers through social exclusion by their mates. Ostracized individuals are often ignored, deliberately not greeted, and voted out of social gatherings at work. Being ostracized is painful for the employees and has detrimental effects for workers' self-esteem (Huertas-Valdivia et al., 2019). It aggravates job tension and can indirectly cause deviant behavior (such as leaving work early and intentionally being late for work) (Hsieh & Karatepe, 2019).

Workplace bullying is a stressor. Among other negative outcomes, it can affect individuals' resilience, and mental and emotional health, indirectly causing emotional exhaustion (Anasori et al., 2020). Its effects differ from person to person. The good news is that emotional intelligence plays a significant (moderating) role in reducing the effects of job stress on job performance (Wu, 2011)

Research has demonstrated that customer-related social stressors aggravate hotel and restaurant employees' emotional exhaustion as well (Ma et al., 2019).

Boukis et al. (2020) note that customer incivility (e.g., verbal aggression, unreasonable demands) is a widespread global phenomenon suffered by front line employees such as fast food workers, although different types of customer incivility may affect frontline workers in different ways. Employees encountering uncivil customers normally suffer role stress, rumination, retaliation, and withdrawal intention, which can result in spirals of negative exchange (Boukis et al., 2020). Customer incivility severely depletes workers' psychological resources and the customer service they deliver.

Various theories, such as ego depletion (Baumeister & Vohs, 2007) and the job demands-resources model (Demerouti et al., 2001), suggest that loss of employees' psychological resources negatively affects daily work. Nevertheless, Boukis et al.'s (2020) findings, which draw on the job demands-resources framework (Demerouti et al., 2001), indicate that managers' leadership style can influence how frontline workers respond to incivility. Boukis argues that empowering (as opposed to laissez-faire) leaders can counteract the negative consequences of abusive treatment from customers.

Top companies (e.g., Four Seasons, Ritz Carlton) attribute their strong position to employee empowerment. Empowering employees psychologically motivates them by fostering active orientation toward their work role; the construct of psychological empowerment encompasses four dimensions: Meaning, competence, self-determination, and impact (Spreitzer, 1995). The lens of Self-Determination Theory (Ryan & Deci, 2000) explains how more empowering practices at work help to increase individuals' basic need for self-determination, also enhancing their emotional tone and raising their self-esteem. What is more, psychological empowerment mitigates the negative effects of workplace ostracism on hospitality employees' self-esteem (Huertas-Valdivia et al., 2019).

Leadership Processes

Understanding leadership as a social process that involves leaders and followers can explain how different methods of leading significantly influence the group's dynamics.

Leaders can shape group dynamics by establishing common goals, organizing work in a particular way, motivating followers, etc. Group dynamics, in turn, can exert social pressure on leaders, influencing them (even unconsciously) to change their tactics based on the group's reaction.

Leaders' actions, behaviors, and attitudes have a significant effect on their followers. Ideally, all tourism organizations would have good leaders with outstanding skills who adapted their leadership style to the team they are overseeing. However, this is not always the case. A large proportion of tourism companies are still managed by traditional authoritative leaders. In fact, Shum et al. (2020b) identified frequent abusive or destructive leadership behaviors among industry managers, especially in hotels and restaurants. Øgaard et al. (2008, p. 669) underscore the need for better and more participative leadership styles and more organic organizational models in the sector. Their results indicate that "the hospitality industry might have a general problem with their managers, who are characterised by traditional leadership styles that fail to make the most of the employee's resources".

The leader's role is of utmost importance for the organization's functioning. From a social psychology perspective, other members of the organization frequently consider leaders as role models and take their actions and behaviors as an example to imitate. Social Learning Theory (Bandura, 1969) explains the process by which people learn from each other through a process of social interaction: Followers tend to observe their leaders, replicate their leaders' behavior, and reproduce similar actions at work.

Research has begun to explore the influence of leaders' attitudes on negative employee behaviors that are harmful for the organization or for other employees (e.g., deviance or counterproductive, antisocial behavior) (Brown & Treviño, 2006). Unfavorable treatment by supervisors has negative consequences, creating a more hostile work climate. Exploitative leaders, for example, display egoistical behavior, taking credit for their subordinates' achievements or underchallenging them. Such leaders tend to manipulate their subordinates in pursuit of self-interest and may even pit co-workers against each other (Schmid et al., 2019). Under these circumstances employees feel that their contributions are not respected and that they are discouraged from collaborating with and trusting each other.

Abusive leaders are inconsiderate and display hostile verbal and nonverbal behaviors (Brown et al., 2005). Abusive supervision not only decreases OCB but exacerbates mental health problems (Peltokorpi & Ramaswami, 2021) and increases counterproductive behavior (Sulea et al., 2013).

Fair treatment, in contrast, fosters trust in the leader and encourages good communication. Employees under the supervision of ethical leaders will take more initiative in contributing to the organization, communicating with management, and proposing ways to solve problems (Brown et al., 2005). Ethical leadership not only discourages employees from negative behavior but also improves workplace climate and organizational values (thus preventing service sabotage) (Yeşiltaş & Tuna, 2018).

Researchers advocate specially for *other-oriented* leadership styles in the service industry, such as servant leadership (Brownell, 2010). Studies have confirmed the positive impact of servant leadership on service quality (Qiu et al., 2020) and customer-oriented OCB (Elche et al., 2020; L.-Z. Wu et al., 2013).

The type of relationship between leader and follower is also extremely important to ensuring good experiences and a positive climate at work. Leader-Member Exchange Theory (Graen & Uhl-Bein, 1995) underscores the benefits of a high-quality relationship between leader and follower in terms of trust, interaction, and support. The leader will be more attentive, and the follower will reap more benefits. Leader-member exchange is proven to generate or facilitate multiple positive employee outcomes, such as OCB and innovative behavior (Dhar, 2016; L.-Z. Wu et al., 2013). Further, strong employee-manager relationships reduce undesired or harmful behavior (Brown & Treviño, 2006).

Social Reinforcement

Social reinforcement is defined as any response individuals "experience" from their surroundings (e.g., praise, recognition, compliments, positive body language) (Lieberman et al., 2001). This experience includes not only external responses but also the individual's own feelings.

In his seminal work *Social behavior as exchange*, Homans (1958, p. 599) states that "the greater the reinforcement, the more often is the reinforced behavior emitted". Homans understands social behavior as transactional. It involves exchanging both tangible and intangible goods, including positive recognition or status. Social exchange theory (Blau, 1964) is one of the most widely applied frameworks in tourism research (Tang, 2014), due to its ability to explain many of workers' attitudinal and behavioral responses.

According to Social Exchange Theory (Blau, 1964), employees who perceive a supportive work environment and believe their company takes good care of them (in terms of job stability, fair compensation, career development, etc.) feel indebted to their organization and tend to reciprocate with positive behavior that is beneficial to the company. Certain human resources management (HRM) practices nurture the employee's sense of belonging to the organization and can help organizations to promote certain positive behaviors among their workforce (Chuang et al., 2016; Huertas-Valdivia et al., 2018).

Receiving fair rewards from the organization is also crucial. Some companies reward outstanding performance. *Marriott Stories of Excellence* (https://www.marriott.com/culture-and-values/awards-of-excellence.mi) reflect these experiences. (*Marriott Awards of Excellence* praise employees who serve customers in challenging situations and/or display proactive behavior to help customers.) Moreover, recognition programs such as "employee of the month" (by combined vote of customers and co-workers) can help to send employees clear messages about desired work attitude and promote the behavior desired.

Unfortunately, tourism businesses sometimes overlook important HRM practices. Low wages and insufficient recognition are common features. In hospitality, for example, the most qualified workers are frequently penalized because the industry rewards talent much less than other sectors (Casado-Díaz & Simón, 2016). Further, the tourism sector is profoundly affected by seasonality and uneven workload throughout the year, and employers frequently adapt to fluctuating demand by hiring part-time or temporary workers. Job instability can compromise workers' psychological availability and even their mental health.

Reviewing the literature on gender equity in the travel, tourism, and hospitality sectors, Remington and Kitterlin-Lynch (2018) identified job training, promotion, and pay as the areas of greatest gender inequity. The results of their Delphi study indicate that the proverbial "glass ceiling" still exists; female industry leaders still face challenges to professional advancement, including "work-life balance", "organizational commitment", "inadequate support systems/mentors", "systematic barriers to advancement", and "lack of female role models".

Final Considerations and Suggestions for Practitioners

Tourism businesses' profitability depends on essential employee attitudes and behaviors toward *external customers* but also among *internal customers* – the workforce.

Delivering superior customer value and satisfaction is crucial to service firms' competitiveness, and such value is grounded in work climate, teamwork chain actions, and cooperation among members. Furthermore, effective communication and positive group dynamics are essential to overall organizational goal achievement.

Employees' attitudes to each other at work are influenced by activity sector, type of job, and working conditions. Individual traits (such as personality, motivation, experience, and emotional intelligence) and team management or leadership can also affect relationships among employees. Because groups and leaders interact with and influence each other, group dynamics are closely tied to leadership processes (Lewin, 1944).

Tourism companies must place special emphasis on identifying and eliminating "bad managers", that is, managers who are unethical and unprofessional; display autocratic management styles; and have low leadership, operational, and technical skills as well as poor decision and delegation skills (Hight et al., 2019). Fostering more participative and employee-oriented leadership styles encourages more collaboration among workers. Nurturing a fair, service-oriented organizational climate is also indispensable for building a more positive work environment in a stressful sector like tourism and hospitality.

In comparing groups dynamics in hospitality to other industries, such as retail chains and banks, Mohanty and Mohanty (2018) found that hotels have better group dynamics among the sectors analyzed. Banks scored highest in group loyalty, and retail chains had higher levels of conflict than hotels and banks. Their study also showed that communication, group dynamics, and teamwork among departments are important to teamwork's effectiveness. It is thus crucial for tourism businesses to develop clear communication systems so that their employees have a clear understanding of the company's goals and objectives.

Mohanty and Mohanty (2018) encourage training programs for hotel employees to improve weaknesses in communication (interpersonal, intergroup, and interlevel), and collaboration, and activities to promote team building in the workforce and improve cooperation.

Workplace misbehavior or poor performance may be the result of intercultural or emotional tension that spreads from one group to another (Chien & Ritchie, 2018). External circumstances can also affect well-being and attitudes among employees at work, for example, a bad attitude toward tourism businesses from *tourism-phobic* host residents (Zerva et al., 2019). Recently, research has demonstrated the negative impact of the Covid-19 pandemic on hospitality workers' mental health (Karatepe et al., 2021). The research also demonstrates that psychological distress during the pandemic made restaurant employees more prone to drug and alcohol use and increased their turnover intentions (Bufquin et al., 2021).

Long-term accumulation of negative emotions (within or outside the workplace) leads employees to experience stress and display negative attitudes at work, impeding the good collaboration required to perform optimally in a people-dependent sector like tourism. Companies in the sector must make every effort to strengthen the resilience of their workforce, especially during crisis management (Aguiar-Quintana et al., 2021). Coaching and training sessions can also help to build a more ethical organizational culture and a healthier work atmosphere.

References

Aguiar-Quintana, T., Nguyen, H., Araujo-Cabrera, Y., & Sanabria-Díaz, J. M. (2021). Do job insecurity, anxiety and depression caused by the COVID-19 pandemic influence hotel employees' self-rated task

performance? The moderating role of employee resilience. *International Journal of Hospitality Management, 94*, 102868. https://doi.org/10.1016/j.ijhm.2021.102868
Anasori, E., Bayighomog, S. W., & Tanova, C. (2020). Workplace bullying, psychological distress, resilience, mindfulness, and emotional exhaustion. *Service Industries Journal, 40*(1–2), 65–89. https://doi.org/10.1080/02642069.2019.1589456
Ashforth, B. E., & Kreiner, G. E. (1999). "How can you do it?": Dirty work and the challenge of constructing a positive identity. *Academy of Management Review, 24*(3), 413–434. https://doi.org/10.5465/AMR.1999.2202129
Bandura, A. (1969). Social-learning theory of identificatory processes. In *Handbook of Socialization Theory and Research*. https://doi.org/10.1080/19371918.2011.591629
Baumeister, R. F., & Vohs, K. D. (2007). Self-regulation, ego depletion, and motivation. *Social and Personality Psychology Compass, 1*(1), 115–128. https://doi.org/10.1111/j.1751-9004.2007.00001.x
Bitner, M. J. (1992). Servicescapes: The impact of physical surroundings on customers and employees. *Journal of Marketing, 56*(2), 57–71. https://doi.org/10.1177/002224299205600205
Blau, P. M. (1964). Exchange and power in social life. In *Exchange and Power in Social Life*. https://doi.org/10.2307/2091154
Boukis, A., Koritos, C., Daunt, K. L., & Papastathopoulos, A. (2020). Effects of customer incivility on frontline employees and the moderating role of supervisor leadership style. *Tourism Management, 77*, 103997. https://doi.org/10.1016/j.tourman.2019.103997
Brown, M. E., & Treviño, L. K. (2006). Ethical leadership: A review and future directions. *Leadership Quarterly, 17*(6), 595–616. https://doi.org/10.1016/j.leaqua.2006.10.004
Brown, M. E., Treviño, L. K., & Harrison, D. A. (2005). Ethical leadership: A social learning perspective for construct development and testing. *Organizational Behavior and Human Decision Processes, 97*(2), 117–134. https://doi.org/10.1016/j.obhdp.2005.03.002
Brownell, J. (2010). Leadership in the service of hospitality. *Cornell Hospitality Quarterly, 51*(3), 363–378. https://doi.org/10.1177/1938965510368651
Bufquin, D., Park, J. Y., Back, R. M., de Souza Meira, J. V., & Hight, S. K. (2021). Employee work status, mental health, substance use, and career turnover intentions: An examination of restaurant employees during COVID-19. *International Journal of Hospitality Management, 93*, 102764. https://doi.org/10.1016/j.ijhm.2020.102764
Casado-Díaz, J. M., & Simon, H. (2016). Wage differences in the hospitality sector. *Tourism Management, 52*, 96–109.
Chen, C. T., Hu, H. H. S., & King, B. (2018). Shaping the organizational citizenship behavior or workplace deviance: Key determining factors in the hospitality workforce. *Journal of Hospitality and Tourism Management, 35*, 1–8. https://doi.org/10.1016/j.jhtm.2018.01.003
Chien, P. M., & Ritchie, B. W. (2018). Understanding intergroup conflicts in tourism. *Annals of Tourism Research, 72*, 177–179. https://doi.org/10.1016/j.annals.2018.03.004
Chuang, C.-H., Jackson, S. E., & Jiang, Y. (2016). Can knowledge-intensive teamwork be managed? Examining the roles of HRM systems, leadership, and tacit knowledge. *Journal of Management, 42*(2), 524–554. https://doi.org/10.1177/0149206313478189
Demerouti, E., Bakker, A. B., Nachreiner, F., & Schaufeli, W. B. (2001). The job demands-resources model of burnout. *The Journal of Applied Psychology, 86*(3), 499–512. https://doi.org/10.1108/02683940710733115
Dennett, A. (2018). Identity construction in transient spaces: Hospitality work on-board cruise ships. *Tourism in Marine Environments, 13*(4), 231–241. https://doi.org/10.3727/154427318X15438502059120
Dhar, R. L. (2016). *Ethical Leadership and Its Impact on Service Innovative Behavior: The Role of LMX and Job Autonomy*. https://doi.org/10.1016/j.tourman.2016.05.011
Elche, D., Ruiz-Palomino, P., & Linuesa-Langreo, J. (2020). Servant leadership and organizational citizenship behavior: The mediating effect of empathy and service climate. *International Journal of Contemporary Hospitality Management, 32*(6), 2035–2053. https://doi.org/10.1108/IJCHM-05-2019-0501
Goffman, E. (1963). *Stigma*. New York: Simon & Schuster.
Graen, G. B., & Uhl-Bein, M. (1995). Relationship based approac to leadership; Development of Leader-Member Exchange (LMX) theory of leadership over 25 years. *Leadership Quarterly, 6*(2)(Lmx), 219–247. https://doi.org/10.1016/1048-9843(95)90036-5
Gursoy, D., Chi, C. G. Q., & Karadag, E. (2013). Generational differences in work values and attitudes among frontline and service contact employees. *International Journal of Hospitality Management, 32*(1), 40–48. https://doi.org/10.1016/j.ijhm.2012.04.002
Harris, L. C., & Ogbonna, E. (2002). Exploring service sabotage. *Journal of Service Research, 4*(3), 163–183. https://doi.org/10.1177/1094670502004003001

Hight, S. K., Gajjar, T., & Okumus, F. (2019). Managers from "Hell" in the hospitality industry: How do hospitality employees profile bad managers? *International Journal of Hospitality Management, 77*, 97–107. https://doi.org/10.1016/j.ijhm.2018.06.018
Homans, G. C. (1958). Social behavior as exchange. *American Journal of Sociology, 63*(6), 597–606. https://doi.org/10.1086/222355
Hsieh, H., & Karatepe, O. M. (2019). Outcomes of workplace ostracism among restaurant employees. *Tourism Management Perspectives, 30*, 129–137. https://doi.org/10.1016/j.tmp.2019.02.015
Huertas-Valdivia, I., Braojos, J., & Lloréns-Montes, F. J. (2019). Counteracting workplace ostracism in hospitality with psychological empowerment. *International Journal of Hospitality Management, 76*, 240–251. https://doi.org/10.1016/j.ijhm.2018.05.013
Huertas-Valdivia, I., Llorens-Montes, F. J., & Ruiz-Moreno, A. (2018). Achieving engagement among hospitality employees: A serial mediation model. *International Journal of Contemporary Hospitality Management, 30*(1), 217–241. https://doi.org/10.1108/IJCHM-09-2016-0538
Jung, H. S., & Yoon, H. H. (2018). Improving frontline service employees' innovative behavior using conflict management in the hospitality industry: The mediating role of engagement. *Tourism Management, 69*, 498–507. https://doi.org/10.1016/j.tourman.2018.06.035
Karatepe, O. M., Saydam, M. B., & Okumus, F. (2021). COVID-19, mental health problems, and their detrimental effects on hotel employees' propensity to be late for work, absenteeism, and life satisfaction. *Current Issues in Tourism, 24*(7), 934–951. https://doi.org/10.1080/13683500.2021.1884665
Lewin, K. (1943). Defining the field at a given time. *Psychological Review, 50*(3), 292–310.
Lewin, K. (1944). The dynamics of group action. *Educational Leadership, 1*(4), 195–200.
Lieberman, M., Gauvin, L., Bukowski, W. M., & White, D. R. (2001). Interpersonal influence and disordered eating behaviors in adolescent girls: The role of peer modeling, social reinforcement, and body-related teasing. *Eating Behaviors, 2*(3), 215–236.
Link, B. G., & Phelan, J. C. (2001). Conceptualizing stigma. *Annual Review of Sociology, 27*, 363–385. https://doi.org/10.1146/annurev.soc.27.1.363
Ma, Z., Kim, H. J., & Shin, K. H. (2019). From customer-related social stressors to emotional exhaustion: An application of the demands–control model. *Journal of Hospitality and Tourism Research, 43*(7), 1068–1091. https://doi.org/10.1177/1096348019849667
Mejia, C., Pittman, R., Beltramo, J. M. D., Horan, K., Grinley, A., & Shoss, M. K. (2021). Stigma & dirty work: In-group and out-group perceptions of essential service workers during COVID-19. *International Journal of Hospitality Management, 93*, 102772. https://doi.org/10.1016/j.ijhm.2020.102772
Mohanty, A., & Mohanty, S. (2018). The impact of communication and group dynamics on teamwork effectiveness: The case of service sector organisations. *Academy of Strategic Management Journal, 17*(4), 1–14. https://www.proquest.com/docview/2124077361?pq-origsite=gscholar&fromopenview=true
Øgaard, T., Marnburg, E., & Larsen, S. (2008). Perceptions of organizational structure in the hospitality industry: Consequences for commitment, job satisfaction and perceived performance. *Tourism Management, 29*(4), 661–671. https://doi.org/10.1016/j.tourman.2007.07.006
Peltokorpi, V., & Ramaswami, A. (2021). Abusive supervision and subordinates' physical and mental health: the effects of job satisfaction and power distance orientation. *International Journal of Human Resource Management, 32*(4), 893–919. https://doi.org/10.1080/09585192.2018.1511617
Qiu, S., Dooley, L. M., & Xie, L. (2020). How servant leadership and self-efficacy interact to affect service quality in the hospitality industry: A polynomial regression with response surface analysis. *Tourism Management, 78*, 104051. https://doi.org/10.1016/j.tourman.2019.104051
Radic, A., Arjona-Fuentes, J. M., Ariza-Montes, A., Han, H., & Law, R. (2020). Job demands–job resources (JD-R) model, work engagement, and well-being of cruise ship employees. *International Journal of Hospitality Management, 88*, 102518. https://doi.org/10.1016/j.ijhm.2020.102518
Remington, J., & Kitterlin-Lynch, M. (2018). Still pounding on the glass ceiling: A study of female leaders in hospitality, travel, and tourism management. *Journal of Human Resources in Hospitality and Tourism, 17*(1), 22–37. https://doi.org/10.1080/15332845.2017.1328259
Ryan, R. M., & Deci, E. L. (2000). Self-determination theory and the facilitation of intrinsic motivation, social development, and well-being. *American Psychologist, 55*(1), 68–78. https://doi.org/10.1037/0003-066X.55.1.68
Schmid, E. A., Pircher Verdorfer, A., & Peus, C. (2019). Shedding light on leaders' self-interest: Theory and measurement of exploitative leadership. *Journal of Management, 45*(4), 1401–1433. https://doi.org/10.1177/0149206317707810
Shigihara, A. M. (2018). "(Not) forever talk": Restaurant employees managing occupational stigma consciousness. *Qualitative Research in Organizations and Management: An International Journal, 13*(4), 384–402. https://doi.org/10.1108/QROM-12-2016-1464

Shum, C., Gatling, A., & Garlington, J. (2020a). All people are created equal? Racial discrimination and its impact on hospitality career satisfaction. *International Journal of Hospitality Management, 89*, 102407. https://doi.org/10.1016/j.ijhm.2019.102407

Shum, C., Gatling, A., & Tu, M. H. (2020b). When do abusive leaders experience guilt? *International Journal of Contemporary Hospitality Management, 32*(6), 2239–2256. https://doi.org/10.1108/IJCHM-05-2019-0474

Spreitzer, G. M. (1995). Psychological empowerment in the workplace: Dimensions, measurement, and validation. *Academy of Management Journal, 38*(5), 1442–1465. https://doi.org/10.2307/256865

Stein, A. A. (1976). Conflict and cohesion: A review of the literature. *Journal of Conflict Resolution, 20*(1), 143–172. https://doi.org/10.1177/002200277602000106

Sulea, C., Fine, S., Fischmann, G., Sava, F. A., & Dumitru, C. (2013). Abusive supervision and counterproductive work behaviors: The moderating effects of personality. *Journal of Personnel Psychology, 12*(4), 196–200. https://doi.org/10.1027/1866-5888/a000097

Tajfel, H. (1982a). Social psychology of intergroup relations. *Annual Review of Psychology, 33*(1), 1–39.

Tajfel, H. (1982b). *Social Identity and Intergroup Relations*. Cambridge: Cambridge University Pres/Paris: Editions de la Maison des Sciences de l' Homme.

Tajfel, H., & Turner, J. C. (1986). The social identity theory of intergroup behavior. In S. Worchel & W. G. Austin (Eds.), *Psychology of Intergroup Relations*, pp. 7–24. Chicago: Nelson-Hall.

Tang, L. R. (2014). The application of social psychology theories and concepts in hospitality and tourism studies: A review and research agenda. *International Journal of Hospitality Management, 36*, 188–196. https://doi.org/10.1016/j.ijhm.2013.09.003

Terry, W. C. (2009). Working on thewater: On legal space and seafarer protection in the cruise industry. *Economic Geography, 85*(4), 463–482. https://doi.org/10.1111/j.1944-8287.2009.01045.x

Terry, W. C. (2014). The perfect worker: discursive makings of Filipinos in the workplace hierarchy of the globalized cruise industry. *Social and Cultural Geography, 15*(1), 73–93. https://doi.org/10.1080/14649365.2013.864781

Wigert, B. & Agrawal, S. (2018). *Employee burnout, Part 1: The 5 main causes*. Retrieved from https:// www.gallup.com/workplace/237059/employee-burnout-part-main-causes.aspx.

Wu, L.-Z., Tse, E. C.-Y., Fu, P., Kwan, H. K., & Liu, J. (2013). The impact of servant leadership on hotel employees' "Servant Behavior." *Cornell Hospitality Quarterly, 54*(4), 383–395. https://doi.org/10.1177/1938965513482519

Wu, Y. C. (2011). Job stress and Job performance among employees in the Taiwanese finance sector: The role of emotional intelligence. *Social Behavior and Personality, 39*(1), 21–31. https://doi.org/10.2224/sbp.2011.39.1.21

Ye, Y., Lyu, Y., Kwan, H. K., Chen, X., & Cheng, X. M. (2021). The antecedents and consequences of being envied by coworkers: An investigation from the victim perspective. *International Journal of Hospitality Management, 94*, 102751. https://doi.org/10.1016/j.ijhm.2020.102751

Yeşiltaş, M., & Tuna, M. (2018). The effect of ethical leadership on service sabotage. *Service Industries Journal, 38*(15–16), 1133–1159. https://doi.org/10.1080/02642069.2018.1433164

Zerva, K., Palou, S., Blasco, D., & Donaire, J. A. B. (2019). Tourism-philia versus tourism-phobia: Residents and destination management organization's publicly expressed tourism perceptions in Barcelona. Tourism Geographies, *21*(2), 306–329. https://doi.org/10.1080/14616688.2018.1522510

Zhu, H., Lyu, Y., Deng, X., & Ye, Y. (2017). Workplace ostracism and proactive customer service performance: A conservation of resources perspective. *International Journal of Hospitality Management, 64*, 62–72. https://doi.org/10.1016/j.ijhm.2017.04.004

26

ATTITUDES (STEREOTYPE AND PREJUDICE) OF LOCAL PEOPLE TOWARDS SEASONAL TOURISM WORKERS

Zanete Garanti and Galina Berjozkina

Introduction

According to the World Economic Forum, tourism is one of the world's largest industries; it provides 10% of the world's GDP, 7% global trade, and as many as one in every 11 jobs globally. Tourism is a seasonal industry characterised by instability of demand and revenues (Krakover, 2000). Tourism seasonality is impacted by many factors, for example, natural resources (sun, sea, snow, rain), holidays, festivals, lifestyle trends, and availability of transportation, among many others. Seasonality can be seen as a weekly, monthly, or yearly trend (Rosselló & Sansó, 2017) of demand for tourism activity in a particular place. Tourism seasonality provides many challenges to industry practitioners, like short business operating seasons, the need to generate revenues of a full year in a short period, problems maintaining supply chain, and relations with tour operators and service providers (Baum & Lundtorp, 2001). Many tourism destinations are working towards a more sustainable tourism model to extend the tourism season and create new tourism activities (Andriotis, 2005) to ensure a more stable demand and supply tourism model. However, worldwide tourism is still seasonal (Tsiotas et al., 2020; Vergori & Arima, 2020), and the degree of seasonality dramatically depends on the location (seasonality is exceptionally high in Southern and Mediterranean Europe) and income in emitting markets (Duro & Turrión-Prats, 2019), which can significantly reduce seasonality.

Two interdependent factors drive tourism employment seasonality: urbanisation and migration of people to live and work in the cities, but on the other side, there is a growing demand for remote, undiscovered, rural tourism (Salazar, 2020). Given the fluctuation in tourism activity, the tourism and hospitality industry demand for a workforce is also unstable. Rural destinations struggling to find employees locally are forced to employ seasonal workers to bring the necessary workforce, skills, and talents to provide tourism and hospitality services. It becomes evident that with a significant inflow of seasonal employees and tourists, the local lifestyle is disturbed (Martínez et al., 2019), and residents might not always perceive the trend positively.

Seasonal tourism employment has both positive and negative effects on multiple stakeholders. First, its employees have to leave their homes, lifestyle, and families to travel to a new destination for work, not always providing good work conditions and pay (McCole, 2015). Nevertheless, some argue that it is not the pay that motivates seasonal employees. Instead, it is experiences and a chance to meet new people (Lundberg et al., 2009). For tourism businesses and destinations, significant challenges relate to the seasonal employees, as seasonality does not allow to maintain product and service quality due to the absence of permanent workers, and short-term employment

DOI: 10.4324/9781003161868-26

rather than long-term employment creates either off-season unemployment or temporary outward migration (Baum & Lundtorp, 2001). Nevertheless, seasonal employees can help knowledge transfer among destinations and be essential drivers of innovation (Ericsson et al., 2020). Moreover, Lundmark (2006) emphasise the long term benefits of attracting sessional and younger population to the region, which, in return, contribute towards "positive development of the local service base, infrastructure and public facilities further making the area attractive to investors, tourists and possible in-migrants" (p. 209).

An area that is less explored in academic literature is the attitudes of residents towards seasonal tourism employees. Generally speaking, the more residents economically benefit from the tourism activity in the area, the more supportive they become towards it (Boley et al., 2018; Ferreira et al., 2020). However, residents benefit from tourism activity in the form of employment and income (Bhat & Mishra, 2020), which can be disturbed by an inflow of seasonal employees, often ready to undertake work for lower pay, job conditions, and security (Cave et al., 2012). Moreover, there are also cultural, environmental, ethical issues and differences (Araslı & Arıcı, 2019; Gülduran & Gürdoğan, 2021) that affect the residents and contribute towards formation of attitude. Therefore, this chapter is a conceptual literature review and mainly focusses on resident attitudes (prejudice and stereotypes) towards seasonal tourism employees.

Seasonal Employment in the Tourism and Hospitality Industry

Within a time series, seasonal variations are defined as the changes in levels of employment recurring in similar timing, duration, and intensity from year to year (Stynes & Pigozzi, 1983). In the tourism and hospitality industry, it is a standard practice to hire and fire employees based on seasonality, and Krakover (2000) suggests that measures of foreign and domestic demand, rates of bed occupancy, and indices of the expected monthly fluctuation and the long-term trend can be significant determinants of adjusting labour needs in tourism and hospitality industry.

The majority of the sessional tourism employees are referred to as peripheral labour, made up of less educated, semi-skilled or unskilled, full-time, part-time, temporary, or short-term contract workers (Krakover, 2000). Therefore, seasonal work is characterised by low pay rates and long working hours, low job security and advantages than full-time employees, and overall seasonal employees are less likely to have their demands and expectations satisfied (Araslı et al., 2020). However, seasonal employment in the tourism industry, while characterised by poor pay and condition, plays an essential role in ensuring tourism and hospitality services and "should not be viewed as qualitatively inferior in all senses" (Ball, 1988, p. 512).

Seasonal employment in the tourism industry can be classified into voluntary and involuntary. Involuntary seasonal workers are those who "involuntarily accept such jobs because there is no alternative regular job opportunity at an acceptable level of remuneration and working conditions" (Ball, 1988, p. 512). In other words, a seasonal job is the only opportunity for these people to work and have any income, even when it is seasonal. The opposite of it is voluntary seasonal workers who accept sessional jobs voluntarily and "participate in the seasonal labour market, perhaps because they have other commitments outside of the seasonal work (…) or they prefer the leisure time that working only seasonally may offer" (Ball, 1988, p. 512). Therefore, seasonal work could be a choice for students, people employed elsewhere, having other commitments, and people who enjoy time off work in the low season.

Among seasonal workers, two specific categories of employees are evident, each with its perception of the seasonal job quality: young workers and workers employed in front-line positions. While young workers enjoy the flexibility of seasonal jobs, they perceive their jobs as "bad" jobs, primarily not satisfying their expectations and not giving opportunities to develop their career paths. Evaluation of job quality improves with age and a chance to work in the front-line, directly interacting with the customer (Guidetti et al., 2020). Other authors emphasise that motivations that drive seasonal employment in the tourism industry are more complex. For example, Tuulentie and Heimtun (2014) classify seasonal workers into five distinct types: (1) migrant tourism workers, who are attached to the mobility of the lifestyle and do not have intentions to stay in the place; instead, they are mobile and travelling around to make money and see different places, (2) working-holiday tourists, who want to make money and see the world, but are in this life stage temporarily without intentions to do it for a long term, (3) hobbyists, who have other jobs and interests elsewhere, but love experiencing and practicality new activities, (4) professional holiday employees who are starting their career and seek to find a permanent job that is suitable for them, (5) local seasonal workers that are familiar with the place, are visiting the home region and working during the visit, but have no intentions to stay due to the lack of the opportunities off-season. The categories are flexible and are situation related. Finally, it is also possible to differentiate seasonal employees on whether they have the option to live in or live out (Lee-Ross, 1995). Those who live within the facilities have less or sometimes even no interactions with locals, while those who prefer to live out have more contact and interaction with locals.

The tourism industry has long been characterised by ethnically and culturally diverse seasonal employment (Vassou et al., 2017), bringing unique challenges in managing a diverse workforce and providing good quality service. Many destinations around the world are lowly populated but frequently visited by tourists, heavily relying on foreign seasonal employment (Chen & Wang, 2015), and paradoxically locality of destination, whether it is local gastronomic, cultural, natural, or other experiences, is provided by global workforce (Lozanski & Baumgartner, 2020). Among all characteristics of seasonal workers, the one that makes the most significant difference when focussing on the perceptions and attitudes towards seasonal employees is the origin of those employees (Ioannides & Zampoukos, 2018). A local student that recently graduated and came back to his home city and undertakes seasonal work to gain his first job experience will be perceived differently from the seasonal employee who travelled from another country due to a lack of job and income in his home country and is not familiar with local customs, language, and history.

One of the most significant factors when choosing seasonal employment is the risk of being unemployed in the off-season (Nukhu & Singh, 2020). However, as Jolliffe and Farnsworth (2003) emphasised, it can be either embraced or perceived as a challenge. Embracing seasonality allows tourism industry businesses to focus on temporary workers, brief orientation and training, and perform specific tasks while offering matching or leading pay compared to competitors and incentives to say all sessions. Moreover, Grobelna and Skrzeszewska (2019), in the study among tourism and hospitality students, concludes that seasonality, when appropriately managed, can be seen as a positive trend rather than negative. It is also reported that tourism employees can be motivated to re-accept the seasonal jobs. Some of the most important motivators leading to this choice are personal characteristics (age, education, experience), wages, job requirements, teamwork and workplace characteristics, and satisfaction with the supervisor and workplace (Šošić et al., 2018). Even more, seasonal workers might bring many potential benefits to rural areas. As many countries are becoming urbanised with the local population migrating towards cities, seasonal workers can become future residents (Möller et al., 2014), as this is one of the motivators to apply for the seasonal job.

Local Residents and Their Attitudes

According to Perloff (1993), an attitude is "a learned, global evaluation of an object (person, place, or issue) that influences thought and action" (p. 39). Attitude is generated from the experience of learning (Rheu, 2020) and guides the present actions, thinking, way of expressing, and all efforts towards the object. Attitude formation is a field in social psychology (Olson & Fazio, 2001), trying to understand and explain how people evaluate things either positively or negatively. Researchers differentiate between effect- and cognition-based attitudes (Edwards, 1990), and tripartite theory proposes that attitudes are cognitive, affective, and behavioural (Park & MacInnis, 2006). It is also argued by Fazio et al. (1984) that attitudes form spontaneously. Not necessarily people have attitude unless they "are directly questioned about their feelings toward the attitude objects and/or perceive some situational cue that implies that it may be functional in the future to know one's attitude toward the objects in question" (p. 231). Nowadays, attitudes are also formed based on media and social networks (Winter, 2020), where other people significantly influence the evaluation process and attitude development among individuals.

Local residents are a vital part of the tourism and hospitality industry, and the tourism industry greatly benefits from their involvement and participation in co-creating tourist experiences (Huber & Gross, 2021). Simply speaking, it is the traditional lifestyle, customs, and traditions, gastronomic experiences, heritage, the environment of residents that tourists came to experience (Belma & Alverez, 2019). However, tourism activity also brings many economic, social, and environmental challenges to the lives of residents (But & Ap, 2017; Gupta & Chomplay, 2021), which contributes towards attitude forming. Economically, tourism provides income and work opportunities for residents (Godovykh & Ridderstaat, 2020; Mamirkulova et al., 2020), which could be challenged by an inflow of lower-cost seasonal and migrant workers employees. The influx of migrant seasonal employees also requires economic adaptation, social integration, cultural acceptance, and psychological integration (Sun et al., 2020). It is the trade-off situation where residents seek to find the balance between benefits (positive aspects of tourism development) and costs (negative elements) that would explain individuals' decision-making towards issues related to tourism, including seasonal employment (Qiu et al., 2019).

Local residents, experiencing both positive and negative effects of tourism development and increased number of seasonal employees to ensure quality service for tourists, learn from these experiences and interactions and be expected to form an attitude towards seasonal employees. Attitude is highly correlated with knowledge – the more an individual learns and knows the subject, the more effects of subjective ambivalence on attitude impact would be expected (Wallace et al., 2020). Attitude varies in its strength, and a strong attitude is defined by its impact and durability (Luttrell & Sawicki, 2020). Formed attitudes can also vary in valence and be classified into positive, negative and neutral (Li et al., 2017). It allows concluding that residents would form an attitude, whether positive, neutral, or negative, and strong or weak based on their knowledge and past experiences with seasonal tourism employees. These attitudes, in return, would affect and guide their behaviour towards the seasonal employees. The attitude that is of particular interest in this chapter is prejudice and stereotypes the host community develops towards seasonal tourism employees.

Prejudice and Stereotypes towards Seasonal Tourism Employees

Prejudice is animus, or negative bias, towards social groups and their putative members (Paluck et al., 2021), and stereotypes are a set of cognitive generalisations about the qualities and characteristics of the members of a group or social category. "Stereotypes [...] simplify and expedite perceptions and judgments, but they are often exaggerated, negative rather than positive, and resistant to revision" (APA Dictionary, 2021). It can be summarised that there is a negative opinion formed about seasonal tourism employees before interacting with particular individuals but drawing the biased attitude based on some group representatives.

Seasonality is, in essence, a highly complex phenomenon. Its prejudice is sometimes conflicting. Quite often, introducing a tourist product to the market results in a spike in seasonal variations and pricing, particularly during the summer. A seasonal increase in demand also accompanies it.

Seasonality is determined by social factors, such as attitudes, most often biased towards certain workers. This attitude is observed both among local residents (e.g. a local resident came to work in another city for the season) and among migrants (local residents are outraged by the arrival of such a large number of migrants). Locals change their values, beliefs, and acceptable norms of behaviour due to tourism, resulting in creating a false culture to identify with tourists. When tourism development plans fail to respect the people's interests, they may feel alienated from their communities (Diniz et al., 2014). The attitude is ambiguous. Residents more often prefer locals to visitors since, in their opinion, residents, albeit from another city, are dearer and more understandable in mentality than newcomers. Residents are afraid of those who have arrived because they do not know what to expect, and stereotypes associated with countries and cultures of the visitors play a significant role too.

Even if the permanent resident population continues to diminish, the villages are occupied seasonally by people who have a long-term connection (e.g. through family connections) and newcomers who arrive from outside. Some rural areas face socioeconomic obstacles, and some studies suggest that the complexity of these issues has been lessened for some groups of individuals. Small-scale tourism enterprises (owned by locals, lifestyle migrants from Europe, or people with an Asian or refugee background) who produce and sell local food in collaboration with tourists, for example, are noted as crucial for rural variety. Different resources, in the best-case situation, can complement rather than compete with one another. Negative pictures of vacant storefronts, decaying infrastructure, and the social shame associated with large-scale decline may make such places less appealing to new migrants, mainly amenity migrants (Carson et al., 2020a; Carson et al., 2020b). Seasonality, transit patterns, and visitor concentration may help safeguard sensitive nature regions from increasing tourism-related environmental stresses (Carson et al., 2020).

Seasonal population injections (which can occur in the summer, winter, or the autumn hunting season) attract seasonal enterprises, especially in tourism and hospitality and agricultural and food production. The externalisation of employment and the growing importance of seasonal habitation and economic activity indicate village residents' increased mobility (Carson et al., 2020). Those who live and work seasonally or temporarily in the region can contribute to the sustainability of an area. There is another approach towards seasonality and work itself that some other countries have undertaken. For example, young people aim to experience the magic of Santa Claus through the seasonal jobs that they are offered. Here we are talking about international students that have less or no knowledge of Christmas at all. Living in a foreign country with a different culture is not always easy and adjustable. However, for some young seasonal employees, particularly overseas students, a lack of (lived) awareness of Christmas customs, as well as the types of duties and positions available, frequently means that the joy of Christmas fades quickly. This implies that while the attraction of the place may not diminish for these young people, the enchantment of Christmas may.

Academic literature has extensively explored racial prejudice and stereotyping (Wang et al., 2011), leading to the conclusion that prejudice is mainly targeted towards seasonal employees that are perceived as "others". In other words, prejudice is focussed on migrant seasonal employees – those that are not from the particular destination, but arrive at the destination for work from outside regions or countries. Spatial proximity is shown to be an essential determinant of attitudes in host-stranger relationships (Lujala et al., 2020). It can be assumed that the closer the seasonal employee comes from, the less prejudice it would face because the employee would be perceived as "almost one of us". When taking Spain as an example, the degree of strength of the prejudice and stereotyping would vary towards French and Indian seasonal employees; it would be expected that Indian employees are facing more negative attitudes because they would be perceived substantially different from locals in terms of demography, culture, and social organisation (Carson & Argent, 2020) or in other words perceived as "others", different, strange, and unknown.

Regarding economic stereotyping and prejudice, Esses et al. (2012) suggest recognising the mutual economic dependency that the host community has with migrant employees, as both sides benefit economically. However, this relationship is subject to various attitudes. The industries will take seasonal workers and temporary labour to sustain jobs in the specific tourism sector for fewer expenses. Migrant workers engage in the work that locals are unwilling to do – low pay and low-status jobs, creating negative biased opinions towards them being more subordinate class employees accepting conditions locals would not. Seasonal tourism employees can also face prejudice when the host community perceives that they are taking away their jobs and income. Also, the vast majority of foreign workers reside in rented housing or quarters provided by their employers. Some employers refuse to offer accommodation for their employees (Suratman et al., 2019). Foreign and seasonal employees in their workplaces have a less healthful physical environment and inadequate housing conditions.

Migrant employees can raise the tension within the local community that creates a negative bias towards them, especially in the cases where a migrant employee does not know the language, does not gain host nation friends that ease the transition, and does not adapt, or even challenge the local culture (Janta et al., 2011). It is also reported that local communities often perceive migrant workers as a threat to the culture and heritage of the host community (Tunon & Baruah, 2012), mainly because of not knowing and respecting local values, traditions, and norms. Suppose a seasonal employee is coming from a country of different religious beliefs. In that case, it can be seen as a threat (Lazaridis & Wickens, 1999), mainly due to stereotypes that people have about different religions.

Overall, seasonal tourism employees are placed in a very unfavourable situation – from one side, there are unfair work practices, lower pay, and worse job conditions, but on the other, there is pressure from the host community to become one of them and adapt, learn the language, and accept and practice cultural, religious, and lifestyle practices. When seasonal employees are not understood, conflicts due to differences arise with the host community (Widyawati, 2006). Moreover, in line with the anti-immigration enforcement and criminalisation of undocumented status among migrant workers in many developed countries like the USA, EU, and others (Becerra, 2020), seasonal migrant employees are also perceived and prejudiced as being criminals. In such a context, the host community develops negative attitudes towards seasonal tourism employees. A study by Kim et al. (2018) reported that seasonal migrant employees even feel that the host community hates them.

Acquiring local language and culture skills is crucial for integrating seasonal employees and acceptance by the host community. Language skills increase acceptance and give opportunities to get better jobs, acquire managerial positions, and increase earnings (Budría et al., 2019). Moreover, learning culture, local customs, norms, and lifestyle also provide many benefits for seasonal employees, both for individuals and employees (Wendt et al., 2020). One of the paradoxes indicated in academic literature is that the tourism industry heavily depends on a seasonal workforce who are "guests" in the host community but are working to provide "local" hospitality experiences to tourists (Linge et al., 2020). Such a situation can give a reason for prejudice by the local community, who would perceive seasonal employees as not qualified to provide such a service. The more the seasonal employees are educated about language, culture, and lifestyle, the better their attitude.

Conclusions

Seasonality in tourism presents several problems to industry practitioners. Many tourism locations are attempting to develop a more sustainable tourism model to extend the tourism season and generate new tourism activities. Tourism is seasonal throughout the world, and it is mainly determined by geography and income-generating markets. Tourism employment seasonality is driven by two interrelated factors: urbanisation and the movement of people to live and work in cities, while on the other hand, there is a growing demand for isolated, unknown, rural tourism.

Among all seasonal worker characteristics, the origin of those personnel makes the most significant impact when concentrating on views and attitudes about seasonal employees. Graduates of the same country as the residents are perceived as more friendly than those forced to travel to a foreign country for a seasonal job and accept the conditions that are sometimes not as good as they should be.

Prejudice and stereotyping are directed toward migrant seasonal employees who are not native to the location but go there for work from other areas or countries. Migrant workers undertake labour that locals are unwilling to do, such as low-paying and low-status employment, which creates unfavourable prejudice against them as lower-class employees who accept circumstances that locals would not. Seasonal tourist workers may encounter discrimination if the host community believes they are stealing their jobs and money.

Rural areas that cannot locate personnel locally are obliged to hire seasonal workers to supply the necessary workforce, skills, and abilities for tourism and hospitality services. It is clear that a substantial influx of seasonal labour and visitors disrupts the local lifestyle, and local inhabitants may not always see the development positively. Seasonal tourist employment has both beneficial and negative consequences for a variety of stakeholders. For starters, its employees must leave their homes, lifestyles, and families to go to a new location for employment, which does not necessarily provide decent working conditions and compensation. Integration in the new environment is also often problematic, due to the prejudice and stereotyping that is targeted towards seasonal migrant employees.

References

Andriotis, K. (2005). Seasonality in Crete: Problem or a way of life?. *Tourism Economics, 11*(2), 207–224.

APA Dictionary (2021). Available at: https://dictionary.apa.org/stereotype

Araslı, H., & Arıcı, H. E. (2019). The art of retaining seasonal employees: Three industry-specific leadership styles. *The Service Industries Journal, 39*(3–4), 175–205.

Araslı, H., Altinay, L., & Arıcı, H. E. (2020). Seasonal employee leadership in the hospitality industry: A scale development. *International Journal of Contemporary Hospitality Management, 32*(6), 2195–2215.

Ball, R. M. (1988). Seasonality: A problem for workers in the tourism labour market?. *The Service Industries Journal, 8*(4), 501–513.

Baum, T., & Lundtorp, S. (2001). Introduction. In Baum, T. and Lundtrop, S. (Eds.), *Seasonality in tourism* (pp. 1–4). Routledge.

Becerra, D. (2020). "They say we are criminals": The stress, fears, and hopes of migrant dairy workers as a result of us immigration policies. *Journal of Poverty, 24*(5–6), 389–407.

Belma, S. U. N. A., & Alverez, M. D. (2019). Gastronomic identity of Gaziantep: Perceptions of tourists and residents. *Advances in Hospitality and Tourism Research (AHTR), 7*(2), 167–187.

Bhat, A. A., & Mishra, R. K. (2020). Demographic characteristics and residents' attitude towards tourism development: A case of Kashmir region. *Journal of Public Affairs, 21*(e2179), 1–13.

Boley, B. B., Strzelecka, M., & Woosnam, K. M. (2018). Resident perceptions of the economic benefits of tourism: Toward a common measure. *Journal of Hospitality & Tourism Research, 42*(8), 1295–1314.

Budría, S., Colino, A., & de Ibarreta, C. M. (2019). The impact of host language proficiency on employment outcomes among immigrants in Spain. *Empirica, 46*(4), 625–652.

But, J. W. P., & Ap, J. (2017). The impacts of casino tourism development on Macao residents' livelihood. *Worldwide Hospitality and Tourism Themes, 9*(3), 260–273.

Carson, D. A., Åberg, K. G., & Prideaux, B. (2020a). Cities of the North: Gateways, competitors or regional markets for hinterland tourism destinations? In L. Lundmark, D.B. Carson & M. Eimermann (Eds.), *Dipping into the North* (pp. 285–310). Singapore: Palgrave Macmillan.

Carson, D. B., & Argent, N. (2020). Who lives in the inland north? Dynamic, diverse, fragile, robust. In L. Lundmark, D.B. Carson & M. Eimermann (Eds.), *Dipping in to the North* (pp. 15–25). Singapore: Palgrave Macmillan.

Carson, D. B., Carson, D. A., Eimermann, M., Thompson, M., & Hayes, M. (2020b). Small villages and socio-economic change in resource peripheries: A view from Northern Sweden. In L. Lundmark, D.B. Carson & M. Eimermann (Eds.), *Dipping in to the North* (pp. 27–53). Singapore: Palgrave Macmillan.

Cave, J., Brown, K. G., & Baum, T. (2012). Human resource management in tourism: A small island perspective. *International Journal of Culture, Tourism and Hospitality Research, 6*(2), 124–132.

Chen, J. S., & Wang, W. (2015). Foreign labours in Arctic destinations: Seasonal workers' motivations and job skills. *Current Issues in Tourism, 18*(4), 350–360.

Diniz, S. R., Falleiro, S. P., & De Barros, M. M. (2014). Local residents' perception of the psycho-social and economic impact of tourism in Goa. *International Journal of Scientific and Research Publications, 4*(11), 1–7.

Duro, J. A., & Turrión-Prats, J. (2019). Tourism seasonality worldwide. *Tourism Management Perspectives, 31*, 38–53.

Edwards, K. (1990). The interplay of affect and cognition in attitude formation and change. *Journal of Personality and Social Psychology, 59*(2), 202.

Ericsson, B., Overvåg, K., & Möller, C. (2020). Seasonal workers as innovation triggers. In *Tourism Employment in Nordic Countries* (pp. 235–256). Cham: Palgrave Macmillan.

Esses, V. M., Brochu, P. M., & Dickson, K. R. (2012). Economic costs, economic benefits, and attitudes toward immigrants and immigration. *Analyses of Social Issues and Public Policy, 12*(1), 133–137.

Fazio, R. H., Lenn, T. M., & Effrein, E. A. (1984). Spontaneous attitude formation. *Social Cognition, 2*(3), 217–234.

Ferreira, F. A., Castro, C., & Gomes, A. S. (2020, October). Positive and negative social-cultural, economic and environmental impacts of tourism on residents. In *International Conference on Tourism, Technology and Systems* (pp. 288–298). Singapore: Springer.

Godovykh, M., & Ridderstaat, J. (2020). Health outcomes of tourism development: A longitudinal study of the impact of tourism arrivals on residents' health. *Journal of Destination Marketing & Management, 17*, 100462.

Grobelna, A., & Skrzeszewska, K. (2019). Seasonality: Is it a problem or challenge facing future tourism employment? Implications for management. *Journal of Entrepreneurship, Management and Innovation, 15*(1), 205–230.

Guidetti, G., Pedrini, G., & Zamparini, L. (2020). Assessing perceived job quality among seasonal tourism workers: The case of Rimini, Italy. *Tourism Economics*. Advance online publication.

Gülduran, Ç. A., & Gürdoğan, A. (2021). Investigation of intercultural sensitivity levels of department managers working in hotel enterprises according to some demographic features. In E. Özen, S. Grima & R.D. Gonzi (Eds.), *New Challenges for Future Sustainability and Wellbeing* (pp. 95–117). Emerald Publishing Limited.

Gupta, V., & Chomplay, P. (2021). Local Residents' Perceptions Regarding the Negative Impacts of Overtourism: A Case of Shimla. In A. Sharma & A. Hassan (Eds.), *Overtourism as Destination Risk* (pp. 69–80). Bingley: Emerald Publishing Limited.

Huber, D., & Gross, S. (2021). Local residents' contribution to tourist experiences: A community perspective from Garmisch-Partenkirchen, Germany. *Tourism Review*. Advance online publication.

Ioannides, D., & Zampoukos, K. (2018). Tourism's labour geographies: Bringing tourism into work and work into tourism. *Tourism Geographies, 20*(1), 1–10.

Janta, H., Brown, L., Lugosi, P., & Ladkin, A. (2011). Migrant relationships and tourism employment. *Annals of Tourism Research, 38*(4), 1322–1343.

Jolliffe, L., & Farnsworth, R. (2003). Seasonality in tourism employment: Human resource challenges. *International Journal of Contemporary Hospitality Management, 15*(6), 312–316.

Kim, H. J., Choi, H. J., Lee, K. W., & Li, G. M. (2018). Acculturation strategies used by unskilled migrant workers in South Korea. *Ethnic and Racial Studies, 41*(9), 1691–1709.

Krakover, S. (2000). Partitioning seasonal employment in the hospitality industry. *Tourism Management, 21*(-5), 461–471.

Lazaridis, G., & Wickens, E. (1999). "Us" and the "Others": Ethnic minorities in Greece. *Annals of tourism research, 26*(3), 632–655.

Lee-Ross, D. (1995). Attitudes and work motivation of subgroups of seasonal hotel workers. *Service Industries Journal, 15*(3), 295–313.

Li, C., Guo, X., & Mei, Q. (2017). Deep memory networks for attitude identification. In *Proceedings of the Tenth ACM International Conference on Web Search and Data Mining* (pp. 671–680). WDSM.

Linge, T. T., Furunes, T., Baum, T., & Duncan, T. (2020). Hospitality through hospitableness: Offering a welcome to migrants through employment in the hospitality industry. In *Tourism Employment in Nordic Countries* (pp. 401–424). Cham: Palgrave Macmillan.

Lozanski, K., & Baumgartner, K. (2020). Local gastronomy, transnational labour: Farm-to-table tourism and migrant agricultural workers in Niagara-on-the-Lake, Canada. *Tourism Geographies*. Advance online publication.
Lujala, P., Bezu, S., Kolstad, I., Mahmud, M., & Wiig, A. (2020). How do host–migrant proximities shape attitudes toward internal climate migrants?. *Global Environmental Change, 65*, 102156.
Lundberg, C., Gudmundson, A., & Andersson, T. D. (2009). Herzberg's two-factor theory of work motivation tested empirically on seasonal workers in hospitality and tourism. *Tourism Management, 30*(6), 890–899.
Lundmark, L. (2006). Mobility, migration and seasonal tourism employment: Evidence from Swedish mountain municipalities. *Scandinavian Journal of Hospitality and Tourism, 6*(3), 197–213.
Luttrell, A., & Sawicki, V. (2020). Attitude strength: Distinguishing predictors versus defining features. *Social and Personality Psychology Compass, 14*(8), e12555.
Mamirkulova, G., Mi, J., Abbas, J., Mahmood, S., Mubeen, R., & Ziapour, A. (2020). New Silk Road infrastructure opportunities in developing tourism environment for residents better quality of life. *Global Ecology and Conservation, 24*, e01194.
Martínez, J. M. G., Martín, J. M. M., Fernández, J. A. S., & Mogorrón-Guerrero, H. (2019). An analysis of the stability of rural tourism as a desired condition for sustainable tourism. *Journal of Business Research, 100*, 165–174.
McCole, D. (2015). Seasonal employees: The link between sense of community and retention. *Journal of Travel Research, 54*(2), 193–205.
Möller, C., Ericsson, B., & Overvåg, K. (2014). Seasonal workers in Swedish and Norwegian ski resorts–potential in-migrants?. *Scandinavian Journal of Hospitality and Tourism, 14*(4), 385–402.
Nukhu, R., & Singh, S. (2020). Perceived Sustainability of seasonal employees on destination and work—a study in the tourism industry. In *Sustainable Human Resource Management* (pp. 213–225). Singapore: Springer.
Olson, M. A., & Fazio, R. H. (2001). Implicit attitude formation through classical conditioning. *Psychological Science, 12*(5), 413–417.
Paluck, E. L., Porat, R., Clark, C. S., & Green, D. P. (2021). Prejudice reduction: Progress and challenges. *Annual Review of Psychology, 72*, 533–560.
Park, C. W., & MacInnis, D. J. (2006). What's in and what's out: Questions on the boundaries of the attitude construct. *Journal of Consumer Research, 33*(1), 16–18.
Perloff, R. M. (1993). *The Dynamics of Persuasion: Communication and Attitudes in the 21st Century*. New York: Routledge.
Qiu, H., Fan, D. X., Lyu, J., Lin, P. M., & Jenkins, C. L. (2019). Analysing the economic sustainability of tourism development: Evidence from Hong Kong. *Journal of Hospitality & Tourism Research, 43*(2), 226–248.
Rheu, M. (2020). Attitude. In Van den Bulck, J. (Ed.), *The International Encyclopedia of Media Psychology*, (pp. 1–12). UK: John Wiley & Sons.
Rosselló, J., & Sansó, A. (2017). Yearly, monthly and weekly seasonality of tourism demand: A decomposition analysis. *Tourism Management, 60*, 379–389.
Salazar, N. B. (2020). Labour migration and tourism mobilities: Time to bring sustainability into the debate. *Tourism Geographies, 22*(3), 1–11.
Šošić, M. M., Bečić, M., & Jasprica, D. (2018). Factors influencing worker's intention to re-accept seasonal employment. In *6th International OFEL Conference on Governance, Management and Entrepreneurship. New Business Models and Institutional Entrepreneurs: Leading Disruptive Change*. Governance Research and Development Centre (CIRU).
Stynes, B. W., & Pigozzi, B. W. (1983). A tool for investigating tourism-related seasonal employment. *Journal of Travel Research, 21*(3), 19–24.
Sun, J., Ling, L., & Huang, Z. J. (2020). Tourism migrant workers: The internal integration from urban to rural destinations. *Annals of Tourism Research, 84*, 102972.
Suratman, R., Samsudin, S., Aminuddin, S. Z., & Saim, Z. (2019). Resident's perception on foreign workers housing in high rise residential building. In *IOP Conference Series: Earth and Environmental Science* (pp. 12–33). Malaysia: IOP Publishing.
Tsiotas, D., Krabokoukis, T., & Polyzos, S. (2020). Detecting interregional patterns in tourism seasonality of Greece: A principal components analysis approach. *Regional Science Inquiry, 12*(2), 91–112.
Tunon, M., & Baruah, N. (2012). Public attitudes towards migrant workers in Asia. *Migration and Development, 1*(1), 149–162.
Tuulentie, S., & Heimtun, B. (2014). New rural residents or working tourists? Place attachment of mobile tourism workers in Finnish Lapland and Northern Norway. *Scandinavian Journal of Hospitality and Tourism, 14*(4), 367–384.

Vassou, C., Zopiatis, A., & Theocharous, A. L. (2017). Intercultural workplace relationships in the hospitality industry: Beyond the tip of the iceberg. *International Journal of Hospitality Management, 61*, 14–25.
Vergori, A. S., & Arima, S. (2020). Transport modes and tourism seasonality in Italy: By air or by road?. *Tourism Economics*. Advance online publication.
Wallace, L. E., Patton, K. M., Luttrell, A., Sawicki, V., Fabrigar, L. R., Teeny, J.,... & Wegener, D. T. (2020). Perceived knowledge moderates the relation between subjective ambivalence and the "impact" of attitudes: An attitude strength perspective. *Personality and Social Psychology Bulletin, 46*(5), 709–722.
Wang, L., Ma, Q., Song, Z., Shi, Y., Wang, Y., & Pfotenhauer, L. (2011). N400 and the activation of prejudice against rural migrant workers in China. *Brain Research, 1375*, 103–110.
Wendt, M., Jóhannesson, G. T., & Skaptadóttir, U. D. (2020). On the move: Migrant workers in Icelandic hotels. In *Tourism Employment in Nordic Countries* (pp. 123–142). Palgrave Macmillan, Cham.
Widyawati, N. (2006). Representations of migrant workers in Malaysian Newspapers. Are we up to the challenge?. *Current Crises and the Asian Intellectual Community*. The work of the 2005/2006 API Fellows. Bangkok: The Nippon Foundation.
Winter, S. (2020). Do anticipated Facebook discussions diminish the importance of argument quality? An experimental investigation of attitude formation in social media. *Media Psychology, 23*(1), 79–106.

27

SOCIAL-PSYCHOLOGICAL BACKGROUND OF DISCRIMINATION AND ITS REFLECTIONS ON TOURISM

Filiz Gümüş Dönmez and Serkan Aylan

Introduction

Although discrimination based on race, color, religion, gender, and nationality was prohibited with the 1964 Civil Rights Act, the issue of discrimination has been increasing day by day through globalization (Cheung et al., 2016). With the removal of borders as a result of globalization, a great many people can go to the transnational geographies to live, work and get an education. However, since they make up a small minority in the places they go, those people may be exposed to discrimination much more easily. In this respect, it is quite possible to say that inequality and discrimination in the world have been triggered more by such a high level of globalization (Centeno, 2010).

Discrimination, as can be seen in all areas of life, observed in business life restricts employees from explicitly expressing their wages, goals, objectives, opinions, career desires, and expectations. Even though the regulations and laws regarding discrimination after industrialization have brought some restrictions, they haven't been enough to eliminate discrimination, which is still a growing problem in business life today (Elei, 2016). And one of the industries that has been affected by the discrimination phenomenon is tourism, as well.

The tourism industry has always been a growing industry apart from the crises such as terrorism, Covid-19, and natural disasters. Approximately 10% of the total employment in the world is currently employed in this sector industry. But this growth reveals the lack of qualified personnel especially due to the fragile and seasonal structure of the tourism industry in many countries. What's more, the industry has a problem of retaining talented and qualified personnel because of the high turnover rate of the labor force. Since the people employed in the industry tend to leave the tourism industry as they get older, businesses in the industry have been recruiting young talents from tourism schools to solve this problem. The job opportunities given to young talents rather than the mature ones and considered a kind of luck actually means low salaries, low status, and this has been continuing with endless career crises (Shum et al., 2020; Wen & Madera, 2013). And the discrimination in the tourism industry actually starts right at this point. Employees who started the industry with such discrimination from the beginning, are discriminated against or are exposed to discrimination and continues with other types of discrimination such as gender, religion, language and color.

Defining Discrimination and Workplace Discrimination

The concept of discrimination is in a close relationship with the concepts of treatment and behavior. For this reason, it differs from attitudinal and emotional biases such as prejudice and cognitive

DOI: 10.4324/9781003161868-27

qualities and beliefs such as stereotypes (Dipboye & Collella, 2005). However, prior to these concepts, it is necessary to look at the concept of attitudes that causes prejudice and stereotypes. Attitude is divided into two categories as cognitive attitude and emotional attitude. Cognitive attitude includes certain thoughts or beliefs about the object to be developed. As for emotional attitude, it includes emotions and feelings concerning the attitude object. Although cognitive and emotional tendencies between individuals are compatible with each other, they may be inconsistent from time to time. More importantly, the place of cognitive and emotional attitude in discrimination varies according to individual differences and group relationships (Dovidio & Hebl, 2005).

Prejudice can be expressed as an individual's negative and unfair attitude towards a person or group without knowing the person or the group (Ramiah et al., 2010). Discrimination arising from prejudice is stated as exhibiting an unfair or unequal attitude and behavior to one or more of the members in the group due to some certain characteristics of theirs (Dion, 2002). The behavior here involves activities, judgments, and decisions that are displayed by the ones within the group against those outside the group (Ramiah et al., 2010). Correll et al. (2010) defined discrimination as a behavior displayed to group members only because they are a member of a certain group/category rather than being a certain right or reciprocity. Accordingly, while discriminators may justify themselves because the other party deserves it, the other party does not agree with this issue (Ramiah et al., 2010).

Stereotype is the generalizations regarding unjustified beliefs about a group or group members (Ramiah et al., 2010). These generalizations include erroneous thoughts, over-generalizations, factual inaccuracies, and disproportionately rigid behavior concerning prejudiced attitudes and discriminatory behaviors.

Discrimination and the perception of discrimination, with various sources and indicators, have been an important problem among individuals, groups, and organizations all over the world. In addition, discrimination occurs at many levels, including social, cultural, institutional, and individual. There are a number of good and bad ways people can discriminate (Weisskopf, 2010). In the essence of discrimination lies the rights of group members that are not sufficiently represented and their being made disadvantaged in terms of social, political, and economical aspects. Discrimination has been both an interesting and frustrating topic for social scientists. The reason of its being interesting is that it has a strong mechanism based on inequality in its past and present structure. The reason of its being annoying is that it is very difficult to measure and understand (Kaas & Manger, 2010). The studies conducted by social scientists to differentiate, define and reveal the effects of discrimination have resulted in several methods so far. The findings obtained show that discrimination is mostly displayed as racism, against women, mentally ill individuals, gays, lesbians, and overweight people (Carr & Friedman, 2005; Smith, 2003).

As for Weisskopf (2010), he stated that the most common discrimination as a social problem is the discrimination displayed by the wealthy and strong group of people in society against the poorer and weak/weak group in the same society. On the other hand, people spend one-third of their everyday lives working, in other words, at the workplace. For this reason, one of the areas where discrimination is considered is the workplace. Workplace discrimination is quite widespread in underdeveloped and developing countries as well as in developed countries. The main reason for this is that countries' opening their doors to foreign migration. Therefore, this situation has nothing to do with economic development. People who come to the host country due to foreign migration may encounter an unequal and discriminatory attitude in matters such as religion, language, and race, as they form a minority group in that country (Banarjee, 2008). Workplace discrimination, which had different forms in the past, have still been continuing as a major problem in the workplace (Dipboye & Collella, 2005).

While job discrimination directly affects health as a certain source of stress, it may also affect promotion and income indirectly. Moreover, unequal behaviors or discrimination have negative consequences on individual and organizational issues such as productivity, organizational commitment, job satisfaction, and well-being (Banarjee, 2008). Besides its organizational effects, discrimination is highly effective in chronic diseases such as stress, depression, burnout, blood pressure, alcohol addiction, heart diseases, and births of babies with low weight (Castro et al., 2008).

The United Nations' International Labor Organization defined discrimination in the workplace as any distinction, exclusion or preference made on the basis of race, color, sex, religion, political opinion, national extraction or social origin, which has the effect of nullifying or impairing equality of opportunity or treatment in employment or occupation. (Cheung et al., 2016: 4). This definition has been criticized from two aspects. First, the definition encompasses general discrimination involving behavior outside of organizational practices such as personal dialogues with customers and social characteristics that are often not legally protected like weight, sexual orientation, physical attractivity. Second, this definition emphasizes equality of opportunity rather than interpersonal experiences within the scope of social identity characteristics (Cheung et al., 2016; Shen & Dhanani, 2015).

According to the International Labor Organization (ILO) (2007), workplace discrimination is an example of human rights violation that negatively affects productivity and economic growth and causes the abilities of the individual to be wasted. In addition, workplace discrimination has an impact hindering social cohesion and solidarity and creating social inequalities in poverty reduction. Workplace discrimination may be displayed at the employee selection stage, during the employment process or while quitting the job. Discrimination stands for some particular and inappropriate attitudes towards individuals due to the reasons stated in the previous sentence, regardless of the individual's ability and characteristics to perform his/her duty. For example, by alleging the relations of an employee with the management as an excuse for recruiting a candidate from outside the company instead of that employee who deserves to be promoted by providing all the necessary conditions creates discrimination.

Although the nature of discrimination is generally characterized by negative attitudes and behaviors, it is possible to observe positive discrimination practices in business life. Negative discrimination is the evaluation of an individual with unequal, incomplete, or inadequate characteristics compared to other members. On the other hand, positive discrimination is based on acting in favor of the individual by ignoring the disadvantages of the individual due to his/her age, gender, religion, language, race, and health status in order to behave equally (Ramiah et al. 2010). For instance, not giving a pregnant woman a promotional and advertising task that requires standing for a long period of time, not giving long-distance travel permission to an elderly company driver, and allowing a Muslim employee from the workplace to attend Friday prayer during the required hours are within the scope of positive discrimination practices.

The discrimination arising from the effects of economic, social, cultural, demographic, or other characteristics creates important outcomes not only at personal level but also at the organizational and social levels. In this context, there are a number of theories that shape the phenomenon of discrimination within a society and connect the issue to a certain basis.

Theories of Discrimination

In the related literature, there are a number of theories that constitute the understanding of intergroup relations, prejudice and discrimination. Some of these theories are social identity theory, behaviors from intergroup affect and stereotypes map (BIAS map), aversive racism, and system justification theory.

Social Identity Theory: This theory explains how ethnic minorities, immigrants, and foreign employees are exposed to discrimination (Reskin, 2000). As a matter of fact, the aspects that lie at the root of the Social Identity Theory can be listed as identification with a group, abiding by the group norms, and discrimination against individuals outside the group (Cheung et al., 2016). In this theory, people unconsciously classify individuals in two categories as in-group and out-of-group individuals. The individuals who are integrated into one of the groups clearly highlight the differences between the groups and minimize the differences of the group that they belong to.

Behaviors from Intergroup Affect and Stereotypes Map (BIAS map): While social identity theory analyzes the basic and general processes that cause in-group discrimination, the BIAS map provides insights into certain ways of discrimination displayed against particular group members. Within this context, BIAS map is an extension of the Stereo Content Model (Ramiah et al., 2010). BIAS Map distinguishes possible behavioral consequences of stereotype clusters from each other. Within the scope of the model, feelings, thoughts and behaviors related to discrimination have been combined under a sole model (Cuddy et al., 2007).

Aversive racism: The theory is based on the conflict between white Americans' denial of personal prejudice and their unconscious negative feelings and beliefs towards black people (Dovidio & Gaertner, 2004). Aversive racists externally reprimand and deny their being prejudiced claiming to have egalitarian values. However, when they encounter individuals outside of their groups, they do not hesitate to feel negative emotions such as fear, anxiety, disgust or indifference (Cheung et al., 2016).

System justification theory: System Legitimation is a trend that is used to defend, support and rationalize the existing social, economic and political order and to motivate the individuals in this direction. The theory was developed with the aim of maintaining the accepted stability, supporting the current social order, and opposing the social change (Jost & Andrews, 2012). There are three main factors in the core structure of System Legitimation Theory: ego, group and system. Ego legitimation is the legitimization of the individual's personal interests. Group legitimation is the individuals' serving to the interests of the group to which they belong for the sake of the interests of the group. System legitimation, on the other hand, means accepting the existing system without opposing the system for the continuity of the system, although it is against the interests of the individual and the group (Ramiah et al., 2010).

Types of Employment Discrimination and Its Reflections on Tourism

Age Discrimination (Ageism)

Ageism, also known as age discrimination, is a type of discrimination against individuals or groups due to their age. According to Low and Ang (2013), ageism can be defined as a combination of three interrelated elements. According to them, the first of these three elements is harmful attitudes towards the elderly, old age, and the aging process. The second element is discriminatory practices against the elderly. The third element is institutional practices and policies that perpetuate stereotypes about the elderly. Age discrimination in the workplace occurs most often in recruitment, promotion, and performance evaluation (Perry & Finkelstein, 1999). The preference of a certain age range in age-based discrimination in businesses varies depending on the employer's or manager's evaluation rather than the characteristics of the employee (Duncan & Loretto, 2004). For example, some businesses prefer younger energetic people as employees, while others may prefer more experienced and even retired people. It has been observed that especially elderly individuals are frequently exposed to age discrimination at work (Low & Ang, 2011) or during the recruitment process (Capelli & Novelli, 2010).

As in many industries, discrimination can also be mentioned in the tourism industry, where labor-intensive employees with different characteristics work (Yeşiltaş et al., 2012). McNair and Flynn (2007) stated that the hotel industry, which is included in the tourism industry, is an important industry in terms of the business and the number of employees it employs. In addition, they emphasized the low average age in the industry and drew attention to the existence of age discrimination in terms of employee selection process in the hotel industry and promotion and wage policies in the business life of the organization. Demir (2011) states in his study that people in a certain age range are preferred for job postings and applications in tourism.

Sex Discrimination (Sexism)

Gender is a position in society attributed to the individual. Individuals have no control over their attributed status. Gender is also a fundamental status (Sullivan, 2003). It is seen that gender discrimination has different definitions in the literature. For example, according to Glick and Fiske (1996), gender discrimination is discrimination against a person only because of their gender. According to Parrillo (2005), apart from negative behaviors, sexual discrimination is the distribution of prestige, power, and property according to gender, not individual characteristics. According to WHO (2001), gender discrimination is the exposure of any discrimination, exclusion, or restriction based on socially structured gender roles and norms that prevent the individual from fully enjoying human rights. It is seen that there are different classifications regarding the types of gender discrimination. These classifications can be listed as "contradictory sexism" (Ambivalent), which consists of hostile and benevolent sexism (Glick & Fiske, 1996), "overt, covert, and subtle sexism" (Benokraities & Feagin, 1995), "direct and indirect sexism" (Demirbilek, 2007), "vertical and horizontal sexism" (Tükeltürk & Perçin, 2008).

Due to the diversity of jobs in the tourism industry and biological differences, there may be discrimination in favor of women and sometimes in favor of men in employee employment and promotions. For example, Segovia-Perez et al. (2019) argue that horizontal and vertical gender discrimination is common in tourism employment. Muñoz-Bullón (2009) found in his study in Spain that male employees in the tourism industry earn 6.7% more than women.

Racial/Ethnic Discrimination (Racism)

Racial discrimination is any kind of discrimination against any person based on racial or ethnic origin (Behera, 2018). Today, some authors suggest that there are four or five basic races, while others suggest that there are up to three dozen basic races. In addition, Count Joseph Arthur de Gobineau argued that there are three basic races: White (Caucasian), Black (Negroid), and Yellow (Mongoloid). Gobineau claimed that the western influence spread to the whole world because the white race has superior intelligence, morality, will and all other positive hereditary characteristics (Sayın & Candan, 2016). Racial discrimination or racism is a phenomenon that can be seen in many areas of life such as education, employment, health, and housing (Li et al., 2020). For example, Marginson et al. (2010) stated that the main factor that makes international students studying in foreign countries other than their home country unhappy is the racism they are exposed to. One of the industries where racial discrimination or racism is seen is tourism. In tourism, which mostly covers tourists from different races, practices involving racial discrimination in both tourists and employment are encountered (Deale & Wilborn, 2006). In their research Li et al. (2020), examining the comments of tourists on many tourism websites around the world, revealed the existence of racism that tourists face on a global scale. The concern that potential tourists will be exposed to racism during their holidays can affect their vacation decisions. For example, Hudson et al. (2020) found that fear of racial discrimination continues to be an important factor preventing African American tourists from taking and enjoying their vacations.

Skin Color Discrimination (Colorism)

Skin color discrimination is defined as a type of discrimination based on skin color in the literature and the term "colorism" is used. Colorism is the subjection of people to different social and economic treatments according to their skin color (Hamilton et al., 2009: 32). Colorism is often confused with racism. Colorism and racism are too intertwined to be easily distinguished from each other, but they are still different from each other. Racism is the name given to people being discriminated against due to their race and ethnicity. However, colorism is the discrimination of individuals due to their phenotype, regardless of the race or ethnicity they are associated with. In discrimination based on color, color is not an indicator of race. Between two people of the same race with different color tones, the lighter-skinned one can be considered superior to the darker-skinned one (Jones, 2000: 1497). It is possible to state that colorism is a common type of discrimination in business life. According to the 2019 report of the EEOC (The Equal Employment Opportunity Commission), 4.7% (3415) of the accusations of discrimination in the workplace in the USA are related to skin color discrimination (EEOC, 2020). People of color who work in different industries stated that they suffer from hostile attitudes due to lack of career opportunities (Wen & Madera, 2013), low social status (DiTomaso et al., 2007), and racial prejudice (Ziegert & Hanges, 2005). Shum et al. (2020) found in their study in the accommodation industry that non-white employees were exposed to a higher level of discrimination than whites. At the same time, Benjamin and Dillette (2021), in their qualitative research in the USA, revealed that black tourists are still exposed to discrimination and racism today.

Disability Discrimination

Disability is any impairment of the body or mind that makes it difficult for a person to do certain activities (activity limitation) and interact with the world around them (participation restrictions) (CDC, 2020). Disability discrimination is a less advantageous treatment of a disabled person than a non-disabled person in the same or similar circumstances (Australian Human Rights Commission, 2012). Disability discrimination can be seen in two different ways. These are defined as direct and indirect disability discrimination. Direct disability discrimination occurs when people with disabilities are treated worse because of their different personal status than others in a similar situation for a prohibited justification. Indirect discrimination, on the other hand, means that laws, policies or practices appear neutral at first glance but have an extremely negative impact on a disabled individual. This type of discrimination occurs when some individuals are excluded from accessing an opportunity that actually appears accessible due to their conditions (United Nations, 2018: 4–5). People with disabilities may face discrimination in many areas such as education, health, and employment. When evaluated in terms of the business world, discrimination occurs not only in the job search process but also in the provision of wages and services (Stewart et al., 2011). Considering the types of discrimination experienced by persons with disabilities in the tourism and accommodation industry, the physical facilities of the enterprise are generally insufficient for persons with disabilities, the fact that these facilities are built without considering the disabled tourists, the services provided to the customers in the enterprises are kept together with the disabled tourists and other non-disabled tourists and their special needs and free it seems that their will is being ignored. In addition, it is seen that there are words and behaviors reflected on the disabled by other non-disabled tourists in the business and which indicate discomfort (Lim, 2020).

Religious Discrimination

Discrimination based on religion refers to individuals being subjected to unfair treatment due to their religious beliefs, preventing or restricting their religious worship in the work environment or during the process of job application (Ghumman et al., 2013: 441). It is observed that people in different countries of the world are exposed to discrimination due to their religion or sect. For example, in the discrimination study conducted by Russell et al. (2008) in Ireland, it is revealed that individuals who live in Ireland and belong to different religious beliefs (Christianity, Islam, etc.) or different sects of the same religion (Catholic, Protestant, and Orthodox) and those who do not believe in any religion are faced discrimination not only in social life but also in business life. In the study conducted by Weller (2011) in the UK, Muslim respondents reported that they were subjected to a lot of religious discrimination and unfair treatment after the September 11, 2001 attack. People working in the tourism industry may be exposed to discrimination because they dress in accordance with the religion they believe. For example, WAFIQ (2018) mentions that three women working in a five-star hotel in Kuala Lumpur were refused to wear headscarves and the difficulties that came with this situation.

Pregnancy and Parenthood Discrimination

Women are exposed to many discriminatory practices in working life. One of these is discrimination regarding pregnancy and parenthood. Pregnancy discrimination involves the unfavorable treatment of a woman (who may be an applicant or an employee) due to her pregnancy, birth, or a medical condition related to pregnancy or childbirth (EEOC, 2021). In Western societies, it is common for women to be discriminated against due to pregnancy in the workplace. At the same time, almost half of working women are exposed to discriminatory practices such as not being given the opportunity to receive education, criticizing their job performance or appearance, and changing the job description (Mäkelä, 2012: 680). In a study conducted in Austria and Croatia, it was seen that although many parents admit that they are discriminated against, a few of them report discrimination during the employment process (Parents@work, 2020). Dalkıranoğlu & Çetinel (2008), in their research on five-star hotel managers in Istanbul, revealed that pregnant women are viewed negatively by both male and female managers when hiring, and that in case of a possible crisis, pregnant women can be dismissed first.

Other Discrimination Types

In addition to the types of discrimination mentioned above, there are other types of discrimination that are relatively new in business life. For example, sexual orientation/LGBT discrimination, discrimination against HIV carriers, genetic discrimination, obesity discrimination, nepotism and political discrimination are some of them. Sexual orientation discrimination is discrimination type that applied to lesbian, gay, bisexual, and transgender individuals in different areas of life due to their sexual choice or sexual orientation. In the report of "Discrimination on grounds of sexual orientation and gender identity in Europe" prepared by the European Commission of Human Rights, it is revealed that LGBT individuals are exposed to some discrimination and barriers by law enforcement authorities in protection, access to health and education opportunities, marriage and family establishment, and employment (Council of Europe Commissioner for Human Rights, 2011).

When looking at genetic discrimination, which is another type of discrimination, practices such as refusal of a person's job application or termination of his job can be seen as a result of the genetic information of the individuals as a result of genetic tests. Genetic screening has important implications for the workplace, as employers may consider excluding or firing employees whose genetic status is prone to a particular disease in the future. Genetic testing can easily lead to unfair dismissal or denial of employment (ILO, 2007).

Another type of discrimination after genetic discrimination is HIV/AIDS discrimination. People living with HIV/AIDS in society and business life often face discrimination. According to a report on the rights violations of people living with HIV in Turkey, 6.6% (out of 103 people) whose rights were violated while working or receiving services in the public industry. It has also been revealed that people living with HIV feel under great threat from workplace harassment and are even fired from work (Surgevil & Mayaturk-Akyol, 2011).

Obesity, another health problem, can also cause individuals to face discrimination. The exposure of individuals to attitudes such as prejudice and discrimination is among the most important causes of social and psychological problems caused by obesity, which is defined by the World Health Organization (WHO) as "abnormal or excessive fat accumulation that can harm health" (Puhl et al., 2014). There are studies showing that obese individuals have a low chance of getting a job offer, their starting salary is lower, and they generally encounter some adversities in the working environment (O'Brien et al., 2008).

Finally, other types of discrimination seen in the business world are nepotism and political discrimination. Nepotism, which is the behavior of hiring and even promoting an employee due to kinship, blood ties, family ties, friendship, and so on connections in an organization or business (Kawo & Torun, 2020: 54), is a common type of discrimination in the business world. Characteristics of the tourism industry and hotel businesses provide a suitable ground for nepotism practices, and as a result, factors such as job satisfaction and organizational commitment of the staff in the organization negatively affect (Pelit et al., 2015).

Political discrimination is discrimination against people's opinions, including political party membership and political, socio-political, and moral attitudes ILO (2009). In the labor market discrimination study conducted by Larja et al. (2012) in Finland, it was revealed that 4% of the employees participating in the study stated that they were exposed to political discrimination. Studies are showing the existence of political discrimination in the tourism industry as well. For example, Yeşiltaş et al. (2012), in their study of 407 employees working in four- and five-star hotel establishments in Ankara, found that the participants thought that there was little political discrimination in employee selection and in the business life of the organization.

Conclusion

In the tourism sector, services are provided by employees from different cultures, communities, and countries. However, the sector, which grows every year, is experiencing a problem of qualified personnel due to the labor shortage. For this reason, employees in my tourism industry are employed with problems such as low wages, low status, and lack of career development. This situation causes the industry to discriminate people. Despite the laws/practices against discrimination in almost all countries, it has not been possible to eliminate this problem and it continues to be a growing problem today. However, different types of discrimination in the tourism industry both in terms of tourists and in employment can be prevented with some practices and precautions. For example, employees working in tourism businesses can be trained on the issue of discrimination as an international problem, types of discrimination, and what behaviors can be understood as discrimination, so that discrimination against foreign tourists of different races and skin colors can be prevented. if we look from the viewpoint of employment, equal numbers of male and female employees in tourism businesses and based on merit in promotions will prevent gender discrimination, albeit a little. In addition, the equality of white and nonwhite personnel in terms of numbers can prevent color discrimination, employment of employees with different beliefs and providing the necessary convenience and opportunity for employees to live their beliefs can also prevent belief discrimination.

Consequently, the concept of discrimination and its implications in tourism are evaluated in this study. In this context, the concepts of discrimination and employment discrimination are explained, and the theories that make up the concept are included. In addition, anti-discrimination laws in leading countries such as America are mentioned. Finally, the study was concluded by mentioning the common issues of discrimination and their reflections in tourism.

References

Australian Human Rights Commission. (2012). *Disability discrimination-know your rights.* https://humanrights.gov.au/sites/default/files/content/pdf/disability_rights/dda_brochure.pdf

Banarjee, R. (2008). An examination of factors affecting perception of workplace discrimination. *Journal of Labor Research, 29,* 380–401. https://doi.org/10.1007/s12122-008-9047-0

Behera, K. R. (2018). Racial discrimination: A social injustice. In Dayal, D. (Ed.), *Complexion Based Discriminations: Global Insights.* Notion Press.

Benjamin, S. & Dillette, A. K. (2021). Black travel movement: Systemic racism informing tourism. *Annals of Tourism Research, 88,* 103–169. 10.1016/j.annals.2021.103169

Benokraities, N. V. & Feagin, J. R. (1995). *Modern Sexism: Blatant, Subtle and Covert Discrimination.* Prentice Hall.

Capelli, P. & Novelli, B. (2010). *Managing the Older Worker: How to Prepare for the New Organizational Order.* Harvard Business School Press.

Carr, D. & Friedman, M. A. (2005). Is obesity stigmatizing? Body weight, perceived discrimination, and psychological well-being in the United States. *Journal of Health and Social Behavior, 46,* 244–259. 10.1177/002214650504600303

Castro, A. B., Gee, G. C. & Takeuchi, D. T. (2008). Workplace discrimination and health among Filipinos in the United States. *American Journal of Public Health,* 98(3), 520–526. https://doi.org/10.2105/AJPH.2007.110163

CDC-Centers for Disease Control and Prevention. (2020, September 16). *Disability and health overview.* https://www.cdc.gov/ncbddd/disabilityandhealth/disability.html#:~:text=A%20disability%20is%20any%20condition, around%20them%20(participation%20restrictions.

Centeno, M. A. (2010). *Discrimination in an Unequal World.* Ed. Centeno, M. A. & Newman, K. S. Oxford University Press.

Cheung, H. K., King, E., Lindsey, A., Membere, A. Markell, H. & Kilcullen, M. (2016). Understanding and reducing workplace discrimination. *Research in Personnel and Human Resources Management, 34,* 1–93. https://doi.org/10.1108/S0742-730120160000034010

Correll, J., Judd, C. M., Park, B. & Wittenbrink, B. (2010). Measuring prejudice, stereotypes and discrimination. In Dovidio, J. F., Hewstone, M. Glick, P., & Esses, V. M. (Eds.), *The Sage Handbook of Prejudice, Stereotyping, and Discrimination.* Sage.

Council of Europe Commissioner for Human Rights (2011). *Discrimination on grounds of sexual orientation and gender identity in Europe.* Council of Europe Publishing. https://rm.coe.int/discrimination-on-grounds-of-sexual-orientation-and-gender-identity-in/16807b76e8

Cuddy, A. J. C., Fiske, S.T. & Glick, P. (2007). The BIAS map: Behaviors from intergroup affect and stereotypes. *Journal of Personality and Social Psychology, 92,* 631–648. https://doi.org/10.1037/0022-3514.92.4.631

Dalkıranoğlu, T. & Çetinel, F. G. (2008). A comparative analysis of female and male managers' approach to gender discrimination in hospitality businesses. *Dumlupınar Üniversitesi Sosyal Bilimler Dergisi,* 20, 277–298. https://dergipark.org.tr/tr/pub/dpusbe/issue/4762/65433

Deale, C. S. & Wilborn, L. R. (2006). Hospitality students and their stereotypes: A pilot study. *Journal of Hospitality & Tourism Education, 18*(1), 33–45. https://doi.org/10.1080/10963758.2006.10696848

Demir, M. (2011). Discrimination in the working life: A sample of tourism sector. *Journal of International Human Science, 8*(1), 762–784. https://www.j-humansciences.com/ojs/index.php/IJHS/article/view/1602

Demirbilek, S. (2007). Cinsiyet ayrımcılığın sosyolojik açıdan incelenmesi. *Finans Politik & Ekonomik Yorumlar Dergisi, 44*(511), 45–49. http://www.ekonomikyorumlar.com.tr/files/articles/152820003974_1.pdf

Dion, K. L. (2002). The social psychology of perceived prejudice and discrimination. *Canadian Psychology/Psychologie Canadienne, 43*(1), 1–10. https://doi.org/10.1037/h0086899

Dipboye, R. L. & Collella, A. (2005). An introduction. In Dipboye, R. L. & Collella, A. (Eds.), *Discrimination at Work: The Psychological and Organizational Bases* (pp. 1–10). Lawrence Erlbaum Associates.

DiTomaso, N., Post, C. & Parks-Yancy, R., (2007). Workforce diversity and inequality: Power, status, and numbers. *Annual Review of Sociology, 33*(1), 473–501. 10.1146/annurev.soc.33.040406.131805

Dovidio, J. F. & Gaertner, S. L. (2004). Aversive racism. In Zanna, M. P. (Ed.), *Advances in Experimental Social Psychology* (pp. 1–51). Academic Press.

Dovidio, J. F. & Hebl, M. R. (2005). Discrimination at the level of the individual: Cognitive and affective factors. In Dipboye, R. L. & Collella, A. (Eds.), *Discrimination at Work: The Psychological and Organizational Bases* (pp. 11–36). Lawrence Erlbaum Associates.

Duncan, C. & Loretto, W. (2004). Never the right age? Gender and age-based discrimination in employment. *Gender, Work and Organization, 11*(1), 95–115. https://doi.org/10.1111/j.1468-0432.2004.00222.x

EEOC. (2020, January 24). *EEOC releases fiscal year 2019 enforcement and litigation data.* https://www.eeoc.gov/newsroom/eeoc-releases-fiscal-year-2019-enforcement-and-litigation-data

EEOC. (2021). *Pregnancy discrimination.* https://www.eeoc.gov/pregnancy-discrimination

EEOC. (2021, April 2). *Disability discrimination.* https://www.eeoc.gov/disability-discrimination

EEOC. (2021, April 2). *The pregnancy discrimination act of 1978.* https://www.eeoc.gov/statutes/pregnancy-discrimination-act-1978

Elei, G. C. U. (2016). Effects of workplace discrimination on employee performance. *Texila International Journal of Management, 2*(2), 1–7. DOI: 10.21522/TIJMG.2015.02.02.Art012

Ghumman, S., Ryan, A. M., Barclay, L. A. & Markel, K. S. (2013). Religious discrimination in the workplace: A review and examination of current and future trends. *Journal of Business and Psychology, 28*, 439–454. https://doi.org/10.1007/s10869-013-9290-0

Glick, P. & Fiske, S. (1996). The ambivalent sexism inventory: Differentiating hostile and benevolent sexism. *Journal of Personality and Social Psychology, 70*(3), 491–512. https://doi.org/10.1037/0022-3514.70.3.491

Hamilton, D., Goldsmith, A. H. & Darity, W. J. (2009). Shedding "light" on marriage: The influence of skin shade on marriage for black females. *Journal of Economic Behavior & Organization, 72*, 30–50. https://doi.org/10.1016/j.jebo.2009.05.024

Hudson, S., So, K. K. F., Meng, F., Cárdenas, D. & Li, J. (2020). Racial discrimination in tourism: The case of African-American travellers in South Carolina. *Current Issues in Tourism, 23*(4), 438–451. https://doi.org/10.1080/13683500.2018.1516743

ILO. (2003, May 23). *ILO: Workplace discrimination, a picture of hope and concern.* https://www.ilo.org/global/-about-the-ilo/mission-and-objectives/features/WCMS_075613/lang--en/index.htm

ILO. (2007). *Equality at work: Tackling the challenges: Report of the director-general global report under the follow-up to the ILO declaration on fundamental principles and rights at work.* https://www.ilo.org/wcmsp5/groups/public/---dgreports/---dcomm/---webdev/documents/publication/wcms_082607.pdf

ILO. (2009). *Eliminating discrimination in the workplace.* http://www.oit.org/wcmsp5/groups/public/---ed_emp/---emp_ent/---multi/documents/publication/wcms_116342.pdf

Jones, T. (2000). Shades of brown: The law of skin color. *Duke Law Journal, 49*, 1487–1557. https://scholarship.law.duke.edu/faculty_scholarship/72

Jost, J. & Andrews, R. (2012). System justification theory. In Christie, D. J. (Ed.), *Encyclopedia of Peace Psychology* (Vol. II, 1092–1096). Wiley-Blackwell.

Kaas, L. & Manger, M. (2010). *Ethnic discrimination in Germany's labour market: A field experiment*, Discussion Paper Series. https://doi.org/10.1111/j.1468-0475.2011.00538.x

Kawo, J. W. & Torun, A. (2020). The relationship between nepotism and disengagement: The case of institutions in Ethiopia. *Journal of Management Marketing and Logistics*, 7(1), 53–65. http://doi.org/10.17261/Pressacademia.2020.1197

Larja, L., Warius, J. Sundback, L., Liebkind, K., Kandolin, I. & Jasinsnkaja-Lahti, I. (2012). *Discrimination in the Finish labor market. An overview and a field experiment on recruitment.* Employment and entrepreneurship. Publication of the Ministry of Employment. https://yhdenvertaisuus.fi/documents/5232670/5376058/Discrimination+in+the+Finnish+Labour+Market

Li, S., Li, G., Law, R. & Paradies, Y. (2020). Racism in tourism reviews. *Tourism Management, 80*, 104100. https://doi.org/10.1016/j.tourman.2020.104100

Lim, J. E. (2020). Understanding the discrimination experienced by customers with disabilities in the tourism and hospitality industry: The case of Seoul in South Korea. *Sustainability, 12*, 7328. *RePEc:gam:jsusta:v:12:y:2020:i:18:p:7328-:d:410031*

Low, K. C. P. & Ang, S. L. (2011). The case study: High time to quit? *Educational Research, 2*(8), 1330–1333. http://www.interesjournals.org/ER

Low, P. K. C. & Ang, S. L. (2013). Ageism. *Encyclopedia of Corporate Social Responsibility.* https://link.springer.com/referenceworkentry/10.1007%2F978-3-642-28036-8_423

Mäkelä, L.A. (2012). A narrative approach to pregnancy-related discrimination and leader–follower relationships. *Gender, Work and Organization, 19*(6), 677–698. https://doi.org/10.1111/j.1468-0432.2010.00544.x

Marginson, S., Nyland, C., Sawir, E. & Forbes-Mewett, H. (2010). *International Student Security.* Cambridge University Press.

Mcnair, S. & Flynn, M. (2007). *Managing an ageing workforce in hospitality: A report for employers.* https://webarchive.nationalarchives.gov.uk/20100623152202/http://www.dwp.gov.uk/docs/man-age-hospitality-v4.pdf

Muñoz-Bullón, F. (2009). The gap between male and female pay in Spanish tourism industry. *Tourism Management, 30*, 638–649. 10.1016/j.tourman.2008.11.007

O'Brien, K. S., Latner, J. D. & Halberstadt, J. (2008). Do antifat attitudes predict antifat behaviors? *Obesity 16*(2), 87–93. 10.1038/oby.2008.456
Parents@work. (2020). *Experiences of discrimination at the workplace based on parenthood summary of case studies conducted in Austria and Croatia.* https://parentsatwork.eu/wp-content/uploads/2020/05/EN_Summary-Case-Studies Parents@work_April2020-1.pdf
Parrillo, V. (2005). *Contemporary Social Problems* (6th ed.). Pearson Education, Inc.
Pelit, E., İstanbullu Dinçer, F. & Kılıç, İ. (2015). The effect of nepotism on organizational silence, alienation and commitment: A study on hotel employees in Turkey. *Journal of Management Research,* 7(4), 82–110. 10.5296/jmr.v7i4.7806
Perry, E. L. & Finkelstein, L. M. (1999). Toward a broader view of age discrimination in employment-related decisions: A joint consideration of organizational factors and cognitive processes. *Human Resource Management Review, 9*(1), 21–49. https://doi.org/10.1016/S1053-4822(99)00010-8
Puhl, R. M., Luedicke, J. & Grilo, C. M. (2014). Obesity bias in training: Attitudes, beliefs, and observations among advanced trainees in professional health disciplines. *Obesity, 22*(4), 1008–1015. 10.1002/oby.20637
Ramiah, A., Hewstone, M., Dovidio, J. F. & Penner, L. A. (2010). The social psychology of discrimination: Theory, measurement, and consequences. In Russell, H., Bond, L. & McGinnity, F. (Eds.), *Making Equality Count: Irish and International Approaches to Measuring Discrimination* (pp. 84–112). Liffey Press.
Reskin, B. (2000). The proximate causes of employment discrimination. *Contemporary Sociology, 29,* 319–328. https://doi.org/10.2307/2654387
Russell, H., Quinn, E. O'Riain, R. K. & McGinnity, F. (2008). *The Experience of Discrimination in Ireland: Analysis of the QNHS Equality Module.* Dublin: The Equality Authority and the Economic and Social Research Institute. https://www.esri.ie/system/files/media/file-uploads/2015-07/BKMNEXT120.pdf
Sayın, E. & Candan, H. (2016). The rising global racism. *Ardahan University Journal of Faculty of Economics and Administrative Sciences, 4,* 35–46. https://www.ardahan.edu.tr/iibfdergi/arsiv/4/03.pdf
Segovia-Perez, M., Figueroa-Domecq, C., Fuentes-Moraleda, L. & Munoz-Mazon, A. (2019). Incorporating a gender approach in the hospitality industry: Female executives' perceptions. *International Journal of Hospitality Management, 76,* 184–193. https://doi.org/10.1016/j.ijhm.2018.05.008
Shen, W. & Dhanani, L. (2015). Measuring and defining discrimination. In Colella, A. J. & King, E. B. (Eds.), *The Oxford Handbook of Workplace Discrimination.* Oxford University Press.
Shum, C., Gatling, A. & Garlington, J. (2020). All people are created equal? Racial discrimination and its impact on hospitality career satisfaction. *International Journal of Hospitality Management, 89,* 1–10. 10.1016/j.ijhm.2019.102407
Smith, S. (2003). Exploring the efficacy of African Americans' job referral networks. *Ethnic and Racial Studies, 26*(6), 1029–1045. 10.1080/0141987032000132478
Stewart, A., Niccolai, S. & Hoskyns, C. (2011). Disability discrimination by association: A case of the double yes? *Social & Legal Studies, 20*(2), 173–190. http://dx.doi.org/10.1177/0964663910391519
Sullivan, T.J. (2003). *Introduction to Social Problems* (6th ed.). Pearson Education.
Surgevil, O. & Mayaturk-Akyol, E. (2011). Discrimination against people living with HIV/AIDS in the workplace: Turkey context. *Equality Diversity and Inclusion: An International Journal 30*(6), 463–481. https://doi.org/10.1108/02610151111157693
Tükeltürk, A. Ş. & Perçin, N. Ş. (2008). Turizm sektöründe kadın çalışanların karşılaştıkları kariyer engelleri ve cam tavan sendromu: cam tavanı kırmaya yönelik stratejiler. *Yönetim Bilimleri Dergisi, 6*(2), 113–128. https://dergipark.org.tr/tr/pub/comuybd/issue/4111/54125
United Nations. (2018, April 26). *Convention on the rights of persons with disabilities- committee on the rights of persons with disabilities.* https://digitallibrary.un.org/record/1626976
WAFIQ. (2018). *Religious discrimination: Banning of Hijab among hotel employees.* https://wafiq.my/2018/01/09/religious-discrimination-banning-of-hijab-among-hotel-employees/
Weisskopf, T. E. (2010). Reflections on globalization, discrimination and affirmative action. In Centeno, M. A. & Newman, K. S. (Eds.), *Discrimination in an Unequal World* (pp. 23–44). Oxford University Press.
Weller, P. (2011). Religious discrimination in Britain: A review of research evidence, 2000–10. Equality and Human Rights Commission Research Report, 73. https://www.equalityhumanrights.com/sites/default/files/research_report_73_religious_discrimination.pdf
Wen, H. & Madera, J. M. (2013). Perceptions of hospitality careers among ethnic minority students. *Journal of Hospitality Leisure Sport & Tourism Education, 13,* 161–167. http://dx.doi.org/10.1016/j.jhlste.2013.09.003
WHO. (2001). *Transforminghealthsystems: Genderandrightsinreproductivehealth.* http://apps.who.int/iris/bitstream/handle/10665/67233/WHO_RHR_01.29.pdf;jsessionid=F8766F9814A398C80E215C169B8BCC40?sequence=1

Yeşiltaş, M., Arslan, Ö. E. & Temizkan, R. (2012). Political discrimination in the process of employee recruitment and work life in the organization: A study at hotel establishments. *Journal of Business Research-Turk, 4*(1), 94–117. https://www.isarder.org/isardercom/2012vol4Issue1/Vol.4_Issue.1-06_full_text.pdf
Ziegert, J. C. & Hanges, P. J. (2005). Employment discrimination: The role of implicit attitudes, motivation, and a climate for racial bias. *Journal of Applied Psychology, 90*(3), 553–562. https://doi.org/10.1037/0021-9010.90.3.553

CONCLUSIONS

Tourism and Social Psychology

Dogan Gursoy and Sedat Çelik

This concluding chapter draws together the issues and concepts discussed in this book to present an overview of the issues and concepts in tourism and social psychology. This chapter also summarizes the issues and aspects highlighted, as well as conclusions formulated by authors in chapters, and provides suggestions and recommendations for hospitality and tourism scholars, practitioners, and destination and business managers to enable them to successfully utilize the concepts discussed in this book.

Considering the fact tourism involves close interactions among tourists, locals, businesses, and other stakeholders, interactions and communications among individuals who speak different languages, belong to different religions and races can have significant impact on tourism experience formation and determine the quality and satisfaction perceptions (Chen, Han, Bilgihan & Okumus, 2021). The consequences of those interactions and communications on tourists' attitudes and behaviors suggest that both tourism scholars and practitioners need to pay a close attention to social and psychological causes and consequences of tourists' attitudes and behaviors (Dewnarain, Ramkissoon & Mavondo, 2021). As a result, over the past several decades, tourism research has witnessed a tremendous growth in the amount of information and knowledge generated by both academics and practitioners (Gursoy, Ouyang, Nunkoo & Wei, 2019). The main reason for this tremendous growth is that researchers and practitioners have been trying to better understand decision-making processes, experience formation and evaluation processes and the factors that may influence these processes (Chi, Wen & Ouyang, 2020). This understanding is even likely to become more critical as the competition among destinations becomes fiercer since many destinations heavily rely heavily on tourism earnings for their economies (Sota, Chaudhry & Srivastava, 2020).

There is no question that there is strong relationship between the tourism field and many disciplines, and tourism scholarship has been significantly influenced by the theories and concepts developed in those disciplines. One of those disciplines that have significantly influenced tourism research over the years is the social psychology. As the tourism industry reaches its maturity and scientific sophistication, it is important that we as researchers and practitioners fully understand the breadth and depth of the influence of social psychology on travelers' experience formation, satisfaction, and loyalty behaviors. This understanding can help us better explain, understand, monitor, and predict factors that can influence travelers' experience formation, satisfaction, and loyalty behaviors. Furthermore, considering the fact that today's travelers are in search of very complex "multi-faceted" and "hybrid" experiences that can help them have unique travel experiences, understanding the effects of social psychological factors on their decision making, planning,

DOI: 10.4324/9781003161868-28

and experiencing hospitality and tourism products and services is a focal challenge for scholars and destination managers in a globalized and highly competitive environment.

Creation of compelling customer experiences requires strong understanding of social psychology since experiences are "deconstructed" products due to the hybrid nature of travel experiences (Horng & Hsu, 2021). Because of this fragmented nature of tourism experiences, scholars and practitioners face various issues and challenges in creating satisfactory experiences. Furthermore, the COVID-19 pandemic that wreaked havoc in the tourism industry made a thorough understanding of social psychology and its impact on developing and delivering tourist experiences, satisfaction, and loyalty behaviors a critical issue for both scholars and practitioners (Atadil & Lu, 2021; Gursoy & Chi, 2020).

These issues and challenges can prove vital for the success of any destination regardless of its product offerings, location, size, and target markets. Ignoring these social psychological issues and challenges can have detrimental effects.

While a few books were published on the subject in the past (Bochner, 1982; Pearce, 1982; Filep & Pearce, 2014; Filep et al., 2017), most of them were published in an earlier time when tourism was still at infancy or development stage of its life cycle. Since then, number of people traveling and the number of destinations have grown significantly. These changes require a new examination of tourism and its impact from a social psychological perspective. Thus, this handbook carefully examines social psychology and tourism issues and challenges that are raised in the contemporary tourism literature and faced by tourism scholars and destination managers in their everyday operations. Defining key social psychology concepts and issues and exploring the type of impacts they may have on travelers' experience formation, satisfaction, and loyalty behaviors is critical for both scholars and practitioners. Furthermore, this exploration can provide critical insights for successful development and implementation of tourist experiences.

The purpose of this handbook is to discuss and analyze the issues related to tourism and travelers' experience formation, satisfaction, and loyalty behaviors from the social psychology perspective. This handbook offers insights into social psychology issues, principles and practices that influence travelers' experiences, satisfaction, and loyalty behaviors. Contributors to this handbook who are experts in the field have employed various methodologies in their contributions, mainly conceptual approaches, literature reviews, and qualitative studies. We discuss the main issues, important aspects, and challenges discussed and highlighted by various contributors in their chapters in the next section.

Chapter 1: Psychology, Sociology, and Social Psychology

In this chapter, the relationship between sociology, psychology, and social psychology is examined. Areas of social psychology are addressed by comparing three main social psychology approaches by focusing on the main concepts and problems the discipline focused on, perspectives, methodology, levels of analysis, and interpretation of the results.

Chapter 2: The Relationship between Tourism and Sociology, Psychology, and Social Psychology

This chapter presents the relationship between tourism and its human side; sociology, psychology, and social psychology that are fundamentally interrelated to shed light on some overlooked characteristics of tourism in conceptual and practical terms. Although the relationship between tourism and social psychology is meaningful, it has not received sufficient attention from tourism scholars. This paper sets the historical context of travel and tourism before examining the importance of the relationship between tourism and sociology, psychology, and social psychology.

The psychological and sociological foundations of tourism and its inseparable connections with social psychology are emphasized to develop a comprehensive understanding of the tourism phenomenon beyond the oversize hats, photographs taken, number of international arrivals, and tourism revenues

Chapter 3: The Effects of Tourism: Economic, Environmental, Social, Cultural, Social Psychological

In this chapter, studies that investigated the economic, social, cultural, environmental, and social psychological effects of tourism over the years are examined. Tourism is generally defined as multifaceted since it requires economic, social, cultural, and environmental inputs. Hence, the effects of tourism are discussed under the headings of social, cultural, economic, and environmental since tourism is not only an economic event but also has social, political, environmental, cultural, and social psychological effects. When the studies on the effects of tourism in the literature are examined, it is seen that most of the studies conducted until 1980 focused on the positive economic effects of tourism, while the studies conducted after 1980 began to focus on the environmental, social, cultural, and social-psychological effects of tourism. However, the number of studies focusing on tourism's psychological and social psychological effects is limited.

Chapter 4: Tourism, Prejudice, Stereotypes, and Personal Contact: Gordon Allport's Contributions to Tourism Research

Unfortunately, tourism scholars have not paid much attention to the problem of racism in tourism and hospitality. In fact, most scholars have argued tourism promotes peace and political stability by reducing the possibilities of inter-ethnic or inter-class conflicts. In this chapter, authors have shown precisely the opposite. Prejudice and racism take a hidden form in service sectors where the hosts are subordinate to guest´s desires. The classic racism, which was based on so-called biological differences, set the pace to a subtle cultural racism camouflaged as cultural diversity. This chapter puts the contributions of Gordon Allport and the social contact theory to the forefront while laying the foundations towards a new radical understanding of racism as well as the hostilities to foreign tourists.

Chapter 5: Mere Exposure Effect and Tourism Relationship

There are many objects or stimuli in daily life, at home, at school, on the way to work, on TV broadcasts, social media, and environment. Over time, people begin to feel familiar with these stimuli they see constantly and begin to recognize them. People also evaluate these familiar stimuli constantly. In this evaluation process, positive approaches emerge in liking, trusting, preferring, and purchasing compared to other stimuli. Visitors who travel outside of their permanent residence also encounter similar situations. When a tourist who goes to the hotel is constantly exposed to similar local dishes, s/he becomes unfamiliar with local dishes. After a while, the local dishes he/she is exposed to begin to look familiar compared to other local dishes.

Chapter 6: Social Exchange Theory and Tourism

In this chapter, social exchange theory (SET) whose roots date back to Homans (1958) and Blau (1964) is reviewed through tourism perspective. SET has been widely used to explain residents' perceptions towards tourism or tourism development. SET suggests that possible costs or benefits obtained from this exchange play important role in residents' evaluation of this exchange process.

From a tourism perspective, individuals who perceive the benefits of exchange, which means interaction between residents and tourists, tend to consider that exchange positive. Thus, residents who benefit from tourism positively consider tourism development. The main factors that affect residents' support towards tourism divided into three groups: economic, social, and environmental. Tourism studies reveal that SET is still the dominant theory explaining residents' attitudes towards tourism and its impacts. However, the widespread use of SET in tourism research also increases the amount of criticism of SET. Whereas previous research based on SET have offered an appropriate theoretical framework to explain local people's attitude towards tourism, recent studies highlight that SET does not have the "theoretical precision" that is needed to explain support for tourism. Thus, researchers have begun to use other theories and conceptual framework to better explain support for tourism development.

Chapter 7: Social Representation Theory and Tourism

The argument about how the social context affects behavior has been an important research question generator in the axis of social psychology. The interaction of individuals and groups create an extensive network of relations, and various common identities and realities may be formed in this pattern. Social representation theory, developed by Serge Moscovici in 1961, refers to the values and figuration of images created by social groups. According to this theory, some collective sense-making processes function in the interaction of individuals and groups, and people hold a view about the social world they live in by building a common sense during this process. Tourism, as a social phenomenon, gives room for individual and social interactions, and creates a productive interrelation field for social research by bringing various tourist groups and host communities together. This chapter attempts to respond to questions about what Social Representation theory is, how representations are formed, and the standing of SR theory in the context of tourism.

Chapter 8: Travel Career Pattern Theory of Motivation

Travel career pattern is an approach to understanding the travel motives of tourists. The concept was originated from travel career ladder developed by Philip L. Pearce, which denotes five different levels of travel motivation. Travel career pattern framework comprises 14 factors that are divided into three layers: the core motives, the middle motives, and the outer motives. The core motives that include novelty-seeking, escaping/relaxing, and relationship building are viewed as very important for travelers regardless of their travel experiences and life-stages. The other motives consist of nature, self-development (host-site involvement), kinship, self-actualization, self-enhancement, stimulation, isolation, nostalgia, autonomy, social status, and romance. Various studies in different settings show that the travel career pattern approach can be applied to a cross-cultural context. This chapter aims to explore the concept of travel career pattern and its usage. More specifically, the first aim is to address the elaboration of travel career pattern theory of motivation, while the second purpose is to depict the use of travel career pattern in tourism research.

Chapter 9: Social Comparison Theory and Tourism

This chapter discusses the importance of social comparison theory in tourist behavior. Since the 1930s, there have been great changes in perspectives on human psychology. With the theory of social comparisons put forward by Festinger, important developments took place in interpreting individual and group changes in human behavior. Subsequent studies made contributions to the development of this theory. Today, making sense of tourist behavior and creating appropriate marketing strategies is one of the most important competitive strategies. In this context, it is important to reveal social comparisons within understanding tourist behavior and tourism marketing.

Chapter 10: Hotel Corporate Social Responsibility (CSR) May Not Always Lead to Positive Outcomes: The Role of Attributions about Motives Behind CSR Initiatives

This chapter aims to discuss three key stakeholders' (employees, customers, and local people) reactions to hotel companies' CSR practices through the lens of attribution theory. Corporate social responsibility practices have become common among hotel companies regardless of their size to gain and force their competitive advantage. This is because some key stakeholders such as customers, employees, and even local people positively respond to these initiatives. To build a good corporate reputation and gain stakeholders' positive responses, hotel companies are engaged in CSR practices and communicate their activities to the stakeholders. However, recent empirical studies indicated that stakeholders may negatively react to CSR if they attribute the motives behind these initiatives as egoistic or self-serving. These findings suggest that it is not easy for companies to reach positive reactions from their stakeholders through social and environmental activities. That created some key questions for companies to respond to while formulating and communicating their initiatives to stakeholders.

Chapter 11: Attitudes in Tourism and Traveling as a Tool/Instrument for Attitude Change

This chapter reviews the studies on the attitudes of host community and tourists to determine whether tourism plays a mediating role in individuals' attitude change and reveal how this issue is approached in current research. Attitudes can be defined as general evaluations of the objects, ideas, and people that an individual encounters in his or her life or, a cognitive process involving positive or negative evaluations of an individual towards a specific object. Attitudes guide the thoughts, behaviors, and even feelings of an individual. In other words, this process can also affect the individual's behavior towards an object. Attitude change refers to the change in the direction and the intensity of the individual's attitude towards an object of attitude. Due to the functional value of attitudes, processes that change attitudes have always been an important topic of social psychology. Travel can be considered as one of these processes. The problem of whether tourism and travel are mediators of attitude change is of great importance in current social psychology research.

Chapter 12: Explaining Intergroup and Intragroup Dynamics in Tourism: A Social Identity Approach

This chapter provides a useful theoretical framework for understanding the intergroup and intragroup dynamics in the tourism ecosystem and offers directions for advancing the tourist/resident behavior research agenda. In international travel, tourists are often required to interact with other tourists and residents of the destination, whose cultures, norms and values might diverge substantially from their own. Intergroup interactions, therefore, represent an integral part of tourism experience and are vital for the sustainable development of a destination. Through the social identity lens, this chapter discusses the psychological mechanisms underlying individuals' intergroup and intragroup behaviors: the construal of ingroup and outgroup in the tourism context, ingroup identification, ingroup bias, intergroup attribution, the black sheep effect, and appraisal of positive/negative intergroup contact. Specifically, the discussion highlights how individuals (e.g., tourists) are likely to respond when facing an ingroup wrongdoing (e.g., misbehavior by compatriot tourists) and threat from an outgroup (e.g., a conflict with residents).

Chapter 13: Travel and Transformation: An Examination of Tourists' Attitude Changes

This chapter adopts a transformative tourism framework to examine tourist transformation processes. This transformation can occur at three levels: existential, cognitive, and behavioral. Travel is associated with transformative benefits that can lead to changes in the attitudes in its participants. While advances have been made to understand and measure attitude change as a proxy for tourist loyalty towards a destination, the novelty of this chapter lies in applying a transformative tourism framework to dissect attitude change as a multifaceted outcome of travel. The tourist transformation model (Pung et al., 2020a) and the transformative travel experience scale (Soulard et al., 2020) were developed to conceptualize and measure the process and outcomes of transformative travel.

Chapter 14: Does Tourism Impact on Prejudice, Discrimination, Assimilation, Genocide, Segregation, Integration?

Travel and tourism prepare an environment for intercultural interaction processes. Due to the different cultural characteristics of tourists from different nationalities, their attitudes towards each other and the results of these attitudes, and the threats they perceive from the outgroup can be determined within the framework of the integrated threat theory. It is possible that different behaviors and attitudes will emerge in the travel and tourism sector, where cultural differences are intensely observed. Tourists of different nationalities and cultures, working together and spending time in common areas, reflect the traces of many personal and cultural factors while meeting their needs. In this context, in addition to demographic factors such as age, gender, marital status as well as psychological factors such as motivation, perception, personality structure, learning, factors of the cultural dimension of nationality, which belong to individual factors, may allow the emergence of a cultural differentiation and a different intergroup contact. For this reason, the traditions, social values, attitudes, and behaviors of each nationality should be examined first. Thus, the different images that tourists have in each other may cause them to approach with subjective evaluations by moving away from an objective approach. Integrated threat theory is important for intercultural research, as it helps to explore social issues and to understand the cognitive processes behind people's attitudes, beliefs, and behaviors. Some of the applicable research topics of the integrated threat theory are, as expressed by Redmond (2009: 3), "religious intolerance", "public attitudes towards immigration", "racial profiling and stereotyping", "public attitudes towards same gender relationships", "support for feminist movements", "diversity" and "national identity", and "different motives in the workplace". In this sense, this chapter aims to examine the question of "Does traveling/tourism impact on prejudice, discrimination, assimilation, genocide, segregation, and integration?" And if the answer is yes, it is aimed to determine the answer of the question of, "How does traveling/tourism impact on prejudice, discrimination, assimilation, genocide, segregation, and integration?"

Chapter 15: Re-examining the Tourism and Peace Nexus: A Social Network Theory Perspective

There are mixed views regarding the relationship between tourism and peace. It is perceived by some researchers that tourism is a force for creating peace and by others that peace is a mechanism to generate tourism. This chapter seeks to re-examine the relationship between tourism and peace from a deductive reasoning approach. To provide a broader understanding of both concepts, the chapter analyzes the model of peace tourism as well the contribution of both tourism and peace to sustainable development. The deduction is that tourism, being a global activity, can generate peace and peace is a catalyst for tourism and its development. Both tourism and peace play vital roles in the global mission of sustainable development. In employing the social network theory to explain the relationship between tourism and peace, the chapter proposes tourism nodes and tourism ties that are critical in tourism networks to attain peace and for peace to generate tourism. The study will be of importance to tourism stakeholders and scholars.

Chapter 16: What Influences Attitude Change? Tourist Satisfaction, Motivation, Personality, Tolerance Level, Contact Situation (Level, Type, Frequency)

Many factors are affecting the attitudes of tourists, who are the main components of the tourism sector. Both the changes in the attitudes of the tourists over time and the changes in marketing activities oblige the stakeholders of the sector to take measures. As argued in the literature, personality, satisfaction levels, motivations of the tourists, and the frequency of interaction with both tourism employees and other tourism stakeholders are the main factors that affect the sector. One of the most important factors that help explain tourist behavior is motivation. Tourist motivation can be the sum of biological and socio-cultural forces that energize and generate people's behavior. Tourist motivation emerges as a very important variable and driving force behind the behaviors of tourists. While it is easy to answer the questions of who, where, how, and when in tourism, it is very difficult to answer the question of why tourists visit. Thus, understating people's motivations for visiting a destination is crucial. The differences between tourists due to their personality can influence their motivations and their level of satisfaction with the service they receive. Satisfaction creates positive feelings towards any brand, product, or service and as a result, this positive feeling leads to the brand, product, or service being repurchased or recommended to others whereas dissatisfaction can cause negative attitudes, negative announcements, and not being preferred. This point emerges as an important factor affecting the attitudes of tourists towards the tourism employees and the service or product they receive. Another important phenomenon underlying the principle of satisfaction and motivation in tourists is the contact situation (level, type, and frequency) impact. From this point of view, in this chapter, the relations between the mentioned topics are discussed and the effects of these issues on the attitude changes of tourists are explained with examples based on the literature.

Chapter 17: Social/Cultural Distance and Its Reflections on Tourism

Developments in the tourism industry increase the flow of people with different social and cultural backgrounds to tourism destinations, and different groups such as domestic and foreign tourists, local people and resident foreigners, and tourism workers and tourism enterprises interact in touristic destinations. Thus, they can have positive and negative social and cultural interactions. This situation affects cultural and social distances between people. Hofstede (1980) defines cultural distance as the degree to which norms and values in one country differ from norms in another country. Social distance is the degree of understanding that exists between individuals, between groups, and between a person and groups (Bogardus, 1940). In this chapter, social and cultural distance is examined within the tourism in the context of Hofstede's theory of cultural dimensions and social distance scale developed by Bogardus, as it is one of the most widely used classifications in the relevant literature.

Chapter 18: Inbound Tourism and Alteration in Social Culture, Norms, and Community Attitudes in the Tourism Industry: The South Asian Experiences

The interaction between host and tourists is a natural phenomenon that causes an exchange of values, norms, and cultural practices, leading to a change in the social atmosphere of a tourist destination. With the continuous development of tourism industry, it has been observed that traditional social cultures have changed vividly in many tourist destinations around the world. These changes include the impact of tourism development on inhabitants' traditional values, lifestyles, and interpersonal relationships both in rural and urban localities. However, sometimes the mixing of rich foreign tourists and relatively poor local residents can increase the likelihood of a community backlash against tourists and tourism development. Therefore, it is extremely important to scrutinize the host residents' perception of tourists and their attitude towards them as the residents of many tourist destinations are the fundamental element of the tourism "product" and have a sizable impact on its sustainability. In this chapter, we develop a conceptual model to explain the vicissitudes in social cultures, values, norms, and attitudes in the emerging South Asian countries. Based on the discussion, we aim to provide policy advice for the policymakers to prepare guidelines for integrated sustainable tourism management and development in the South Asian region.

Chapter 19: Culture Shock Experiences of Tourists: A Transformative Perspective

Curiosity in culture shock and its outcomes has increased, but there has not yet been a clear comprehension of the transformative process of this phenomenon. This study attempts to clarify the culture shock experiences of tourists from the perspective of transformative learning theory and to explore related permanent changes in tourists' attitudes, behaviors, beliefs, and feelings. Through a hermeneutic approach, this study reviews and discusses the main components of tourists' culture shock and its transformative process. This study proposes that culture shock experienced at the tourist destination triggers transformation. Transformative process of the culture shock of tourists is modeled in six steps, by illustrating with a model: fancy, disorienting dilemma, self-evaluation, liminality, coping strategies, and transformation. This proposed model presents a conceptual foundation for further investigation and is proper for planning tourists' transformative culture shock experiences.

Chapter 20: Tourist-to-Tourist Interaction (TTI): A Social Distance Perspective

The basis of tourism is the interactions of tourist-local community, tourist-service personnel, and tourist-tourist. A critical issue for the tourism literature is to understand the effects of these encounters. Social psychology, which draws attention to the social context of the encounters in question and the social factors affecting them, is a discipline that has been applied in the field of tourism since the 1980s. The concept of social distance, which was put forward by Robert Park in sociology and developed by Emory Bogardus, is of great importance in understanding the attitude changes of tourists within the scope of social psychology. The concept of social distance is the perceived affinity and mutual sympathy between different groups and describes the extent to which these groups are willing to interact with each other. Most of the pioneering studies were used social distancing to understand the interaction between the local community and tourists. Fundamentally, tourist-tourist interaction has always been the subject of less study than local community-tourist interaction. However, it is seen that tourist-tourist interaction (TTI) is the subject of studies in areas such as cruising, backpacking, group tours, and festivals. This chapter aims to explain tourists' encounters with tourists from a social psychological perspective and explain the possible positive-negative effects of the mentioned encounters on the social distance between tourists.

Chapter 21: Value Typology in the Context of the Tourism Sector

The value philosophy, which has been put forward in parallel with the developments in marketing science, has been discussed by many researchers since the 1950s. In addition, the relationship between businesses and customers has been explained. It can be said that Holbrook (1994, 1996, 2002) contributed greatly to the in-depth examination of the theory of value and its incorporation into marketing literature. Topics such as the processing of the value theme, its importance, antecedents, and methods of use in businesses have been discussed in many studies by the author. The importance of the tourism sector in the service industry is increasing day by day. Considering that it has contributed greatly to the development of countries in the world not only in sociocultural terms but also in economic terms, the tourism sector is an important sector for almost every country. Value typology is an important point to understand the service industry. However, there has never been any research on value typology in the tourism sector when the literature is examined. Therefore, value perceptions that lead the tourists to develop attitude-behavior is explained in this study. Value typology presented in the literature by Holbrook (2002) is used to reveal the value perceptions of tourists. Within the scope of the research, sectorial examples are included in each typology. In addition, the value typology is detailed by examining different perspectives depending on the different conditions of the sector.

Chapter 22: Impact of Overtourism on Residents

Recent discussions on overtourism (OT) have demonstrated tourisms' negative effects on residents of frequently visited destinations. In many places such as Barcelona, Venice, or Palma de Mallorca, the masses of tourists led to resistance among the residents of visited destinations. The reactions involve more than just irritation, as highlighted in Doxey's well-known irritation index model (1975). In more and more places, rather small and spontaneous neighborhood actions have developed into (international) coordinated social movements demanding substantiated changes as far as turning away from the dominating growth-driven capitalist system. Those demands towards degrowth demonstrate the politicizing effect of OT and suggest that OT may even have transformative effects on residents. To study these effects, Mezirow's theory of transformative learning may be a useful framework. The chapter suggests that investigating the impact of OT on residents is highly relevant and that further research should be done on the transformative power of tourism.

Chapter 23: The Dyadic Influence of Personal and Cultural Factors on Tourism and Hospitality

As tourism involves many social interactions of individuals, both as customers and service providers, with groups, a study of tourism and hospitality from the lens of social psychology is extremely important. As tourism and hospitality covers intense and continuous social contact and interaction, and especially increasingly in multicultural service environments, a study of personal and social factors is important for understanding human behavior in tourism and hospitality, both from the perspectives of customers and service providers. By analyzing aspects of psychology (personal characteristics such as gender and personality) and sociology (social characteristics such as culture) this chapter offers a framework for understanding behavior in tourism and hospitality.

Chapter 24: Social-Psychological Issues in Tourism Business

Tourism is a multidimensional and complex phenomenon. During tourism activities, there is an ongoing communication network with non-profit and profit-seeking institutions/organizations.

Relations between tourists, employees, and local people and the results of these relations are evaluated within the scope of social psychology. The fact that each segment has different characteristics within the specified groups increases the complexity. Which tourist? Which employee? Which people? According to demographic and cultural features, the answers to the questions indicate how difficult it is to measure in this regard. For example, the communication between a tourist group classified according to their nationality and the employees of different cultures and local people grouped according to their income from tourism is far from general evaluations. An interdisciplinary approach is possible to achieve positive/negative outcomes of this communication and interactions, which will bring very different results. Some theories and concepts are used to explain these relationships. This chapter discusses social psychology in the tourism industry from the employee's perspective.

Chapter 25: Attitudes and Behaviors of Tourism Employees at Work and among Co-Workers

Tourism is a labor-intensive industry that employs a diverse, multicultural workforce. Due to seasonality, tourism activity is uneven throughout the year and frequently requires irregular working hours and changing shifts. Unforeseen contingencies can also affect service delivery. Service performance – and thus customer satisfaction – are not only conditioned by employees' attributes but also highly dependent on teamwork effectiveness, cohesion, and collaboration. Diverse factors can thus affect employees' attitudes, behaviors, and performance.

This chapter first identifies the unique features of working in the tourism industry. It then discusses group dynamics and social identity construction in different subsectors. Next, the chapter explains how specific difficulties and situations at work can generate tensions among employees. The importance of leadership and social reinforcement is also addressed. The final section provides further recommendations for tourism practitioners.

Chapter 26: Attitudes (Stereotype and Prejudice) of Local People towards Seasonal Tourism Workers

In the economic growth of many regions worldwide, tourism plays a significant role; however, seasonality tends to be the leading factor in employment in the tourism industry. Migration towards seasonal destinations happens internally and externally and enables seasonal employees to move and travel around to work in the destinations in high demand. Many tourism destinations are characterized by high seasonality and small capacity of local employees to ensure demand in high season, leading to a large share of seasonal and mostly foreign employment in the industry. The chapter reviews related literature to explore the relationships between the stereotypical and prejudicial attitudes of local people towards seasonal tourism workers. The findings show that the literature on contemporary immigrant-host relations has been largely explored in the last decades. Literature suggests that due to the work conditions seasonal employees accept (low paid and casual work, no social insurance, etc.), they have been suspected of prejudice and stereotypes, where local people relegate those employees in the category of "others", seeing them as a threat rather than a benefit. Circumstances in the job market, stereotyping and prejudice rendered migrant employees more vulnerable than local ones, especially where migrant workers have little to no experience or training. There is an increased risk of exploitation. This chapter provides theoretical insights and valuable, practical implications for managing stereotypes and prejudice of local people towards seasonal tourism workers.

Chapter 27: Social-Psychological Background of Discrimination and Its Reflections on Tourism

Discrimination can arise from various factors such as age, gender, race, ethnic origin, sexual orientation, physical appearance, and social class. As in other parts of society, discrimination is common in modern workplaces even in developed countries as well as in developing countries. Thus, workplace discrimination is a persistent problem all over the world although precautions and legislation are designed to prevent discrimination practices. One of the workplaces where discrimination has been adversely observed is tourism industry. In tourism industry, it is strongly possible to see employees from different cultures, societies, and countries. The growing industry-tourism worsens the skilled labor because of number of job openings in the sector. In this context, this chapter first covers social-psychological background of discrimination and theories concerning the subject. Second, it examines the level of discrimination such as age, gender, race, ethnicity, skin color, nationality, mental or physical disability, and other discriminations such as nepotism. Third, it evaluates the consequences of discrimination by focusing of tourism perspective.

This handbook includes a number of chapters that focus on the issues and aspects related to tourism and social psychology. Each chapter formulates specific research questions and provides suggestions and recommendations for hospitality and tourism scholars, practitioners, and destination and business managers to enable them to successfully utilize the concepts discussed in this book. It is worth noting that there is no magic recipe to guarantee the successful creation and delivery of satisfactory tourism experiences. All involved actors, stakeholders, planners, managers, and marketers must be aware of the challenges, obstacles, and difficulties in creating and delivering a satisfactory tourism experience. They have to devote energy and resources to surmount them and achieve a successful partnership with all stakeholders in for successful creation and delivery of satisfactory experiences (Wu, Gursoy & Zhang, 2021). It is quite clear that the challenges, problems, and opportunities will continue to evolve as all tourism destinations and businesses strive for delivering better and improved experiences.

In creation and delivery of satisfactory tourism experiences, understanding tourists, employees, and other stakeholders' needs and wants, and collaboration among all stakeholders is an imperative and crucial for successful creation and delivery of tourism experiences. Thus, investing in understanding needs and wants of all stakeholders is vital for development of successful creation and delivery of tourism experiences at destination and business level. All actors involved, at any level and field, must utilize the adequate strategies and approaches to deliver satisfactory experiences to tourists while keeping all other stakeholders happy.

We hope that this book will generate great interest and discussion of tourism and social psychology and provide a foundation for a much greater research contribution from both scholars and business practitioners.

References

Atadil, H. A., & Lu, Q. (2021). An investigation of underlying dimensions of customers' perceptions of a safe hotel in the COVID-19 era: Effects of those perceptions on hotel selection behavior. *Journal of Hospitality Marketing & Management*, *30*(6), 655–672.

Blau, P. M. (1964). *Exchange and Power in Social Life*. New York: John Wiley and Sons. https://doi.org/10.4324/9780203792643

Bochner, S. (1982). *Cultures in Contact*. Pergamon Press.

Bogardus, E. S. (1940). Scales in social research. *Sociology and Social Research*, *24*, 69–75.

Chen, S., Han, X., Bilgihan, A., & Okumus, F. (2021). Customer engagement research in hospitality and tourism: A systematic review. *Journal of Hospitality Marketing & Management*, *30*(7), 871–904.

Chi, C. G. Q., Wen, B., & Ouyang, Z. (2020). Developing relationship quality in economy hotels: The role of perceived justice, service quality, and commercial friendship. *Journal of Hospitality Marketing & Management, 29*(8), 1027–1051.

Dewnarain, S., Ramkissoon, H., & Mavondo, F. (2021). Social customer relationship management: A customer perspective. *Journal of Hospitality Marketing & Management, 30*(6), 673–698.

Doxey, G. V. (1975). A Causation theory of visitor–resident irritants, methodology and research inferences: The impact of tourism. *Travel Research Association, 6th Annual Conference Proceedings*, 195–198.

Filep, S., Laing, J., & Csikszentmihalyi, M. (eds.). (2017). *Positive Tourism.* Routledge, Taylor & Francis Group.

Filep, S., & Pearce, F. (2014). *Tourist Experience and Fulfilment: Insights from Positive Psychology.* Routledge.

Gursoy, D., & Chi, C. G. (2020). Effects of COVID-19 pandemic on hospitality industry: Review of the current situations and a research agenda. *Journal of Hospitality Marketing & Management, 30*(5), 527–529.

Gursoy, D., Ouyang, Z., Nunkoo, R., & Wei, W. (2019). Residents' impact perceptions of and attitudes towards tourism development: A meta-analysis. *Journal of Hospitality Marketing & Management, 28*(3), 306–333.

Hofstede, G. (1980). Culture and organizations. *International Studies of Management & Organization, 10*(4), 15–41. doi:10.1080/00208825.1980.11656300.

Holbrook, M. B. (1994). The nature of customer value: An axiology of services in the consumption experience. In R. T. (ed.), *Service Quality: New Directions in Theory and Practice* (pp. 21–71). Sage Publications.

Holbrook, M. B. (1996). Customer value-a framework for analysis and research. In K. P. Corfman and J. G. Lynch (eds.), *Association for Consumer Research-23 ed.*(pp. 138–142). *Provo,* UT.

Holbrook, M. B. (2002). *Introduction to Consumer Value in Consumer Value.* UK-London: Routledge.

Homans, G. C. (1958). Social behavior as exchange. *American Journal of Sociology, 63*(6), 597–606.

Horng, J. S., & Hsu, H. (2021). Esthetic dining experience: The relations among aesthetic stimulation, pleasantness, memorable experience, and behavioral intentions. *Journal of Hospitality Marketing & Management, 30*(4), 419–437.

Pearce, P. L. (1982). *The Social Psychology of Tourist Behaviour.* Oxford: Pergamon.

Pung, J. M., Gnoth, J., & Del Chiappa, G. (2020a). Tourist transformation: Towards a conceptual model. *Annals of Tourism Research, 81,* Article 102885. https://doi.org/10.1016/j.annals.2020.102885

Redmond, Brian. (2009). Lesson 8 commentary: Intergroup theories: How do the people around me influence me? *Work Attitudes and Motivation* (pp. 1–14). The Pennsylvania State University World Campus.

Sota, S., Chaudhry, H., & Srivastava, M. K. (2020). Customer relationship management research in hospitality industry: A review and classification. *Journal of Hospitality Marketing & Management, 29*(1), 39–64.

Soulard, J., McGehee, N., & Knollenberg, W. (2020). Developing and testing the Transformative Travel Experience Scale (TTES). *Journal of Travel Research, 60*(5), 923–946.

Wu, X., Gursoy, D., & Zhang, M. (2021). Effects of social interaction flow on experiential quality, service quality and satisfaction: Moderating effects of self-service technologies to reduce employee interruptions. *Journal of Hospitality Marketing & Management, 30*(5), 571–591.

INDEX

Note: Bold page numbers refer to tables; *italic* page numbers refer to figures and page numbers followed by "n" denote endnotes.